SELF-HEALING at the NANOSCALE

Mechanisms and Key Concepts of Natural and Artificial Systems

SELF-HEALING at the NANOSCALE

Mechanisms and Key Concepts of Natural and Artificial Systems

Edited by
Vincenzo Amendola • Moreno Meneghetti

Foreword by **Francesco Stellacci**

CRC Press
Taylor & Francis Group
Boca Raton London New York

CRC Press is an imprint of the
Taylor & Francis Group, an **informa** business

CRC Press
Taylor & Francis Group
6000 Broken Sound Parkway NW, Suite 300
Boca Raton, FL 33487-2742

First issued in paperback 2018

ISBN-13: 978-1-4398-5473-0 (hbk)
ISBN-13: 978-1-138-37447-8 (pbk)

Library of Congress Cataloging-in-Publication Data

Self-healing at the nanoscale : mechanisms and key concepts of natural and artificial systems / edited by Vincenzo Amendola and Moreno Meneghetti.
p. cm.
Includes bibliographical references and index.
ISBN 978-1-4398-5473-0 (hardback)
1. Self-healing materials. 2. Nanostructured materials. I. Amendola, Vincenzo. II. Meneghetti, Moreno.

TA418.9.S62S435 2012
620.1'15--dc23 2011035860

Visit the Taylor & Francis Web site at
http://www.taylorandfrancis.com

and the CRC Press Web site at
http://www.crcpress.com

Contents

Foreword

Materials have characterized the development of mankind. From the Stone Age to the Bronze Age, from wood to the present silicon age, materials with their properties have determined what civilization could and could not dream and do. It is reasonable to ask what will come next. What new material will enable a new leap forward in civilization? At present, it is hard to believe that we will unveil a property, and consequently a material, that we cannot foresee now. The next leap forward in materials will be in moving from a "property"-dominated vision toward a "function"-dominated vision.

In nature, materials (or, better, materials assemblies) have functions more than properties. They grow, renew, and, yes, they self-repair. This book addresses all aspects of self-healing from a broad viewpoint. It starts with biological systems and then covers devices, nanomaterials, and complex assembly processes. The link is the conquer of this important function. The ability to study different systems and materials and find a common thread will give the readers an edge in understanding and mastering this important concept of self-healing materials and/or systems.

Francesco Stellacci
Constellium Professor
Institute of Materials
École Polytechnique Fédérale de Lausanne
Lausanne, Switzerland

Preface

It has been more than 30 years since Gerd Binnig and Heinrich Rohrer invented the scanning tunneling microscope at IBM Zurich in 1981, which many regard as the founding stone of science at the nanoscale. Since then, the science of nanostructures has become a mature and widely diffused research field. A common feature of well-established disciplines is the shift toward problems with high complexity, often bordering other scientific disciplines. Nanoscience is no exception to this rule and has recently expanded its frontiers in the study of more advanced systems such as, for example, nanomaterials that can self-repair.

A self-healing material is a material that is capable of repairing its components when they are damaged by wear and tear or by accidental external events. In general, self-repair represents a sophisticated strategy to increase the robustness and extend the life span of a material, especially in cases where repair or replacement of materials is economically detrimental, dangerous, or impossible. This is the case in materials that are exposed to extreme physical and chemical conditions, such as corrosive environments; high-intensity light irradiation, ionizing radiation, or high temperatures; and materials that are used in areas inaccessible to humans, such as outer space and the deep sea.

In the case of nanomaterials, however, the self-repairing ability may be needed even in ordinary conditions. For purely thermodynamic reasons, nanomaterials are generally more fragile materials than their micro- and macroscopic equivalents. In fact, nanostructures are characterized by a high surface-atom-to-bulk-atom ratio, and the creation of interfaces has high costs in energy. It thus follows that, frequently, the excellent functional properties of nanomaterials are balanced by fast degradation and structural damage during use. The creation of nanosystems that are able to repair themselves can be the solution whenever it is too expensive or not possible to create more robust materials with the same functions.

The field of self-healing materials and nanomaterials is still in its early stage, and there are very few examples of such systems. Due to this, self-healing materials are classified into systems that are capable of autonomous self-repair, i.e., that do not require any external intervention to restore the damage (and are a minority), and systems that are capable of nonautonomous self-repair, i.e., that require an external trigger to start the repair process, such as increasing the temperature or applying a magnetic field or an external force.

One of the reasons that makes the design of self-healing systems a very ambitious target is the need for expertise in different fields like material science, chemistry, engineering, and biology. Biologists' contributions are particularly valuable because nature provides sophisticated, elegant, and efficient mechanisms of damage repair from the molecular to the macroscopic scale, many of which have been studied extensively. Even if they far exceed the self-healing mechanisms used in material science in terms of efficiency and complexity, it is worth noting that biological building blocks, like proteins and nucleic acids, are of nanometric size; therefore, most of the biological machinery takes place at the same size scale as that of nanomaterials. If nature has been successful in creating extraordinary self-repairing systems using components with nanoscale dimensions, then humans, in principle, should also be able to create self-repairing artificial nanosystems using nanomaterials. In addition, hybrid materials composed of engineered nanomaterials and biological components can provide an extra degree of freedom for designing self-repairing systems. Therefore, self-healing at the nanoscale is an area that has enormous potential for the future. Indeed, a few brilliant examples in this direction have appeared recently (and are also discussed in this book).

BASIC FEATURES OF SELF-HEALING SYSTEMS

For many years, self-repairing strategies of nanoscale materials included almost exclusively the spontaneous or externally triggered atomic-scale processes that are driven by a net gain of free energy. These are well-known processes in materials science and have been studied for a long time.

In the category of processes driven by a gain of enthalpy are the phenomena of passivation and surface reconstruction. The binding energy of individual atoms composing the material increases during passivation and reconstruction and can be considered as simple cases of autonomous self-healing systems. For example, when the surface of metals like Al, Cr, Ti, Zr, and Cu is exposed to air, a passivating oxide layer forms in order to decrease surface free energy. This layer is quite impermeable and protects the underlying metal from further oxidation. In case of a mechanical removal of a part of this layer, air oxidation of the exposed metal surface readily restores the protective film (Hultquist et al. 2001). A similar effect is observed in some systems by surface reconstruction. For instance, CdSe and CdTe semiconductor quantum dots undergo surface reconstruction in the presence of a shell of passivating ligands on the surface (Puzder et al. 2004). The interface reconstruction allows the preservation of the inner crystalline structure and composition and has great importance in opening the optical gap in these nanoparticles.

In the category of self-repair processes driven by a gain of entropy are shape memory materials that have the ability to restore their initial form after a plastic deformation. For instance, shape memory materials can recover their form after a bulk deformation on their surface integrity after slight bumps, indentations, or scratches. Shape recovery requires an external trigger, in most cases a temperature increase above a transition threshold. Therefore, these materials fall into the category of non-autonomic self-healing materials. Several metal alloys, ceramics, polymers, and hydrogels with shape memory behavior are known. Shape memory basically exploits an entropy-driven phase transition, for instance, from martensite to austenite in the case of the NiTi alloy, from tetragonal to monoclinic in the case of ZrO_2, from glassy state to rubber–elastic state in the case of polymers, and from hydrophilic to hydrophobic state in the case of hydrogels (Mather et al. 2009; Müller and Seelecke 2001).

In recent years, new systems with higher levels of complexity have appeared among self-repairing materials. These systems are based, in particular, on autonomous assembly or on responsive chemical reactions. In the first case, the damage is repaired thanks to the self-assembling ability of some components in the material. In the second case, the repair process uses a chemical reaction between appropriate compounds included in the material that are self-activated after the damage or are activated by an external stimulus.

Some of the self-healing materials that have appeared in recent years have been inspired by natural repair strategies. Although natural and artificial self-healing systems are different in terms of complexity, sophistication, functions, and components, it is still possible to identify some general characteristics that are common in both cases (Table P.1):

1. Modularity: The system consists of several components with distinct functions.
2. Self-assembly: The different components or building blocks of the system are able to assemble autonomously.
3. Redundancy: The system is able to perform its function even when a part of it is damaged and is being repaired.
4. Transport: The damaged components are transferred from the functional site to the repair site.
5. Specialized repairing components: Some of the components of the system have functions dedicated exclusively to transport and repair of the damaged components.
6. Hierarchical structure: Repairing components have a higher level of complexity and are larger than the damaged unit. Sometimes, the repair can take place in sites other than those in which the system performs its function.

By increasing the level of complexity and efficiency of the self-healing system, the number of aforementioned characteristics that can be identified in the system also increases. To date, only a few natural repair systems have all these six properties.

TABLE P.1
Frequent Features of Self-Healing Systems

1. Modularity
2. Self-assembly
3. Redundancy
4. Transport
5. Specialized repairing components
6. Hierarchical structure

OUTLINE OF THE BOOK

The book includes contributions about selected representative examples in the diverse and multidisciplinary field of self-repairing systems. Several self-healing systems found in natural processes and others created by man-made activity are reported with a special emphasis on the key concepts, strategies, and mechanisms at the atomic, molecular, and nanometric scales. The contributions of the book are grouped into three parts.

Part I: Natural Self-Healing Systems—covers paradigmatic self-repair systems developed by nature in living organisms.

Solar radiation is both a source of energy and a possible cause of damage. In Chapter 1, Nixon and Komenda describe how plant cells are able to selectively regenerate only those components affected by light damage among all those involved in the photosynthesis. In Chapter 2, Ma et al. show how self-repair is a prerequisite for the proper functioning of the visual cycle in humans, namely, a process with different stages in which the photoactive material is transported from the site of operation to the repair site and back again continuously.

In living organisms, the repair of larger systems inevitably requires the involvement of a greater number of components, which indicates a rise in the level of complexity of the repair mechanism. In Chapter 3, Sinigaglia presents an overview of the biological mechanisms of repair on an increasing size scale, highlighting the importance of the presence of a hierarchy between the different components.

Part II: Artificial Self-Healing Systems—describes various materials whose structures have been engineered at the micro- or nanoscale to obtain the self-repair ability. Also, some artificial nano or nanostructured materials that unexpectedly revealed the ability to repair themselves are described.

Chapter 4 shows that a nanosystem with only three components can reveal a complex network of interactions sufficient to allow the spontaneous repair of its functional nanostructures, as happens in the case of the self-healing of gold nanoparticles during laser irradiation. In Chapter 5, Lackinger shows that the capacity for self-assembly at the nanoscale can guarantee the self-repair ability for purely thermodynamic reasons, as happens in two-dimensional molecular monolayers.

Extracting energy from the sun with the maximum efficiency and the minimum waste of resources is one of the most appealing targets for researchers. In Chapter 6, Bonchio et al. describe the state of the art and the working principles of inorganic nanocatalysts for water splitting, which can either self-repair or withstand the harmful conditions in which water oxidation is carried out. In Chapter 7, Strano et al. present a sophisticated way to solve the problem of photodamage in photovoltaic hybrid systems, incorporating the properties of carbon nanotubes with those of natural photosystems.

In Chapter 8, Alessandri provides a comprehensive overview of nanocomposite materials in which nanostructures were introduced specifically to obtain self-healing through autonomous or nonautonomous mechanisms. In Chapter 9, Picchioni et al. describe the sound results and the interesting perspectives offered by synthetic chemistry for the construction of macromolecular and polymeric systems with functional groups that grant self-repairing ability to the system.

Change in microscopic structure can confer self-healing properties to traditional materials as well. In Chapter 10, Shchukin et al. discuss the state of the art for the use of microscopic smart containers in obtaining surfaces that can self-repair the damages of corrosion, which is one of the oldest problems in materials science. In Chapter 11, Méar et al. demonstrate the possibility of conferring the self-repairing ability even to a traditionally fragile material like glass, when the structure of the material is modified on a very small size scale. Finally, in Chapter 12, Hager and Schubert cover the most significant strategies for the repair of polymers, ceramics, concrete, and metals by the engineering of their structure on a microscopic scale.

Part III: Frontiers of Self-Healing Systems—includes contributions on systems that were studied in recent years and that have shown good potential for developing or inspiring new self-healing nanomaterials in the future.

Nanoscale self-healing processes have been studied not only by experimental scientists, but they have also been studied using computational methods, with striking results. An excellent example is provided in Chapter 13 by Balazs et al. describing a repair-and-go system for site-specific healing at the nanoscale.

If we consider the founding event of nanotechnology, more than 50 years ago, to be Richard Feynman's famous speech "There's plenty of room at the bottom" during the annual meeting of the American Physical Society, then we soon go to the vision of nanotechnology as the discipline that would have allowed us to obtain "a hundred tiny hands" able of self-replication down to the atomic scale. In Chapter 14, Stano shows how self-healing and self-replication are connected abilities in living organisms; hence, the realization of self-healing nanomaterials could be a pivotal step in the process of fabricating self-replicating nanomachines. Moreover, in Chapter 15, Månsson et al. describe the fascinating case of a molecular machine optimized by nature that has a relevant role in a biological nanoscale self-healing process.

To date, only one type of artificial system has reached a complexity level that is comparable to nature—that is, the information technology system. In Chapter 16, Tempesti et al. discuss how this area opens interesting opportunities for the development of nanoscale electronic devices capable of self-healing.

The organization of the book in three parts not necessarily is the best possible, and some readers may prefer a cross-reading path. For example, Chapters 1, 6, and 7 are connected by the theme of the need to convert sunlight into energy by systems that are able to resist photodamage. In Chapters 3, 12, and 13, a comparison between the natural healing strategies of tissues, the best repair strategies for structural materials achieved by man so far, and the strategies that theoreticians indicate as the most promising for the development of more efficient self-repairing artificial systems is provided. Chapters 5 and 14 show that self-assembly is an effective strategy for self-repair in natural and artificial molecular systems. It is likely that the readers will be able to find even more cross-correlations and alternative reading paths through the book.

In conclusion, we hope that this book will provide a panorama of the concepts, mechanisms, and types of self-healing systems at the nanoscale in order to be a source of inspiration for the development of the self-healing systems of the future.

Vincenzo Amendola

Moreno Meneghetti

REFERENCES

Hultquist, G., Tveten, B., Hörnlund, E., Limbäck, M., and Haugsrud, R. 2001, Self-repairing metal oxides, *Oxidation of Metals*, 56(3), 313–346.

Mather, P.T., Luo, X., and Rousseau, I.A. 2009, Shape memory polymer research, *Annual Review of Materials Research*, 39, 445–471.

Müller, I. and Seelecke, S. 2001, Thermodynamic aspects of shape memory alloys, *Mathematical and Computer Modelling*, 34(12–13), 1307–1355.

Puzder, A., Williamson, A.J., Gygi, F., and Galli, G. 2004, Self-healing of CdSe nanocrystals: First-principles calculations, *Physical Review Letters*, 92, 217401.

Acknowledgments

This book was made possible thanks to the valuable work and the scientific expertise of all the contributors. We would therefore gratefully acknowledge all the contributors.

Editors

Vincenzo Amendola has been an assistant professor of physical chemistry in the Department of Chemical Sciences of Padua University since 2008. He received his PhD in material science and engineering in 2007 and his master's degree in material science in 2004, both from Padua University, working in the group of Professor M. Meneghetti. In 2007, he was a 6-month visiting PhD student at the Massachusetts Institute of Technology, as part of the group of Professor F. Stellacci, and in 2011 he was a 7-month academic visitor at Cambridge University, as part of the group of Professor A.C. Ferrari. He was honored with the Levi prize and the Semerano prize from the Italian Chemical Society. He is the author of several peer-reviewed papers, one patent, and has given invited talks at several conferences, including the first conference about *Laser Irradiation and Nanoparticles Generation in Liquids (ANGEL2010).* He serves as a referee for several leading journals of Physical Chemistry, Material Science and Nanotechnology.

Amendola's main research topics are the laser ablation synthesis in solution (LASiS) of functional and functionalizable nanostructures, the study of plasmonic properties of nanoparticles, and, in general, the interaction of nanosecond laser pulses with matter for the generation of new materials and of new phenomena (like self-healing processes). He is currently collaborating with various research groups in Europe and Asia. More information about his research activities can be found at www.chimica.unipd.it/vincenzo.amendola.

Moreno Meneghetti received his doctor's degree in 1979 and is now a full professor of physical chemistry at the University of Padova. He focused his research interest on the optical properties of materials as tools for understanding their behavior. He used both experimental and theoretical approaches for characterizing phase transitions in conducting and superconducting organic materials focusing especially on electron–phonon couplings involving molecular units. He developed full diagonalization and non-mean field approaches to study the optical properties (vibrational and electronic) of low-dimensional solids. His research interests include nonlinear optical properties of molecular systems and, in particular, dynamics of nonlinear transmission of excited states with applications like optical limiting. Nonlinear optical properties and Raman properties of nanostructures based on carbon-like fullerenes and nanotubes were also investigated for their reactivity and for the preparation of new materials. Models for the electronic properties of single-wall nanotubes based on nonconventional methodologies are also the focus of his current research. Laser ablation of metals in solution for the production of metal nanoparticles is another aim of his research, focusing especially on the SERS effect. In particular, gold nanoparticles are used as nanobioconjugates with applications in nanomedicine for targeting and imaging at the subcell level.

Meneghetti's research activity has been published in more than 130 papers in peer-reviewed journals.

Contributors

Ivano Alessandri
Interuniversity Consortium on Science and Technology of Materials
and
Chemistry for Technologies Laboratory
University of Brescia
Brescia, Italy

Vincenzo Amendola
Department of Chemical Sciences
University of Padua
Padua, Italy

Anna C. Balazs
Chemical Engineering Department
University of Pittsburgh
Pittsburgh, Pennsylvania

Serena Berardi
Department of Chemical Sciences
University of Padua
Padua, Italy

Ardemis A. Boghossian
Department of Chemical Engineering
Massachusetts Institute of Technology
Cambridge, Massachusetts

Marcella Bonchio
Institute on Membrane Technology
Italian National Council of Research
Padua, Italy

A.A. Broekhuis
Chemical Engineering Department (Product Technology)
University of Groningen
Groningen, the Netherlands

Mauro Carraro
Department of Chemical Sciences
University of Padua
Padua, Italy

Jong Hyun Choi
Birck Nanotechnology Center
Bindley Bioscience Center
School of Mechanical Engineering
Purdue University
West Lafayette, Indiana

Daniel Coillot
Calalysis and Solid State Chemistry Unit
Université Lille Nord de France
Lille, France

Alfred J. Crosby
Polymer Science and Engineering Department
University of Massachusetts
Amherst, Massachusetts

Todd Emrick
Polymer Science and Engineering Department
University of Massachusetts
Amherst, Massachusetts

Dmitry O. Grigoriev
Department of Interfaces
Max-Planck Institute of Colloids and Interfaces
Potsdam, Germany

Martin D. Hager
Laboratory of Organic and Macromolecular Chemistry (IOMC)
Jena Center for Soft Matter (JCSM)
Friedrich-Schiller-University Jena
Jena, Germany

Moon-Ho Ham
Department of Chemical Engineering
Massachusetts Institute of Technology
Cambridge, Massachusetts

German V. Kolmakov
Department of Physics
New York City College of Technology
City University of New York
Brooklyn, New York

Josef Komenda
Laboratory of Photosynthesis
Institute of Microbiology
Trebon, Czech Republic

Markus Lackinger
TUM School of Education
Technical University Munich
Munich, Germany

and

Deutsches Museum
Munich, Germany

Jian-Xing Ma
Department of Physiology
University of Oklahoma Health Science Center
Oklahoma City, Oklahoma

Alf Månsson
School of Natural Sciences
Linnaeus University
Kalmar, Sweden

François O. Méar
Calalysis and Solid State Chemistry Unit
Université Lille Nord de France
Lille, France

Moreno Meneghetti
Department of Chemical Sciences
University of Padua
Padua, Italy

Franck Michoux
Division of Biology
Department of Life Sciences
Imperial College London
London, United Kingdom

Gennadiy Moiseyev
Department of Physiology
University of Oklahoma Health Science Center
Oklahoma City, Oklahoma

Lionel Montagne
Calalysis and Solid State Chemistry Unit
Université Lille Nord de France
Lille, France

Peter J. Nixon
Division of Biology
Department of Life Sciences
Imperial College London
London, United Kingdom

Marlene Norrby
School of Natural Sciences
Linnaeus University
Kalmar, Sweden

F. Picchioni
Chemical Engineering Department (Product Technology)
University of Groningen
Groningen, the Netherlands

Renaud Podor
Institut de Chimie Séparative de Marcoule
Bagnols-sur-Cèze, France

D. De Reus
Chemical Engineering Department (Product Technology)
University of Groningen
Groningen, the Netherlands

Joël Rossier
Reconfigurable & Embedded Digital Systems (REDS) Institute
Haute Ecole d'Ingénierie et de Gestion du Canton de Vaud
Yverdon-les-Bains, Switzerland

Thomas P. Russell
Polymer Science and Engineering Department
University of Massachusetts
Amherst, Massachusetts

Andrea Sartorel
Department of Chemical Sciences
University of Padua
Padua, Italy

Ulrich S. Schubert
Laboratory of Organic and Macromolecular Chemistry (IOMC)
Jena Center for Soft Matter (JCSM)
Friedrich-Schiller-University Jena
Jena, Germany

Dmitry G. Shchukin
Department of Interfaces
Max-Planck Institute of Colloids and Interfaces
Potsdam, Germany

Alessandro Sinigaglia
Department of Histology, Microbiology, and Medical Biotechnologies
University of Padua
and
Venetian Oncological Institute
Padua, Italy

Pasquale Stano
Biology Department
University of Roma Tre
Rome, Italy

André Stauffer
Logic Systems Laboratory
Ecole Polytechnique Fédérale de Lausanne
Lausanne, Switzerland

Michael S. Strano
Department of Chemical Engineering
Massachusetts Institute of Technology
Cambridge, Massachusetts

Sven Tågerud
School of Natural Sciences
Linnaeus University
Kalmar, Sweden

Gianluca Tempesti
Department of Electronics
University of York
York, United Kingdom

C. Toncelli
Chemical Engineering Department (Product Technology)
University of Groningen
Groningen, the Netherlands

Part I

Natural Self-Healing Systems

1 Keeping the Green World Alive

The Repair Cycle of Photosystem II

Josef Komenda, Franck Michoux, and Peter J. Nixon

CONTENTS

1.1 INTRODUCTION

Oxygenic photosynthesis is a key biological process on Earth: it gives rise to much of its organic matter and is essential for the maintenance of most forms of life. During this process, light energy is converted into chemical energy, which is stored in the form of highly reduced organic compounds. The initial events of this transformation involve absorption of solar radiation by various pigments (predominantly chlorophylls [Chls] and carotenoids [Cars] but also by phycobilins in some organisms), and subsequent trapping of absorbed energy by a group of specialized pigment molecules (the so-called reaction center [RC] pigments) bound within membrane protein complexes termed photosystems or RC complexes. Excitation of these RC pigments leads to transmembrane charge separation and initiation of electron flow within the membrane to ultimately produce reduced nicotinamide adenine dinucleotide phosphate (NADPH), a source of reducing power for the cell. The electron transport chain consists of a system of electron carriers ranging from simple organic compounds

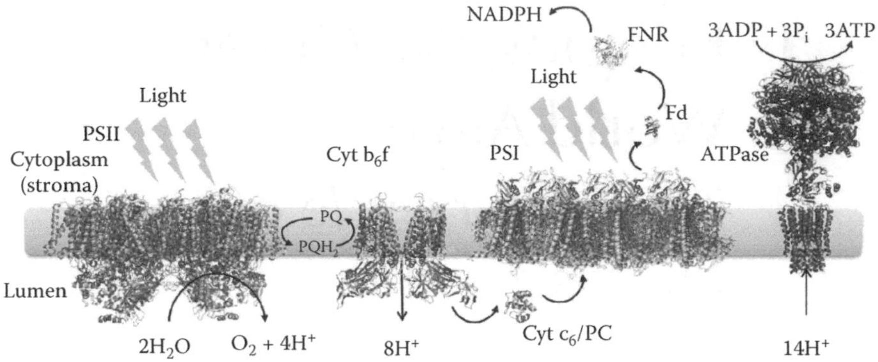

FIGURE 1.1 **(See color insert.)** Protein components of the photosynthetic electron transport chain. Electron transport from water to $NADP^+$ is driven by PSII extracting electrons from water and transferring them to plastoquinone (PQ) to produce plastoquinol (PQH_2). PSI catalyzes the reduction of ferredoxin and, subsequently, formation of NADPH via ferredoxin-$NADP^+$ reductase (FNR). Oxidized PSI is re-reduced by electrons from the cytochrome b_6f complex (Cyt b_6f) via cytochrome c_6 (Cyt c_6) or plastocyanin (PC), whereas PQH_2 produced by PSII reduces oxidized Cyt b_6f. Electron transport is coupled to the translocation of protons from the cytoplasm in cyanobacteria (or stroma in chloroplasts) to the lumen to produce a proton-motive force. In the case of Cyt b_6f, the Q-cycle operates (not shown in the figure to aid clarity) to give a stoichiometry of eight H^+ deposited in the lumen per four electrons transported to PSI. Additional translocation of protons across the membrane is mediated by cyclic electron flow around PSI (data not shown). The proton-motive force coupled to movement of protons back across the membrane through the ATP synthase (ATPase) drives synthesis of ATP from ADP and inorganic phosphate (P_i). The protein complexes shown are the crystal structures derived from various cyanobacteria.

(e.g., quinones) and small redox-active proteins up to multisubunit pigment-binding protein complexes, all of which are embedded in specialized photosynthetic membranes (termed thylakoids) found within cyanobacteria and the chloroplasts of plants and algae (Figure 1.1). Electron transfer is coupled to formation of a gradient of protons across the thylakoid membrane, which is utilized for ATP synthesis by the ATP synthase. NADPH and ATP generated by the "light reactions" are used in the "dark reactions" for synthesis of reduced organic compounds, mostly sugars, from CO_2.

In oxygenic photosynthesis, linear electron transport is driven by two RC complexes called Photosystem I (PSI) and Photosystem II (PSII). PSII has the unique ability to extract electrons from water and so plants, algae, and cyanobacteria are called oxygenic phototrophs since oxidation of water results in oxygen release. There also exist anoxygenic photosynthetic bacteria that use different electron donors but their importance for the overall photosynthetic productivity is relatively small.

1.2 STRUCTURE AND FUNCTION OF PHOTOSYSTEM II

The structure of the PSII complex from thermophilic cyanobacteria has been determined to high resolution by x-ray crystallography (Ferreira et al. 2004; Guskov et al. 2009; Umena et al. 2011). The crystallized complex contains 16 or 17 intrinsic membrane proteins, depending on the preparation, three extrinsic membrane proteins, and a large number of cofactors including Chls, pheophytins (Pheos), β-carotenes, plastoquinones, lipids, and ions including those forming the oxygen-evolving Mn_4Ca cluster (Ferreira et al. 2004; Guskov et al. 2009; Umena et al. 2011). Unfortunately, high-resolution structural models have not yet been obtained for plant PSII, but, based on low-resolution models (Barber 2002), it is highly likely that the structures will be very similar, except for the set of extrinsic proteins on the lumenal side of the complex which differ (Enami et al. 2008).

The most important protein constituents of PSII are homologous, five transmembrane helix–containing proteins termed D1 and D2. These proteins form the so-called D1–D2 heterodimer that binds six Chls, two Pheos, and two β-carotenes. The N-termini of both proteins are exposed to the cytoplasmic side (stromal side in chloroplasts) of the thylakoid membrane, whereas the C-termini protrude into the lumen (inner part of thylakoids). D1 is synthesized as a precursor protein (pD1) with a cleavable C-terminal extension (Marder et al. 1984). The extension is present in all oxygenic phototrophs with the exception of *Euglena* and some dinoflagellates and its length is species-dependent; in plants the extension usually consists of 9 amino acid residues, whereas in cyanobacteria the extension is usually 16 residues. Removal of the extension by a cut on the carboxyl side of residue Ala344 (Takahashi et al. 1988) is required for the assembly of the functional oxygen-evolving complex (Nixon et al. 1992). The extension is removed by a specific processing endoprotease termed CtpA (Anbudurai et al. 1994), which cleaves the plant extension in a single step, whereas the cleavage in cyanobacteria seems to occur in two consecutive steps with a primary cut occurring close to the middle of the extension (Komenda et al. 2007a). The precise role of the D1 extension is not yet clear, although it is needed for optimal photosynthetic performance under high irradiance (Ivleva et al. 2000; Kuviková et al. 2005). The extension interacts with a PSII assembly factor, YCF48, required for the efficient formation of PSII (Komenda et al. 2008), so lower fitness of the extension-less cyanobacterial mutant under high irradiance might be due to an attenuation of this interaction.

The D1–D2 heterodimer is surrounded by CP47 and CP43, two homologous proteins that bind a number of Chls and β-carotenes and function as an inner or proximal light-harvesting antenna system delivering energy to the RC pigments bound to the D1 and D2 proteins. CP47 and CP43 possess six transmembrane helices and a large lumenal loop joining transmembrane helices 5 and 6 that is involved in the binding of the lumenal extrinsic proteins (in the cases of both CP47 and CP43) or in coordinating one of the Mn ions of the oxygen-evolving complex (in the case of CP43; Ferreira et al. 2004). Both proteins are located symmetrically either side of the D1–D2 heterodimer: CP47 next to D2 and CP43 next to D1 (Figure 1.2).

The PSII complex also contains a number of small membrane protein subunits that are assumed to assist the correct assembly of the complex and to optimize PSII electron flow (for review, see Müh et al. 2008). On the lumenal side of the PSII, there is an inorganic metal cluster (Mn_4CaO_5), consisting of a single Ca and four Mn ions linked by oxo bridges, which forms the oxygen-evolving center (OEC), the catalytic site at which water binds and is oxidized (Umena et al. 2011) (Figure 1.2B). The OEC is

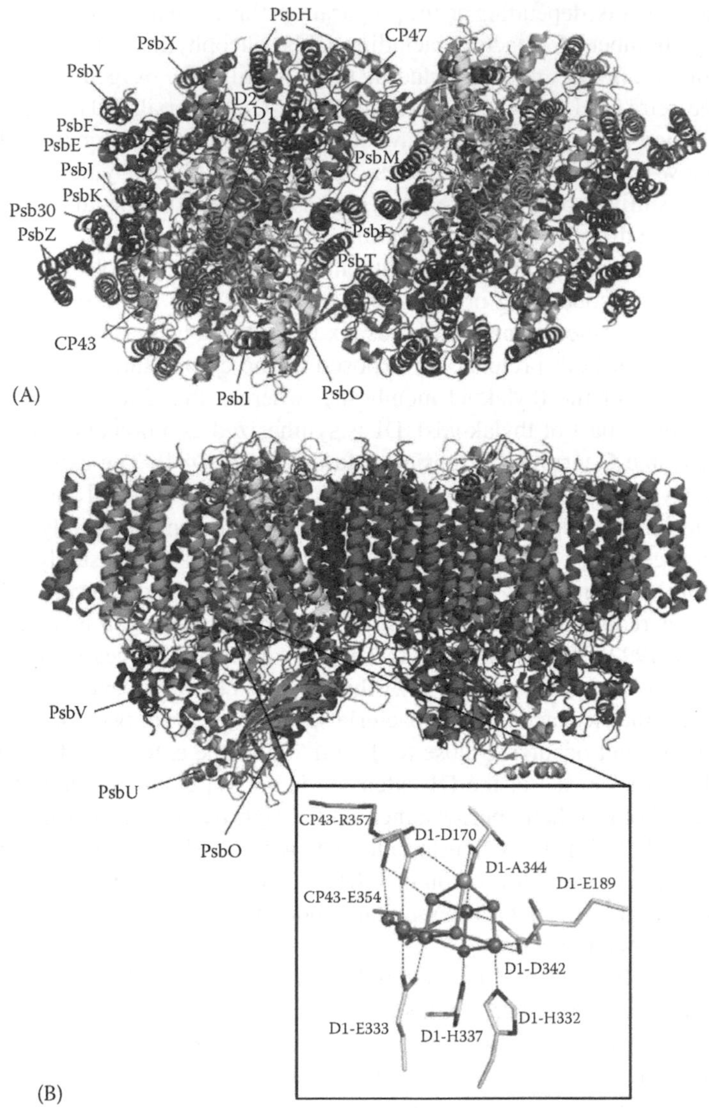

FIGURE 1.2 **(See color insert.)** Crystal structure of the PSII complex isolated from thermophilic cyanobacteria. View of the homodimeric complex from *Thermosynechococcus elongatus* from the cytoplasmic side of the thylakoid membrane (panel A) and perpendicular to the membrane normal (panel B). The 20 subunits have been annotated in panel A and color-coded: D1 (yellow), D2 (orange), CP43 (green), CP47 (red), cytochrome b-559 (purple), PsbO (violet), PsbV (dark blue), and PsbU (light blue). The remaining 11 small transmembrane subunits are shown in gray. For clarity, pigments have been omitted. The inset in panel B shows the structure of the Mn_4CaO_5 cluster involved in water oxidation, with coordinating amino acid residues, determined for *Thermosynechococcus vulcanus* (Umena et al. 2011). The Ca ion is shown in gray, the Mn ions in violet, and the oxo bridges in red. For clarity, bound waters have been omitted. The figure was created with the software Pymol (http://pymol.sourceforge.net, version 0.99) and the PDB files 3BZ1 and 3BZ2 (Guskov et al. 2009) and 3ARC (Umena et al. 2011).

stabilized by a set of extrinsic proteins that cap the cluster shielding it from the aqueous phase. Significant differences can be found in this part of PSII between plants and other oxygenic phototrophs (for review, see Roose et al. 2007a). The common protein for all groups of organisms is the 33 kDa polypeptide, the product of the *psbO* gene. Cyanobacteria and red algae contain, in addition, the 15-kDa *psbV* (cytochrome c-550) and 12-kDa *psbU* gene products. In higher plants, the PsbU and PsbV proteins are missing and their function is fulfilled by the 23-kDa PsbP and 16-kDa PsbQ subunits. Cyanobacterial homologues of PsbP and PsbQ have been identified (Thornton et al. 2004) but their precise roles in PSII function remain unclear (Summerfield et al. 2005a,b; Roose et al. 2007b).

The outer or distal light-harvesting systems also differ significantly among oxygenic phototrophs (Green et al. 2003). Large peripheral phycobilisomes, consisting of water-soluble proteins with covalently bound chromophores (phycobilins), are attached to the cytoplasmic/stromal side of the thylakoid membrane of cyanobacteria and some algal groups (e.g., Rhodophyta) (Figure 1.3). In other algae and in higher plants, the antenna complexes are embedded in the membrane and bind various types of Chl and Cars.

Despite the presence of many proteins in PSII, the cofactors involved in transferring electrons from water to plastoquinone are bound primarily by the D1–D2 heterodimer (Diner and Babcock 1996) (Figure 1.3). In brief, light energy captured by pigments within the antenna system is delivered to the Chls that constitute the primary electron donor, P680, to generate an excited singlet state, P680*, which initiates electron flow by reducing a nearby Pheo, a molecule structurally similar to Chl but lacking the central Mg ion, to form the primary radical pair $P680^+Pheo^-$. $Pheo^-$ then reduces the primary plastoquinone electron acceptor, Q_A, bound to the D2 protein. Unlike regular quinone molecules, Q_A is normally able to accept only a single electron due to its special protein environment. The electron is then transferred to the secondary plastoquinone electron acceptor, Q_B, bound to D1, which becomes doubly reduced after another charge separation event and is then protonated to form the plastoquinol, Q_BH_2. This molecule leaves PSII and is replaced by an oxidized plastoquinone molecule from a pool of plastoquinone molecules located in the lipid bilayer. The oxidized primary donor, $P680^+$, is reduced by an electron from a redox-active tyrosine residue, Tyr161 of the D1 protein, termed Y_Z (Debus et al. 1988; Metz et al. 1989), which in turn is reduced by the Mn_4Ca cluster, coordinated by aspartate, glutamate, and histidine residues in the D1 protein, including the C-terminus of mature D1, and by a glutamate residue of CP43 (Ferreira et al. 2004; Guskov et al. 2009; Umena et al. 2011) (Figure 1.2B). The Mn_4Ca cluster exists in five distinct oxidation or S states, termed S_0–S_4, where the subscript indicates the number of accumulated oxidizing equivalents. Oxidation of two molecules of water to one molecule of dioxygen occurs after formation of the S_4 state, resetting the enzyme to the S_0 state, and so occurs with a periodicity of four. The recent structural model of the cluster determined at atomic resolution has identified potential binding sites for substrate waters plus channels for the movement of incoming water molecules and exiting protons (Umena et al. 2011).

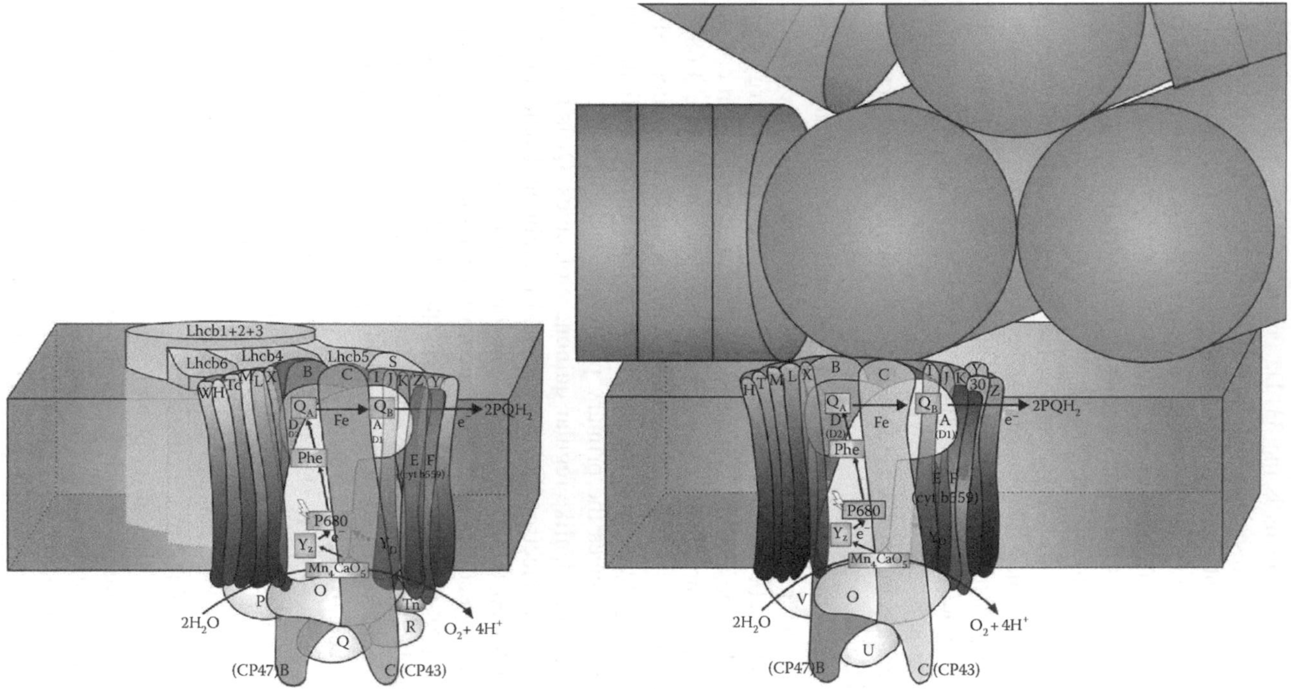

FIGURE 1.3 **(See color insert.)** Comparison of PSII in chloroplasts (left panel) and cyanobacteria (right panel). Excitation of the primary electron donor, P680, leads to stepwise reduction of the pheophytin electron acceptor (Phe) and the plastoquinones, Q_A and Q_B, located close to a non-heme iron (Fe). PQH_2 is produced after two photoacts. On the donor side, P680$^+$ oxidizes tyrosine, Y_z, which in turn oxidizes the Mn_4CaO_5 oxygen-evolving center. Other redox-active components not involved directly in water oxidation include a second redox-active tyrosine, Y_D, located within the D2 subunit and the heterodimeric Cyt b-559 complex. Subunits are annotated so that PsbA is labeled A, and so on. Chloroplasts contain an integral light-harvesting system (Lhcb subunits), whereas cyanobacteria contain the extremely large phycobilisome (large blue–green cylinders) that docks on to the cytoplasmic surface of the thylakoid membrane. The figure was kindly provided by Dr. Jon Nield (http://www.queenmaryphotosynthesis.org//nield//).

1.3 PHOTODAMAGE TO PHOTOSYSTEM II

Despite the existence of a plethora of photoprotective mechanisms *in vivo* to prevent oxidative damage (Niyogi 1999), a characteristic feature of PSII is its relative vulnerability to photodamage. The quantum yield of photodamage has been estimated to be 7×10^{-8} in pumpkin leaves which, although a deceptively low value, still means that approximately half the PSII complexes are damaged in a leaf during a 2h exposure to an irradiance of 1500 μmol of photons/m^2/s (Tyystjärvi and Aro 1996). Indeed, damage to PSII is thought to be an important obstacle for optimal photosynthetic performance in the field and its mitigation is a prime target for crop improvement (Murchie and Niyogi 2011).

Given the complexity of electron transfer events within PSII, it has proved difficult to determine where and how the inactivation to PSII occurs. Indeed, after almost three decades of research, there is still intense debate regarding the dominant mechanism of damage. Several mechanisms have been proposed, which differ in the specific light conditions and the functional state of PSII (reviewed by Tyystjärvi 2008; Vass 2011). The so-called acceptor-side mechanism was proposed to occur in fully functional PSII under high light conditions when the plastoquinone pool becomes overreduced. Under these conditions, the primary quinone acceptor, Q_A, becomes singly reduced or even leaves its binding site on the D2 protein if it becomes doubly reduced and protonated (Vass et al. 1992). This leads to a block in the oxidation of reduced Pheo (hence why it is termed an acceptor-side mechanism of photoinactivation), which increases the probability of charge recombination to form the triplet state of the primary electron donor, ^{3}P680, which in turn can react with triplet oxygen to generate the highly damaging singlet oxygen species (Hideg et al. 1994). Another proposed type of photoinactivation called donor side–induced inactivation is considered to be a meaningful event when the oxygen-evolving machinery is not fully functional and donation of electrons to oxidized P680 is too slow. Under these conditions, oxidative events interrupt the electron flow from oxygen-evolving complex to the primary donor P680 (Chen et al. 1992). The third proposed pathway of PSII photoinactivation occurs at extremely low light intensities when reactive singlet oxygen is generated from ^{3}P680 formed via charge recombination from the singly reduced secondary quinone acceptor, Q_B^-, and the oxidized Mn_4Ca cluster (Keren et al. 1997). All of the aforementioned mechanisms have been demonstrated to occur *in vitro* but there is still uncertainty about which one dominates *in vivo*. Recent studies showing a higher quantum yield of inactivation for blue light, a region of the spectrum that is directly absorbed by the Mn ions in the Mn_4Ca cluster, would seem to speak for a modified version of the donor-side mechanism in which the initial damaging event is the cluster itself (Hakala et al. 2005; Ohnishi et al. 2005). A similar mechanism is also assumed for inactivation of PSII by UV-B radiation (Szilárd et al. 2007). Damage to the Mn_4Ca cluster is also supported by the finding that the rate of photoinactivation of PSII is directly proportional to the light intensity (with exception of extremely low light as mentioned earlier) (Tyystjärvi and Aro 2006), which seems to be at odds with inactivation mechanisms that are dependent on a specific redox state of certain electron carriers (overreduction of acceptor-side electron carriers or overoxidation of the donor-side electron carriers).

1.4 PHOTOSYSTEM II REPAIR CYCLE

Regardless of the mechanism of PSII photodamage, it is accepted that PSII damage results in irreversible damage to protein subunits of PSII, in particular to the D1 subunit (Adir et al. 2003). The high frequency of damage has necessitated the development of a sophisticated mechanism to repair the inactive PSII complex and fully restore its photochemical activity. Such a mechanism is termed the "PSII repair cycle." It is known to consist of a number of distinct steps but a detailed understanding of the process is still lacking (Figure 1.4). It is assumed that the inactivation of the native dimeric PSII core complex is accompanied by a monomerization of the complex. Subsequently, the CP43 antenna and the lumenal extrinsic proteins stabilizing oxygen evolution are detached from the monomer. In this way, the damaged D1 protein within PSII becomes accessible for proteolytic removal and subsequent replacement by a newly synthesized copy of D1. The last phase of the cycle consists of rebinding of CP43 and the lumenal extrinsic proteins, assembly of the Mn_4Ca cluster, reactivation of oxygen evolution, and dimerization of the complex; the precise order of the last steps remains unknown.

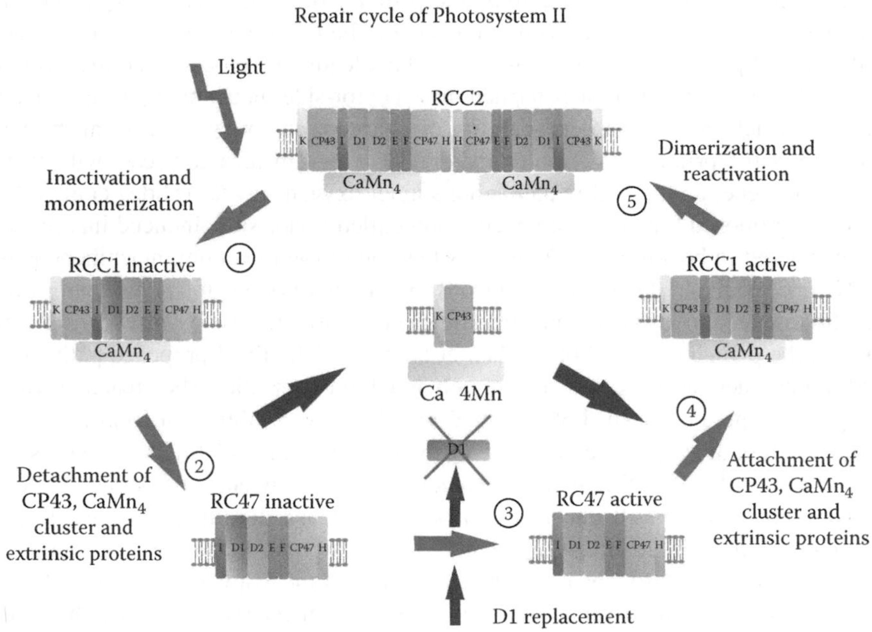

FIGURE 1.4 The repair cycle of Photosystem II. PSII repair cycle starts with light-induced inactivation of the PSII dimer, RCC2, and its monomerization (step 1). Then CP43, the $CaMn_4$ cluster, and extrinsic proteins (displayed collectively as a vanilla-colored shape) are detached (step 2) and the damaged D1 subunit is selectively replaced by a newly synthesized D1 copy at the level of the RC47 complex (step 3). Finally, CP43, Ca and Mn ions, and extrinsic proteins are reattached (step 4), and the repaired PSII monomer is dimerized with concomitant reactivation of the oxygen-evolving machinery (step 5). For clarity, many of the smaller PSII subunits have been omitted.

1.4.1 Initiation of the Repair Cycle

It is generally accepted that light-driven damage to PSII initiates the process of repair although the details remain unclear. Many early reports suggested a crucial role for the Q_B-binding pocket in signaling damage and triggering repair (Kirilovsky et al. 1988; Gong and Ohad 1991; Komenda and Barber 1995). However, relatively normal PSII repair in mutants of the cyanobacterium *Synechocystis* PCC 6803 lacking up to 20 amino acid residues in the vicinity of the Q_B-binding site argued against participation of this region (Nixon et al. 1995; Mulo et al. 1997). Instead, recent data obtained with cyanobacterial mutants having an impaired oxygen-evolving complex are in agreement with a triggering role for the lumenal part of PSII surrounding the Mn_4Ca cluster (Komenda et al. 2010), in line with the concept of direct photoinactivation of the Mn_4Ca cluster as the dominant mechanism *in vivo*. Since the cluster is coordinated by residues in D1 and CP43, we assume that damage and disruption of the cluster could lead to destabilization of the surrounding D1 structure, perturbed binding of extrinsic proteins, and even possibly an altered interaction between D1 and CP43, which may be recognized by the protease.

A key step of the PSII repair process is the selective removal of the D1 protein and its replacement by the newly synthesized D1 copy. This process can be followed experimentally by a radioactive pulse-chase method, which involves labeling of cells with radioactively labeled amino acid and subsequent chasing of the radiolabeled protein in the presence of non-radioactive ("cold") amino acid. Using this approach, the rate of protein turnover can be estimated after electrophoretic separation of membrane proteins and autoradiography. Comparable but not fully equivalent information can also be obtained by following the amount of the protein after inhibition of cellular protein synthesis. Both these approaches revealed much faster turnover/degradation of the D1 protein when compared with other PSII subunits (Edelman and Reisfeld 1978; Schuster et al. 1988). D1 turnover occurs at all light intensities and its rate is similar to the rate of PSII photoinactivation and proportional to the light intensity (Mattoo et al. 1984; Komenda and Barber 1995; Tyystjärvi and Aro 1996). Nevertheless, a detailed kinetic study of D1 degradation in *Spirodela* plants revealed that D1 degradation is a low light–associated process and irradiance as low as 5 μmol photons/m^2/s already elicits about 25% of the total D1 degradation response observed at full sunlight (about 1500 μmol photons/m^2/s) (Jansen et al. 1999).

The crystal structures of cyanobacterial PSII (Ferreira et al. 2004; Guskov et al. 2009; Umena et al. 2011) show that the D1 protein is buried within the center of the large PSII complex and is not easily accessible to the machinery that catalyzes its selective replacement. Therefore, to allow access to the protein, the neighboring inner antenna CP43 must be temporarily detached from the core complex (Barbato et al. 1992; Komenda and Masojídek 1995) and also most probably the proteins stabilizing the Mn_4Ca cluster must be at least partly relocated. How these proteins are detached and whether specific protein factors are involved in this event remain unknown. The exposed D1 protein still bound in the complex lacking CP43 (termed the RC47 complex) becomes a target of the protease (Adir et al. 1990; Komenda et al. 2006).

1.4.2 Selective D1 Degradation and the Identity of the Protease

The identity of the protease involved in selective D1 degradation remained an enigma for nearly 30 years after the first detection of fast D1 turnover (Edelman and Reisfeld 1978). To simplify the search for the protease, early studies initially focused on the use of isolated membranes (Ohad et al. 1985) and later on isolated PSII complexes in which exposure to high light intensities induced formation of specific D1 fragments (Virgin et al. 1990). The inhibitory effect of serine protease inhibitors on D1 fragmentation in various types of isolated PSII complex led to the proposal that either the inner antenna CP43 (Salter et al. 1992) or the RC complex consisting of D1, D2, and additional small polypeptides (Shipton and Barber 1991) possesses a serine protease activity participating in D1 degradation. However, later it appeared that the fragments detected in many *in vitro* studies may represent products of chemical cleavage caused by the direct action of reactive oxygen species (ROS) rather than the result of enzymatic proteolysis (Miyao et al. 1995; Lupínková and Komenda 2004). Therefore, attention has switched more recently to the analysis of physiologically relevant proteases and, in particular, homologues of the HtrA/DegP and FtsH bacterial proteases that are present in cyanobacteria and chloroplasts (Adam 1996).

Initially, an ATP-independent Deg-type serine protease (Deg2) was proposed to be responsible for the initial cleavage of damaged D1, in the loop between the fourth and fifth transmembrane helix, leading to the generation of a 23 kDa N-terminal D1 fragment (Haussühl et al. 2001) that had been detected both *in vitro* (De Las Rivas et al. 1992; Salter et al. 1992) and *in vivo* (Greenberg et al. 1987). Subsequent proteolysis of the breakdown fragments was proposed to be catalyzed by an FtsH-type protease based on the dependence of its degradation on the presence of ATP and Zn ions (Lindahl et al. 2000). However, later studies cast doubts upon the strict cooperation of DegP and FtsH during D1 degradation as inactivation of the *ftsH2* gene in the cyanobacterium *Synechocystis* 6803 (Silva et al. 2003; Komenda et al. 2006) led to the strong inhibition of D1 degradation while no such effect was observed after inactivation of all three genes coding for Deg proteases (Barker et al. 2006). Moreover, a mutant lacking FtsH2 showed stabilization of damaged D1 protein, and no evidence of fragmentation, despite loss of PSII function. This has led to the present prevailing opinion that FtsH homologues are the primary candidates for the enigmatic D1-degrading protease.

The genome of the model plant *Arabidopsis thaliana* contains a number of FtsH-related genes and two of them, FtsH2 (yellow variegated 2 or *VAR2*; Bailey et al. 2002) and FtsH5 (*VAR1*; Sakamoto et al. 2002) have also been found to be important for D1 degradation in analogy to the effect of the *ftsH2* deletion in the cyanobacterium *Synechocystis* 6803.

In chloroplasts, there have been a number of suggestions that Deg proteases play a physiological role in D1 degradation (Edelman and Mattoo 2008; Huesgen et al. 2009). Chloroplast of *Arabidopsis* contains a set of Deg proteases with Deg2 and Deg7 being present in the stroma and Deg1, Deg5, and Deg8 in the lumen. Deg2 is able to cleave the D1 protein in isolated thylakoids giving rise to the aforementioned N-terminal 23-kDa fragment (Haussühl et al. 2001) but *in vivo* is not essential for D1 degradation (Huesgen et al. 2006). Neither the lumenal Deg5 and/or Deg8, nor the stromal Deg7 proteases are required for D1 degradation under standard growth light conditions (Sun et al. 2007, 2010a), Nevertheless, under high irradiance D1 degradation is slowed down in *deg5/deg8* or *deg7* mutants which at first sight would indicate that under extreme conditions when the FtsH-mediated pathway of D1 degradation is insufficient, Deg5, Deg7, and Deg8 may additionally help to cope with high levels of PSII damage. However, the data currently available do not exclude an indirect role for these proteases in D1 replacement, such as ensuring that sufficient ATP is available. The remaining lumenal protease Deg1 has also been suggested to participate in D1 degradation (Kapri-Pardes et al. 2007), but more recent data showed that D1 turnover was relatively unaffected in its absence; instead, Deg1 appears to be important for PSII assembly via an interaction with the D2 protein (Sun et al. 2010b). In summary, the current mutant data support the view that the chloroplast Deg proteases are not essential for D1 degradation *in vivo* but might play an auxiliary role at higher light intensities, possibly cleaving interhelical loops, to facilitate and accelerate the FtsH-mediated processive degradation of D1.

1.4.3 Mechanism of D1 Replacement and Its Dynamics

FtsH homologues in cyanobacteria and chloroplasts are predicted to have two transmembrane helices close to the N-terminus, connected by a lumenal loop, followed by an AAA^+ (ATPase associated with diverse cellular activities) module and a Zn^{2+}-metalloproteinase domain both exposed to the stroma/cytoplasm (Nixon et al. 2005). Studies in non-photosynthetic systems suggest that FtsH subunits are able to assemble into either homo/hexameric complexes or hetero/hexameric complexes consisting of two types of subunit, with the N-terminal region of FtsH important for oligomerization. The FtsH complex is hypothesized to use ATP hydrolysis to pull target membrane protein substrates into an inner cavity where proteolysis occurs (Langer 2000). The FtsH-mediated proteolysis of membrane proteins in *Escherichia coli* most frequently starts from either the N- or C-terminus and continues in a processive way without generation of large polypeptide fragments (Chiba et al. 2002). Alternatively, FtsH can also cleave in membrane-exposed protein loops and processively degrade the resulting fragments (Okuno et al. 2006). With the proteolytic cavity exposed to the cytoplasmic/stromal side, the most attractive FtsH targets in the D1 structure are the long N-terminus and the loop between the fourth and fifth transmembrane helix (Ferreira et al. 2004). Proteolysis from the N-terminus is particularly attractive, as this part of the protein protrudes from the overall structure of the complex and is sufficiently long to be attacked by FtsH. Indeed, a cyanobacterial D1 protein lacking the first 20 amino acid residues is degraded very slowly (Komenda et al. 2007b), whereas mutants lacking most of the large stromal part between the fourth and fifth transmembrane helix of D1 still maintains a fast rate of its degradation (Nixon et al. 1995; Mulo et al. 1997). Moreover, D1 degradation from the N-terminus might explain why N-terminal phosphorylation of D1 in chloroplasts affects its degradation rate (Koivuniemi et al. 1995).

After removal of damaged D1 protein, insertion of the new D1 molecule must occur promptly to minimize the time when the complex is disassembled and when the probability of ROS generation is increased. Therefore, it is important that the new D1 copy or a synthesis intermediate is available for insertion after the degradation of the "old" D1. Data from isolated chloroplasts indicate that the insertion of the new D1 copy into the D1-less PSII subcomplex occurs cotranslationally (Zhang and Aro 2002). On the other hand, the observation that cyanobacterial mutants with inhibited D1 degradation (due to the absence of the FtsH2 protease or due to the N-terminal truncation of D1; Komenda et al. 2006, 2007b) accumulate unassembled full-length D1 protein rather speaks for the quick posttranslational integration of the whole protein into the complex, although cotranslational insertion in wild type cannot be totally excluded.

Degradation of damaged D1 and insertion of the new D1 protein into the PSII complex are highly coordinated processes that seem to stimulate each other. This has been well documented in cyanobacteria where it has been shown that ongoing D1 synthesis is necessary for the maximal rate of the D1 degradation (Komenda and Barber 1995; Komenda et al. 2000), and that mutants displaying a high rate of D1 degradation, for instance, those with an impaired oxygen-evolving complex, also stimulate the synthesis of D1 (Komenda et al. 2010). The mechanism of this synchronization remains unclear.

The cyanobacterial photosynthetic membrane complexes are located in thylakoid membranes that usually form several concentric layers more or less parallel with the cellular surface. The synthesis of D1 occurs on ribosomes bound to membranes but it is not clear yet if these are regular thylakoids, a specific subpopulation of thylakoids, or even cytoplasmic membranes. Nickelsen and colleagues have recently proposed a model in which PSII biogenesis occurs in specific, so-called intermediate membranes that are in contact with both the thylakoid and cytoplasmic membranes (Nickelsen et al. 2010). These membranes might be identical to the so-called "thylakoid centers" identified by electron microscopy in proximity of the cytoplasmic membrane (Kunkel 1982; van de Meene et al. 2006). These centers may represent the site of not only PSII assembly but also PSII repair. If so, PSII complexes in need of repair would have to migrate to this region of the membrane or into its vicinity. However, there are currently no compelling data to confirm this movement. PSII seems to be unusually immobile as suggested by "fluorescence recovery after photobleaching" (FRAP) studies and only strong light of specific wavelength induced some rearrangement of the complex (Sarcina et al. 2006).

The dynamics of PSII repair is potentially even more complicated in higher plants. PSII dimers and its associated light-harvesting system are located in semi-ordered arrays within the appressed membranes of the granal stacks (Dekker and Boekema 2005), which poses logistical problems for access of the FtsH protease to damaged D1 during PSII repair. Current models suggest that phosphorylation of light-harvesting proteins and PSII core proteins, including D1, facilitates PSII repair by enhancing the mobility of damaged PSII complexes within the appressed membranes to allow migration to the granal margins where PSII repair is likely to occur (Tikkanen et al. 2008; Fristedt et al. 2009; Goral et al. 2010; Yoshioka et al. 2010). Whether D1 phosphorylation at the N-terminus plays a role in the synchronization of D1 degradation and synthesis (Rintamäki et al. 1996) is currently unclear as higher plant mutants specifically lacking this phosphorylation site have not yet been studied. In the case of cyanobacteria, D1 is not phosphorylated, even though the phosphorylatable threonine residue is conserved, consistent with the idea that phosphorylation is a specific adaptation associated with the formation of grana.

1.4.4 Fate of Cofactors during D1 Replacement

The degradation of damaged D1 protein is accompanied by the release of cofactors such as Chls, Cars, quinones, and ions from the Mn_4Ca cluster, which then have to rebind again during, or soon after, the insertion of the newly synthesized protein. However, it is not known to what extent the cofactors present in the freshly repaired PSII are identical to the released ones and to what extent they are replaced by the new, externally delivered ones. If the cofactors are to be reused, the replacement process must occur in the vicinity of transient carriers that are able to immediately catch them and promptly deliver to the newly repaired PSII complex. Alternatively, the replacement of D1 may occur without assistance of carriers in a compartment with high local concentration of the cofactors due to their effective transport into a spatially separated compartment in which the cofactors may quickly bind to the newly synthesized D1 protein as soon as it is synthesized. Such a compartment may represent the aforementioned thylakoid centers. However, the high local concentration model cannot apply to Chl molecules as their permanent binding to proteins is essential for preventing generation of reactive singlet oxygen, which occurs when Chl remains free. The newly synthesized or released Chl molecules must therefore immediately be bound to transient carriers or to their permanent binding sites in the Chl-binding proteins including the D1 protein. The candidates for the carrier function in cyanobacteria are the so-called high light–induced proteins (HLIPS) (Dolganov et al. 1995), which are also termed small CAB-like proteins (SCPs) (Funk and Vermaas 1999). They belong to a large family of proteins, which in plants participate in light harvesting (e.g., LHCII) as well as in protection against excessive light (e.g., the PsbS subunit). All HLIPs identified in *Synechocystis* 6803 contain a putative Chl-binding site similar to that found in plant light-harvesting complexes (Funk and Vermaas 1999) and are predicted to contain carotenoids (Xu et al. 2004). Recent data on the biogenesis of the inner PSII antennae, CP43 and CP47, in the cyanobacterium *Synechocystis* 6803 show that binding of Chls and Cars to these apoproteins is able to occur before incorporation of CP47 and CP43 into larger complexes during PSII assembly (Boehm et al. 2011). It seems reasonable to assume that binding of pigments occurs during or very soon after translation of the corresponding protein but convincing experimental evidence is still missing. This binding should occur immediately after release of the Chl molecule from the last enzymes of the Chl biosynthesis pathway or there might be a transient Chl carrier(s) that deliver it to the correct protein in the membrane. There is no information about whether the newly synthesized D1 protein already contains Chl bound when it is either inserted into the D1-less RC47 complex or binds to the D2 protein during formation of new RC subcomplexes.

Assembly of the Mn_4Ca cluster requires light to drive the photooxidation of Mn^{2+} ions in a process known as photoactivation (reviewed by Burnap 2004; Dasgupta et al. 2008). The fact that D1 provides amino acid ligands to all four Mn ions and the Ca ion means that the cluster must be assembled de novo following selective D1 replacement during PSII repair. Early kinetic models have suggested a two quantum mechanism for photoactivation in which an Mn^{2+} ion binds at a high-affinity site in an Mn-depleted PSII complex and is oxidized by turnover of the PSII RC, following absorption of the first photon, to generate an unstable Mn^{3+} state. After a dark step, thought to involve some sort of structural rearrangement, a second Mn^{2+} ion is photooxidized by turnover of the PSII RC following absorption of the second photon to give a relatively stable binuclear Mn center (Tamura and Cheniae 1987; Zaltsman et al. 1997) with an oxidation state equivalent to a super-reduced S state equivalent to S_{-3} (Burnap 2004). Binding and oxidation of two further Mn^{2+} by the PSII RC leads to an Mn_4 cluster with oxidation state (III, III, IV, IV) in the dark-stable S_1 state (Kulik et al. 2007). Mutant studies indicate that the Mn ion ligated by D1-Asp170 and D1-Glu333 in the crystal structure (Figure 1.2B) is the first one that is bound and oxidized during photoactivation (Nixon and Diner 1992; Cohen et al. 2007) and that the C-terminal region of D1 is already folded back close to D1-Asp170 in the absence of bound Mn (Cohen et al. 2007). Although assembly of the Mn_4Ca cluster can occur *in vitro*, it is likely that assembly *in vivo* is highly regulated to minimize photodamage. This could include the participation of "assembly factors" to deliver Mn^{2+} to PSII, and auxiliary proteins to synchronize C-terminal processing of D1, photoactivation of the cluster, and attachment of the lumenal extrinsic proteins. Indeed, the recently discovered Psb27 protein would appear to play such a regulatory role, although the mechanism is currently unclear (Roose and Pakrasi 2008).

1.5 CONCLUSIONS AND PERSPECTIVES

Relatively little is known about how multisubunit membrane proteins are repaired following damage (Daley 2008). The relatively high rates of damage and repair observed with PSII make it an excellent system to study fundamental aspects of protein recognition, selective protein degradation, and replacement and insertion of pigment and metal cofactors. Given that FtsH proteases, which are considered housekeeping proteases in the cell, with broad substrate range, play such an important role in D1 degradation, it seems likely that many of the lessons learned with PSII might be applicable to other complexes in the thylakoid membrane.

A combination of molecular biology and biochemistry has identified a number of auxiliary proteins that are important for PSII repair (Nixon et al. 2010). The next challenge will be to determine when and where they act in the PSII repair cycle and how they engage with PSII. It should even be possible, with the appropriate genetic background, to isolate PSII repair complexes for detailed structural studies and so gain insights at the sub-nano level.

ACKNOWLEDGMENT

We would like to thank Dr. Jon Nield (Queen Mary, University of London) for his kind generosity in providing Figure 1.4 and for fruitful scientific discussions.

REFERENCES

Adam, Z. 1996. Protein stability and degradation in chloroplasts. *Plant Mol. Biol.* 32:773–783.

Adir, N., Shochat, S., Ohad, I. 1990. Light-dependent D1 protein synthesis and translocation is regulated by reaction center II. Reaction center II serves as an acceptor for the D1 precursor. *J. Biol. Chem.* 265:12563–12568.

Adir, N., Zer, H., Shochat, S., Ohad, I. 2003. Photoinhibition—A historical perspective. *Photosynth. Res.* 76:343–370.

Anbudurai, P.R., Mor, T.S., Ohad, I., Shestakov, S.V., Pakrasi, H.B. 1994. The *CtpA* gene encodes the C-terminal processing protease for the D1 protein of the photosystem-II reaction-center complex. *Proc. Natl. Acad. Sci. USA* 91:8082–8086.

Bailey, S., Thompson, E., Nixon, P.J. et al. 2002. A critical role for the Var2 FtsH homologue of *Arabidopsis thaliana* in the photosystem II repair cycle in vivo. *J. Biol. Chem.* 277:2006–2011.

Barbato, R., Friso, G., Rigoni, F., Vecchia, F.D., Giacometti, G.M. 1992. Structural changes and lateral redistribution of photosystem II during donor side photoinhibition of thylakoids. *J. Cell Biol.* 119:325–335.

Barber, J. 2002. Photosystem II: A multisubunit membrane protein that oxidises water. *Curr. Opin. Struct. Biol.* 12:523–530.

Barker, M., de Vries, R., Nield, J., Komenda, J., Nixon, P.J. 2006. The Deg proteases protect *Synechocystis* sp. PCC 6803 during heat and light stresses but are not essential for removal of damaged D1 protein during the photosystem two repair cycle. *J. Biol. Chem.* 281:30347–30355.

Boehm, M., Romeo, E., Reisinger, V. et al. 2011. Investigating the early stages of photosystem II assembly in *Synechocystis* sp. PCC 6803: Isolation of CP47 and CP43 complexes. *J. Biol. Chem.* 286:14812–14819.

Burnap, R.L. 2004. D1 protein processing and Mn cluster assembly in light of the emerging photosystem II structure. *Phys. Chem. Chem. Phys.* 6:4803–4809.

Chen, G.X., Kazimir, J., Cheniae, G.M. 1992. Photoinhibition of hydroxylamine-extracted photosystem II membranes: Studies of the mechanism. *Biochemistry* 31:11072–11083.

Chiba, S., Akiyama, Y., Ito, K. 2002. Membrane protein degradation by FtsH can be initiated from either end. *J. Bacteriol.* 184:4775–4782.

Cohen, R.O., Nixon, P.J., Diner, B.A. 2007. Participation of the C-terminal region of the D1-polypeptide in the first steps in the assembly of the Mn_4Ca cluster of photosystem II. *J. Biol. Chem.* 282:7209–7218.

Daley, D.O. 2008. The assembly of membrane proteins into complexes. *Curr. Opin. Struct. Biol.* 18:420–424.

Dasgupta, J., Ananyev, G.M., Dismukes, G.C. 2008. Photoassembly of the water-oxidizing complex in photosystem II. *Coord. Chem. Rev.* 252:347–360.

Debus, R.J., Barry, B.A., Sithole, I., Babcock, G., McIntosh, L. 1988. Directed mutagenesis indicates that the donor to P680 in photosystem II is tyrosine-161 of the D1 polypeptide. *Biochemistry* 27:9071–9074.

De Las Rivas, J., Andersson, B., Barber, J. 1992. Two sites of primary degradation of the D1-protein induced by acceptor or donor side photoinhibition in PSII core complexes. *FEBS Lett.* 301:246–252.

Dekker, J.P., Boekema, E.J. 2005. Supramolecular organization of thylakoid membrane proteins in green plants. *Biochim. Biophys. Acta* 1706:12–39.

Diner, B.A., Babcock, G.T. 1996. Structure, dynamics, and energy conversion efficiency in photosystem II. In: *Oxygenic Photosynthesis: The Light Reactions*, eds. Ort, D.R., Yocum, C.F., pp. 213–247. Dordrecht, the Netherlands: Kluwer Academic Publishers.

Dolganov, N.A.M., Bhaya, D., Grossmann, A.R. 1995. Cyanobacterial protein with similarity to the chlorophyll a/b binding-proteins of higher-plants—Evolution and regulation. *Proc. Natl. Acad. Sci. USA* 92:636–640.

Edelman, M., Mattoo, A.K. 2008. D1-protein dynamics in photosystem II: The lingering enigma. *Photosynth. Res.* 98:609–620.

Edelman, M., Reisfeld, A. 1978. Characterization, translation and control of the 32,000 dalton chloroplast membrane protein in *Spirodela*. In: *Chloroplast Development*, eds. Akoyunoglou, G., Argyroudi-Akoyunoglou, J.H., pp. 641–652. New York: Elsevier.

Enami, I., Okamura, A., Nagao, R., Suzuki, T., Iwai, M., Shen, J.R. 2008. Structures and functions of the extrinsic proteins of photosystem II from different species. *Photosynth. Res.* 98:349–363.

Ferreira, K.N., Iverson, T.N., Maglaoui, K., Barber, J., Iwata, S. 2004. Architecture of the photosynthetic oxygen-evolving center. *Science* 303:1831–1838.

Fristedt, R., Willig, A., Granath, P., Crevecoeur, M., Rochaix, J.D., Vener, A.V. 2009. Phosphorylation of photosystem II controls functional macroscopic folding of photosynthetic membranes in *Arabidopsis*. *Plant Cell* 21:3950–3964.

Funk, C., Vermaas, W. 1999. A cyanobacterial gene family coding for single helix proteins resembling part of the light-harvesting proteins from higher plants. *Biochemistry* 38:9397–9404.

Gong, H., Ohad, I. 1991. The PQ/PQH_2 ratio and occupancy of photosystem II Q_B site by plastoquinone control the degradation of the D1 protein during photoinhibition in vivo. *J. Biol. Chem.* 266:21293–21252.

Goral, T.K., Johnson, M.P., Brain, A.P., Kirchhoff, H., Ruban, A.V., Mullineaux, C.W. 2010. Visualising the mobility and distribution of chlorophyll-proteins in higher plant thylakoid membranes: Effects of photoinhibition and protein phosphorylation. *Plant J.* 62:948–959.

Green, B.R., Anderson, J.M., Parson, W.W. 2003. Photosynthetic membranes and their light-harvesting antennas. In: *Light-Harvesting Antennas in Photosynthesis*, eds. Greeen, B.R., Parson, W.W., pp. 1–28. Dordrecht, the Netherlands: Kluwer Academic Publishers.

Greenberg, B.M., Gaba, V., Mattoo, A.K., Edelman, M. 1987. Identification of a primary in vivo degradation product of the rapidly-turning-over 32 kD protein of photosystem II. *EMBO J.* 6:2865–2869.

Guskov, A., Kern, J., Gabdulkhakov, A., Broker, M., Zouni, A., Saenger, W. 2009. Cyanobacterial photosystem II at 2.9-Å resolution and the role of quinones, lipids, channels and chloride. *Nat. Struct. Biol.* 16:334–342.

Hakala, M., Tuominen, I., Keranen, M., Tyystjaervi, T., Tyystjaervi, E. 2005. Evidence for the role of the oxygen-evolving manganese complex in photoinhibition of photosystem II. *Biochim. Biophys. Acta* 1706:68–80.

Haussühl, K., Andersson, B., Adamska, I. 2001. A chloroplast DegP2 protease performs the primary cleavage of the photodamaged D1 protein in plant photosystem II. *EMBO J.* 20:713–722.

Hideg, É., Spetea, C., Vass, I. 1994. Singlet oxygen and free radical production during acceptor- and donor- induced photoinhibition. Studies with spin trapping EPR spectroscopy. *Biochim. Biophys. Acta* 1186:143–152.

Huesgen, P.F., Schuhmann, H., Adamska, I. 2006. Photodamaged D1 protein is degraded in *Arabidopsis* mutants lacking the Deg2 protease. *FEBS Lett.* 580:6929–6932.

Huesgen, P.F., Schuhmann, H., Adamska, I. 2009. Deg/HtrA proteases as components of a network for photosystem II duality control in chloroplasts and cyanobacteria. *Res. Microbiol.* 160:726–732.

Ivleva, N.B., Shestakov, S.V., Pakrasi, H.B. 2000. The carboxyl-terminal extension of the precursor D1 protein of photosystem II is required for optimal photosynthetic performance of the cyanobacterium *Synechocystis* sp. PCC 6803. *Plant Physiol.* 124:1403–1411.

Jansen, M.A.K., Mattoo, A.K., Edelman, M. 1999. D1–D2 protein degradation in the chloroplast. Complex light saturation kinetics. *Eur. J. Biochem.* 260:527–532.

Kapri-Pardes, E., Naveh, L., Adam, Z. 2007. The thylakoid lumen protease Deg 1 is involved in the repair of photosystem II from photoinhibition in *Arabidopsis. Plant Cell* 19:1039–1047.

Keren, N., Berg, A., van Kann, P.J.M., Levanon, H., Ohad, I. 1997. Mechanism of photosystem II photoinactivation and D1 protein degradation at low light: The role of back electron flow. *Proc. Natl. Acad. Sci. USA* 94:1579–1584.

Kirilovsky, D., Vernotte, C., Astier, C., Etienne, A.-L. 1988. Reversible and irreversible photoinhibition in herbicide-resistant mutants of *Synechocystis* 6714. *Biochim. Biophys. Acta* 933:124–131.

Koivuniemi, A., Aro, E.-M., Andersson, B. 1995. Degradation of the D1 and D2 proteins of photosystem II in higher plants is regulated by reversible phosphorylation. *Biochemistry* 34:16022–16029.

Komenda, J., Barber, J. 1995. Comparison of *psb*O and *psb*H deletion mutants of *Synechocystis* PCC 6803 indicates that degradation of the D1 protein is regulated by the Q(B) site and dependent on protein synthesis. *Biochemistry* 32:1454–1465.

Komenda, J., Barker, M., Kuviková, S., DeVries, R., Mullineaux, C.W., Tichý, M., Nixon, P.J. 2006. The FtsH protease slr0228 is important for quality control of photosystem II in the thylakoid membrane of *Synechocystis* PCC 6803. *J. Biol. Chem.* 281:1145–1151.

Komenda, J., Hassan, H.A.G., Diner, B.A., Debus, R.J., Barber, J., Nixon, P.J. 2000. Degradation of the photosystem II D1 and D2 proteins in different strains of the cyanobacterium *Synechocystis* PCC 6803 varying with respect to the type and level of *psbA* transcript. *Plant Mol. Biol.* 42:635–645.

Komenda, J., Knoppová, J., Krynická, V., Nixon, P.J., Tichý, M. 2010. Role of FtsH2 in the repair of photosystem II in mutants of the cyanobacterium *Synechocystis* PCC 6803 with impaired assembly or stability of the $CaMn_4$ cluster. *Biochim. Biophys. Acta* 1797:566–575.

Komenda, J., Kuviková, S., Granvogl, B. et al. 2007a. Cleavage after residue Ala352 in the C-terminal extension is an early step in the maturation of the D1 subunit of photosystem II in *Synechocystis* PCC 6803. *Biochim. Biophys. Acta* 1767:829–837.

Komenda, J., Masojídek, J. 1995. Functional and structural changes of the photosystem II complex induced by high irradiance in cyanobacterial cells. *Eur. J. Biochem.* 233:677–682.

Komenda, J., Nickelsen, J., Eichacker, L.A. et al. 2008. The cyanobacterial homologue of HCF136/YCF48 is a component of an early photosystem II assembly complex and is important for both the efficient assembly and repair of photosystem II in *Synechocystis* sp. PCC6803. *J. Biol. Chem.* 283:22390–22399.

Komenda, J., Tichy, M., Prasil, O. et al. 2007b. The exposed N-terminal tail of the D1 subunit is required for rapid D1 degradation during photosystem II repair in *Synechocystis* sp. PCC 6803. *Plant Cell* 19:2839–2854.

Kulik, L.V., Epel, B., Lubitz, W., Messinger, J. 2007. Electronic structure of the Mn_4O_xCa cluster in the S_0 and S_2 states of the oxygen-evolving complex of photosystem II based on pulse ^{55}Mn-ENDOR and EPR spectroscopy. *J. Am. Chem. Soc.* 129:13421–13435.

Kunkel, D.D. 1982. Thylakoid centers: Structures associated with the cyanobacterial photosynthetic membrane system. *Arch. Microbiol.* 133:97–99.

Kuviková, S., Tichý, M., Komenda, J. 2005. A role of the C-terminal extension of the photosystem II D1 protein in sensitivity of the cyanobacterium *Synechocystis* PCC 6803 to photoinhibition. *Photochem. Photobiol. Sci.* 4:1044–1048.

Langer, T. 2000. AAA proteases: Cellular machines for degrading membrane proteins. *Trends Biochem. Sci.* 25:247–251.

Lindahl, M., Spetea, C., Hundal, T., Oppenhaim, A.B., Adam, Z., Andersson, B. 2000. The thylakoid FtsH protease plays a role in the light-induced turnover of the PSII D1 protein. *Plant Cell* 12:419–431.

Lupínková, L., Komenda, J. 2004. Oxidative modifications of the photosystem II D1 protein by reactive oxygen species: From isolated protein to cyanobacterial cells. *Photochem. Photobiol.* 79:152–162.

Marder, J.B., Goloubinoff, P., Edelman, M. 1984. Molecular architecture of the rapidly metabolized 32-kilodalton protein of photosystem II—Indications for COOH-terminal processing of a chloroplast membrane polypeptide. *J. Biol. Chem.* 259:3900–3908.

Mattoo, A.K., Hoffman-Falk, H., Marder, J.B., Edelman, M. 1984. Regulation of protein metabolism; coupling of photosynthetic electron transport to in vivo degradation of the rapidly metabolized 32-kDa protein of chloroplast membranes. *Proc. Natl. Acad. Sci. USA* 81:1380–1384.

Metz, J.G., Nixon, P.J., Rögner, M., Brudvig, G.W., Diner, B.A. 1989. Directed alteration of the D1 polypeptide of photosystem II: Evidence that tyrosine-161 is the redox component, Z, connecting the oxygen-evolving complex to the primary electron donor, P680. *Biochemistry* 28:6960–6969.

Miyao, M., Ikeuchi, M., Yamamoto, N., Ono, Y. 1995. Specific degradation of the D1 protein of photosystem II by treatment with hydrogen peroxide in darkness: Implications for the mechanism of degradation of the D1 protein under illumination. *Biochemistry* 34:10019–10026.

Müh, F., Renger, T., Zouni, A. 2008. Crystal structure of cyanobacterial photosystem II at 3.0Å resolution: A closer look at the antenna system and the small membrane-intrinsic subunits. *Plant Physiol. Biochem.* 46:238–264.

Mulo, P., Tyystjärvi, T., Tyystjärvi, E., Govindjee, Mäenpää, P., Aro, E.-M. 1997. Mutagenesis of the D–E loop of photosystem II reaction center protein D1. Function and assembly of photosytem II. *Plant Mol. Biol.* 33:1059–1071.

Murchie, E.H., Niyogi, K.K. 2011. Manipulation of photoprotection to improve photosynthesis. *Plant Physiol.* 155:86–92.

Nickelsen, J., Rengstl, B., Stengel, A., Schottkowski, M., Soll, J., Ankele, E. 2010. Biogenesis of the cyanobacterial thylakoid membrane system—An update. *FEMS Microbiol. Lett.* 315:1–5.

Nixon, P.J., Barker, M., Boehm, M., de Vries, R., Komenda, J. 2005. FtsH-mediated repair of the photosystem two complex in response to light stress in the cyanobacterium *Synechocystis* sp. PCC 6803. *J. Exp. Bot.* 56:347–356.

Nixon, P.J., Diner, B.A. 1992. Aspartate 170 of the photosystem II reaction center polypeptide D1 is involved in the assembly of the oxygen-evolving manganese cluster. *Biochemistry* 31:942–948.

Nixon, P.J., Komenda, J., Barber, J., Deak, Z., Vass, I., Diner, B.A. 1995. Deletion of the PEST-like region of photosystem II modifies the Q_B-binding pocket but does not prevent rapid turnover of D1. *J. Biol. Chem.* 270:14919–14927.

Nixon, P.J., Michoux, F., Yu, J., Boehm, M., Komenda, J. 2010. Recent advances in understanding the assembly and repair of photosystem II. *Annals Botany* 106:1–16.

Nixon, P.J., Trost, J.T., Diner, B.A. 1992. Role of the carboxy terminus of polypeptide D1 in the assembly of a functional water-oxidizing manganese cluster in photosystem II of the cyanobacterium *Synechocystis* sp. PCC 6803—Assembly requires a free carboxyl group at C-terminal position 344. *Biochemistry* 31:10859–10871.

Niyogi, K.K. 1999. Photoprotection revisited: Genetic and molecular approaches. *Annu. Rev. Plant Physiol. Plant Mol. Biol.* 50:333–359.

Ohad, I., Kyle, D.J., Hirchberg, J. 1985. Light-dependent degradation of the Q_B protein in isolated pea thylakoids. *EMBO J.* 4:1655–1659.

Ohnishi, N., Allakhverdiev, S.I., Takahashi, S. et al. 2005. Two-step mechanism of photodamage to photosystem II: Step 1 occurs at the oxygen-evolving complex and step 2 occurs at the photochemical reaction center. *Biochemistry* 44:8494–8499.

Okuno, T., Yamanaka, K., Ogura, T. 2006. An AAA protease FtsH can initiate proteolysis from internal sites of a model substrate, apo-flavodoxin. *Genes Cells* 11:261–268.

Rintamäki, E., Kettunen, R., Aro, E.-M. 1996. Differential D1 dephosphorylation in functional and photodamaged photosystem II centers. Dephosphorylation is a prerequisite for degradation of damaged D1. *J. Biol. Chem.* 271:14870–14875.

Roose, J.L., Kashino, Y., Pakrasi, H.B. 2007b. The PsbQ protein defines cyanobacterial photosystem II complexes with highest activity and stability. *Proc. Natl. Acad. Sci. USA* 104:2548–2553.

Roose, J.L., Pakrasi, H.B. 2008. The Psb27 protein facilitates Mn cluster assembly in photosystem II. *J. Biol. Chem.* 283:4044–4050.

Roose, J.L., Wegener, K.M., Pakrasi, H.B. 2007a. The extrinsic proteins of photosystem II. *Photosynth. Res.* 92:369–387.

Sakamoto, W., Tamura, T., Hanba-Tomita, Y., Sodmergen, Murata, M. 2002. The VAR1 locus of *Arabidopsis* encodes a chloroplastic FtsH and is responsible for leaf variegation in the mutant alleles. *Genes Cells* 7:769–780.

Salter, A.H., Virgin, I., Hagman, A., Andersson, B. 1992. On the molecular mechanism of light-induced D1 protein degradation in PSII core particles. *Biochemistry* 31:3990–3998.

Sarcina, M., Bouzovitis, N., Mullineaux, C.W. 2006. Mobilization of photosystem II induced by intense red light in the cyanobacterium *Synechococcus* sp. PCC7942. *Plant Cell* 18:457–464.

Schuster, G., Timberg, R., Ohad, I. 1988. Turnover of thylakoid photosystem II proteins during photoinhibition of *Chlamydomonas reinhardtii. Eur. J. Biochem.* 177:403–410.

Shipton, C.A., Barber, J. 1991. Photoinduced degradation of the D1-polypeptide in isolated reaction centers of photosystem-II—Evidence for an autoproteolytic process triggered by the oxidizing side of the photosystem. *Proc. Natl. Acad. Sci. USA* 88:6691–6695.

Silva, P., Thompson, E., Bailey, S. et al. 2003. FtsH is involved in the early stages of repair of photosystem II in *Synechocystis* sp. PCC 6803. *Plant Cell* 15:2152–2164.

Summerfield, T.C., Shand, J.A., Bentley, F.K., Eaton-Rye, J.J. 2005a. PsbQ (Sll1638) in *Synechocystis* sp. PCC 6803 is required for photosystem II activity in specific mutants and in nutrient-limiting conditions. *Biochemistry* 44:805–815.

Summerfield, T.C., Winter, R.T., Eaton-Rye, J.J. 2005b. Investigation of a requirement for the PsbP-like protein in *Synechocystis* sp. PCC 6803. *Photosynth. Res.* 84:263–268.

Sun, X., Fu, T., Chen, N., Guo, J., Ma, J., Zou, M., Lu, C., Zhang, L. 2010a. The stromal chloroplast Deg7 protease participates in the repair of photosystem II after photoinhibition in *Arabidopsis. Plant Physiol.* 152:1263–1273.

Sun, X., Ouyang, M., Guo, J., Ma, J., Lu, C., Adam, Z., Zhang, L. 2010b. The thylakoid protease Deg1 is involved in photosystem II assembly in *Arabidopsis thaliana. Plant J.* 62:240–249.

Sun, X., Peng, L., Guo, J., Chi, W., Ma, J., Lu, C., Zhang, L. 2007. Formation of DEG5 and DEG8 complexes and their involvement in the degradation of photodamaged photosystem II reaction center D1 protein in *Arabidopsis. Plant Cell* 19:1388–1402.

Szilárd, A., Sass, L., Deák, Z., Vass, I. 2007. The sensitivity of photosystem II to damage by UV-B radiation depends on the oxidation state of the water-splitting complex. *Biochim. Biophys. Acta* 1767:876–882.

Takahashi, M., Shiraishi, T., Asada, K. 1988. COOH-terminal residues of D1 and the 44-kDa CPa-2 at spinach photosystem II core complex. *FEBS Lett.* 240:6–8.

Tamura, N., Cheniae, G. 1987. Photoactivation of the water-oxidizing complex in photosystem II membranes depleted of Mn and extrinsic proteins. 1. Biochemical and kinetic characterization. *Biochim. Biophys. Acta* 890:179–194.

Thornton, L.E., Ohkawa, H., Roose, J.L., Kashino, Y., Keren, N., Pakrasi, H.B. 2004. Homologs of plant PsbP and PsbQ proteins are necessary for regulation of photosystem II activity in the cyanobacterium *Synechocystis* 6803. *Plant Cell* 16:2164–2175.

Tikkanen, M., Nurmi, M., Kangasjarvi, S., Aro, E.M. 2008. Core protein phosphorylation facilitates the repair of photodamaged photosystem II at high light. *Biochim. Biophys. Acta* 1777:1432–1437.

Tyystjärvi, E. 2008. Photoinhibition of photosystem and photodamage of the oxygen evolving manganese cluster. *Coord. Chem. Rev*. 252:361–376.

Tyystjärvi, E., Aro, E.-M. 1996. The rate constant of photoinhibition, measured in lincomycin-treated leaves, is directly proportional to light intensity. *Proc. Natl. Acad. Sci. USA* 93:2213–2218.

Umena, Y., Kawakami, K., Shen, J.-R., Kamiya, N. 2011. Crystal structure of oxygen-evolving photosystem II at a resolution of 1.9 Å. *Nature* 473:55–60.

van de Meene, A.M.L., Hohmann-Marriott, M.F., Vermaas, W.F.J., Roberson, R.W. 2006. The three-dimensional structure of the cyanobacterium *Synechocystis* sp. PCC 6803. *Arch. Microbiol.* 184:259–270.

Vass, I. 2011. Molecular mechanisms of photodamage in the photosystem II complex. *Biochim. Biophys. Acta*. doi:10.1016/j.bbabio.2011.04.014.

Vass, I., Styring, S., Hundal, T., Koivuniemi, A., Aro, E.-M., Andersson, B. 1992. Reversible and irreversible intermediates during photoinhibition of photosystem II: Stable reduced Q_A species promote chlorophyll triplet formation. *Proc. Natl. Acad. Sci. USA* 89:1408–1412.

Virgin, I., Ghanotakis, D.F., Andersson, B. 1990. Light-induced D1-protein degradation in isolated photosytem II core complexes. *FEBS Lett.* 287:125–128.

Xu, H., Vavilin, D., Funk, C., Vermaas, W. 2004. Multiple deletions of small Cab-like proteins in the cyanobacterium *Synechocystis* sp. PCC 6803. *J. Biol. Chem.* 279:27971–27979.

Yoshioka, M., Nakayama, Y., Yoshida, M. et al. 2010. Quality control of photosystem II. FtsH hexamers are localized near photosystem II at grana for swift repair of damage. *J. Biol. Chem.* 285:41972–41981.

Zaltsman, L., Ananyev, G.M., Bruntrager, E., Dismukes, G.C. 1997. Quantitative kinetic model for photoassembly of the photosynthetic water oxidase from its inorganic constituents: Requirements for manganese and calcium in the kinetically resolved steps. *Biochemistry* 36:8914–8922.

Zhang, L., Aro, E.-M. 2002. Synthesis, membrane insertion and assembly of the chloroplast-encoded D1 protein into photosystem II. *FEBS Lett.* 512:13–18.

2 Regeneration of 11-*cis*-Retinal in the Retinoid Visual Cycle

Gennadiy Moiseyev and Jian-Xing Ma

CONTENTS

2.1 INTRODUCTION

In vertebrates, both rod and cone visual pigments require 11-*cis*-retinal as a chromophore [1]. Upon absorption of photon, 11-*cis*-retinal photoisomerizes to all-*trans*-retinal, which triggers the conformational change of opsin, activating G-protein transducin and initiating vision [1,2]. The recycling of 11-*cis*-retinal termed the visual cycle is a key process in the regeneration of visual pigments (for reviews, see [3–5]; Figure 2.1). All-*trans*-retinal generated by photoactivation is dissociated from opsin and converted to all-*trans*-retinol by all-*trans*-retinol dehydrogenase [6,7]. The all-*trans*-retinol is then exported from photoreceptors to the retinal pigment epithelium (RPE) through the interphotoreceptor matrix [3,5]. In the RPE cells, all-*trans*-retinol is esterified to all-*trans*-retinyl esters [8]. The key enzyme of the visual cycle—isomerohydrolase—processes all-*trans*-retinyl esters into 11-*cis*-retinol [9,10]. The chemical nature of the isomerohydrolase was unknown for a long time. Recently, we and two other groups reported that the protein RPE65 possesses isomerohydrolase activity [11–13] (see the following text). Our recent study shows that purified RPE65 constituted into liposomes displays robust isomerohydrolase activity and thus provides conclusive evidence that RPE65 is the isomerohydrolase in the visual cycle [14]. There are two possible pathways for 11-*cis*-retinol in the RPE: the first pathway is to be esterified and stored as retinyl ester in the RPE [3,5] and the second is for

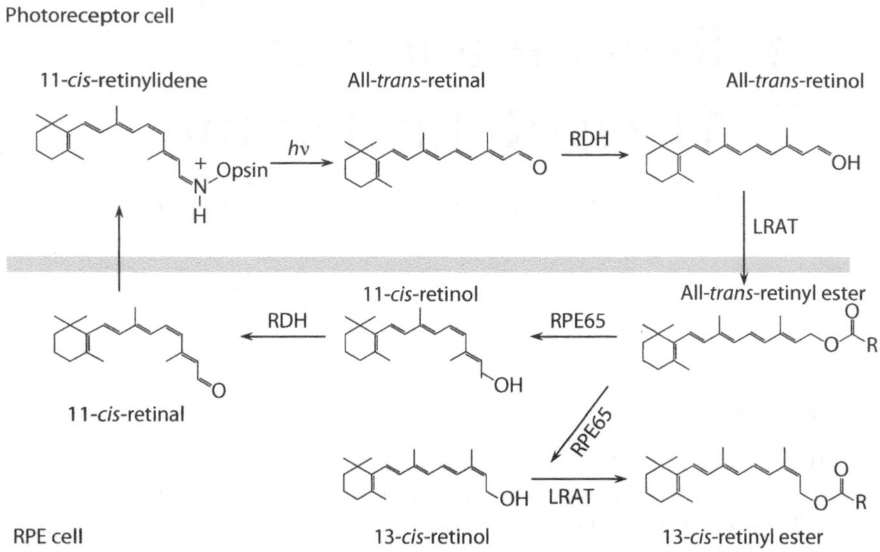

FIGURE 2.1 Overall scheme of regeneration of 11-*cis*-retinal in the visual cycle.

11-*cis*-retinol to be oxidized to become 11-*cis*-retinal by 11-*cis*-retinol dehydrogenase [3,15]. The 11-*cis*-retinal produced is then transported back to the photoreceptor cell, and recombines with opsin to form visual pigments, completing the visual cycle. In this review, we will discuss novel aspects of the enzymology of the visual cycle and mechanism of action of the enzymes catalyzing its reactions.

2.2 TRANSPORTATION OF ALL-*TRANS*-RETINAL OUT OF ROD OUTER SEGMENTS

After the decay of meta II rhodopsin following photoisomerization, all-*trans*-retinal dissociates from opsin and initially localizes in the membrane interior or outer segment disk. Inside the membrane, all-*trans*-retinal readily and reversibly reacts with phosphatidylethanolamine (PE), an abundant lipid of membranes, and forms *N*-retinylidene-PE (*N*-ret-PE) [16]. Formation of *N*-ret-PE delays the delivery of all-*trans*-retinal outside of the interior of the disks. Biochemical and animal model studies suggest that *N*-ret-PE is a substrate for ATP-binding cassette transporter (ABCA4 or ABCR), which is specifically expressed in the rim of photoreceptor outer segment disks [17–20]. Mice with mutations that disrupt ABCA4 function have a diminished rate of all-*trans*-retinal clearance after light exposure [21]. The slower removal of all-*trans*-retinal from the membranes of outer segments results in the second condensation of *N*-ret-PE and another molecule of all-*trans*-retinal and formation of toxic nondegradable A2E compound. Since photoreceptors permanently shed their outer segment disks, which enter RPE cells via phagocytosis, A2E eventually accumulates inside RPE cells and can cause various deleterious effects such as destabilization of the membranes [22,23], inhibiting respiration in mitochondria [24,25], increasing blue-light photodamage [26,27], impairing lysosomal acidification [28], and impairing degradation of phospholipids from phagocytosed outer segments [29]. The oxiran products of A2E irradiation by blue light may induce DNA fragmentation in cultured RPE cells [30,31]. It was also suggested that A2E oxidation products may trigger the complement system that predisposes the macula to disease and contributes to chronic inflammation [32]. It has also been shown that A2E is a ligand of the retinoic acid receptor, and this interaction upregulates VEGF expression in the RPE and choroid [33]. Recently, we have demonstrated that A2E inhibits the visual cycle through binding with RPE65 isomerohydrolase [34]. This is interesting, since autosomal recessive Stargardt disease has been associated with mutations in the ABCA4 gene [18], suggesting that loss of ABCA4 function may lead to macular degeneration by causing a delay in atRAL clearance that leads to the generation of A2E, a toxic derivative that may induce inflammation and also impairs many aspects of RPE cell function, including the visual cycle itself. *In vitro* biochemical experiments showed that ABCA4 protein translocates all-*trans*-retinal and *N*-ret-PE from the luminal to the cytoplasmic side of rod outer segments for subsequent conversion to all-*trans*-retinol. It was reported earlier that mice with mutations disrupting ABCA4 function have diminished rate of all-*trans*-retinal clearance after light exposure, although later this finding was not confirmed. Also, 11-*cis*-retinal synthesis was not decreased in *abca4*$^{-/-}$ mice.

2.3 ALL-*TRANS*-RETINAL REDUCTION IN PHOTORECEPTORS

The first enzymatic event in the regeneration of 11-*cis*-retinal after bleaching of rhodopsin is the conversion of all-*trans*-retinal dissociated from opsin to all-*trans*-retinol, which is further directed to the RPE. This reaction is catalyzed by retinol dehydrogenase (RDH) acting as a reductase (Figure 2.2). The molecular identity of this retinal reductase has been ambiguous. There are two major RDHs characterized in photoreceptors: RDH8 [35] and RDH12 [36]. These dehydrogenases belong to the superfamily of short-chain dehydrogenases/reductases (SDRs) [37]. SDR includes enzymes of ~250 amino acids, which have common structural motifs, including YXXXK, which contains the catalytic tyrosine and the nucleotide-binding site GXXXGXG [38,39]. In addition to retinoids, steroids [38] and short-chain aldehydes

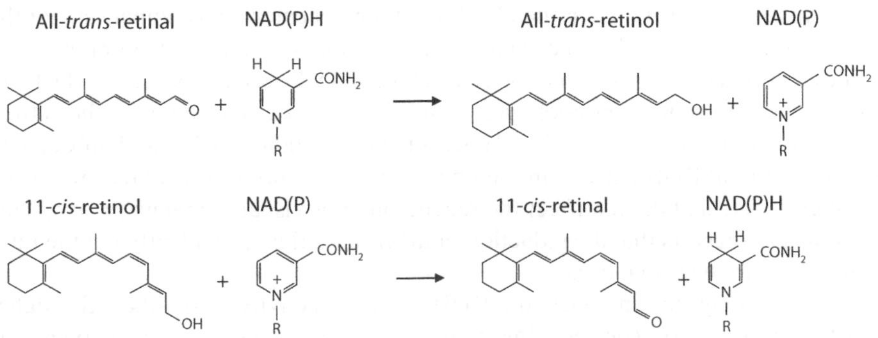

FIGURE 2.2 Reduction of all-*trans*-retinal and oxidation of 11-*cis*-retinol by retinol dehydrogenases.

can also be substrates for some members of the SDR family. RDH8 and RDH12 both require NADPH as a preferred cofactor [35,36,40]. Importantly, photoreceptors efficiently produce NADPH cofactor from NADP through the hexose monophosphate pathway to provide a necessary coenzyme for reduction of all-*trans*-retinal [41].

RDH12 has been suggested as an important enzyme in the visual cycle, since mutations of RDH12 were found in patients with autosomal recessive retinal dystrophy (arRD) [42] and Leber congenital amaurosis (LCA) [43,44]. *In vitro* studies using photoreceptor homogenates from *rdh12*$^{-/-}$ mice showed a significant (30%) contribution of RDH12 enzyme to overall RDH activity [45]. However, the mouse model for RDH12 deficiency did not demonstrate marked differences in retinoid flow or visual cycle function and exhibited normal retinal histology [45,46]. RDH12 intracellular localization in the inner segment of photoreceptors also argues against the specific role of RDH12 in the visual cycle [36]. Besides retinal, RDH12 can also catalyze the reduction of medium-chain aldehydes, the products of lipid peroxidation, which may cause cell apoptosis [40,47]. Therefore, it has been proposed that RDH12 plays a protective role against photoreceptor apoptosis during persistent illumination. Indeed, human patients with RDH12 inactivating mutations are not blind in early childhood and develop a legal blindness only in early adulthood (18–25 years) [42,43]. Therefore, it is possible that vision loss associated with RDH12 mutations develops at an older age mostly due to the effects of chronic light damage.

RDH8 appears to be the major retinal-reducing enzyme in the visual cycle. RDH8 is located in outer segments of photoreceptors and it efficiently reduces all-*trans*-retinal *in vitro*. RDH8 accounts for about 70% of the total RDH activity in the eye according to *in vitro* assays [45]. It requires NADPH as a cofactor and shows 20 times higher activity toward all-*trans*-retinal as compared with 11-*cis*-retinal [35]. However, *rdh8*$^{-/-}$ mice manifest only very slightly delayed clearance of all-*trans*-retinal released from light-activated rhodopsin [48], and retinal degeneration was not observed in *rdh8*$^{-/-}$ mice. Surprisingly, even double knockout *rdh8*$^{-/-}$*rdh12*$^{-/-}$ animals have nearly normal regeneration of visual pigments [45]. This mouse model showed only mild light-dependent retinal degeneration, delayed dark adaptation, and reduced processing of all-*trans*-retinal after the bleach. No mutations in the RDH8 gene have been linked to eye diseases. The absence of severe phenotype in the regeneration of 11-*cis*-retinal is likely due to the redundancy of RDH members expressed in the eye, and other yet unidentified RDHs in photoreceptors can compensate for the ablation of *rdh8* and *rdh12* genes and reduce all-*trans*-retinal. It is well known that besides short-chain RDHs, some medium-chain dehydrogenases such as ADH can reduce retinal [49,50]. In physiological conditions, their contribution in the retinal reduction may not exceed 2%–3%. However, the deletion of *rdh8* and subsequent reduction of total RDH activity in the photoreceptor do not significantly impair the visual cycle rate and do not affect the regeneration of visual pigments *in vivo*. This phenomenon suggests that the reduction of all-*trans*-retinal is evidently not the rate-limiting step in the visual cycle.

It is interesting that in cones, the RDH-reducing activity is significantly higher than that in rods [51]. The reducing activity in outer segments of carp cones was >30 times higher than rod RDH activity [51]. This high activity was attributed to a higher content of RDH8 in cones. It is likely that a higher abundance and a higher total activity of RDH are necessary for a faster cone response and the ability of cones to operate under bright day light. Interestingly, RDH12 expression in cones was not high. Instead, RD8L2, an isoform of RDH8, predominantly localized in the cone inner segment, can adopt the function of RDH12 in cone-dominant species [51].

In addition to providing all-*trans*-retinol for further synthesis of 11-*cis*-retinal chromophore, the reductase function of RDH in photoreceptors has another physiological significance. All-*trans*-retinal can react with PE in the photoreceptor membranes and form diretinal adduct A2E [52,53]. A2E is nondegradable and accumulates with age and causes age-related macular degeneration (AMD). The delay in clearance of retinal from photoreceptors due to ablation of the RDH8 and RDH12 genes results in accumulation of toxic bisretinoid A2E in RPE cells [45].

2.4 RETINOID TRANSFER BETWEEN PHOTORECEPTORS AND RPE

All-*trans*-retinol produced by RDH in the retina is released from photoreceptors and reaches the RPE through the interphotoreceptor matrix. This transfer is facilitated by interphotoreceptor-binding protein (IRBP). IRBP is a large (140 kDa) glycosylated protein abundantly presented in the intercellular space between the photoreceptor and RPE [54–56]. It can bind both all-*trans*-retinol and 11-*cis*-retinal with high affinity [57]. *In vitro*, IRBP accelerates the clearance of all-*trans*-retinol from photoreceptors [58]. IRBP also protects and stabilizes retinoids, preventing their thermal isomerization and oxidation [59]. However, $Irbp^{-/-}$ mice display surprisingly mild phenotypes and regeneration of rhodopsin was similar to that found in wild-type (wt) mice [60]. A possible explanation of the lack of severe phenotype is that some alternative pathways may play a role in the transfer of retinol.

2.5 ESTERIFICATION OF ALL-*TRANS*-RETINOL IN THE RPE BY LECITHIN: RETINOL ACYLTRANSFERASE

Lecithin:retinol acyltransferase (LRAT) plays an important role in the visual cycle, providing a substrate for isomerohydrolase [9,10] and creating storage form of vitamin A in the RPE [61,62]. LRAT catalyzes the transfer of fatty acid moiety from *sn*-1 position of phosphatidylcholine (lecithin) to vitamin A [63] (Figure 2.3). Lecithin was found to be a preferred phospholipid as a cosubstrate, and thus, the major product of esterification is retinyl palmitate or retinyl stearate. Although LRAT was initially resistant to purification [64], many features of LRAT reaction were described before its chemical characterization. It was shown that all-*trans*-retinol is a substrate at least 10 times more efficient than 11-*cis*-retinol [65]. This LRAT property seems logical since 11-*cis*-retinol formed in the RPE should not be esterified and should be converted to 11-*cis*-retinal, the chromophore for regeneration of rhodopsin and thus, 11-*cis*-retinol esterification would be counterproductive for vision. It has been demonstrated that the kinetics of LRAT follows through an ordered ping-pong mechanism, suggesting that the lecithin is bound first and donates the acyl group to the enzyme, releasing the lysophospholipid [66]. Next, all-*trans*-retinol binds to LRAT, accepts fatty acid from acyl enzyme, and dissociates from the active site as retinyl ester [66]. This enzymatic mechanism requires a nucleophile at the active

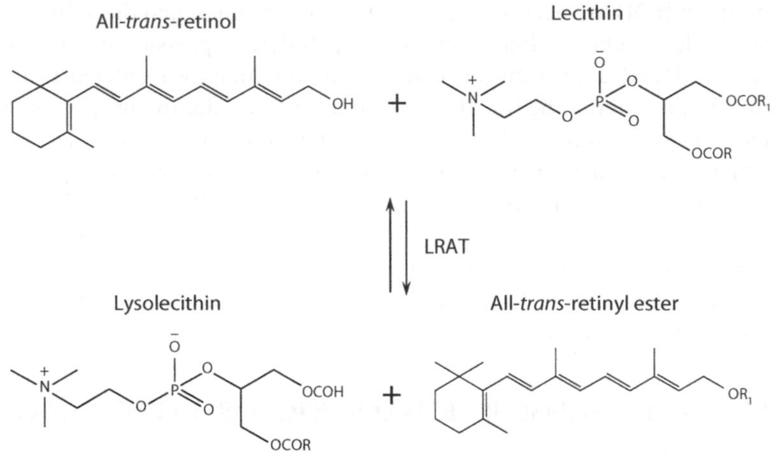

FIGURE 2.3 Transesterification of all-*trans*-retinol catalyzed by LRAT.

site of LRAT to serve as acceptor of acyl moiety. By analogy with thiol-dependent proteases, it was suggested that cysteine residue can function as nucleophile, and this was later confirmed by site-directed mutagenesis of LRAT [67]. LRAT preferentially uses all-*trans*-retinol in the complex with cellular retinol-binding protein (CRBP-1) as a substrate [68]. Interestingly, apo-CRBP strongly inhibits LRAT activity, and in this way it regulates the available all-*trans*-retinol inside the cells [69].

Chemical characterization of LRAT was performed in elegant studies by Robert Rando's group [70]. LRAT was estimated to constitute <0.1% of total RPE membrane proteins. Given the enormous difficulty associated with purifying minor membrane proteins, Ruiz et al. used biotinylated retinoid reagent for affinity labeling and subsequent isolation of the enzyme [70]. After partial sequencing of LRAT peptides, the enzyme has been successfully cloned. LRAT is a polypeptide of 230 amino acids and with molecular mass 25.3 kDa. Its amino acid sequence did not show homology to any known protein at that time. Later, some other homologous proteins such as tumor suppressors and others were identified. Hydropathy analysis suggested that there are two putative α-helical transmembrane sequences: the N-terminal and C-terminal domains. Deletion of these helixes results in a soluble form of LRAT (tLRAT) with catalytical activities, when expressed in *Escherichia coli* [71]. However, more detailed study of LRAT subcellular localization demonstrated that in eukaryotic cells LRAT assumes a single membrane–spanning topology with N-terminal cytoplasmic/C-terminal luminal orientation [72]. Site-directed mutagenesis demonstrated that His60 and Cys161 form an essential catalytic dyad of LRAT [67,73]. Cys161 serves as a nucleophile that forms an acyl enzyme through thioester bond [74]. His60 in a deprotonated form most likely plays a role as an acceptor of hydrogen from the hydroxyl group of retinol. Tyr154 was also found to be essential for catalysis since its substitution for phenylalanine abolishes LRAT activity [74]. This residue is also responsible for $pK_a = 9.95$ observed in pH-dependent profile. These three catalytic residues are conserved in the LRAT family [73].

Loss of function mutations in the human LRAT gene has been identified in patients with LCA [75]. These individuals are otherwise healthy, which suggests that LRAT is not essential for total retinoid metabolism when sufficient dietary retinol is available. *lrat*$^{-/-}$ mouse photoreceptors lack functional rhodopsin and the RPE does not contain all-*trans*-retinyl ester or all-*trans*-retinol [62]. The electroretinogram responses are severely depressed in *lrat*$^{-/-}$ mice, presumably because of absence of 11-*cis*-retinal chromophore, which generates nonfunctional visual pigments [62]. Pupillary response was also decreased by three orders of magnitude in *lrat*$^{-/-}$ mice [62]. This mouse phenotype confirms that LRAT is a key component of the visual cycle. The histological analysis showed a 35% decrease in length of ROS at the age of 6–8 weeks [62]. Only trace amounts of retinyl esters were found in the liver, eye, and blood; but the level of all-*trans*-retinol was only slightly decreased as compared with wt mice. Deletion of the LRAT gene in mice makes animals more susceptible to vitamin A deficiency [76]. Interestingly, *lrat*$^{-/-}$ mice have increased levels of retinyl esters in adipose tissue [77]. Previously, it has been suggested that there is another retinyl ester synthase, acyl CoA:retinol acyl transferase (ARAT) in the retina of cone-dominant species [78]. However, this enzyme has never been identified. Acyl CoA:diacylglycerol acyl transferase (DGAT1) catalyzes palmitoyl CoA–dependent synthesis of all-*trans*-retinyl esters but it is expressed mostly in the intestine and contributes to retinyl ester synthesis when larger pharmacological doses of vitamin A are administered [79]. Thus, other palmitoyl CoA–dependent esterification enzymes such as ARAT are not likely to contribute significantly to esterification of retinol, at least in mammalian species. The absence of retinyl esters in the RPE and retina of *lrat*$^{-/-}$ mice suggests that LRAT is the major enzyme responsible for the synthesis of retinyl esters in the eye.

2.6 GENERATION OF 11-*CIS*-RETINOL IN THE RPE

Retinol isomerase in the RPE converts all-*trans*-retinoids into 11-*cis*-retinol, which requires a sterically unfavorable conformation change. Therefore, this reaction is endergonic and requires energy input. Rando proposed that this energy can be provided by simultaneous hydrolysis of the retinyl ester bond. Indeed, it has since been demonstrated that retinyl ester rather than all-*trans*-retinol is the substrate for retinol isomerase. Rando proposed the term isomerohydrolase, which underlines that isomerization and hydrolysis occur in concerted action. Although the retinol isomerization activity was first discovered >20 years ago, the chemical nature of the retinol isomerase was revealed only recently. RPE65 was identified and cloned independently by two groups [80–82]. Hamel et al. [82] identified a 65-kDa protein using an RPE-specific monoclonal antibody RPE9. This protein was preferentially expressed in the RPE and named RPE65 because of its apparent molecular weight. This protein was shown to associate with the cell membranes and it was detected preferentially in the microsomal fraction. It was relatively abundant in the RPE with 11 μg/bovine RPE [83]. Several lines of evidence suggest that this protein is not glycosylated [82,84]. The Eriksson group independently cloned this protein (p63) as it had an affinity for the plasma retinol–binding protein (RBP) [81] and therefore, initially it was speculated to be the RBP receptor. However, this proposal has not been confirmed later. It was found to have a molecular mass of 63 kDa and to be glycosylated [80]. In spite of its distinct properties, the cloning of this protein later showed that it has the same deduced amino acid sequence as RPE65, suggesting that p63 and RPE65 are the same protein.

The RPE65 cDNA and gene have been cloned from multiple species [84,85]. The human RPE65 gene consists of 14 exons interrupted by 13 introns and is localized in human chromosome 1p31 [85,86]. The 3′ noncoding region contains multiple AU-rich elements, a characteristic of unstable and rapid-degrading RNA, suggesting that it may be tightly downregulated posttranscriptionally [85,87]. The RPE65 gene promoter has also been characterized [88]. The sequence of RPE65 is highly conserved across species. RPE65 shares low but significant sequence homology with β-carotene monooxygenases, which were originally classified as β-carotene dioxygenases [89–91]. Later, they were renamed as β-carotene monooxygenases based on the mechanism of action [92]. These β-carotene monooxygenases are a family of enzymes, which cleave β-carotene to generate vitamin A [90].

The deduced amino acid sequence of RPE65 contains 533 amino acid residues. It does not have any predicted transmembrane helices, and thus it is not an ordinary integral membrane protein. This conclusion is supported by the observation that this protein exists in multiple extractable forms and in complex with other proteins in the RPE [82]. Recombinant RPE65 failed to insert into the cell membrane, further supporting the conclusion that it is not a typical integral membrane protein [93]. The tissue-specific expression, high abundance, evolutionary conservation, developmental regulation, and posttranscriptional down-regulation suggest that RPE65 is a functionally important protein [85].

The homozygous RPE65 knockout mouse (*Rpe65*$^{-/-}$) displays photoreceptor degeneration and diminished rod and cone response in the ERG [94]. Retinoid analysis demonstrated disturbed retinoid profiles in the RPE and retina of the *Rpe65*$^{-/-}$ mice. 11-*cis*-Retinoids are absent in the retina and thus, rhodopsin regeneration is abolished, although free opsin is available. The lack of rhodopsin was supported by the finding that *Rpe65*$^{-/-}$ mice were completely protected from light-induced damage to the retina [95]. In contrast, retinyl ester accumulated in the RPE of the *Rpe65*$^{-/-}$ mouse. Ester saponification showed that all of the retinyl ester was in the all-*trans* form while the 11-*cis*-retinyl ester was absent [94]. The results suggest that in the *Rpe65*$^{-/-}$ mice, the regeneration of 11-*cis*-retinal is interrupted at the isomerization–hydrolysis step, supporting the hypothesis that this protein is an essential component of the isomerohydrolase complex. Although there exists a trace of visual response in the rod in *Rpe65*$^{-/-}$ mice [96], our recent studies showed that it can be ascribed to isorhodopsin, which contains 9-*cis*-retinal as the chromophore, and that there is no 11-*cis*-retinal in *Rpe65*$^{-/-}$ mice [97]. It has been demonstrated that a supplement of 9-*cis*-retinal partially reversed the phenotypes of the *Rpe65*$^{-/-}$ mouse [98,99], suggesting that lack of 11-*cis*-retinal is responsible for the functional abnormalities.

RPE65 is essential for normal vision. Mutations in the RPE65 gene have been reported to be associated with several forms of inherited retinal dystrophies [100–104], including most notably LCA, which is an inherited disease characterized by blindness at birth [101]. RPE65 mutations have been suggested to cause the abnormal visual function in the LCA patients, although the impacts of these mutations on the structure and function of the protein are unknown at this time. In addition, RPE65 mutations have been found to associate with autosomal recessive childhood-onset severe retinal dystrophy [100] and early-onset severe rod–cone dystrophies [103]. RPE65 mutations have also been identified in patients with some forms of retinitis pigmentosa (RP) [102]. In addition, RPE65 gene defects were linked to retinal dystrophy and congenital stationary night blindness in dogs [105,106]. Recently, a number of studies using AAV-mediated gene delivery have demonstrated that RPE65 gene delivery can restore vision in a canine model of childhood blindness and in *Rpe65*$^{-/-}$ mice [107–111]. These findings indicate that a functional RPE65 protein is essential for maintaining physiological functions of the RPE and retina.

Although RPE65 was identified 17 years ago, its physiological function was not revealed until recently. The initial hypothesis that RPE65 could be a receptor for retinoid-binding protein [80,81,93] has been disapproved, since RPE65 has no extracellular or transmembrane domains. The severe phenotype of *Rpe65*$^{-/-}$ mice, the accumulation of retinyl esters in their RPE, and the absence of 11-*cis*-retinal in their retina initially suggested that RPE65 participates in processing retinoids and is essential for regeneration of rhodopsin [94]. Later, it was found that RPE65 has low, yet significant sequence homology with plant dioxygenases such as VP14 [112]. The finding that RPE65 has four conserved histidine residues that coordinate an iron ion provided evidence that RPE65 could also be an enzyme. However, some other studies were in disagreement with this hypothesis. It was reported that almost complete depletion of RPE65 from bovine RPE membranes did not change the isomerase activity [113]. Moreover, purified RPE65 did not show isomerase activity, although it bound retinyl ester with high affinity [114,115]. Both Rando's and Travis's group have demonstrated that RPE65 binds to all-*trans*-retinyl ester and thus, they suggested that RPE65 is a retinyl ester–binding protein [114,115]. Therefore, until 2005 the dominant hypothesis was that RPE65 is a chaperon-like protein that binds retinyl ester and presents it to a yet unknown isomerohydrolase. A major breakthrough in the understanding of RPE65 function was achieved in 2005, when we and others demonstrated that co-expression of RPE65 and LRAT in nonvisual mammalian cell lines created retinol isomerase activity [11–13]. Expression of LRAT was necessary to synthesize insoluble retinyl ester substrate *in situ* from added all-*trans*-retinol. RPE65 extracted retinyl ester from the membrane and converted it to 11-*cis*-retinol. This was the first time that retinol isomerase activity was observed in non-RPE cells. Despite similar reports from three groups, some doubts still remain about RPE65 isomerase function because the activity in our previous work relied upon cell crude lysates and purified RPE65 protein did not show any activity. This last controversy was solved by the work of Nikolaeva et al., which demonstrated the isomerase activity of purified RPE65 after binding with liposomes containing all-*trans*-retinyl ester [14]. This study also concluded that the association of RPE65 with the lipid membrane was critical for restoring its catalytic activity. Most likely, RPE65 changes its conformation upon binding to lipid membrane and becomes catalytically active. In agreement with this finding, PLA2 treatment of RPE microsomal membranes significantly decreases the retinoid isomerization activity [116]. Disruption of the membranes may change the orientation of the retinyl ester substrate inside the phospholipid bilayer in such a way that it could not be extracted by RPE65. The demonstration of isomerohydrolase activity of purified RPE65 provides final proof that RPE65 is the isomerohydrolase in the RPE [12,117].

Previously, we have shown that both recombinant and native RPE65 proteins have two forms with different molecular weights, a membrane associate form and a soluble form [83]. These two forms were initially found to be different in palmitoylations of the protein [118]. Three palmitoylated Cys residues (Cys231, Cys329, Cys330) were identified in RPE65 by Rando's group [118] and were implicated in binding with the membrane. However, later it was demonstrated that these cysteines are not palmitoylated and a new palmitoylation site (C112) was discovered [119]. The crystal structure of RPE65 has been recently solved, and provided some new insights in the interaction of RPE65 with the membrane [120]. Bovine RPE65 structure contains a seven-bladed β-propeller chain fold similar to that found in water-soluble proteins. However, there is a cluster of hydrophobic and aromatic residues (Phe, Leu, and Ile) on a surface of the protein, which can potentially interact with the hydrophobic core of the membrane. Unfortunately, the α-helical amphipathic fragment 109–126, which is a good candidate for membrane binding, was highly disordered in the crystal structure. The hydrophobic patch surrounds the entrance to the long channel, which leads to the catalytic site of RPE65. It is likely that binding of RPE65 through the amphipathic helix 109–126 facilitates the extraction of retinyl ester dissolved in the membrane and the delivery of the substrate to the active site. An alternative model of RPE65 interaction with the membrane was proposed by Yuan et al. [121]. Based on the interaction of RPE65 with paper strips containing various lipids, the authors suggested that the negatively charged lipids are essential for binding. It was reported by the authors that RPE65 exclusively reacts with the substrate incorporated in the negatively charged liposomes. However, the results of this work are contradictory to our measurements, which showed a robust catalytic activity of RPE65 toward retinyl ester embedded in neutral liposomes [14]. Therefore, even if negatively charged residues (354–359 peptide) may contribute to binding energy, most likely they are not critical for membrane interaction and RPE65 function. We have shown that iron 2+ cofactor is necessary for catalytic activity of RPE65 and cannot be substituted by iron 3+ or other metal ions [117]. The crystal structure of RPE65 demonstrated that Fe^{2+} ion is coordinated by four conserved histidine residues: His180, His241, His313, and His527, with average bond length 2.2 Å (Figure 2.4) [13,120,122]. The geometry of iron coordination is octahedral with two vacant coordination sites. Three of four His residues (His241, His313, and His527) form hydrogen bonds with three highly conserved glutamic residues (Glu148, Glu417, and Glu469). This second coordination sphere is also essential for enzymatic activity, particularly, the negative charge of the glutamates [13,122,123]. It is likely that E417 negative charge affects the p*K* of H313 and modulates its iron-binding capacity [123]. The exact function of iron ion in the catalytic mechanism is still unclear.

The mechanism underlying RPE65 isomerization reaction as well as the function of catalytic residues at the active site is currently unknown. Most likely, the reaction proceeds through SN1 mechanism. This includes the cleavage of alkyl oxygen bond to generate a relatively stable carbocation intermediate [124]. The formation of carbocation is promoted by the interaction of some Lewis acid (e.g., metal ion) with

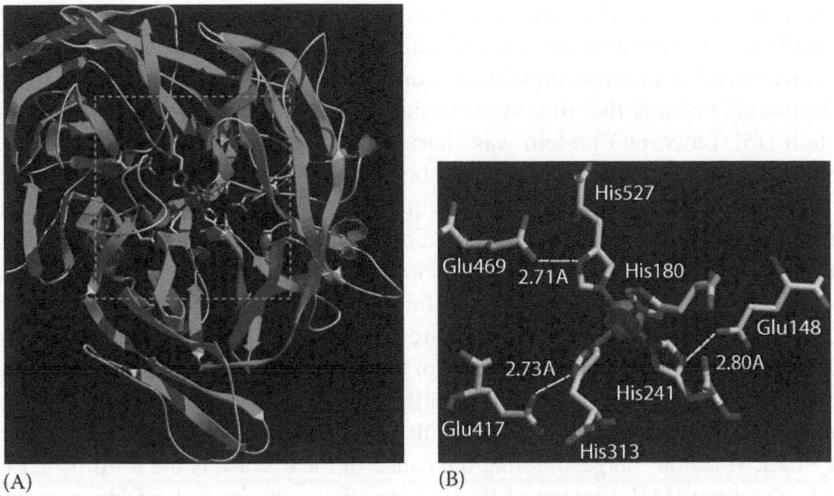

FIGURE 2.4 (See color insert.) Molecular model of the bovine RPE65 iron-binding site. (A) Overall 3D structure of RPE65, where the conserved Fe^{2+} site marked with purple dashed line. (B) A detailed view of the RPE65 structure in the vicinity of iron-binding site. The conserved residues His180, His241, His313, and His527 coordinate the catalytically essential iron ion.

ester acyl oxygen. Iron ion at the active site of RPE65 can play a role of a Lewis acid. When carbocation is formed, the bond order in the polyene chain decreases due to partial delocalization of electron density, which results in substantial decrease in the activation energy of isomerization [124]. After isomerization of the carbocation, the nucleophile attack of water molecule at C15 position produces 11-*cis*-retinol. The generation of 13-*cis*-retinol as a side product of the isomerase reaction supports the suggested carbocation mechanism [125]. Indeed, Redmond et al. have recently reported that both 13-*cis*-retinol and 11-*cis*-retinol are produced by RPE65 in a cell culture model [125]. Although in many previous studies we did not observe the formation of 13-*cis*-retinol, this may be due to the presence of LRAT that efficiently converts 13-*cis*-retinol to 13-*cis*-retinyl ester [117,122,126,127]. This 13-*cis*-retinyl ester thermally isomerizes to all-*trans*-retinyl ester, which again serves as a substrate for RPE65 (Figure 2.1). It is also known that 11-*cis*-retinol is an inefficient substrate for LRAT [71]. In addition, CRALBP binds 11-*cis*-retinol produced by RPE65 and protects it from the esterification by LRAT [128]. This likely explains why the formation of 13-*cis*-retinol in the RPE has not been detected *in vivo*, and that it is probably not physiologically important.

2.7 OXIDATION OF 11-*CIS*-RETINOL IN THE RPE

The final step of 11-*cis*-retinal production is the oxidation of 11-*cis*-retinol to generate 11-*cis*-retinal in the RPE. The RDH5 gene encodes the dehydrogenase, which is responsible for the majority of retinol oxidation activity in the RPE [15,129]. In humans, mutations in this gene have been found in patients with fundus albipunctatus, a disease characterized by delayed dark adaptations of both types of photoreceptors [130]. 11-*cis*-Retinol-specific dehydrogenase (RDH5) localized in the RPE was the first enzyme of the visual cycle identified at the molecular level. It was discovered as a partner protein co-immunoprecipitated using antibodies against p63 protein, which at that time was considered to be a receptor for retinol-binding protein [15]. Later, p63 protein was characterized as RPE65, isomerohydrolase of the visual cycle. This is not surprising because RPE65 generates 11-*cis*-retinol, a substrate for RDH5. The amino acid sequence of RDH5 was found to be homologous to SDR [129]. The enzyme contains hydrophobic sequences at the N- and C-terminus, and associates with the membrane. The primary structure of RDH5 also contains the putative cofactor–binding site and essential residues involved in catalysis included in an invariant signature sequence YXXXK [131]. Enzyme assays showed that RDH5 oxidizes 11-*cis*-retinol to 11-*cis*-retinal using NAD as a cofactor [129] (Figure 2.2). RDH5 mutations identified in patients with fundus albipunctatus, a form of stationary night blindness, result in decreased protein stability and loss of enzymatic function, suggesting that the cause of the disease is the diminished level of 11-*cis*-retinal [132]. Sixteen of the 18 identified mutations of RDH5 were localized in the catalytic domain that resides inside the lumen of ER and the rest were found in C-terminal transmembrane domain [132]. Among the mutants studied in patients, only one mutant, A294P, showed high 11-*cis*-RDH activity when expressed in COS cells [132]. The pathological phenotype for this mutation was attributed to the dominant-negative effect interfering with the formation of functional dimers of the enzyme. Despite inactivity of RDH5 enzyme, patients with null mutations in RDH5 eventually recover their visual pigments. Therefore, 11-*cis*-retinal can be produced by some alternative enzymes that compensate the absence of RDH5. In agreement with this notion, the mice with *Rdh5* gene knockout also display delayed dark adaptation, but only at very high bleach level [133]. Importantly, the rate of regeneration of 11-*cis*-retinal was indistinguishable between $Rdh5^{-/-}$ and wt mice [133]. An interesting phenotype for $Rdh5^{-/-}$ mice is the accumulation of 11-*cis*-retinol and 13-*cis*-retinol and retinyl esters in their RPE [133]. The accumulation of 13-*cis*-retinoids could not be interpreted as thermal isomerization of 11-*cis*-retinol, since it would require double isomerization at different double bonds, once at a time. We believe that these results can be explained only by the fact discovered by Redmond et al. that RPE65 can produce not only 11-*cis*-retinol but also 13-*cis*-retinol from all-*trans*-retinyl ester [125]. In the absence of RDH5, most of 13-*cis*-retinol is rapidly converted to 13-*cis*-retinyl esters by LRAT, probably, because 13-*cis*-retinol is a good substrate for LRAT. 11-*cis*-Retinol is a poor substrate for LRAT [71]; therefore, it is eventually oxidized to 11-*cis*-retinal by some other RDH. The molecular characterization of other enzymes that can oxidize 11-*cis*-retinol in the RPE has not been performed yet. Enzymatic assay using RPE protein fractions from $Rdh5^{-/-}$ mice showed that dehydrogenase activity was most efficient with pro-S-NADPH cofactor in contrast to the highest activity with pro-S-NADH for wt mice [129]. This suggests that an unidentified enzyme in the RPE that can process *cis*-retinoids should have strong preference toward NADP as a cofactor.

RDH11 has been suggested to serve as an alternative enzyme, which can oxidize 11-*cis*-retinol in the RPE [134]. RDH11 localized in the RPE and Müller cells in the retina and demonstrated dual-substrate specificity *in vitro* for both *cis*- and *trans*-retinoid substrates and has a very low activity with 13-*cis*-retinol as substrate, suggesting that it can play a role in retinoid homeostasis [36]. However, *Rdh11*$^{-/-}$ mice retinoid profile did not show significant differences from that in wt mice [134]. For *Rdh11*$^{-/-}$ or *Rdh5*$^{-/-}$*Rdh11*$^{-/-}$ mice, EM did not reveal any morphological abnormalities in the photoreceptors or RPE cells. Light stimulation caused 73% more *cis*-retinyl esters in *Rdh5*$^{-/-}$*Rdh11*$^{-/-}$ mice compared with *Rdh5*$^{-/-}$ [134]. Although RDH11 has NADPH specificity, it catalyzes the reduction of all-*trans*-retinal 50-fold more efficiently than oxidation of retinol *in vitro*. However, *in vivo* in the RPE, RDH11 may act as oxidative enzyme. First, the ratio of $NADP^+$:NADPH is about 4:1 in the retina [135] and second, oxidation of 11-*cis*-retinol is promoted by cellular retinaldehyde-binding protein (CRALBP) because CRALBP has much higher affinity toward 11-*cis*-retinal than toward 11-*cis*-retinol shifting equilibrium in the oxidation way [136,137].

RDH10 is also expressed abundantly in the RPE in Müller cells [138,139] and has a preference for NADP as a coenzyme; therefore, it can function as a supplemental enzyme that oxidizes 11-*cis*-retinol in the RPE. Initially, it was thought that RDH10 is specific toward all-*trans*-retinol; however, later it has been shown that 11-*cis*-retinol is also a good substrate for RDH10 [140,141]. RDH10 is a unique enzyme among SDR proteins because the disruption of the RDH10 gene was found to be lethal for mice [142]. It has been suggested that at a certain stage of embryo development, RDH10 is indispensable for providing all-*trans*-retinal, the precursor of retinoic acid synthesis. For this reason, evidently RDH10 mutations have not been found to be associated with retinal diseases since they are incompatible with survival. Further studies including conditional knockout mice are necessary for determining the physiological role of RDH10 in 11-*cis*-retinal synthesis in the RPE.

Some researchers indicate that cytosolic alcohol dehydrogenases (ADH) can metabolize retinol to retinal [143]. Soluble ADH4 was found to be expressed in the RPE and to catalyze reaction with *cis*-retinoids [144]. However, soluble fraction of bovine RPE did not show any significant retinol oxidation activity, which makes them improbable candidate for retinoid visual cycle [129]. Membrane-associated SDRs have an important advantage over soluble ADH, since they can directly extract hydrophobic retinoids from cellular membranes.

2.8 SUMMARY AND CONCLUSION

In this chapter, we presented a current knowledge of 11-*cis*-retinal regeneration through the reactions of the retinoid visual cycle. As a result of significant progress made by researchers in recent years, we have a better understanding of a process of regeneration of 11-*cis*-retinal. In the visual cycle, the retinoid *trans–cis* isomerization is the key catalytic step, and the most challenging to study, since the identification and characterization of RPE65 as the isomerohydrolase was fraught with setbacks. The major difficulty for studying the isomerase reaction is that both the enzyme and substrate are not soluble in water. However, the recent successes in developing a new liposome-based assay and obtaining a crystal structure of RPE65 will facilitate further studies of RPE65-catalyzed isomerization. Obtaining a detailed understanding of visual cycle reactions yields new possibilities for the treatment of inherited eye diseases caused by mutations of visual cycle genes.

REFERENCES

1. Baylor, D. 1996. How photons start vision. *Proc Natl Acad Sci USA* 93: 560–565.
2. McBee, J.K., K. Palczewski, W. Baehr, and D.R. Pepperberg. 2001. Confronting complexity: The interlink of phototransduction and retinoid metabolism in the vertebrate retina. *Prog Retin Eye Res* 20: 469–529.
3. Rando, R.R. 2001. The biochemistry of the visual cycle. *Chem Rev* 101: 1881–1896.
4. Saari, J.C., D.L. Bredberg, and N. Noy. 1994. Control of substrate flow at a branch in the visual cycle. *Biochemistry* 33: 3106–3112.
5. Crouch, R.K., G.J. Chader, B. Wiggert, and D.R. Pepperberg. 1996. Retinoids and the visual process. *Photochem Photobiol* 64: 613–621.
6. Palczewski, K., S. Jager, J. Buczylko et al. 1994. Rod outer segment retinol dehydrogenase: Substrate specificity and role in phototransduction. *Biochemistry* 33: 13741–13750.
7. Jang, G.F., J.K. McNee, A.M. Alekseev, F. Haeseleer, and K. Palczewski. 2000. Stereoisomeric specificity of the retinoid cycle in the vertebrate retina. *J Biol Chem* 275: 28128–28138.
8. Bernstein, P.S. and R.R. Rando. 1986. In vivo isomerization of all-*trans*- to 11-*cis*-retinoids in the eye occurs at the alcohol oxidation state. *Biochemistry* 25: 6473–6478.
9. Deigner, P.S., W.C. Law, F.J. Canada, and R.R. Rando. 1989. Membranes as the energy source in the endergonic transformation of vitamin A to 11-*cis*-retinol. *Science* 244: 968–971.
10. Moiseyev, G., R.K. Crouch, P. Goletz et al. 2003. Retinyl esters are the substrate for isomerohydrolase. *Biochemistry* 42: 2229–2238.
11. Jin, M., S. Li, W.N. Moghrabi, H. Sun, and G.H. Travis. 2005. Rpe65 is the retinoid isomerase in bovine retinal pigment epithelium. *Cell* 122: 449–459.
12. Moiseyev, G., Y. Chen, Y. Takahashi, B.X. Wu, and J.X. Ma. 2005. RPE65 is the isomerohydrolase in the retinoid visual cycle. *Proc Natl Acad Sci USA* 102: 12413–12418.
13. Redmond, T.M., E. Poliakov, S. Yu et al. 2005. Mutation of key residues of RPE65 abolishes its enzymatic role as isomerohydrolase in the visual cycle. *Proc Natl Acad Sci USA* 102: 13658–13663.
14. Nikolaeva, O., Y. Takahashi, G. Moiseyev, and J.X. Ma. 2009. Purified RPE65 shows isomerohydrolase activity after reassociation with a phospholipid membrane. *FEBS J* 276: 3020–3030.
15. Simon, A., U. Hellman, C. Wernstedt, and U. Eriksson. 1995. The retinal pigment epithelial-specific 11-*cis* retinol dehydrogenase belongs to the family of short chain alcohol dehydrogenases. *J Biol Chem* 270: 1107–1112.
16. Anderson, R.E. and M.B. Maude. 1970. Phospholipids of bovine outer segments. *Biochemistry* 9: 3624–3628.
17. Beharry, S., M. Zhong, and R.S. Molday. 2004. *N*-Retinylidene-phosphatidylethanolamine is the preferred retinoid substrate for the photoreceptor-specific ABC transporter ABCA4 (ABCR). *J Biol Chem* 279: 53972–53979.
18. Allikmets, R., N. Singh, H. Sun et al. 1997. A photoreceptor cell-specific ATP-binding transporter gene (ABCR) is mutated in recessive Stargardt macular dystrophy. *Nat Genet* 15: 236–246.
19. Illing, M., L.L. Molday, and R.S. Molday. 1997. The 220-kDa rim protein of retinal rod outer segments is a member of the ABC transporter superfamily. *J Biol Chem* 272: 10303–10310.
20. Sun, H., R.S. Molday, and J. Nathans. 1999. Retinal stimulates ATP hydrolysis by purified and reconstituted ABCR, the photoreceptor-specific ATP-binding cassette transporter responsible for Stargardt disease. *J Biol Chem* 274: 8269–8281.
21. Weng, J., N.L. Mata, S.M. Azarian et al. 1999. Insights into the function of Rim protein in photoreceptors and etiology of Stargardt's disease from the phenotype in abcr knockout mice. *Cell* 98: 13–23.

22. De, S. and T.P. Sakmar. 2002. Interaction of A2E with model membranes. Implications to the pathogenesis of age-related macular degeneration. *J Gen Physiol* 120: 147–157.
23. Sparrow, J.R., B. Cai, Y.P. Jang, J. Zhou, and K. Nakanishi. 2006. A2E, a fluorophore of RPE lipofuscin, can destabilize membrane. *Adv Exp Med Biol* 572: 63–68.
24. Sparrow, J.R. and B. Cai. 2001. Blue light-induced apoptosis of A2E-containing RPE: Involvement of caspase-3 and protection by Bcl-2. *Invest Ophthalmol Vis Sci* 42: 1356–1362.
25. Suter, M., C. Reme, C. Grimm et al. 2000. Age-related macular degeneration. The lipofusion component *N*-retinyl-*N*-retinylidene ethanolamine detaches proapoptotic proteins from mitochondria and induces apoptosis in mammalian retinal pigment epithelial cells. *J Biol Chem* 275: 39625–39630.
26. Sparrow, J.R., K. Nakanishi, and C.A. Parish. 2000. The lipofuscin fluorophore A2E mediates blue light-induced damage to retinal pigmented epithelial cells. *Invest Ophthalmol Vis Sci* 41: 1981–1989.
27. Schutt, F., S. Davies, J. Kopitz, F.G. Holz, and M.E. Boulton. 2000. Photodamage to human RPE cells by A2-E, a retinoid component of lipofuscin. *Invest Ophthalmol Vis Sci* 41: 2303–2308.
28. Holz, F.G., F. Schutt, J. Kopitz et al. 1999. Inhibition of lysosomal degradative functions in RPE cells by a retinoid component of lipofuscin. *Invest Ophthalmol Vis Sci* 40: 737–743.
29. Finnemann, S.C., L.W. Leung, and E. Rodriguez-Boulan. 2002. The lipofuscin component A2E selectively inhibits phagolysosomal degradation of photoreceptor phospholipid by the retinal pigment epithelium. *Proc Natl Acad Sci USA* 99: 3842–3847.
30. Sparrow, J.R., H.R. Vollmer-Snarr, J. Zhou et al. 2003. A2E-epoxides damage DNA in retinal pigment epithelial cells. Vitamin E and other antioxidants inhibit A2E-epoxide formation. *J Biol Chem* 278: 18207–18213.
31. Sparrow, J.R., J. Zhou, and B. Cai. 2003. DNA is a target of the photodynamic effects elicited in A2E-laden RPE by blue-light illumination. *Invest Ophthalmol Vis Sci* 44: 2245–2251.
32. Zhou, J., Y.P. Jang, S.R. Kim, and J.R. Sparrow. 2006. Complement activation by photooxidation products of A2E, a lipofuscin constituent of the retinal pigment epithelium. *Proc Natl Acad Sci USA* 103: 16182–16187.
33. Iriyama, A., R. Fujiki, Y. Inoue et al. 2008. A2E, a pigment of the lipofuscin of retinal pigment epithelial cells, is an endogenous ligand for retinoic acid receptor. *J Biol Chem* 283: 11947–11953.
34. Moiseyev, G., O. Nikolaeva, Y. Chen et al. 2010. Inhibition of the visual cycle by A2E through direct interaction with RPE65 and implications in Stargardt disease. *Proc Natl Acad Sci USA* 107: 17551–17556.
35. Rattner, A., P.M. Smallwood, and J. Nathans. 2000. Identification and characterization of all-*trans*-retinol dehydrogenase from photoreceptor outer segments, the visual cycle enzyme that reduces all-*trans*-retinal to all-*trans*-retinol. *J Biol Chem* 275: 11034–11043.
36. Haeseleer, F., G.F. Jang, Y. Imanishi et al. 2002. Dual-substrate specificity short chain retinol dehydrogenases from the vertebrate retina. *J Biol Chem* 277: 45537–45546.
37. Bray, J.E., B.D. Marsden, and U. Oppermann. 2009. The human short-chain dehydrogenase/reductase (SDR) superfamily: A bioinformatics summary. *Chem Biol Interact* 178: 99–109.
38. Lukacik, P., K.L. Kavanagh, and U. Oppermann. 2006. Structure and function of human 17beta-hydroxysteroid dehydrogenases. *Mol Cell Endocrinol* 248: 61–71.
39. Kallberg, Y., U. Oppermann, H. Jornvall, and B. Persson. 2002. Short-chain dehydrogenases/reductases (SDRs). *Eur J Biochem* 269: 4409–4417.

40. Belyaeva, O.V., O.V. Korkina, A.V. Stetsenko et al. 2005. Biochemical properties of purified human retinol dehydrogenase 12 (RDH12): Catalytic efficiency toward retinoids and C9 aldehydes and effects of cellular retinol-binding protein type I (CRBPI) and cellular retinaldehyde-binding protein (CRALBP) on the oxidation and reduction of retinoids. *Biochemistry* 44: 7035–7047.
41. Hsu, S.C. and R.S. Molday. 1994. Glucose metabolism in photoreceptor outer segments. Its role in phototransduction and in NADPH-requiring reactions. *J Biol Chem* 269: 17954–17959.
42. Thompson, D.A., A.R. Janecke, J. Lange et al. 2005. Retinal degeneration associated with RDH12 mutations results from decreased 11-*cis* retinal synthesis due to disruption of the visual cycle. *Hum Mol Genet* 14: 3865–3875.
43. Janecke, A.R., D.A. Thompson, G. Utermann et al. 2004. Mutations in RDH12 encoding a photoreceptor cell retinol dehydrogenase cause childhood-onset severe retinal dystrophy. *Nat Genet* 36: 850–854.
44. Sun, W., C. Gerth, A. Maeda et al. 2007. Novel RDH12 mutations associated with Leber congenital amaurosis and cone-rod dystrophy: Biochemical and clinical evaluations. *Vision Res* 47: 2055–2066.
45. Maeda, A., T. Maeda, W. Sun et al. 2007. Redundant and unique roles of retinol dehydrogenases in the mouse retina. *Proc Natl Acad Sci USA* 104: 19565–19570.
46. Kurth, I., D.A. Thompson, K. Ruther et al. 2007. Targeted disruption of the murine retinal dehydrogenase gene *Rdh12* does not limit visual cycle function. *Mol Cell Biol* 27: 1370–1379.
47. Marchette, L.D., D.A. Thompson, M. Kravtsova et al. 2010. Retinol dehydrogenase 12 detoxifies 4-hydroxynonenal in photoreceptor cells. *Free Radic Biol Med* 48: 16–25.
48. Maeda, A., T. Maeda, Y. Imanishi et al. 2005. Role of photoreceptor-specific retinol dehydrogenase in the retinoid cycle in vivo. *J Biol Chem* 280: 18822–18832.
49. Duester, G. 1996. Involvement of alcohol dehydrogenase, short-chain dehydrogenase/reductase, aldehyde dehydrogenase, and cytochrome P450 in the control of retinoid signaling by activation of retinoic acid synthesis. *Biochemistry* 35: 12221–12227.
50. Gallego, O., O.V. Belyaeva, S. Porte et al. 2006. Comparative functional analysis of human medium-chain dehydrogenases, short-chain dehydrogenases/reductases and aldo–keto reductases with retinoids. *Biochem J* 399: 101–109.
51. Miyazono, S., Y. Shimauchi-Matsukawa, S. Tachibanaki, and S. Kawamura. 2008. Highly efficient retinal metabolism in cones. *Proc Natl Acad Sci USA* 105: 16051–16056.
52. Sparrow, J.R. and M. Boulton. 2005. RPE lipofuscin and its role in retinal pathobiology. *Exp Eye Res* 80: 595–606.
53. Sparrow, J.R., N. Fishkin, J. Zhou et al. 2003. A2E, a byproduct of the visual cycle. *Vision Res* 43: 2983–2990.
54. Adler, A.J., C.D. Evans, and W.F. Stafford, 3rd. 1985. Molecular properties of bovine interphotoreceptor retinol-binding protein. *J Biol Chem* 260: 4850–4855.
55. Adler, A.J. and C.D. Evans. 1985. Some functional characteristics of purified bovine interphotoreceptor retinol-binding protein. *Invest Ophthalmol Vis Sci* 26: 273–282.
56. Redmond, T.M., B. Wiggert, F.A. Robey et al. 1985. Isolation and characterization of monkey interphotoreceptor retinoid-binding protein, a unique extracellular matrix component of the retina. *Biochemistry* 24: 787–793.
57. Shaw, N.S. and N. Noy. 2001. Interphotoreceptor retinoid-binding protein contains three retinoid binding sites. *Exp Eye Res* 72: 183–190.
58. Tsina, E., C. Chen, Y. Koutalos et al. 2004. Physiological and microfluorometric studies of reduction and clearance of retinal in bleached rod photoreceptors. *J Gen Physiol* 124: 429–443.
59. Crouch, R.K., E.S. Hazard, T. Lind et al. 1992. Interphotoreceptor retinoid-binding protein and alpha-tocopherol preserve the isomeric and oxidation state of retinol. *Photochem Photobiol* 56: 251–255.

60. Palczewski, K., J.P. Van Hooser, G.G. Garwin et al. 1999. Kinetics of visual pigment regeneration in excised mouse eyes and in mice with a targeted disruption of the gene encoding interphotoreceptor retinoid-binding protein or arrestin. *Biochemistry* 38: 12012–12019.
61. Imanishi, Y., M.L. Batten, D.W. Piston, W. Baehr, and K. Palczewski. 2004. Noninvasive two-photon imaging reveals retinyl ester storage structures in the eye. *J Cell Biol* 164: 373–383.
62. Batten, M.L., Y. Imanishi, T. Maeda et al. 2004. Lecithin–retinol acyltransferase is essential for accumulation of all-*trans*-retinyl esters in the eye and in the liver. *J Biol Chem* 279: 10422–10432.
63. Saari, J.C. and D.L. Bredberg. 1989. Lecithin:retinol acyltransferase in retinal pigment epithelial microsomes. *J Biol Chem* 264: 8636–8640.
64. Barry, R.J., F.J. Canada, and R.R. Rando. 1989. Solubilization and partial purification of retinyl ester synthetase and retinoid isomerase from bovine ocular pigment epithelium. *J Biol Chem* 264: 9231–9238.
65. Rando, R.R. 2002. Membrane-bound lecithin–retinol acyltransferase. *Biochem Biophys Res Commun* 292: 1243–1250.
66. Shi, Y.Q., I. Hubacek, and R.R. Rando. 1993. Kinetic mechanism of lecithin retinol acyl transferase. *Biochemistry* 32: 1257–1263.
67. Mondal, M.S., A. Ruiz, D. Bok, and R.R. Rando. 2000. Lecithin retinol acyltransferase contains cysteine residues essential for catalysis. *Biochemistry* 39: 5215–5220.
68. Ong, D.E., P.N. MacDonald, and A.M. Gubitosi. 1988. Esterification of retinol in rat liver. Possible participation by cellular retinol-binding protein and cellular retinol-binding protein II. *J Biol Chem* 263: 5789–5796.
69. Herr, F.M. and D.E. Ong. 1992. Differential interaction of lecithin–retinol acyltransferase with cellular retinol binding proteins. *Biochemistry* 31: 6748–6755.
70. Ruiz, A., A. Winston, Y.H. Lim et al. 1999. Molecular and biochemical characterization of lecithin retinol acyltransferase. *J Biol Chem* 274: 3834–3841.
71. Bok, D., A. Ruiz, O. Yaron et al. 2003. Purification and characterization of a transmembrane domain-deleted form of lecithin retinol acyltransferase. *Biochemistry* 42: 6090–6098.
72. Moise, A.R., M. Golczak, Y. Imanishi, and K. Palczewski. 2007. Topology and membrane association of lecithin: Retinol acyltransferase. *J Biol Chem* 282: 2081–2090.
73. Jahng, W.J., L. Xue, and R.R. Rando. 2003. Lecithin retinol acyltransferase is a founder member of a novel family of enzymes. *Biochemistry* 42: 12805–12812.
74. Xue, L. and R.R. Rando. 2004. Roles of cysteine 161 and tyrosine 154 in the lecithin–retinol acyltransferase mechanism. *Biochemistry* 43: 6120–6126.
75. Thompson, D.A., Y. Li, C.L. McHenry et al. 2001. Mutations in the gene encoding lecithin retinol acyltransferase are associated with early-onset severe retinal dystrophy. *Nat Genet* 28: 123–124.
76. Liu, L. and L.J. Gudas. 2005. Disruption of the lecithin:retinol acyltransferase gene makes mice more susceptible to vitamin A deficiency. *J Biol Chem* 280: 40226–40234.
77. O'Byrne, S.M., N. Wongsiriroj, J. Libien et al. 2005. Retinoid absorption and storage is impaired in mice lacking lecithin:retinol acyltransferase (LRAT). *J Biol Chem* 280: 35647–35657.
78. Mata, N.L., A. Ruiz, R.A. Radu, T.V. Bui, and G.H. Travis. 2005. Chicken retinas contain a retinoid isomerase activity that catalyzes the direct conversion of all-*trans*-retinol to 11-*cis*-retinol. *Biochemistry* 44: 11715–11721.
79. Wongsiriroj, N., R. Piantedosi, K. Palczewski et al. 2008. The molecular basis of retinoid absorption: A genetic dissection. *J Biol Chem* 283: 13510–13519.
80. Bavik, C.O., U. Eriksson, R.A. Allen, and P.A. Peterson. 1991. Identification and partial characterization of a retinal pigment epithelial membrane receptor for plasma retinol-binding protein. *J Biol Chem* 266: 14978–14985.

81. Bavik, C.O., C. Busch, and U. Eriksson. 1992. Characterization of a plasma retinol-binding protein membrane receptor expressed in the retinal pigment epithelium. *J Biol Chem* 267: 23035–23042.
82. Hamel, C.P., E. Tsilou, E. Harris et al. 1993. A developmentally regulated microsomal protein specific for the pigment epithelium of the vertebrate retina. *J Neurosci Res* 34: 414–425.
83. Ma, J., J. Zhang, K.L. Othersen et al. 2001. Expression, purification, and MALDI analysis of RPE65. *Invest Ophthalmol Vis Sci* 42: 1429–1435.
84. Hamel, C.P., E. Tsilou, B.A. Pfeffer et al. 1993. Molecular cloning and expression of RPE65, a novel retinal pigment epithelium-specific microsomal protein that is post-transcriptionally regulated in vitro. *J Biol Chem* 268: 15751–15757.
85. Nicoletti, A., D.J. Wong, K. Kawase et al. 1995. Molecular characterization of the human gene encoding an abundant 61 kDa protein specific to the retinal pigment epithelium. *Hum Mol Genet* 4: 641–649.
86. Hamel, C.P., N.A. Jenkins, D.J. Gilbert, N.G. Copeland, and T.M. Redmond. 1994. The gene for the retinal pigment epithelium-specific protein RPE65 is localized to human 1p31 and mouse 3. *Genomics* 20: 509–512.
87. Liu, S.Y. and T.M. Redmond. 1998. Role of the 3′-untranslated region of RPE65 mRNA in the translational regulation of the RPE65 gene: Identification of a specific translation inhibitory element. *Arch Biochem Biophys* 357: 37–44.
88. Nicoletti, A., K. Kawase, and D.A. Thompson. 1998. Promoter analysis of RPE65, the gene encoding a 61-kDa retinal pigment epithelium-specific protein. *Invest Ophthalmol Vis Sci* 39: 637–644.
89. Wyss, A., G. Wirtz, W.D. Woggon et al. 2000. Cloning and expression of beta,beta-carotene 15,15′-dioxygenase. *Biochem Biophys Res Commun* 271: 334–336.
90. Wyss, A. 2004. Carotene oxygenases: A new family of double bond cleavage enzymes. *J Nutr* 134: 246S–250S.
91. von Lintig, J. and K. Vogt. 2000. Filling the gap in vitamin A research. Molecular identification of an enzyme cleaving beta-carotene to retinal. *J Biol Chem* 275: 11915–11920.
92. Woggon, W.D. 2002. Oxidative cleavage of carotenoids catalyzed by enzyme models and beta-carotene 15,15′-monooxygenase. *Pure Appl Chem* 74: 1397–1408.
93. Bavik, C.O., F. Levy, U. Hellman, C. Wernstedt, and U. Eriksson. 1993. The retinal pigment epithelial membrane receptor for plasma retinol-binding protein. Isolation and cDNA cloning of the 63-kDa protein. *J Biol Chem* 268: 20540–20546.
94. Redmond, T.M., S. Yu, E. Lee et al. 1998. Rpe65 is necessary for production of 11-*cis*-vitamin A in the retinal visual cycle. *Nat Genet* 20: 344–351.
95. Grimm, C., A. Wenzel, F. Hafezi et al. 2000. Protection of Rpe65-deficient mice identifies rhodopsin as a mediator of light-induced retinal degeneration. *Nat Genet* 25: 63–66.
96. Seeliger, M.W., C. Grimm, F. Stahlberg et al. 2001. New views on RPE65 deficiency: The rod system is the source of vision in a mouse model of Leber congenital amaurosis. *Nat Genet* 29: 70–74.
97. Fan, J., B. Rohrer, G. Moiseyev, J.X. Ma, and R.K. Crouch. 2003. Isorhodopsin rather than rhodopsin mediates rod function in RPE65 knock-out mice. *Proc Natl Acad Sci USA* 100: 13662–13667.
98. Van Hooser, J.P., Y. Liang, T. Maeda et al. 2002. Recovery of visual functions in a mouse model of Leber congenital amaurosis. *J Biol Chem* 277: 19173–19182.
99. Ablonczy, Z., R.K. Crouch, P.W. Goletz et al. 2002. 11-*cis*-Retinal reduces constitutive opsin phosphorylation and improves quantum catch in retinoid-deficient mouse rod photoreceptors. *J Biol Chem* 277: 40491–40498.
100. Gu, S.M., D.A. Thompson, C.R. Srikumari et al. 1997. Mutations in RPE65 cause autosomal recessive childhood-onset severe retinal dystrophy. *Nat Genet* 17: 194–197.
101. Marlhens, F., C. Bareil, J.M. Griffoin et al. 1997. Mutations in Rpe65 cause Lebers congenital amaurosis. *Nat Genet* 17: 139–141.

102. Morimura, H., G.A. Fishman, S.A. Grover et al. 1998. Mutations in the RPE65 gene in patients with autosomal recessive retinitis pigmentosa or Leber congenital amaurosis. *Proc Natl Acad Sci USA* 95: 3088–3093.
103. Lorenz, B., P. Gyurus, M. Preising et al. 2000. Early-onset severe rod–cone dystrophy in young children with RPE65 mutations. *Invest Ophthalmol Vis Sci* 41: 2735–2742.
104. Thompson, D.A., P. Gyurus, L.L. Fleischer et al. 2000. Genetics and phenotypes of RPE65 mutations in inherited retinal degeneration. *Invest Ophthalmol Vis Sci* 41: 4293–4299.
105. Aguirre, G.D., V. Baldwin, S. Pearce-Kelling et al. 1998. Congenital stationary night blindness in the dog: Common mutation in the RPE65 gene indicates founder effect. *Mol Vis* 4: 23.
106. Veske, A., S.E. Nilsson, K. Narfstrom, and A. Gal. 1999. Retinal dystrophy of Swedish briard/briard-beagle dogs is due to a 4-bp deletion in RPE65. *Genomics* 57: 57–61.
107. Acland, G.M., G.D. Aguirre, J. Ray et al. 2001. Gene therapy restores vision in a canine model of childhood blindness. *Nat Genet* 28: 92–95.
108. Narfstrom, K., M.L. Katz, M. Ford et al. 2003. In vivo gene therapy in young and adult RPE65–/– dogs produces long-term visual improvement. *J Hered* 94: 31–37.
109. Narfstrom, K., R. Bragadottir, T.M. Redmond et al. 2003. Functional and structural evaluation after AAV.RPE65 gene transfer in the canine model of Leber's congenital amaurosis. *Adv Exp Med Biol* 533: 423–430.
110. Bennett, J. 2004. Gene therapy for Leber congenital amaurosis. *Novartis Found Symp* 255: 195–202; discussion 202–207.
111. Lai, C.M., M.J. Yu, M. Brankov et al. 2004. Recombinant adeno-associated virus type 2-mediated gene delivery into the Rpe65–/– knockout mouse eye results in limited rescue. *Genet Vaccines Ther* 2: 3.
112. Tan, B.C., S.H. Schwartz, J.A. Zeevaart, and D.R. McCarty. 1997. Genetic control of abscisic acid biosynthesis in maize. *Proc Natl Acad Sci USA* 94: 12235–12240.
113. Choo, D.W., E. Cheung, and R.R. Rando. 1998. Lack of effect of RPE65 removal on the enzymatic processing of all-*trans*-retinol into 11-*cis*-retinol in vitro. *FEBS Lett* 440: 195–198.
114. Gollapalli, D.R., P. Maiti, and R.R. Rando. 2003. RPE65 operates in the vertebrate visual cycle by stereospecifically binding all-*trans*-retinyl esters. *Biochemistry* 42: 11824–11830.
115. Mata, N.L., W.N. Moghrabi, J.S. Lee et al. 2004. Rpe65 is a retinyl ester binding protein that presents insoluble substrate to the isomerase in retinal pigment epithelial cells. *J Biol Chem* 279: 635–643.
116. Golczak, M., P.D. Kiser, D.T. Lodowski, A. Maeda, and K. Palczewski. 2010. Importance of membrane structural integrity for RPE65 retinoid isomerization activity. *J Biol Chem* 285: 9667–9682.
117. Moiseyev, G., Y. Takahashi, Y. Chen et al. 2006. RPE65 is an iron(II)-dependent isomerohydrolase in the retinoid visual cycle. *J Biol Chem* 281: 2835–2840.
118. Xue, L., D.R. Gollapalli, P. Maiti, W.J. Jahng, and R.R. Rando. 2004. A palmitoylation switch mechanism in the regulation of the visual cycle. *Cell* 117: 761–771.
119. Takahashi, Y., G. Moiseyev, Z. Ablonczy et al. 2009. Identification of a novel palmitylation site essential for membrane association and isomerohydrolase activity of RPE65. *J Biol Chem* 284: 3211–3218.
120. Kiser, P.D., M. Golczak, D.T. Lodowski, M.R. Chance, and K. Palczewski. 2009. Crystal structure of native RPE65, the retinoid isomerase of the visual cycle. *Proc Natl Acad Sci USA* 106: 17325–17330.
121. Yuan, Q., J.J. Kaylor, A. Miu et al. 2010. Rpe65 isomerase associates with membranes through an electrostatic interaction with acidic phospholipid headgroups. *J Biol Chem* 285: 988–999.

122. Takahashi, Y., G. Moiseyev, Y. Chen, and J.X. Ma. 2005. Identification of conserved histidines and glutamic acid as key residues for isomerohydrolase activity of RPE65, an enzyme of the visual cycle in the retinal pigment epithelium. *FEBS Lett* 579: 5414–5418.
123. Nikolaeva, O., Y. Takahashi, G. Moiseyev, and J.X. Ma. 2010. Negative charge of the glutamic acid 417 residue is crucial for isomerohydrolase activity of RPE65. *Biochem Biophys Res Commun* 391: 1757–1761.
124. McBee, J.K., V. Kuksa, R. Alvarez et al. 2000. Isomerization of all-*trans*-retinol to *cis*-retinols in bovine retinal pigment epithelial cells: Dependence on the specificity of retinoid-binding proteins. *Biochemistry* 39: 11370–11380.
125. Redmond, T.M., E. Poliakov, S. Kuo, P. Chander, and S. Gentleman. 2010. RPE65, visual cycle retinol isomerase, is not inherently 11-*cis*-specific: Support for a carbocation mechanism of retinol isomerization. *J Biol Chem* 285: 1919–1927.
126. Moiseyev, G., Y. Takahashi, Y. Chen, S. Kim, and J.X. Ma. 2008. RPE65 from cone-dominant chicken is a more efficient isomerohydrolase compared with that from rod-dominant species. *J Biol Chem* 283: 8110–8117.
127. Takahashi, Y., G. Moiseyev, Y. Chen, and J.X. Ma. 2006. The roles of three palmitoylation sites of RPE65 in its membrane association and isomerohydrolase activity. *Invest Ophthalmol Vis Sci* 47: 5191–5196.
128. Winston, A. and R.R. Rando. 1998. Regulation of isomerohydrolase activity in the visual cycle. *Biochemistry* 37: 2044–2050.
129. Jang, G.F., J.P. Van Hooser, V. Kuksa et al. 2001. Characterization of a dehydrogenase activity responsible for oxidation of 11-*cis*-retinol in the retinal pigment epithelium of mice with a disrupted RDH5 gene: A model for the human hereditary disease fundus albipunctatus. *J Biol Chem* 20: 20.
130. Yamamoto, H., A. Simon, U. Eriksson et al. 1999. Mutations in the gene encoding 11-*cis* retinol dehydrogenase cause delayed dark adaptation and fundus albipunctatus. *Nat Genet* 22: 188–191.
131. Driessen, C.A., B.P. Janssen, H.J. Winkens et al. 1995. Cloning and expression of a cDNA encoding bovine retinal pigment epithelial 11-*cis* retinol dehydrogenase. *Invest Ophthalmol Vis Sci* 36: 1988–1996.
132. Liden, M., A. Romert, K. Tryggvason, B. Persson, and U. Eriksson. 2001. Biochemical defects in 11-*cis*-retinol dehydrogenase mutants associated with fundus albipunctatus. *J Biol Chem* 276: 49251–49257.
133. Driessen, C.A., H.J. Winkens, K. Hoffmann et al. 2000. Disruption of the 11-*cis*-retinol dehydrogenase gene leads to accumulation of *cis*-retinols and *cis*-retinyl esters. *Mol Cell Biol* 20: 4275–4287.
134. Kim, T.S., A. Maeda, T. Maeda et al. 2005. Delayed dark adaptation in 11-*cis*-retinol dehydrogenase-deficient mice: A role of RDH11 in visual processes in vivo. *J Biol Chem* 280: 8694–8704.
135. Matschinsky, F.M. 1968. Quantitative histochemistry of nicotinamide adenine nucleotides in retina of monkey and rabbit. *J Neurochem* 15: 643–657.
136. Saari, J.C. and D.L. Bredberg. 1987. Photochemistry and stereoselectivity of cellular retinaldehyde-binding protein from bovine retina. *J Biol Chem* 262: 7618–7622.
137. Saari, J.C., L. Bredberg, and G.G. Garwin. 1982. Identification of the endogenous retinoids associated with three cellular retinoid-binding proteins from bovine retina and retinal pigment epithelium. *J Biol Chem* 257: 13329–13333.
138. Wu, B.X., Y. Chen, J. Fan et al. 2002. Cloning and characterization of a novel all-*trans* retinol short-chain dehydrogenase/reductase from the RPE. *Invest Ophthalmol Vis Sci* 43: 3365–3372.
139. Wu, B.X., G. Moiseyev, Y. Chen et al. 2004. Identification of RDH10, an all-*trans* retinol dehydrogenase, in retinal Muller cells. *Invest Ophthalmol Vis Sci* 45: 3857–3862.

140. Farjo, K.M., G. Moiseyev, Y. Takahashi, R.K. Crouch, and J.X. Ma. 2009. The 11-*cis*-retinol dehydrogenase activity of RDH10 and its interaction with visual cycle proteins. *Invest Ophthalmol Vis Sci* 50: 5089–5097.
141. Belyaeva, O.V., M.P. Johnson, and N.Y. Kedishvili. 2008. Kinetic analysis of human enzyme RDH10 defines the characteristics of a physiologically relevant retinol dehydrogenase. *J Biol Chem* 283: 20299–20308.
142. Sandell, L.L., B.W. Sanderson, G. Moiseyev et al. 2007. RDH10 is essential for synthesis of embryonic retinoic acid and is required for limb, craniofacial, and organ development. *Genes Dev* 21: 1113–1124.
143. Duester, G. 2001. Genetic dissection of retinoid dehydrogenases. *Chem Biol Interact* 130–132: 469–480.
144. Martras, S., R. Alvarez, S.E. Martinez et al. 2004. The specificity of alcohol dehydrogenase with *cis*-retinoids. Activity with 11-*cis*-retinol and localization in retina. *Eur J Biochem* 271: 1660–1670.

3 Biological Repair Mechanisms
A Short Overview

Alessandro Sinigaglia

CONTENTS

3.1 INTRODUCTION

In current times, material scientists and engineers are often challenged with the development and the management of new materials designed to satisfy peculiar technological needs. Due to the great richness of natural existing materials, destined to extremely variegate functions and endowed with incredibly heterogeneous characteristics, it may be very tempting for scientists to take inspiration from nature and try to find good models inside it before starting on new projects or when working them out to get the better performances they can give.

The same is true when dealing specifically with the theme of repair (and self-repair) systems.

Of course, there are many differences between natural materials and the way they are generated and are designed to work and artificial materials, including the so-called biomimetics, and it may be a good scientific exercise to take a brief look at some examples in natural repair (or better, damage-response) mechanisms and to understand the logics that are behind them, because these logics may be, more than the chemical–physical features of natural materials, a good inspiration for scientists and engineers working on self-repair.

In this brief introductive chapter, we will have a look at a few examples of how biological entities (in this case, animals, just for a simplification choice) manage the adverse event that gives rise to reparation in all its forms: damage.

As we know, repair systems in living beings are targeted at the kind, and the gravity, of the damage caused. The way in which this happens is strictly regulated along its steps, thanks to evolutionary pressure that selected specific mechanisms and refined them across the generations and the rising of new species. From this point of view, it is not surprising that such processes are generally well conserved across living beings of even very different kinds.

In biology, hierarchies are of fundamental importance. In this case, we can better understand how living beings counteract damages and chronic stresses by looking at how they organize themselves, since the repair mechanisms, or, we should better say, the responses to damages, are differentiated on a dimensional scale.

From a simplified point of view, we can see every living being as a whole organism, but also as a finely structured organization of smaller, interdependent, pieces, which in turn are composed of smaller units and so on until reaching the molecular level, which is usually the deepest insight reached by biological studies.

To briefly exemplify the hierarchical organization of biological entities, we can have a look at vertebrate animals, the class humans belong to, and obviously the first to be regarded as a useful model by nonbiological researchers when observing natural mechanisms.

Other kinds of living beings, such as plants, bacteria, yeasts, insects, are organized with different hierarchical structures, yet the concept remains the same to that we choose as an example: a sequential progression of structures with increasing complexity from single molecules to entire organisms (Table 3.1).

In conclusion, what should really be kept as a reminder is that living beings are not materials, nor are they made of materials in proper sense. But, they can act as materials when we consider them from technological points of view, at any level of their organizational hierarchy. So, they can, or actually should, be studied by material scientists and engineers to get inspiration and insights not only from their biochemical components, but especially from their ability to manage complexity under good and bad circumstances through simple, yet finely tuned, mechanisms.

TABLE 3.1
Hierarchical Organization of Biological Structures

Hierarchy	Features	Dimension Scale	Examples
Organism	Made of several billions cells, hundreds of organs, dozens of apparatuses. Capable of self-repair, not of self-duplication	cm–10 m	Human, mammal, fish, bird
Apparatus	Complex of all functionally related organs. Organ–organ communications to perform coordinate work and face damages	cm–m	Skeletal, locomotive, gastrointestinal, reproductive, cardiocirculatory, immune system
Organ	Anatomical working unit made by different organized tissues. Capable of self-repair in some occasion, of self-renewal in sporadic cases	mm–m	Skin, bones, brain, muscles, liver, stomach, heart
Tissue	Local organization of noncellular components and cells with specific functions and morphology, coordinated and in communication to perform physiological tasks. Generally capable of self-renewal	mm–cm	Blood, connective tissue (cartilage, bone), epithelium (skin, mucosae), neural tissue
Cell	Basic unit of life, capable of autoduplication, self-repair, self-killing Contains genetic inheritable instructions	μm–mm	Blood cell, neuron, gamete (reproductive cell), keratinocyte (skin cell), muscle cell
Cellular organelle	Functional and/or topological structures included in cells. Made by aggregation of macromolecules, in some case capable of self-duplication and repair	nm–μm	Membranes, ribosomes, nucleus, cytoplasm, cytoskeleton, mitochondria
Macromolecule	Complexes (often polymers) of simpler molecules with or without sequence and spatial specificity	nm	DNA, RNA, lipids, proteins, carbohydrates
Molecule	Organic compounds organized with defined symmetry properties	A–nm	Nucleotides, fatty acids, amino acids
Atom	Basic unit of biological chemistry	pm–A	C, O, N, Ca, S

In the next section, we will focus on four different situations of biological response to stress and damage:

1. DNA integrity defense against various kinds of offence, as an example of molecular-level self-repair (or self-destruction)
2. Liver physiological and pathological self-renewal, as and example of both cell and organ-level mechanisms
3. Wound healing in skin, as a very simplified model of tissue-level repair and regeneration process
4. Bone repair, as an example of both tissue and organ-level self-repair and regeneration

3.2 CELL RESPONSE TO DNA DAMAGE

3.2.1 DNA Structure

When looking at damage-response mechanisms at microscopic level, it is always important to remember that biological molecules are responsible for the functional efficiency of the living being they are part of. In the case of DNA, in the chemical structure of a biological molecule, the genetic information lays, which must be preserved undamaged through the generations.

From a chemical point of view, DNA is a macromolecule, a polymeric double chain (double-strand) of modular units called nucleotides, which in turn are simpler molecules made up by an invariant part (a polyphosphate molecule linked to a monosaccharide) and a variable part (a nitrogenous base), available in four different chemical structures (adenine, guanine, cytosine, thymine), which confers the molecule its "encoding" capabilities (Figure 3.1). Hydrogen bonds between bases keep the two strands together and are the key for maintaining DNA sequence specificity through replication, transcription, and repair processes. The physiological H-bond pairing are adenine–thymine (uracyl, in RNA only) and guanine–cytosine.

DNA forms long chains that wind up and coil tightly around proteins called histones to form discrete structures known as chromosomes; DNA is stored inside a subcellular membranous structure called nucleus (the presence of nucleus is common to all eukaryote cells), organized in 23 couples of chromosomes (so that, a part from sex chromosomes X and Y, all DNA is present in double copy).

DNA is the most precious content inside a cell, and, thus, good part of the cell ordinary maintenance activities are focused on DNA safety and integrity. There is a simple motivation: losing a part of DNA sequence or allowing modifications to be introduced in the sequence is absolutely incompatible with cell biological rules, so it must be avoided. And, if good repairs cannot be made, a cell can autonomously start a self-destruction program (called apoptosis).

A cell must choose upon living or dying in every moment, and has a complex organization of check points appointed to take this decision (in cell biology, a check point is a coordinated group of sensors, mediators, and effectors, usually proteins, that act like a switch in particular temporal or conditional situations of cell life.

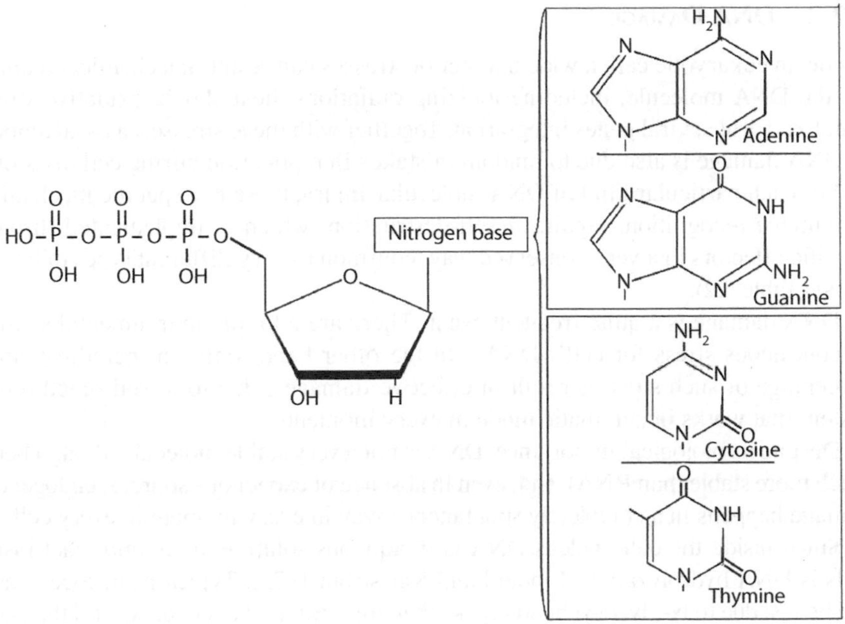

FIGURE 3.1 Structure of nucleotides, DNA monomeric units. Nucleotides are formed by a sugar molecule (deoxyribose) bound to an organic triphosphate group and to a nitrogen base. The four possible canonical nitrogen bases are shown on the right panel, subdivided into purines (adenine, guanine) and pyrimidines (cytosine and thymine).

They are called up to check the status of DNA integrity and replicative capability and, improperly said, "take decisions" on whether proceeding through cell cycle or stopping) (Weinert 1998).

Sometimes, death of a cell occurs, when it is no longer required, or must be eliminated if it becomes potentially dangerous. Sometimes, death is the end point of a long life of chronical stress, and is part of a continuous flow of turnover. But other times, death is an incident occurring to a cell when something inside it, or around it, goes awfully wrong. Chemical or physical stresses, microbial infections, or spontaneous genetic mistakes are the main causes of such incidents.

Fortunately, due to various DNA repair mechanisms, the vast majority of such situations are solved without further complications.

Though DNA itself is not able to self-repair from a chemical point of view, it encodes for the production of virtually everything is required to its repair. So, every cell is virtually self-sufficient for DNA repair duties and we could speculate that, in a biological perspective, DNA is actually able to recognize damages and lesions on its molecule, and repair itself.

Lots of different DNA lesion sources can produce lots of different damages (we will only mention a few), but for each kind of damage at least one specific repair mechanism is available, with the aim to fix genetic errors before DNA replicates or is transcripted (to produce proteins via an RNA messenger).

3.2.2 DNA Damage

Inside an eukaryotic cell, a wide number of stresses can result in a chemical damage for the DNA molecule, including ionizing radiations, heat shock, oxidative stress, genotoxic drugs, viral genes integration. Together with these stresses, a basal amount of DNA damage is also due to random mistakes in replication during cell division.

For each particular kind of DNA molecular injury, there is a specific mechanism of damage recognition, repair, or self-destruction, which is implemented through specific effectors in a very conserved way (common to very different types of organisms) (Table 3.2).

DNA damage is a quite frequent event. There are a lot of environmental sources of continuous stress for cell's DNA. On the other hand, only an incredible small percentage of such stresses result in effective damages, due to a well-tuned repair system that works in automatic mode in every moment.

Despite its biological importance, DNA is not a very stable molecule (though being much more stable than RNA), and, even in absence of exogenous sources, endogenous damage happens in a completely spontaneous way in every moment in every cell.

Since inside the cell nucleus DNA is in aqueous solution, a common fact inside cells is DNA hydrolyzation (Lindahl and Karlström 1973). Typically, nitrogen bases can be lost due to N-glycosyl bond breaks that are estimated to occur several thousand times for each cell in a day. Or, to a smaller but still relevant extent, nitrogen bases can undergo spontaneous deamination (Lindahl and Nyberg 1974), especially on cytosine (so becoming uracyl, a base typical of RNA, not present in DNA) or methylated cytosine (so becoming thymine; 5-methylcytosine is commonly present in DNA, because of enzymatic reactions of cytosine methylation, which have specific biological functions). In turn, if not repaired, such kind of damage can introduce a point mutation in the DNA sequence when DNA is replicated, because the improperly formed thymine bases will be matched to adenine bases instead of guanine, and the original genetic information will be lost. Due to this mechanism, methylated cytosines are considered to be preferential sites for clinically relevant mutations (Jones et al. 1992).

On the other hand, also spontaneous or induced oxidation by oxygen-reactive species (ROS) can result in important damages to DNA molecules. Due to the oxidative metabolism of all eukaryote cells, exposure of DNA to oxidation is a chronic feature (hence the commonly used term "oxidative stress"). An example may be the formation of 8-OHdG (hydroxyguanine) by hydroxyl radicals (Kasai et al. 1984), typically seen in clinical studies of chronic diseases with strong relations with oxidative stress (liver, stomach, esophagus, etc.). Other DNA lesions caused by oxidation by ROS can happen on cytosine or thymine nitrogen bases, whose planar structure is disrupted, or can lead to the formation of covalent bonds between sugar-phosphate backbone and nitrogen bases (Dirksen et al. 1988), or between adjacent nucleotide bases (adenine or guanine dimers). DNA peroxidation can also result in less frequent but very harmful molecular lesions.

DNA can also be subject to chemical modifications other than oxidation. As already mentioned, alkylation, in particular, methylation, is one of these. Nonenzymatic methylation can result in modifications on nitrogen bases with different effects (e.g., the formation of 3-methyladenine results in a DNA replication block and leads to cell death) (Rydberg and Lindahl 1982).

TABLE 3.2
Sources of DNA Damage and Related Repair Systems

Source of Damage	Features	Molecular Damage	Repair Mechanism
ROS (endogenous or exogenous)	DNA is chemically attacked by oxygen-reactive radicals or ions. Covalent modification of bases	Base loss; base mismatch; intrastrand covalent links. Double-strand breaks	Mismatch repair, nucleotide excision, homologous or nonhomologous double-strand repair
Ionizing radiations (x-rays, γ-rays)	High-energy excitation of DNA molecule	DNA intrastrand breaks; base loss; double-strand breaks	Excision repair, nonhomologous joining
UV radiation	Mid-energy excitation of DNA molecule	Base dimerization (often thymine dimers)	Direct repair (not in human), nucleotide-excision repair
Heat	Heat disrupts chemical bonds	DNA breaks	
DNA-binding molecules	Covalent binding of molecules to DNA	Replicative stop; interstrand links (cross-links)	Cross-link repair, excision repair
DNA-intercalating molecules	Insertion of a planar molecule between the two strands	Stabilization and deformation of double helix, replicative stop, repair inhibition	Cross-link repair
Toxins or viral proteins	See ROS, binding or intercalating agents. May also inhibit repair	Various	Excision repair, cross-link repair
Random misincorporation during replication	A certain amount of base incorporation mistakes are made continuously	Base insertion, base deletion, base mismatch	Mismatch repair
Viral DNA	Viral genomes insertion in host cell's DNA	Insertion of exogenous sequences inside genome; frameshift mutation	No specific mechanism
Water	DNA in aqueous solution can be spontaneously degraded	Base loss (especially guanine and adenine); base deamination (transition cytosine–uracyl, methylcytosine–thymine)	Mismatch repair, base excision, nucleotide excision
Methyl groups	Methyl groups are transferred to DNA bases	Inhibition of replication; base mismatch	Direct repair, mismatch repair, base or nucleotide-excision repair

Another example is the production of O6-methylguanine by means of alkylating drug temozolomide that is used in anticancer therapy. We will return to it because of its peculiar repair mechanism.

A very high number of molecules either endogenously produced inside the cell or artificially introduced by diet or medical or environmental administration can give rise to modification of DNA bases with toxic effects by binding DNA (so forming DNA adducts), or can create a heavy interference with DNA replication (and repair) machinery by intercalating inside DNA double helix. Many genotoxic drugs are intentionally designed to form DNA adducts or to intercalate inside DNA in order to stop DNA replication and cause cell death. Obviously, since DNA repair systems are also inhibited by such drugs, this can lead to mutagenic effects.

Apart from chemical compounds, physical agents are a main source of DNA direct and indirect lesions. It is well known that ionizing radiations carry sufficient energy to randomly disrupt chemical bonds inside the cell, including DNA backbone (single or double strand) or nitrogen base bonds (Rupert 1961). So, ionizing radiations must be considered a very dangerous form of genotoxic agent, which can induce mutations and cell death. Naturally, exposition to intense ionizing radiations is not a common experience for living cells, but exposition to UV rays from natural or artificial sources is much more frequent.

Low-energy UV rays (UV-A) are not considered to be directly able to induce lesions in DNA molecules, though they could produce reactive intermediate molecules (they can, e.g., generate ROS, which in turn can cause oxidative damage as mentioned earlier), whereas mid-energy UV radiations (UV-B) are able to generate intrastrand DNA base covalent bonds (typically, pyrimidine dimers) (Setlow et al. 1965), which are responsible for replicative stops or point mutations.

In the case of UV radiations, not only DNA repair mechanism is deployed to prevent genetic alterations, but also the chemical structure of DNA itself has biophysical properties that can prevent most of UV potential harms, thanks to a photoprotective-like mechanism that dissipates energy into heat by rapid internal conversion.

Thermal stress is also responsible for chemical bond breaks inside DNA molecule (loss of nitrogen bases on single-strand breaks), similarly to what described earlier for spontaneous hydrolysis.

Viruses or bacteria can be a source for DNA damage due to DNA-binding toxins or proteins, or by elicitation of ROS production (directly or, most commonly, by the host immune system).

Some viruses, namely, retroviruses, can also have a direct effect on DNA of the infected cells, by producing a DNA copy of their RNA genome that inserts itself inside the host DNA by homologous recombination. This is not a real damage, so it is not generally subject to repair.

Of course, among the DNA damages must be included, though improperly, the spontaneous base misincorporation that happen, to a determinate rate, during DNA synthesis and replication. As in the previous case, this may be considered to be not a real damage; nevertheless, it is subject to efficient repair mechanisms.

Finally, DNA damage may also result from the partial repair of a smaller lesion. Typical example of that is the formation of double-strand breaks resulting from the enzymatic recognition of a single-strand break (Hamilton et al. 1975).

A double-strand break is much more dangerous for the cell, because it can lead to major genetic or even chromosomic alterations that can result in cell death or senescence (complete and permanent replicative stop).

3.2.3 Repair Systems

There are a large number of DNA repair systems spread across all living species. For simplicity, we can organize DNA repair systems that are also present in human cells in five groups, based on their general model of action.

The first group is the one with the simpler mechanism: direct repair, which means that dedicated enzymes are produced to sense for specific damages and, when they find them, they can immediately repair the lesions by themselves. A typical example present in many organisms (bacteria, fungi, plants, invertebrates, and some vertebrates, but not humans) is the enzyme photolyase (Setlow 1972) that is able to find and repair thymine dimers and thymine–cytosine adducts generated by UV-B radiations. Photolyases are evolutionally very ancient and well-conserved enzymes that act by scanning continuously the DNA molecules and binding to UV-generated dimers with high affinity (Figures 3.2 and 3.3).

Each photolyase molecule contains two different chromophores, one being a light harvester (5,10-methenyltetrahydrofolate or 8-hydroxy-5-deazariboflavin, both excited by visible blue light) and the second being a catalytic cofactor (reduced flavin–adenine dinucleotide, FADH-). After blue light photons absorption, the light harvester is excited, and the excitation energy is transferred to FADH- that, in the presence of a UV-induced dimer, donates an electron to it, breaking the bond, allowing the correct bases to be restored, and getting the electron in return. This reaction is called photoreactivation (Sinha and Häder 2002). Another example is the ubiquitous

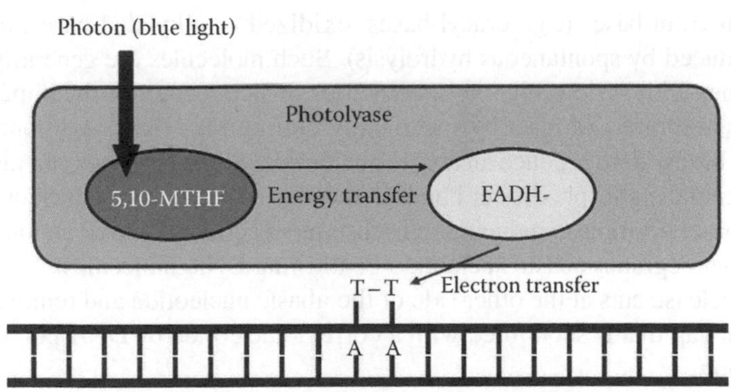

FIGURE 3.2 General mechanism of photoreactivation by enzyme photolyase. When an intrastrand base–base dimer (typically a thymine–thymine dimer) is sensed by the enzyme while randomly binding to DNA molecules in search of damages, the dimer is inserted in the enzyme's catalytic site where the reaction takes place. The enzyme cofactor 5,10-MTHF, in presence of light, absorbs a blue-wavelength photon, gaining energy that is soon transferred to the other cofactor FADH-. After being excited, FADH- transfers an electron to the DNA dimer, disrupting the base–base bond and restoring the correct DNA structure. The electron is eventually given back to the cofactor so that the enzyme is again ready to repair other damages.

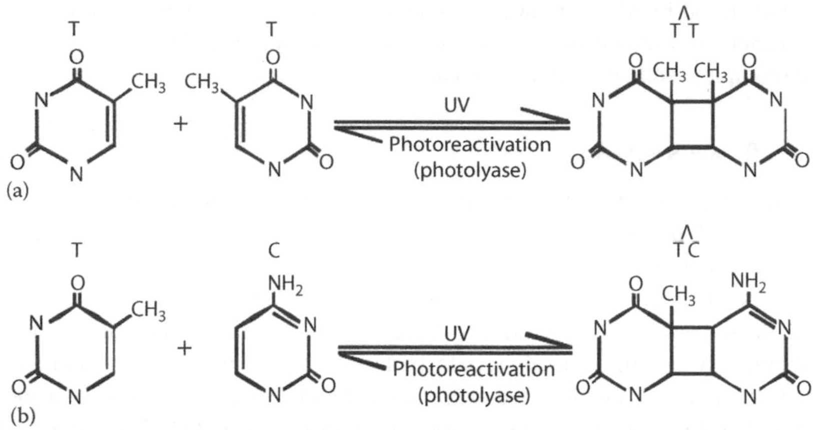

FIGURE 3.3 Thymine–thymine (a) and thymine–cytosine dimers (b) can arise from exposition to UV radiations. Such dimers can be repaired to restore the original structure by means of photoreactivation by enzyme photolyase. (From Sinha, R.P. and Häder, D.P., UV-induced DNA damage and repair: A review, *Photochem. Photobiol. Sci.*, 1, 225–236, 2002. By permission of The Royal Society of Chemistry.)

enzyme MGMT (methylguanine DNA-methyltransferase), which is responsible for specific methylated guanine dealkylation, as described in the previous paragraph.

Unfortunately, since anticancer drugs can produce such methylated bases, MGMT work can result in less effective therapies (Isowa et al. 1991).

A second option for DNA repair is the so-called excision repair system, which in turn is subdivided into base-excision and nucleotide-excision repair (Rupert 1961). In the first case, one or more molecules scan the DNA strands to search for the presence of aberrant bases (e.g., uracyl bases, oxidized or alkylated bases, deaminated bases produced by spontaneous hydrolysis). Such molecules are generally glycosylase enzymes with various binding specificities, which recognize the imperfect DNA backbone structure and react by specifically cutting the N-glycosyl bonds between sugar and bases, so to produce an abasic nucleotide. Some enzymes can also proceed into breaking the phosphodiester bonds at the side of the abasic (a nucleotide with no nitrogen base) position so inducing the subsequent action of an AP endonuclease (an enzyme that degrades abasic nucleotide-containing DNA molecules).

This nuclease cuts at the other side of the abasic nucleotide and removes it, forming a point gap that is soon filled with a correct nucleotide by DNA polymerase and ligase action.

In other cases, endonucleases can cut off a longer stretch of DNA, but it will anyway immediately be resynthesized by DNA polymerase.

The second mechanism of excision repair is nucleotide excision, which is the method of first choice for eliminating bulky DNA adducts, cross-linking, base dimers, and other kind of molecular abnormalities. The lesion is recognized by DNA-binding proteins sensing aspecific double helix symmetry perturbations. They in turn gather up other proteic factors that bind cooperatively (so conferring kinetic specificity to repair reaction) and start nicking DNA both upstream and downstream the damage, causing a stretch of 20–30 nucleotides to be cut. The resulting gap is then filled by polymerase/ligase action as seen for base-excision repair. During DNA transcription (to obtain RNA for gene expression), such machinery is in direct contact with transcription complex, so that DNA lesions that are recognized and stop transcription process can be readily repaired. Such variant of excision repair is called transcription-coupled repair.

Another mechanism of DNA aberrations repair is called mismatch repair (Wildenberg and Meselson 1975). When new DNA is synthesized, base mismatch can be present due to random misincorporation or due to chemical modifications in the template strand. Such mispairing is recognized by specific proteins that in turn start recruiting other effectors in a mechanism close to nuclear excision repair. The main difference is that the newly synthesized strand is recognized (due to a less amount of methylation on specific DNA motifs) and is selectively targeted for being degraded producing a gap of some dozen nucleotides length which is sealed by the usual polymerase–ligase mechanism.

When the template strand is damaged, such mechanism is not sufficient to fix the situation, and a loop of defective repair can start, eventually resulting in replication stop and cell death.

Much more complex repair mechanisms are required when both DNA strands are involved in the damage, that is, when there is a double-strand break or an interstrand bond (cross-link).

In the case of double-strand breaks, which can arise from ionizing radiations, heat shock, heavy oxidation or defective single-strand nick repair, there are two main repair systems, optioned on the basis of damage nature and extension (Resnick and Moore 1979).

In the first case, the break is sealed using a recombination-like machinery (recombination is a natural process of DNA exchange between homologous chromosomes). If a DNA molecule identical to the injured one is readily available inside the cell (during DNA synthesis a perfect copy is produced, but anyway in every time a homologous chromosome can be found), both its strands are used as direct templates for broken strands homolog pairing and ligation to fix the breaks. In the end, four strand structures are generated (called Holliday's junctions) and then rearranged into two identical (and repaired) DNA double-strand molecules.

In the second case, if a "replacement template" is not available, the cut ends of DNA double strand are joined together by specific proteins (Ku dimers) (Chen et al. 1996) that act by recruiting other effectors, resulting in a direct ligation of juxtaposed ends, regardless they are or not part of the same original molecule.

The aforementioned method is error prone, since part of the original information can be lost or altered, so it is used only if no other ways can be found, for example, in case of heavy radiation damage.

On the other hand, this mechanism is essential for a non-repair process such as genetic somatic rearrangements in lymphocytes to obtain specific antibody production (this process is called VDJ recombination).

As previously seen, drugs or radiations can induce interstrand bonds (cross-links), which prevent DNA replication and can lead to cell death. In similar situations, repair can be achieved by a double cycle of excision repair, in order to separate strands and repair both involved nucleotides. When cross-link is recognized during replication, a slightly different mechanism provides both damage repair and new DNA error-free synthesis.

3.3 LIVER REGENERATION

In ancient Greece, physiology knowledge was rather poor; nevertheless, the epic myth of Prometheus (whose liver was eaten every day by vultures, and every day was regenerated anew) shows us how liver self-repair properties were already known in those ancient days.

This peculiar ability of liver is mainly due to the combined effect of the presence of a population of liver-specific stem cells (called oval cells) and of the capability of common liver parenchymal cells (called hepatocytes) to start proliferating with rapidity and efficiency when the surrounding conditions require them to do it. Clinical reports state that after surgical resection of >70% of liver mass, the remaining tissue still retains the capability to regenerate to original dimensions. To current knowledge, no other organ in vertebrate animals behaves the same way.

It is anyway not completely correct to define this process as regeneration, since the anatomical structure of the organ is not necessarily maintained whereas the internal histological structure is restored starting from the undamaged portions. The process should be more properly defined as compensatory growth or compensatory hyperplasia (Fausto et al. 2006).

In the absence of stress conditions, hepatocytes are in a cell cycle condition called phase G_0, meaning that they are not able to divide and proliferate, nor to replicate their DNA. But when it is needed, for example, when a large amount of liver tissue has been removed or has died of acute poisoning, the remaining hepatocytes immediately start to proliferate and stop when the original organ dimension is restored, probably due to the release of strong antiproliferative molecule, tumor growth factor beta (TGF-β) (Strain 1988). Soon after hepatocytes, the other liver cell types (bile duct cells, Kupffer cells, stellate cells, and endothelial cells) also start to proliferate, thus permitting restoration of not only the pre-damage hepatocytes number, but also the physiological tissue and organ architecture. The first step of this regenerative process is the production of signaling molecules by the nonhepatocytic cells of the damaged liver, such as, for example, proinflammatory interleukin (IL)-6 or tumor necrosis factor alpha (TNF-α, produced by the macrophage-like Kupffer cells), which act as primers to hepatocytes proliferation (Iwai et al. 2001). Anyway, after cytokines action to bring hepatocytes out of quiescence and back to the cell cycle, the action of growth factors is strictly required to ensure that proliferation really takes place (in terms of cell cycle, while TNF-α and IL-6 move cells from G_0 to G_1, those growth factors can lead them to phase S, with DNA replication, and finally to mitosis). Hepatic growth factor (HGF), secreted by nonhepatocytic liver cells (mainly fibroblast-like stellate cells), is considered to be the strongest activator of hepatocytes proliferation (Nakamura 1991). Other growth factors involved in the process are transforming growth factor alpha (TGF-α) (produced by the hepatocytes themselves in an autocrine loop) and epidermal growth factor (EGF) (Fleig 1988).

Together with hepatocytes, also oval cells (liver stem cells) seem to be able to proliferate in case of liver damage and differentiate into hepatocytes or colangiocytes (cells lining bile ducts, also part of liver tissue) (Golding et al. 1995).

It has also been hypothesized that circulating hematopoietic stem cells, under certain circumstances, can migrate to liver and differentiate into colangiocytes and hepatocytes (Petersen et al. 1999), but this is still under debate.

Of course, to restore the physiological architecture of liver and to maintain its metabolic activities, it is necessary that new blood vessels and sinusoids (highly permeable capillary vessels, typical of liver) be formed in the correct places, so that all hepatocytes are correctly connected to blood circulation. Blood vessels are formed in regenerating liver by means of specific growth factors, such as vascular endothelial growth factor (VEGF) and angiopoietins, but they can grow with the correct geometrical disposition inside the liver parenchyma only following the new deposition of extracellular matrix by stellate cells, which takes place a few days after the starting of the regenerative process (Martinez-Hernandez and Amenta 1995) (Figure 3.4).

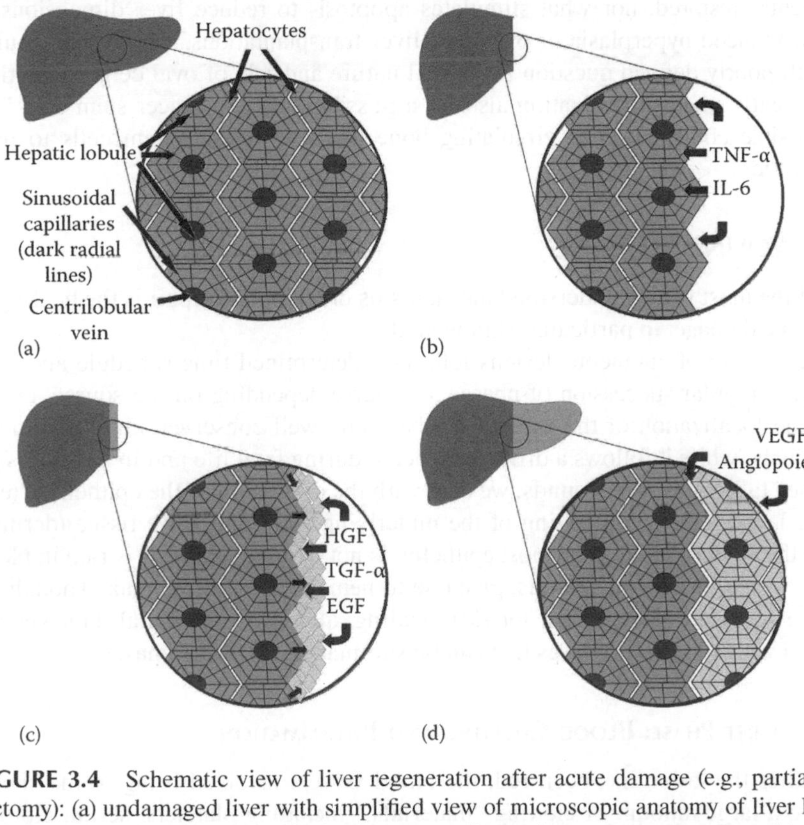

FIGURE 3.4 Schematic view of liver regeneration after acute damage (e.g., partial hepatectomy): (a) undamaged liver with simplified view of microscopic anatomy of liver lobules, including sinusoidal blood capillaries between adjacent hepatocytes, represented by radial dark lines; (b) partial hepatectomy immediately followed by release of TNF-α and IL-6 from surrounding tissue to induce the remaining hepatocytes to enter cell cycle; (c) in a few hours, secretion of growth factors (HGF, EGF, TGF-α) by surrounding tissue induces massive proliferation of healthy hepatocytes (new hepatocytes are represented in lighter color) until the pre-damage parenchyma mass is restored; (d) in the following days, production of connective tissue and new sinusoid vessels between newly formed hepatocytes restores the histological architecture and functionality of the liver.

Needless to say, aberrant liver regeneration can also occur. Typical example of this is cirrhosis: when damaging factors (i.e., alcohol intoxication, metabolic steatosis, or chronic viral hepatitis) are chronically present inside the liver, the regenerative process aimed at replacing poisoned or infected hepatocytes goes on without coming to a stop. In this continuous, excessive turnover, liver architecture is not restored with the same pace, and in mid-to-long terms it is completely deregulated, generating abnormal reactions such as collagen overproduction (fibrosis) or even disseminated liver scarification (cirrhosis) with loss of metabolic functionality and heavy problems in blood perfusion. On the other hand, excessive, never-stopping hepatocytes proliferation can result in accumulation of genetic errors and, in long terms, in hepatic cancer.

Though research in the field is in continuous progress, there is still a certain amount of unsolved questions about liver repair process. For example, it is still not clear which signal can stop hepatocytes from proliferating when the liver mass is sufficiently restored, nor what stimulates apoptosis to reduce liver dimensions in case of induced hyperplasia or oversized liver transplantations. Another intriguing and still poorly defined question is the real nature and role of oval cells, since they are currently under investigation also for a possible role as "cancer stem cells," or the possible contribution of circulating bone marrow–derived stem cells to liver regeneration.

3.4 WOUND HEALING

One of the most widely understood mechanisms of biological repair is the healing of soft tissue damage, in particular, skin wounds.

The healing of cutaneous lesions follows a determined time schedule and goes through a regular succession of phases, of course depending on the source, extension, and localization of the wound, but basically well conserved among all adult individuals (while it follows a different scheme during fetal life and in newborns).

When talking of skin wounds, we deal with the disruption of the epithelium (epidermic layer) and the damaging of the underlying soft connective tissue (dermis). While the epidermic layer, as most epithelia, is not irrorated, dermis is rich in blood vessels which, in case of wounds, give rise to hemorrhagic phenomena. Though we propose schematic mechanism for skin healing, almost all epithelial damages are repaired with similar processes that can be summarized in three phases.

3.4.1 First Phase: Blood Clotting and Inflammation

Soon after the lesion has occurred, the blood vessels involved bring to the site of damage a large amount of clotting "materials," including platelets, serum clotting factors, and other plasma proteins (i.e., fibrin), with the aim of immediately isolating the underlying exposed tissue and stopping the blood leakage restoring hemostasis. Though being not properly a mechanism of wound healing, this first phase of cloth formation is indispensable to the following processes. In facts, platelets embedded in the clot are a source for growth factors and cytokines that will play a major role in starting the following repair mechanisms. Among those secreted factors, EGF, platelet-derived growth factor (PDGF), and TGF-β (Frank et al. 1996) are involved in both recruiting inflammatory blood cells and activating surrounding fibroblasts and keratinocytes (basal, stem-like, epidermal cells).

In the immediately following hours, the bloodstream brings to the wound site a large number of blood white cells, namely, neutrophils and monocytes–macrophages, which are responsible for the inflammation phase, thanks to both their direct phagocytic role (direct to contaminants microorganisms and to necrotic cells or extracellular materials involved in the tissue damage), but also, to some extent, especially, the production of a large number of other inflammation mediators (cytokines), including TGF-α, TGF-β (Rappolee et al. 1988), fibroblast growth factor (FGF), IL-1, and TNF-α (Hübner et al. 1996), which will bring the action of the aforementioned, platelet-produced cytokines in promoting inflammation at wound site.

As studies on animal models showed in recent years, the presence and correct working of neutrophils and especially monocytes–macrophages is critical for the effectiveness of wound healing.

3.4.2 Second Phase: Re-Epithelialization

In a time ranging from a few hours to a couple of days post-injury, the effective skin repair slowly starts to happen with the following phase, the formation of granulation tissue in the damaged dermis, and the migration of epidermis keratinocytes from the healthy margins of the lesion (or, if present in the affected area, from hair follicles). Granulation tissue is formed by means of fibroblasts coming from the surrounding dermis, which are stimulated by the previously mentioned, locally released, cytokines (mainly the FGF family) to produce extracellular matrix (fibronectin, hyaluronic acid, and later collagen) in large quantities, while from the local blood vessels, similarly by means of secreted growth factors (mainly VEGF secreted by macrophages and surrounding epidermal cells) (Brown et al. 1992), starts the production of new capillaries inside the repairing dermis, in turn bringing on-site macrophages and new soluble cytokines and nutrients useful for the reparation process. In the meanwhile, the surrounding basal keratinocytes, the stem-like cells of skin, undergo a heavy phenotypic change, completely losing their cell–cell and cell–basal membrane adhesivity (which in physiological conditions is granted by an array of transmembrane proteins, mainly integrin family members or components of the desmosomal complex), and re-organizing their cytoskeletons to permit crawl-like movement.

Those cells migrate from the edges of the cut epidermis inside the granulation tissue (in this case using secreted proteins called proteases to cut the matrix and gain way through it), to form in a few days a new basal layer of epidermis, and then start to proliferate to produce the new skin upper layers. This process is called re-epithelialization, and is largely regulated by the local secretion of growth factors (EGF, TGF-α, keratinocyte growth factor [KGF]; secreted by fibroblasts) (Werner et al. 1992). In a few days, depending on the wound extension, the gap is completely filled, the epithelial continuity is restored; the upper, differentiated layers of epidermis are also being restored, and make their way through the granulation tissue (that, by this point, is mainly extracellular matrix).

It is noteworthy that though a large network of new capillary blood vessels is formed in the granulation tissue starting from endothelial cells of the surrounding vessels (attracted on-site by a chain of biochemical signaling stimulated by cytokines secreted by fibroblasts, platelets, and macrophages, as seen before), all the new capillaries are to be dissolved by apoptosis of their endothelial cells, since in epidermis, that replaces the granulation there will be no vascularization.

3.4.3 Third Phase: Scar Formation

In the second week after the trauma, the regeneration of the underlying dermis takes place.

As seen before, large numbers of fibroblasts migrate into the granulation tissue, where they carry out the formation of strong collagen fibers. By this point, part of those cells differentiates into a smooth muscle-like phenotype (they are called myofibroblasts

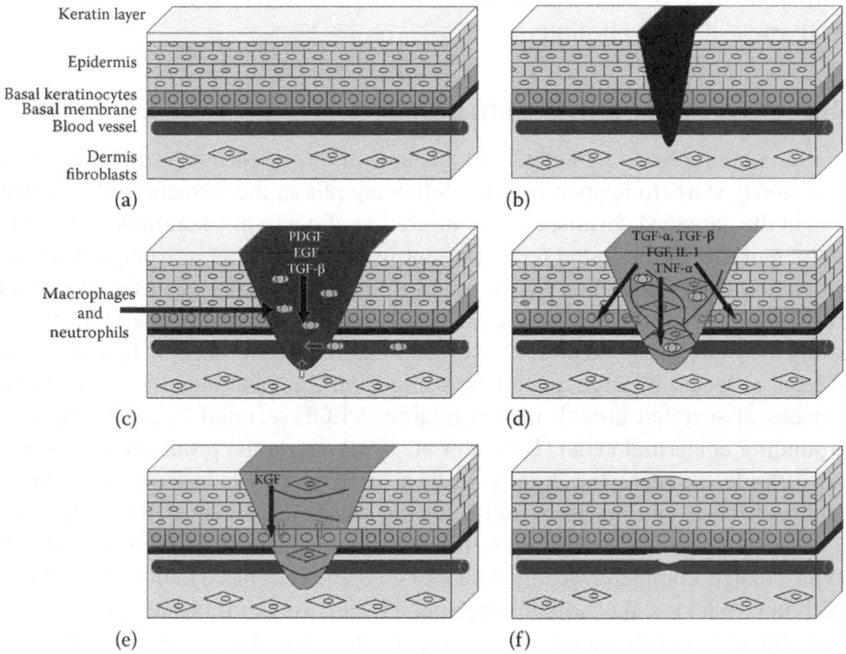

FIGURE 3.5 Schematic view of cutaneous wound healing: (a) undamaged skin is composed by epidermis (multilayer epithelium with basal membrane) and dermis, mainly composed by connective tissue with blood vessels; (b) skin is damaged by a deep lesion; (c) blood vessels in the dermis layer are damaged, causing a hemorrhage, and subsequently the formation of a clot by platelets. Many blood white cells (neutrophils and macrophages) and fibroblasts from surrounding tissue are attracted on-site; (d) fibroblasts lay down fibrous matrix and many new blood capillaries are formed giving rise to granulation tissue (shown in lighter color), replacing the blood clot; basal keratinocytes from undamaged epidermis proliferate and migrate inside granulation tissue to fill the gap; (e) basal keratinocytes repair the gap and proliferate to form the suprabasal epidermis, while granulation tissue slowly loses cells and vessels; (f) in the following weeks, epidermis is almost entirely restored, while granulation tissue completely disappears and new dermis is formed, though containing less fibroblasts and blood vessels (and being less resistant) than before the damage. The newly formed skin portion is called scar.

because of their contractile properties), whose contraction allows the reduction in size of the forming scar and thus accelerates the time for epidermis reconstruction (Walter 1976). During this phase, on the other hand, the collagen fibers produced by fibroblasts in the deeper part of the granulation tissue, which is to give rise to the scarred dermis, are degraded by metalloproteinases action and replaced by stronger fibers, contributing to the (though only partial) restoration of skin mechanical strength.

In a time ranging from a few days to a few weeks, the last phase brings to the scar maturation, with a continuous layer of epidermis completely covering a newly formed dermis with very scarce fibroblasts, very few blood vessels, no glands nor hair bulbs, and very few nervous terminations, which will have not the same functional "performances" of the pre-lesion dermis, being less resistant to mechanical stress (only up to 70% strength can be restored) (Levenson et al. 1965), irradiation, thermal changes, but will maintain skin integrity and, in general, will reestablish the pre-wound anatomical outfit of the damaged area (Figure 3.5).

3.5 BONE CRACKS AND FRACTURES REPAIR

3.5.1 Bone Structure

The technological properties of bones have always led the engineers to consider it as a possible source of inspiration for new composite materials. Elasticity and stress (especially compressive) resistance, together with very low weight, are typical features of bones that could be of interest when seeking to design new materials, but the main peculiarity in bones is their dynamic life, the possibility to be rearranged to follow (and lead) the organism's growth, to maintain or alter the mineral homeostasis in the bloodstream and, of course, to undergo reparation in case of damage (Figure 3.6).

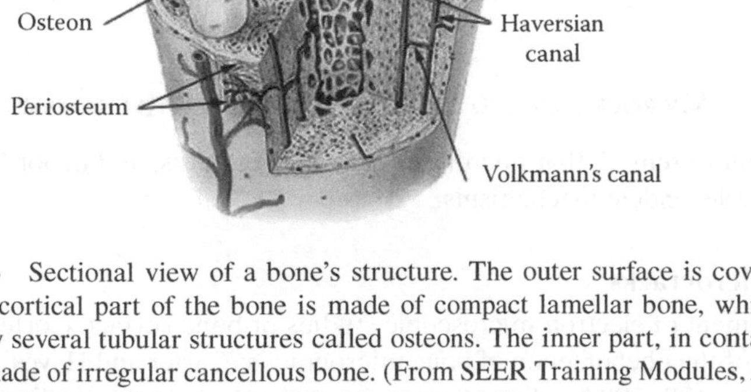

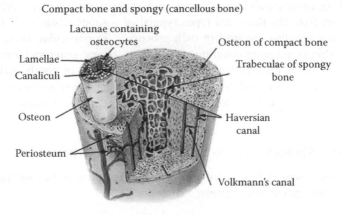

FIGURE 3.6 Sectional view of a bone's structure. The outer surface is covered by periosteum; the cortical part of the bone is made of compact lamellar bone, which in turn is composed by several tubular structures called osteons. The inner part, in contact with bone marrow, is made of irregular cancellous bone. (From SEER Training Modules, *Anatomy and Physiology*, U.S. National Institutes of Health, National Cancer Institute, http://training.seer.cancer.gov//anatomy//skeletal//tissue.html.)

Osseous tissue (also called bone tissue), the major component of human bones (together with marrow, endosteum and periosteum, nerves, blood vessels, and cartilage), is a particularly organized structure made up by specialized cells (osteocytes, osteoblasts, osteoclasts) that are included in a network of noncellular components, with the fundamental contribution of a continuous and widespread stream of blood which is granted by a network of blood vessels.

The three cell types present in bone tissue have specific functions: osteoblasts are progenitor cells, able to lay down extracellular matrix, but also to produce enzymes and hormones. When they are completely embedded in the matrix they secreted, they undergo a maturation process that transforms them into osteocytes. These cells are adhibited to bone maintenance functions and calcium homeostasis through bone deposition and resorption.

Osteoclasts are immune-derivative cells with lytic and phagocytic capabilities aimed to perform bone remodeling both in physiological conditions and in advanced phases of fracture repair processes.

The main extracellular components are proteic fibers (type I collagen), which are secreted by osteoblasts during bone formation but also during lifetime turnover or reparation. In compact bone tissue, collagen fibers are usually disposed in a lamellar shape to form thin cylindrical structures (disposed in parallel with the bone longitudinal axis) called osteons or Haversian systems. The second noncellular component is represented by crystals of hydroxyapatite ($Ca_5(PO_4)_3(OH)$), the same mineral that we find as main constituent in dental enamel. These crystals are literally embedded in the fibrous matrix, filling its gaps and the combination of the two materials confers to bone its peculiar technological properties, while the presence of cells embedded as well permits the bone to turn over its extracellular components, thus maintaining their good performances, and to face the very frequent microfractures that mechanical stress produces.

The self-repairing capabilities of bones are largely dependent on the cellular component. Apart from the three cell types typical of osseous tissue, cells from other parts of bones are also involved in such processes: in particular, periosteum and endosteum, thin membranes covering, respectively, bones' outer and inner surfaces (the inner surface is the one facing the marrow) are rich in poorly differentiated progenitor cells that give birth to osteoblasts and, upon particular conditions (e.g., in case of fracture), to chondroblasts, progenitor cells of cartilage tissues.

3.5.2 Bone Microcracks and Stress Breaks Repair Mechanisms

It is important to remark that microlesions, stress fractures, and major fractures are repaired in independent mechanisms.

3.5.2.1 Microcracks

The development of electron microscopic studies of bone surface (cortex) led to the observation of the phenomenon of bone microcracks (Carter and Hayes 1977), small breaks of some 100–500 μm diameter that are probably caused by ordinary mechanical stress (torsion, stretching, compression).

Though it is still partially unclear if all or only a small part of such breaks undergo repair, it is known that this operation is routinely performed by bones' cells. Osteocytes embedded in bone matrix have several dozen cell processes (cell extroflections with sensor functions) each, thus forming a dense network inside bone tissue, and it is believed by researchers that microcracks interrupt such network causing an alarm system to be deployed by osteocytes: through the secretion of soluble factors, osteocytes may induce the recruiting in the break site of osteoblasts and osteoclasts, that, working synergically, act in remodeling bone tissue, repairing cracks, and refreshing the tissue components by resorption of old matrix and deposition of new one (Hazenberg et al. 2009).

3.5.2.2 Stress Fractures

Under particular conditions of fatigue and work load (as is for professional athletes, especially distance runners), some bones, especially long ones, are subjected to what is believed to be an extreme consequence of microcracking and consequent repair mechanisms (Uthgenannt et al. 2007).

Fracture stress causes bones to start a self-repair process that mainly results in the formation of the so-called "hard callus" starting from the periosteum at break's site: periosteal progenitor cells are differentiated into osteoblasts that start immediately to lay down new bone matrix with nonregular spatial organization, forming what is known as woven bone, continuing in this action until the gap in the bone cortex is completely covered by the woven bone callus, which is much thicker than the original bone, though less strong and dense until a good amount of mineralization is restored. In midterms, osteoclasts will perform resorption of this woven bone and osteoblasts will replace it with new lamellar bone (with regular longitudinally oriented collagen–hydroxyapatite structures).

3.5.3 Bone Major Fractures Repair

The healing process for major bone breaks (fractures) can be schematically subdivided into four phases.

3.5.3.1 Inflammation Phase

Soon after fracture, similar to any other irrorated tissue, bone undergoes a hemorrhagic process that is repaired by canonical blood clotting. The recruiting of blood white cells is responsible for inflammatory and antiseptic processes, but also for secretion of soluble growth factors and cytokines required for bone regeneration (PDGF, VEGF, FGF, TGF-β, bone morphogenetic proteins [BMP]) (Bolander 1992). After clot formation, nearby cells (fibroblasts from surrounding soft tissue) depose a loose connective tissue in which small blood vessels are formed. This new structure is called granulation tissue. In general terms, this first phase is common to soft tissue wound repair mechanisms.

3.5.3.2 Soft Callus Formation

In the days following fracture, the periosteum cells in proximity of the fracture site proliferate and differentiate into chondroblasts, and chondrocytes which in turn start laying down hyaline cartilage inside the gap. In the same way, in the inner part of

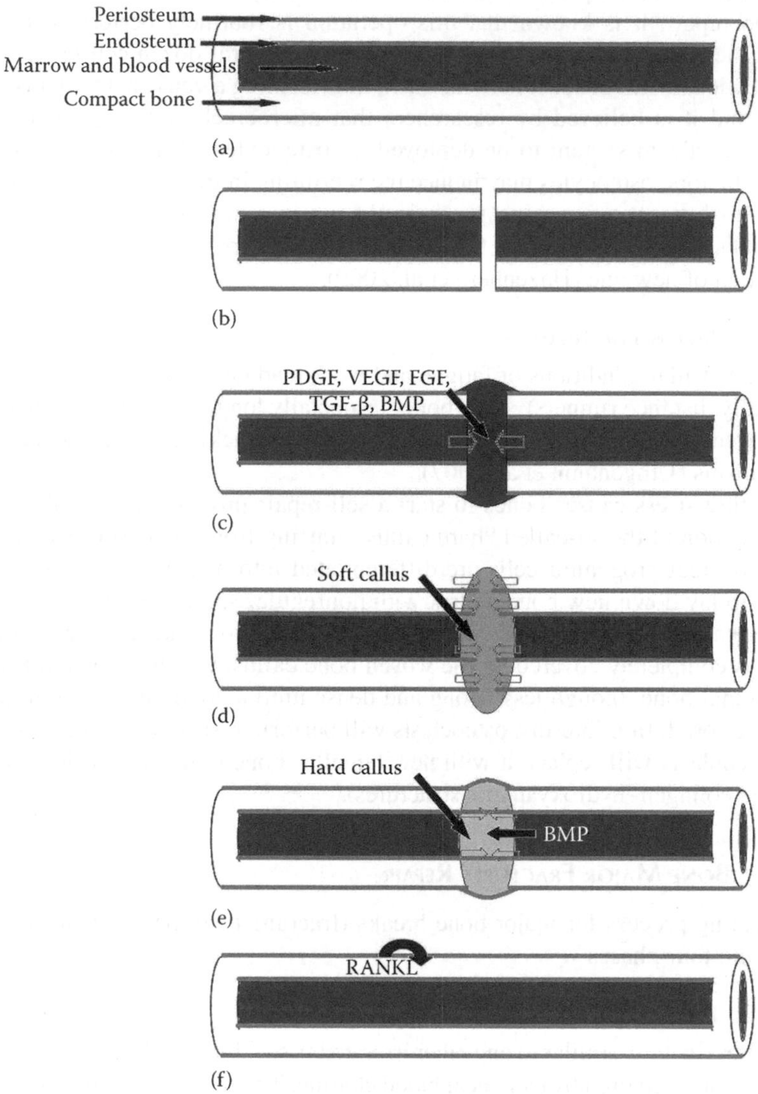

FIGURE 3.7 Schematic view of bone fractures' repair: (a) undamaged bone, longitudinal section; (b) the bone is broken (fracture); (c) blood vessels from inside the broken bone are damaged, causing a hemorrhage and consequently a hematoma to be formed. In the following days, granulation tissue replaces the blood clot; (d) new cells are produced from progenitors in the periosteum and endosteum. These cells lay down a matrix of cartilage and fibrous connective tissue that fills the gap between the broken bone's ends, and is called soft callus; (e) in the following weeks, osteoblasts from the edges of the fracture produce large amounts of woven bone that replaces the mineralized soft callus. This formation is called hard callus and causes the bone to be completely repaired; (f) in the following months, osteoclasts and osteoblasts act in replacing the woven bone with new lamellar compact bone completely restoring bone's original structure and function.

the broken bone chondrocytes and fibroblasts are generated and/or activated and lay down hyaline cartilage or less organized fibrous connective tissue. In a few days, the granulation tissue formed in the middle of the bone break is completely replaced by a vascularized and (in later phases) partially mineralized form of fibrocartilage tissue called "soft callus" that fills the gap and forms a bridge between the two broken extremities. In this phase, the aforementioned growth factors (FGF, VEGF, TGF-β, BMP) are required for cells proliferation, differentiation, and activation.

3.5.3.3 Hard Callus Formation

In the 3–4 following weeks, osteoblasts from the fracture edges (in particular, but not only, from periosteum and endosteum) proliferate and produce large amounts of bone matrix inside the already formed soft callus, generating a mass of irregular (woven) bone with increasing degrees of mineralization called "hard callus," as is for stress fractures. A new network of blood vessels is formed inside the callus to bring on-site the required minerals and growth factors, especially BMP (Deckers et al. 2002). This process is very similar to the embryogenic bone formation process, or osteogenesis (Gerstenfeld et al. 2003), and as in that process, there can be new woven bone formation even in absence of a cartilage callus (most bones are generated via a cartilage model, but some, such as skull bones, are formed via the so-called intramembranous ossification directly on mesenchymal structures, soft tissues, during fetal life). Usually after 4 weeks from fracture, the hard callus completely seals the break with sufficiently mineralized and resistant woven bone.

3.5.3.4 Bone Remodeling

In the following months, the new woven bone is slowly replaced with highly organized lamellar bone by means of osteoclasts (Teitelbaum 2000). As previously seen, these cells act by dissolving the mineralized matrix with acid secretion and proteolytic enzymes. Osteoblasts, soon after that, produce new lamellar bone instead of the dissolved one. Osteoblasts also produce cytokines indispensable for osteoclasts, for example, RANKL. After remodeling, the bone is absolutely similar to the prefracture one. From this point of view, bone fracture repair can be considered as a form of actual regeneration (Figure 3.7).

3.6 CONCLUSIONS

Living beings have evolved through centuries to gain the capability of handling adverse situations, thus ensuring their continuous existence. The central question in this long process has been how to manage the equilibrium between natural, or induced, damages and repair systems.

In particular, living organisms always had the need to keep safe their genetic information, encoded in nucleic acids (DNA). Because of that, all living beings, from the less evolved microorganisms up to humans, have developed mechanism to recognize genetic damages at molecular level and, when possible, to repair it or, when not possible, to minimize worse consequences by cellular self-destruction strategies. DNA itself, encoding the genes that rule the production of all the factors required for the repair systems, is, from a particular perspective, responsible for its own safety, which is philosophically intriguing.

But, scaling up the point of view, it is possible to understand that complex living organisms are made by a hierarchical and modular organization of structures that retain, to some extent, some self-repairing, self-renewal, or regenerative abilities. So, the simple phenomenon of the healing of a skin wound can be seen as the capability of a small part of tissue to produce a new, repaired, version of itself (a scar) when the original form is damaged, while the surprising ability of liver to regenerate completely after a radical loss of tissue mass is an almost unique example of a coordinated cell proliferation mechanism aimed to restore the physiology of an entire organ.

Last, the examples of a complex system such as bones, organs with a technologically evolved, specialized tissue (osseous tissue), which are routinely engaged in a production–resorption cycle to grant tissue with good performances and, in case of failure (when a bone is broken), are autonomously able to start a process ending with the restoration of original tissue functionality and organ architecture.

REFERENCES

Bolander ME: Regulation of fracture repair by growth factors. *Proc Soc Exp Biol Med.* 1992;200:165–170.

Brown LF, Yeo KT, Berse B et al.: Expression of vascular permeability factor (vascular endothelial growth factor) by epidermal keratinocytes during wound healing. *J Exp Med.* 1992;176(5):1375–1379.

Carter DR, Hayes WC: Compact bone fatigue damage: A microscopic examination. *Clin Orthop Relat Res.* 1977;(127):265–274.

Chen F, Peterson SR, Story MD, Chen DJ: Disruption of DNA-PK in Ku80 mutant xrs-6 and the implications in DNA double-strand break repair. *Mutat Res.* 1996;362(1):9–19.

Deckers MM, van Bezooijen RL, van der Horst G et al.: Bone morphogenetic proteins stimulate angiogenesis through osteoblast-derived vascular endothelial growth factor A. *Endocrinology.* 2002;143:1545–1553.

Dirksen ML, Blakely WF, Holwitt E, Dizdaroglu M: Effect of DNA conformation on the hydroxyl radical-induced formation of 8,5′-cyclopurine 2′-deoxyribonucleoside residues in DNA. *Int J Radiat Biol.* 1988;54(2):195–204.

Fausto N, Campbell JS, Riehle KJ: Liver regeneration. *Hepatology.* 2006;43(Suppl. 1):45–53.

Fleig WE: Liver-specific growth factors. *Scand J Gastroenterol Suppl.* 1988;151:31–36.

Frank S, Madlener M, Werner S: Transforming growth factors beta1, beta2, and beta3 and their receptors are differentially regulated during normal and impaired wound healing. *J Biol Chem.* 1996;271(17):10188–10189.

Gerstenfeld LC, Cullinane DM, Barnes GL, Graves DT, Einhorn TA: Fracture healing as a post-natal developmental process: Molecular, spatial, and temporal aspects of its regulation. *J Cell Biochem.* 2003;88:873–884.

Golding M, Sarraf CE, Lalani EN et al.: Oval cell differentiation into hepatocytes in the acetylaminofluorene-treated regenerating rat liver. *Hepatology.* 1995;22(4 Pt 1): 1243–1253.

Hamilton LD, Mahler I, Grossman L: Enzymatic repair of UV-irradiated DNA in vitro. *Basic Life Sci.* 1975;5A:235–243.

Hazenberg JG, Hentunen TA, Heino TJ, Kurata K, Lee TC, Taylor D: Microdamage detection and repair in bone: Fracture mechanics, histology, cell biology. *Technol Health Care.* 2009;17(1):67–75.

Hübner G, Brauchle M, Smola H, Madlener M, Fässler R, Werner S: Differential regulation of pro-inflammatory cytokines during wound healing in normal and glucocorticoid-treated mice. *Cytokine*. 1996;8(7):548–556.

Isowa G, Ishizaki K, Sadamoto T et al.: O_6-Methylguanine–DNA methyltransferase activity in human liver tumors. *Carcinogenesis*. 1991;12(7):1313–1317.

Iwai M, Cui TX, Kitamura H, Saito M, Shimazu T: Increased secretion of tumor necrosis factor and interleukin 6 from isolated, perfused liver of rats after partial hepatectomy. *Cytokine*. 2001;13:60–64.

Jones PA, Rideout WM 3rd, Shen JC, Spruck CH, Tsai YC: Methylation, mutation and cancer. *Bioessays*. 1992;14(1):33–36.

Kasai H, Tanooka H, Nishimura S: Formation of 8-hydroxyguanine residues in DNA by X-irradiation. *Gann*. 1984;75(12):1037–1039.

Levenson SM, Geever EF, Crowley LV, Oates JF 3rd, Berard CW, Rosen H: The healing of rat skin wounds. *Ann Surg*. 1965;161:293–308.

Lindahl T, Karlström O: Heat-induced depyrimidination of deoxyribonucleic acid in neutral solution. *Biochemistry*. 1973;12(25):5151–5154.

Lindahl T, Nyberg B: Heat-induced deamination of cytosine residues in deoxyribonucleic acid. *Biochemistry*. 1974;13(16):3405–3410.

Martinez-Hernandez A, Amenta PS: The extracellular matrix in hepatic regeneration. *FASEB J*. 1995;9:1401–1410.

Nakamura T: Structure and function of hepatocyte growth factor. *Prog Growth Factor Res*. 1991;3(1):67–85.

Petersen BE, Bowen WC, Patrene KD et al.: Bone marrow as a potential source of hepatic oval cells. *Science*. 1999;284(5417):1168–1170.

Rappolee DA, Mark D, Banda MJ, Werb Z: Wound macrophages express TGF-alpha and other growth factors in vivo: Analysis by mRNA phenotyping. *Science*. 1988;241(4866):708–712.

Resnick MA, Moore PD: Molecular recombination and the repair of DNA double-strand breaks in CHO cells. *Nucleic Acids Res*. 1979;6(9):3145–3160.

Rupert CS: Repair of ultraviolet damage in cellular DNA. *J Cell Comp Physiol*. 1961;58(3 Pt 2):57–68.

Rydberg B, Lindahl T: Nonenzymatic methylation of DNA by the intracellular methyl group donor *S*-adenosyl-L-methionine is a potentially mutagenic reaction. *EMBO J*. 1982;1(2):211–216.

Setlow JK: Photorepair of biological systems. *Res Prog Org Biol Med Chem*. 1972;3(Pt 1):335–355.

Setlow RB, Carrier WL, Bollum FJ: Pyrimidine dimers in UV-irradiated poly dI:dC. *Proc Natl Acad Sci USA*. 1965;53(5):1111–1118.

Sinha RP, Häder DP: UV-induced DNA damage and repair: A review. *Photochem Photobiol Sci*. 2002;1:225–236.

Strain AJ: Transforming growth factor beta and inhibition of hepatocellular proliferation. *Scand J Gastroenterol Suppl*. 1988;151:37–45.

Teitelbaum SL: Bone resorption by osteoclasts. *Science*. 2000;289:1504–1508.

Uthgenannt BA, Kramer MH, Hwu JA, Wopenka B, Silva MJ: Skeletal self-repair: Stress fracture healing by rapid formation and densification of woven bone. *J Bone Miner Res*. 2007;22(10):1548–1556.

Walter JB: Wound healing. *J Otolaryngol*. 1976;5(2):171–176.

Weinert T: DNA damage and checkpoint pathways: Molecular anatomy and interactions with repair. *Cell*. 1998;94(5):555–558.

Werner S, Peters KG, Longaker MT, Fuller-Pace F, Banda MJ, Williams LT: Large induction of keratinocyte growth factor expression in the dermis during wound healing. *Proc Natl Acad Sci USA*. 1992;89(15):6896–6900.

Wildenberg J, Meselson M: Mismatch repair in heteroduplex DNA. *Proc Natl Acad Sci USA*. 1975;72(6):2202–2206.

Part II

Artificial Self-Healing Systems

4 Self-Healing of Gold Nanoparticles during Laser Irradiation

An Embryonic Case of "Systems Nanotechnology"

Vincenzo Amendola and Moreno Meneghetti

CONTENTS

4.1 INTRODUCTION

The high interface-to-volume ratio is an intrinsic property of nanomaterials (Feynman 1959). In most cases, the creation of an interface has an energy cost that makes nanostructures less stable than their bulk equivalents, especially in critical conditions such as intense light irradiation, high temperature, or oxidative environments (Drexler 1992; Amendola and Meneghetti 2009b). The design of nanoscale self-repairing systems can solve the problem of structural and functional damages and, therefore, it is an outstanding requirement in nanoscience and nanotechnology. However, a very limited number of nanomaterials with self-healing ability are known. To date, most of the self-healing strategies developed by researchers are devoted to the reparation of structural materials (Amendola and Meneghetti 2009b). Moreover, the strategies developed for the self-healing of micro- or macro-sized materials cannot be easily extended to nanomaterials because of the extreme size confinement (Hager et al. 2010). Different approaches are required, which may be directly or indirectly inspired to natural systems, where several components interact together for the self-repair of functional units (Fratzl 2007).

We studied a blend of gold nanoparticles (AuNPs) and zinc phthalocyanines (ZnPcs) that epitomize and overcome the aforementioned issues. Noble metal nanoparticles (NMNPs) are known for excellent nonlinear absorption properties, superior to that of the best organic optical limiters such as fullerenes or phthalocyanines (Francois et al. 2000, 2001; Ispasoiu et al. 2000; Philip et al. 2000; Bozhevolnyi et al. 2003; West et al. 2003; Porel et al. 2005; Sun et al. 2005; Wang et al. 2005; Elim et al. 2006). However, NMNPs lose much of these properties after few intense laser pulses due to photoinduced fragmentation (Francois et al. 2000, 2001; Ispasoiu et al. 2000; Philip et al. 2000; Bozhevolnyi et al. 2003; West et al. 2003; Porel et al. 2005; Sun et al. 2005; Wang et al. 2005; Elim et al. 2006). During the investigation of optical limiting (OL) in colloidal solutions, we found that a blend of AuNPs and ZnPcs in tetrahydrofuran (THF) shows the ability of self-heal photofragmented AuNPs simultaneously to laser irradiation (Amendola et al. 2009). This self-repairing ability results in enhanced OL performances, that is, lower light limitation threshold and longer durability. Therefore, the blend of AuNPs and ZnPcs represents an infrequent example of a functional nanomaterial that self-heals during its use.

In the following sections, we will briefly outline the main concepts about the study of NMNPs for OL applications, we will show the experimental evidences and the proposed mechanism for the self-healing of AuNPs and, finally, we will discuss the analogy with some processes typical of systems chemistry, in order to show that the self-healing of AuNPs blended with ZnPcs under laser irradiation can be considered as a simple case of "systems nanotechnology."

4.2 OPTICAL LIMITING WITH AuNPs

Controlling the intensity of light is an important requirement in laser and optical technology. Optical limiters are the most common examples of passive devices for the control of light intensity (Sun and Riggs 1999; Chen et al. 2005). The main applications of optical limiters are the protection of optical sensors and of eyes from high-intensity light beams, but they can be used also for all-optical passive switching. OL devices are based on nonlinear optical materials whose transmittance decreases significantly for increasing light fluence. The typical plot of output (i.e., transmitted) vs. input (i.e., incident) fluence in an OL device shows linear behavior until a certain fluence threshold and a clearly nonlinear behavior after this threshold (Figure 4.1). In the nonlinear regime, the output energy reaches a plateau until the point of breakdown. The OL threshold is the parameter characterizing the nonlinear behavior of the material and it is usually defined as the input fluence at which the transmittance is 50% of the linear transmittance (Chen et al. 2005). It is desirable that optical limiters have low OL threshold in the entire spectral range of operation.

Nonlinear absorption and nonlinear scattering are the two main mechanisms exploited in OL devices. Nonlinear absorption is possible when the absorption cross section in a material is higher in the excited states than in the ground state. Nonlinear scattering is usually observed as a consequence of the photoinduced heating and consequent formation of a refractive index discontinuity around the material, such

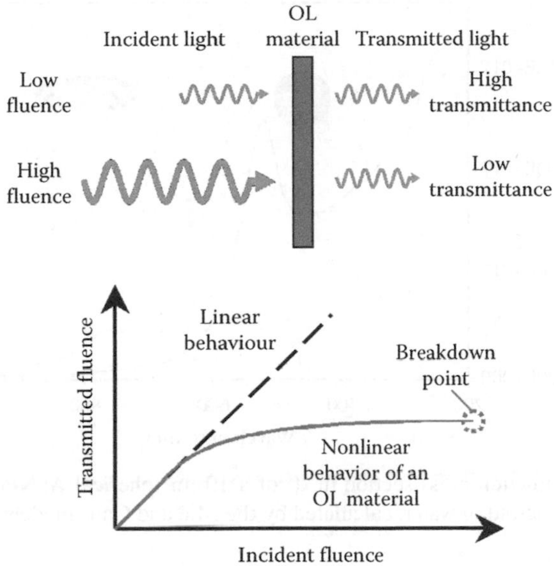

FIGURE 4.1 Sketch of the behavior of an optical limiter (top) and typical curve of output vs. input fluence for an optical limiter (bottom).

as a gas bubble. Among the most efficient nonlinear absorption systems, there are metalloporphyrins and metallophthalocyanines, organometallic complexes, and fullerene derivatives (Sun and Riggs 1999; Chen et al. 2005). Nonlinear scattering is observed in molecular aggregates or nanostructures such as carbon black or carbon nanotube bundles (Sun and Riggs 1999; Chen et al. 2005).

The development of optical limiters that can resist to high laser fluences is a challenging task. In recent years, nanoparticles composed of gold, silver, copper and their alloys were intensely investigated for OL because they show superior performances than best organic optical limiters, with nanoseconds (Sun and Riggs 1999; Francois et al. 2000; Ispasoiu et al. 2000; Bozhevolnyi et al. 2003; Tom et al. 2003; West et al. 2003; Porel et al. 2005; Sun et al. 2005; Wang and Sun 2006), picoseconds (Francois et al. 2000, 2001; Philip et al. 2000), and femtoseconds (Elim et al. 2006) laser pulses. The main OL mechanism of NMNPs has been attributed to free carriers multiphoton absorption processes (Sun and Riggs 1999; Tom et al. 2003; West et al. 2003; Sun et al. 2005; Elim et al. 2006). Nonlinear scattering is usually excluded in well-dispersed NMNPs because several studies did not show differences between open-aperture and closed-aperture OL measurements (Sun and Riggs 1999; Tom et al. 2003; West et al. 2003; Sun et al. 2005; Elim et al. 2006; Wang and Sun 2006).

In principle, Au, Ag, and Cu NPs have high linear and nonlinear susceptibilities, due to the presence of highly polarizable electrons, namely, 1 *sp*-band conduction electron and 10 d-band electrons for each atom in the particle (Kreibig and Vollmer 1995). Moreover, the free electrons give localized surface plasmon modes that can be collectively excited by optical wavelengths (Kreibig and Vollmer 1995). The plasmon resonance absorption cross section of NMNPs is very high. For instance, the

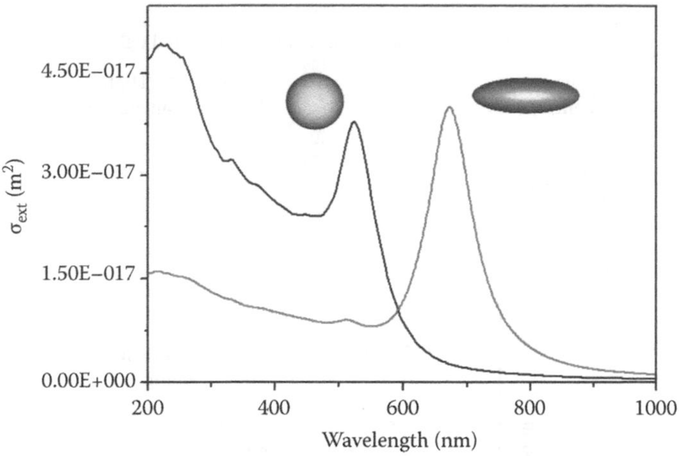

FIGURE 4.2 Extinction cross section (text) of a 10 nm spherical AuNP and of a 15 nm × 5 nm × 5 nm Au spheroid in water, calculated by the Mie and Gans models, respectively.

extinction cross section in AuNPs is on the order of 10^{-17} cm^2 per single atom, that is, on the order of 10^{-12} cm^2 for a 20 nm nanoparticle (Kreibig and Vollmer 1995; Amendola and Meneghetti 2009c). The plasmon absorption band of NMNPs can be tuned from the visible to the near infrared range by alloying Au with Ag or with Cu or by controlling the shape of the nanostructures (Amendola et al. 2010). For instance, spherical AuNPs have one plasmon resonance at about 520 nm, whereas rod-shaped AuNPs have two plasmon modes, one at wavelengths characteristic of spherical AuNPs and a second one at longer wavelengths, depending on the aspect ratio of the rod (Figure 4.2). Similarly, NMNPs aggregation induces the redshift and the broadening of the most intense plasmon bands (Messina et al. 2011).

Resonant excitation of the plasmon modes induces sensible heating of nanoparticles (Link and El-Sayed 2000; Amendola and Meneghetti 2009a). Indeed, due to ultrahigh extinction cross section and ultrafast relaxation dynamics of the plasmon resonance, NMNPs have extremely high conversion efficiency of photons into phonons for unit volume, namely, NMNPs are among the best converter of visible light into heat provided by nature. On the one hand, nanoparticles heating is an efficient way for populating the empty states above the Fermi energy level, which can facilitate excited states multiphoton absorption processes (Boyd 2008). On the other hand, particles heating can originate thermal or electrostatic fragmentation of nanoparticles (Link and El-Sayed 2000; Amendola and Meneghetti 2009a). At moderate laser intensity, metal particles undergo melting, boiling, and vaporization, whereas nanoparticles fragmentation takes place at higher laser intensity as a consequence of photoionization and the following Coulomb explosion (Yamada et al. 2006, 2007). Therefore, OL and photofragmentation of NMNPs are strictly connected, as shown in several experimental works (Francois et al. 2000, 2001; Ispasoiu et al. 2000; Philip et al. 2000; Bozhevolnyi et al. 2003; West et al. 2003; Porel et al. 2005; Sun et al. 2005; Wang et al. 2005; Elim et al. 2006). In particular, both OL performances and photoinduced fragmentation processes are enhanced for larger metal nanoparticles (Francois et al. 2000, 2001; Ispasoiu et al. 2000; Philip et al. 2000; Bozhevolnyi et al. 2003; West et al. 2003; Porel et al. 2005; Sun et al. 2005; Wang et al. 2005; Elim et al. 2006). This is a relevant drawback for OL, because NMNPs loose their OL properties after few laser pulses due to photoinduced fragmentation. The fragmentation process has lower fluence threshold for large particles (20–80 nm), because the absorption cross section per single atom is higher and the heat dissipation rate is lower in large nanoparticles (Francois et al. 2000, 2001; Ispasoiu et al. 2000; Philip et al. 2000; Bozhevolnyi et al. 2003; West et al. 2003; Porel et al. 2005; Sun et al. 2005; Wang et al. 2005; Elim et al. 2006). This can be understood considering that the plasmon absorption cross section per single atom reaches a maximum for nanoparticles of ca. 50 nm, but the cooling efficiency of AuNPs has a $1/d$ dependence (where d is the AuNP diameter). Therefore, the resulting photothermal heating rate per single gold atom has a maximum for nanoparticles of ca. 80 nm (Figure 4.3) (Richardson et al. 2006).

Sometimes, NMNPs were blended with other optical limiters with complementary nonlinear absorption mechanisms and activation thresholds, aiming at enhanced OL performances. Dendrimer-encapsulated nanoparticles and nanoparticles conjugated to fulleropyrrolidine-pyridine derivatives or to a fulleropyrrolidine moiety all showed better OL performances than the isolated moieties (Sun and Riggs 1999; Ispasoiu et al. 2000; Fang et al. 2002; Qu et al. 2002; West et al. 2003; Goodson et al. 2004; Amendola et al. 2005). However, all these studies did not face the main problem associated to OL with plasmonic nanostructures, namely, their fast photoinduced fragmentation.

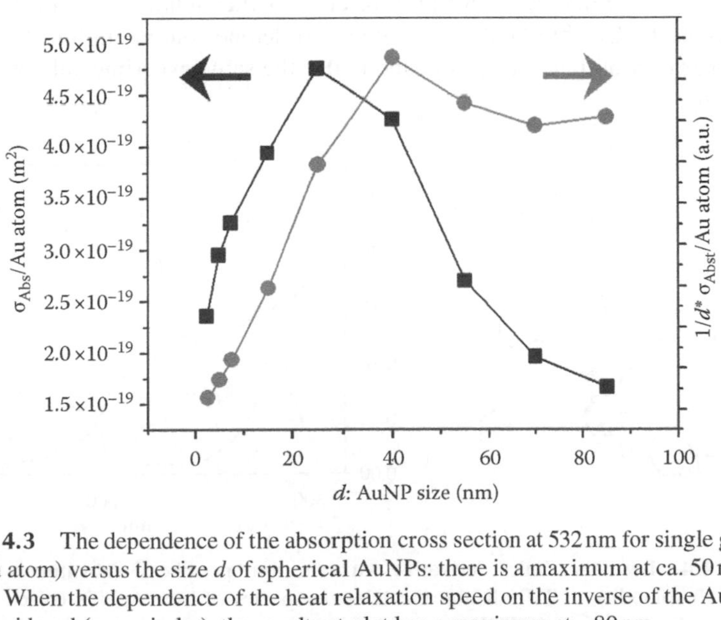

FIGURE 4.3 The dependence of the absorption cross section at 532 nm for single gold atom (sAbs/Au atom) versus the size d of spherical AuNPs: there is a maximum at ca. 50 nm (black squares). When the dependence of the heat relaxation speed on the inverse of the AuNPs size $1/d$ is considered (gray circles), the resultant plot has a maximum at ~80 nm.

4.3 SELF-HEALING OF AuNPs DURING LASER IRRADIATION

We studied a blend of AuNPs and ZnPcs in THF solution that showed the ability to repair photofragmented nanoparticles during laser irradiation, with enormous benefits on the OL performances (Amendola et al. 2009).

AuNPs with average size of 16 nm were obtained by laser ablation synthesis in solution (LASiS) in THF. In a typical LASiS of AuNPs, 1064 nm (9 ns) laser pulses were focused on a 99.99% gold plate placed at the bottom of a cell filled with pure THF. LASiS does not require any other chemical or stabilizing agent around particles surface, because a colloidal solution of charged nanoparticles is obtained (Amendola et al. 2006, 2007; Amendola and Meneghetti 2009a). By Z-spectroscopy, we measured a positive *Z* potential of +42 mV for AuNPs obtained by LASiS in THF.

The ZnPcs molecules were also dissolved in THF and added to AuNPs solution prior to the OL measurements. We used a zinc 2(3)-tetra-[2-(*tert*-butoxy)-ethyloxy]-phthalocyanine obtained from the cyclization of 4-[2-(*tert*-butoxy)-ethyloxy]-phthalonitrile with $ZnCl_2$. The so obtained ZnPc derivative has four peripheral substituents that impart high solubility in THF and completely prevents intermolecular aggregation (Figure 4.4).

The AuNPs–ZnPcs blend showed better performances than isolated AuNPs during a single OL measurement with 532 nm (9 ns) laser pulses and also after multiple repeated measurements. The good durability of the AuNPs–ZnPcs blend is evident especially when compared with the fast degradation of the OL performances of bare AuNPs (Figure 4.5). The OL measurement on bare AuNPs confirms that extensive degradation of the gold nanoparticles takes place under high fluences as discussed earlier.

The AuNPs–ZnPcs blend had an OL threshold of 316 mJ/cm^2 for the first OL cycle and 391 mJ/cm^2 for the eighth cycle, whereas for bare AuNPs we measured 472 mJ/cm^2 for the first cycle and 1485 mJ/cm^2 for the eighth cycle. For the sake of comparison, the OL threshold of a reference fullerene solution in toluene with the same linear transmittance at 532 nm and under the same experimental conditions is 511 mJ/cm^2.

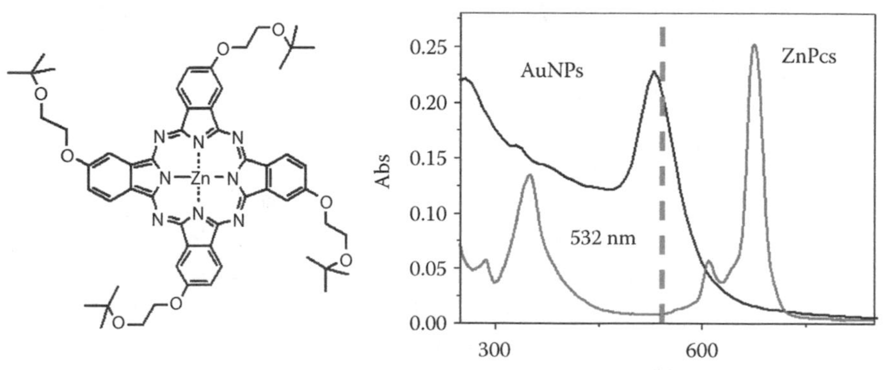

FIGURE 4.4 Left: the ZnPcs derivative with the four peripheral substituents added to impart high solubility in THF. Right: the UV–vis spectra of bare AuNPs (black line) and ZnPcs (gray line) solutions in THF.

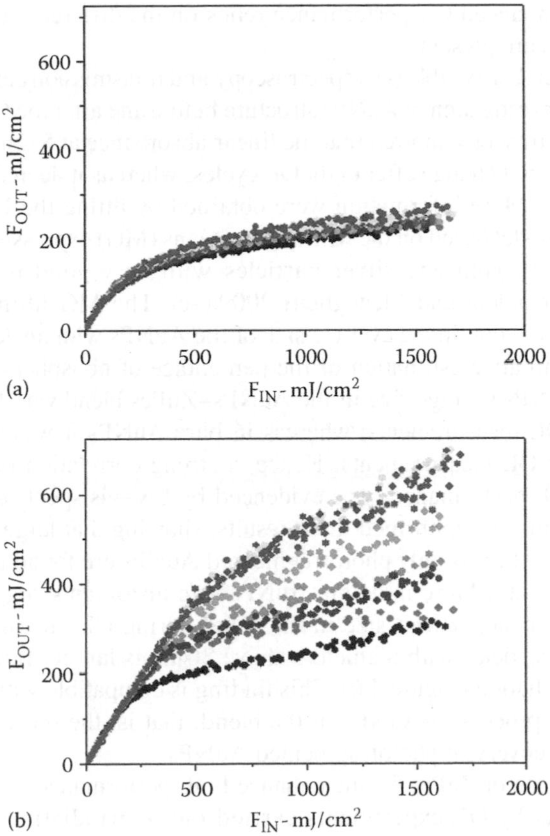

FIGURE 4.5 OL curves at 532 nm (9 ns pulses) for the (a) AuNPs–ZnPcs blend and (b) bare AuNPs in THF. The superior OL performances of the blend is evident during the first OL measurements and is retained after eight OL cycles, whereas bare AuNPs undergo a rapid loss of OL properties. (Reprinted with permission from Amendola, V., Dini, D., Polizzi, S., Shen, J., Kadish, K.M., Calvete, M.J.F., Hanack, M., and Meneghetti, M., Self-healing of gold nanoparticles in the presence of zinc phthalocyanines and their very efficient nonlinear absorption performances, *J. Phys. Chem. C*, 113(20), 8688–8695, 2009. Copyright 2009 American Chemical Society.)

By performing nonlinear transmittance and pump and probe nanosecond spectroscopy on blended and isolated components, we confirmed that the enhanced OL performances of the AuNPs–ZnPcs blend were not arising from the nonlinear behavior of the ZnPcs moiety. In particular, ZnPcs are well-known reverse saturable absorbers, with higher absorption cross section in the triplet state than in the singlet ground state in the visible range (Perry et al. 1996; Chen et al. 2005). However, we did not observe any difference between isolated and blended molecules for both the absorption of the triplet state or the fluorescence from the first excited singlet state. This finding is a clear indication that the presence of AuNPs did not affect the population of the excited states of ZnPcs and their nonlinear absorption properties;

therefore, the enhanced OL performance relies on the different behavior of AuNPs when the ZnPcs are present.

We performed UV–visible (vis) spectroscopy and transmission electron microscopy (TEM) for the investigation of AuNPs structure before and after the OL measurements.

UV–vis spectroscopy showed that the linear absorbance at 532 nm of the AuNPs–ZnPcs blend did not change after eight OL cycles, whereas it decreased continuously for bare AuNPs. More information were obtained by fitting the UV–vis spectra of AuNPs with a model based on the Mie and the Gans (MG) expressions, as previously reported for both gold and silver particles with very good results (Amendola et al. 2007; Amendola and Meneghetti 2009a–c). The MG fitting of the UV–vis spectrum allows to obtain the average size of the AuNPs with an accuracy of ca. 5% and a semiquantitative estimation of the percentage of nonspherical nanoparticles. In our case, AuNPs average size in the AuNPs–ZnPcs blend was 16 nm both before and after the OL measurements, whereas in bare AuNPs it was 16 nm before and only 9 nm after OL measurements. Hence, a strong correlation of average AuNPs size with the OL performances was evidenced by UV–vis spectroscopy for the two samples. TEM images confirmed these results, showing that large AuNPs are present in the blend, whereas only photofragmented AuNPs are found in the not blended solution (Figure 4.6). More in detail, AuNPs size histograms suggested a dynamic evolution of the nanoparticles size in the blend during OL measurements, because the fraction of particles with diameters above 30 nm is larger after laser irradiation than before irradiation (Figure 4.6). This finding is compatible with a fragmentation and a regrowth process of AuNPs in the blend, that is, the presence of the ZnPcs promoted the recovery of photofragmented AuNPs.

The importance of ZnPcs for the enhanced OL performances of AuNPs was further stressed out by OL experiments carried out by irradiating the sample with 350 consecutive laser pulses at a constant fluence of 1100 mJ/cm^2. After the first laser pulses, blended and bare AuNPs have equivalent nonlinear transmittance values (Figure 4.7a). However, the OL efficiency of the bare particles rapidly degraded after few pulses, whereas the blend retained its OL performances. The concentration of ZnPcs in the blend is a relevant parameter for the retention of OL properties. We found that the OL efficiency after 350 laser pulses is entirely retained for an AuNPs:ZnPcs molar ratio above 1:2000 and the preservation effect is still present even for AuNPs:ZnPcs ratio as low as 1:500. Also in this case, we found that the preservation of the OL performances was associated with the preservation of linear absorbance at 532 nm.

A better understanding of the role of ZnPcs was possible by studying the addition of the phthalocyanines to bare AuNPs, instead of starting with the already blended components, during a multiple laser pulses irradiation experiment. After 400 consecutive laser pulses at 532 nm and 1100 mJ/cm^2, the bare AuNPs solution loose most of its OL properties. Interestingly, when we added the ZnPcs to such a solution and prosecuted the irradiation for 200 more laser shots, we observed the recovery of the OL properties after only few laser pulses and up to the end of the irradiation experiment (Figure 4.7b). Moreover, by UV–vis spectroscopy we observed that the addition of ZnPcs to AuNPs irradiated with 400 laser pulses produced the immediate redshift

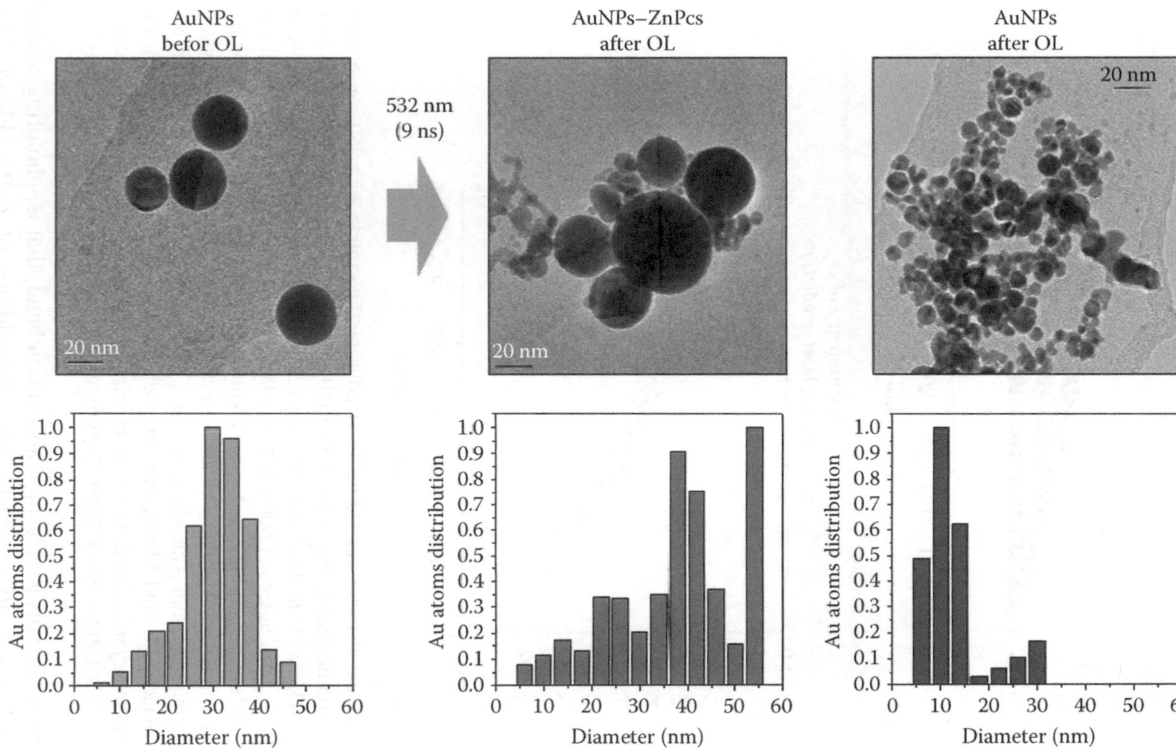

FIGURE 4.6 Top (from left to right): TEM images of AuNPs before laser OL measurement (left) and after OL measurements for the AuNPs–ZnPcs blend (middle) and bare AuNPs (right). Bottom (from left to right): histograms of gold atoms distribution versus AuNPs size before (left) and after the OL experiments for blended (middle) and non-blended (right) AuNPs. (Reprinted with permission from Amendola, V., Dini, D., Polizzi, S., Shen, J., Kadish, K.M., Calvete, M.J.F., Hanack, M., and Meneghetti, M., Self-healing of gold nanoparticles in the presence of zinc phthalocyanines and their very efficient nonlinear absorption performances, *J. Phys. Chem. C*, 113(20), 8688–8695, 2009. Copyright 2009 American Chemical Society.)

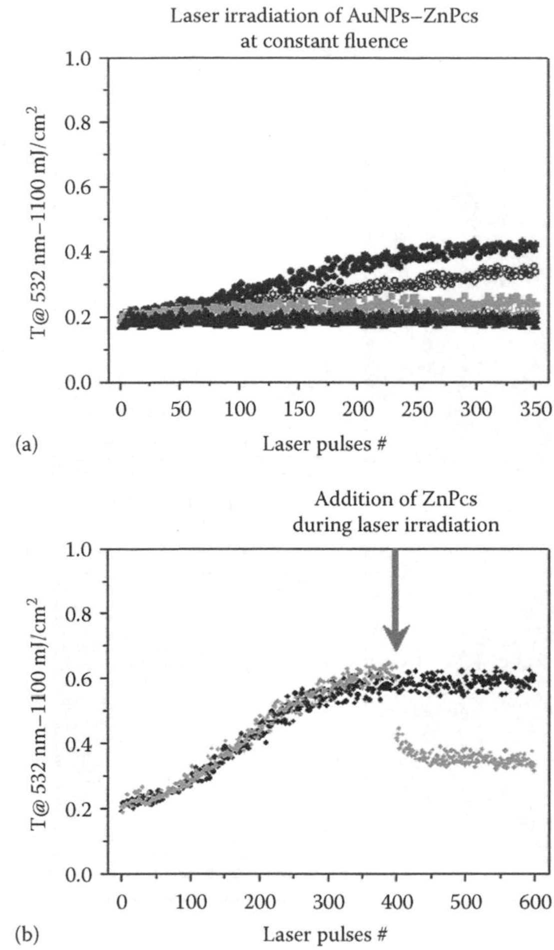

FIGURE 4.7 (a) OL experiments carried out by irradiating the sample with 350 consecutive laser pulses at 532 nm (9 ns) and a constant fluence of 1100 mJ/cm^2. Different AuNPs:ZnPcs ratio were tested: black circles = 1:0, hollow black circles = 1:500, gray squares = 1:1000, hollow gray squares = 1:2000, black triangles = 1:6000, hollow black triangles = 1:15,000. (b) Nonlinear transmittances at 532 nm for multiple pulses irradiation with 1100 mJ/cm^2 reported for the bare AuNPs (black) and bare AuNPs to which ZnPcs were added after 400 pulses (gray). (Reprinted with permission from Amendola, V., Dini, D., Polizzi, S., Shen, J., Kadish, K.M., Calvete, M.J.F., Hanack, M., and Meneghetti, M., Self-healing of gold nanoparticles in the presence of zinc phthalocyanines and their very efficient nonlinear absorption performances, *J. Phys. Chem. C*, 113(20), 8688–8695, 2009. Copyright 2009 American Chemical Society.)

and a broadening of the surface plasmon absorption band, that is, photofragmented AuNPs assembled into small aggregates after the addition of ZnPcs (Figure 4.8). Discrete dipole approximation (DDA) calculations of the plasmon resonance in a cluster formed by 16 small (9 nm) AuNPs aggregated together showed a result similar to that observed experimentally, thus confirming that the redshift and broadening of the surface plasmon absorption band is due to AuNPs aggregation.

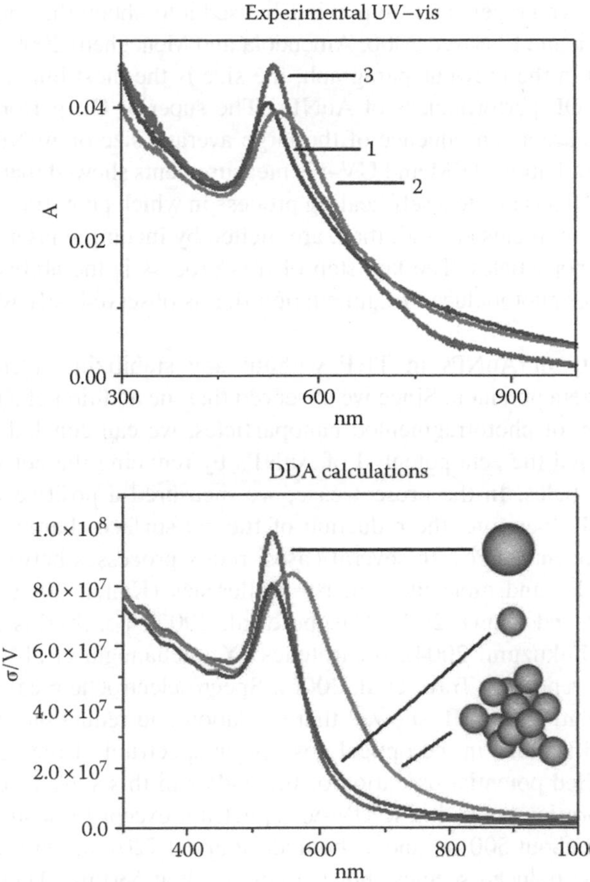

FIGURE 4.8 Left: experimental UV–vis spectra of AuNPs after 400 laser pulses at 532 nm–1100 mJ/cm^2 (spectrum 1), after the addition of ZnPcs (spectrum 2) and after 200 more laser pulses (spectrum 3). The ZnPcs contribute to the UV–vis spectra was subtracted for clarity in the latter two cases. Right: DDA simulations for a gold sphere of diameter 9 nm, for a cluster formed by 16 touching 9 nm AuNPs and for a 20 nm sphere. (Reprinted with permission from Amendola, V., Dini, D., Polizzi, S., Shen, J., Kadish, K.M., Calvete, M.J.F., Hanack, M., and Meneghetti, M., Self-healing of gold nanoparticles in the presence of zinc phthalocyanines and their very efficient nonlinear absorption performances, *J. Phys. Chem. C*, 113(20), 8688–8695, 2009. Copyright 2009 American Chemical Society.)

After the last 200 consecutive laser pulses, we observed another modification of the UV–vis absorption spectra of the sample, namely, the sharpening and the blueshift of the plasmon resonance of aggregated AuNPs (Figure 4.8). The MG fit of the absorption spectrum clearly indicated that the pristine aggregates were transformed into spherical particles with average size of about 20 nm, while a size of 9 nm was measured on the not blended AuNPs after 400 and after 600 pulses. The transformation of AuNPs aggregates into larger particles by laser irradiation through a photomelting process is a very well-known process and we previously used it to obtain the controlled growth of AuNPs (Link and El-Sayed 2000; Amendola and Meneghetti 2009a).

As recalled in the previous paragraph, the size is the most important parameter governing the OL performances of AuNPs. The superior OL performances of the AuNPs–ZnPcs are a consequence of the large average size of AuNPs in the blend during laser irradiation. TEM and UV–vis measurements showed that the large average size of AuNPs is due to a self-healing process in which photofragmented AuNPs reassemble in small clusters and, then, are melted by incoming laser pulses to form again large nanoparticles. The key step of this process is the ability of AuNPs to reassemble after photoinduced fragmentation that is observed only when ZnPcs are present.

The stability of AuNPs in THF without any stabilizing agent depends on nanoparticles zeta potential. Since we observed that the addition of ZnPcs produced the aggregation of photofragmented nanoparticles, we can conclude that phthalocyanines affected the zeta potential of AuNPs by reducing the net surface charge of the nanoparticles. In the present case, we measured a positive Z potential on AuNPs in THF; therefore, the reduction of the net surface charge can happen by the oxidation of the ZnPcs. In several cases, redox processes between gold or silver nanoparticles and molecules such as fullerenes (Kamat 2002; Sudeep et al. 2002; Thomas and Kamat 2003; Hasobe et al. 2005), porphyrins and fullerenes (Imahori and Fukuzumi 2004), β-carotenes (Yakuphanoglu et al. 2006), or with Er^{3+} ions were reported (Trave et al. 2006). Spectroelectrochemical measurements on a ZnPcs solution in THF showed that oxidation and reduction of the molecule brings to modifications in the optical absorption spectrum of the solution. In particular, controlled potential oxidation of the ZnPcs in this solvent results in a loss of all the absorption peaks in the UV–vis spectrum except for a small increase of a broadband at about 500 nm and a shoulder at about 720 nm. On the contrary, the spectrum of the reduced species shows a new peak at 580 nm. Therefore, UV–vis spectroscopy provides a way to support the presence of redox processes during the laser irradiation of the AuNPs–ZnPcs blend. The UV–vis spectra of the AuNPs–ZnPcs blend shows the decrease of the linear absorption of ZnPcs after irradiation with 350 laser pulses at 532 nm and 1100 mJ/cm^2 and no new bands are observed at 580 nm. The bare ZnPcs solution, when subjected to the same experiment, did not show any significant modification of the UV–vis spectrum. This finding definitively suggests that ZnPcs oxidation occurred during laser irradiation in the presence of AuNPs. Therefore, the surface charge of AuNPs was lowered by a redox process with nearby ZnPcs. It is likely that the oxidation process is favored by the higher temperature of the solution in the nearby of photofragmented AuNPs and by the higher local concentration of AuNPs after photofragmentation. Moreover, in the case of AuNPs obtained by LASiS in THF, the electron transfer process from the ZnPcs to AuNPs is favored because these AuNPs are not coated with a shell of interfering ligands (Hirsch et al. 2007; Amendola and Meneghetti 2009a).

In summary, from the aforementioned data, we concluded that enhanced OL performances like lower OL threshold and higher durability for multiple pulse irradiation of the AuNPs–ZnPcs blend are the result of the self-healing of AuNPs via the following mechanism (Figure 4.9):

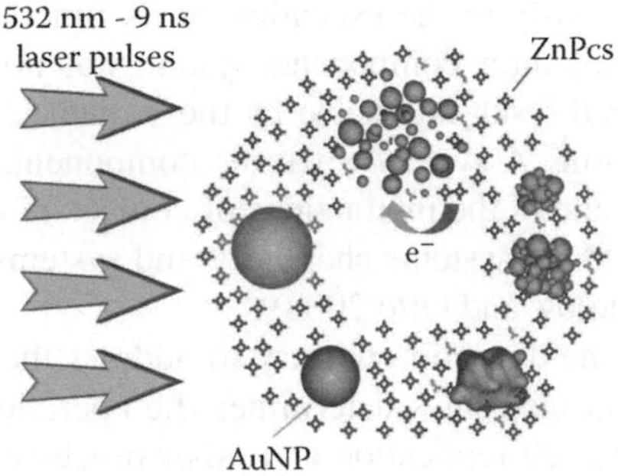

FIGURE 4.9 **(See color insert.)** Sketch of the self-healing mechanism for AuNPs in the presence of ZnPcs. From left in clockwise order: AuNPs absorb photons and heat up over the fragmentation threshold; some of the positive charges of photofragmented AuNPs are neutralized by the oxidation of ZnPcs, promoting their aggregation; aggregates are photomelted into new spherical AuNPs, that are ready to efficiently limit light again. (Reprinted with permission from Amendola, V., Dini, D., Polizzi, S., Shen, J., Kadish, K.M., Calvete, M.J.F., Hanack, M., and Meneghetti, M., Self-healing of gold nanoparticles in the presence of zinc phthalocyanines and their very efficient nonlinear absorption performances, *J. Phys. Chem. C*, 113(20), 8688–8695, 2009. Copyright 2009 American Chemical Society.)

1. AuNPs irradiated with 532 nm and 9 ns laser pulses show excellent OL performances but also suffer from fragmentation into smaller particles, that are not efficient optical limiters.
2. In the absence of ZnPcs, these smaller particles repel each other due to their positive surface charge, whereas in the presence of ZnPcs, the formation of small aggregates of AuNPs is favored by the reduction of their surface charge due to the oxidation of ZnPcs.
3. After the aggregation process, incoming laser pulses produce the photomelting of the aggregates into large spherical particles.
4. Large nanoparticles are ready to absorb light again with high efficiency and to repeat the cycle indefinitely.

4.4 SELF-HEALING OF GOLD NANOPARTICLES AS A SIMPLE CASE OF "SYSTEMS NANOTECHNOLOGY"

In the AuNPs–ZnPcs blend, one can identify three components that cooperate to the healing process: AuNPs, ZnPcs, and the laser pulses. These three components are all included in a matrix, the THF solution that can be considered inert. The interaction of AuNPs and ZnPcs without laser pulses only results in a slow aggregation and sedimentation of nanoparticles over a time of days or weeks. The interaction of bare AuNPs with 9 ns laser pulses at high fluence (i.e., 1100 mJ/cm^2) produces AuNPs photofragmentation. The interaction of ZnPcs molecules with the same laser pulses does not produce any effect. Only the combined action of the three components originates the cascade of events, which leads to the self-healing of photofragmented AuNPs. Therefore, AuNPs, ZnPcs, and the laser pulses can be considered as a whole interacting "system" that has the property of self-repairing its functional components (i.e., AuNPs) simultaneously to the execution of its function (i.e., OL). The result of the interactions of this three components system was not predictable, because it is different than the final result attainable by the sequence of interactions amongst two components at a time. A system where its components cooperate to yield different results than the sum of the interactions of couples of its components is a concept familiar in the fields of systems chemistry and systems biology (Kitano 2002; Guarise et al. 2008; Ludlow and Otto 2008).

Systems biology is the discipline that try to address the issue of how the interaction between different molecules determines the operation of biological systems (Kitano 2002), included self-replication and other processes involved in the origin of life on Earth (Chiarabelli et al. 2009). Systems chemistry deals with the products originated by the chemical reactions and physicochemical interactions of molecules in complex mixtures (Ludlow and Otto 2008) and provides simpler models for the study of biological processes. The study of complex biological and chemical systems is a hard task because, usually, the free energy of complex systems cannot be determined by considering their isolated components, due to the large number of possible interactions that should be considered (Ludlow and Otto 2008). Moreover, these systems often are kinetically controlled rather than thermodynamically controlled (Kitano 2002; Ludlow and Otto 2008).

Complex chemical mixtures can originate a rich panorama of unpredictable behaviors, as molecular self-replication and self-assembling, formation of reaction patterns, or amplification of unconventional products. The case of the dynamic combinatorial amplification of peptides is interesting because it has some analogies with the self-healing mechanism of AuNPs. For instance, in a simple artificial system setup by Gleason and colleagues (Cheeseman et al. 2002), an ensemble of peptides (a peptide library) was added to two reaction chambers separated by a dialysis membrane and containing protease and carbonic hydrase, respectively. Protease is an enzyme with the function of protein catabolism by hydrolysis of the peptide bonds, whereas carbonic hydrase is an enzyme that shows different binding affinity for different peptides of the library. Peptides were small enough to migrate from one reaction chamber to the other (Figure 4.10).

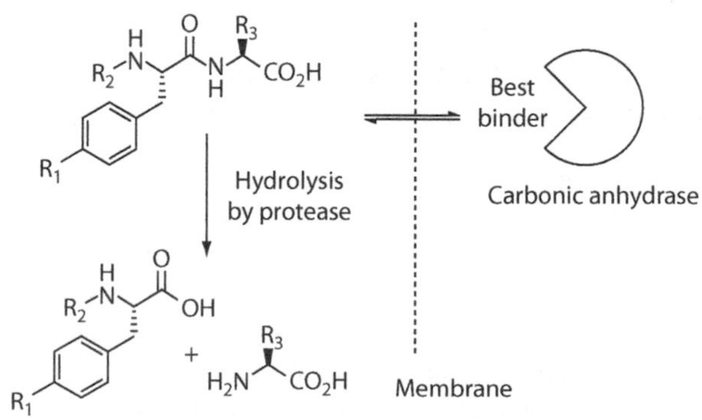

FIGURE 4.10 Sketch of the experiment with two reaction chambers used to demonstrate the kinetically controlled amplification of peptide with best binding affinity among a peptide library. (From Ludlow, R.F. and Otto, S., Systems chemistry, *Chem. Soc. Rev.*, 37(1), 101, 2008. Reproduced by permission of The Royal Society of Chemistry.)

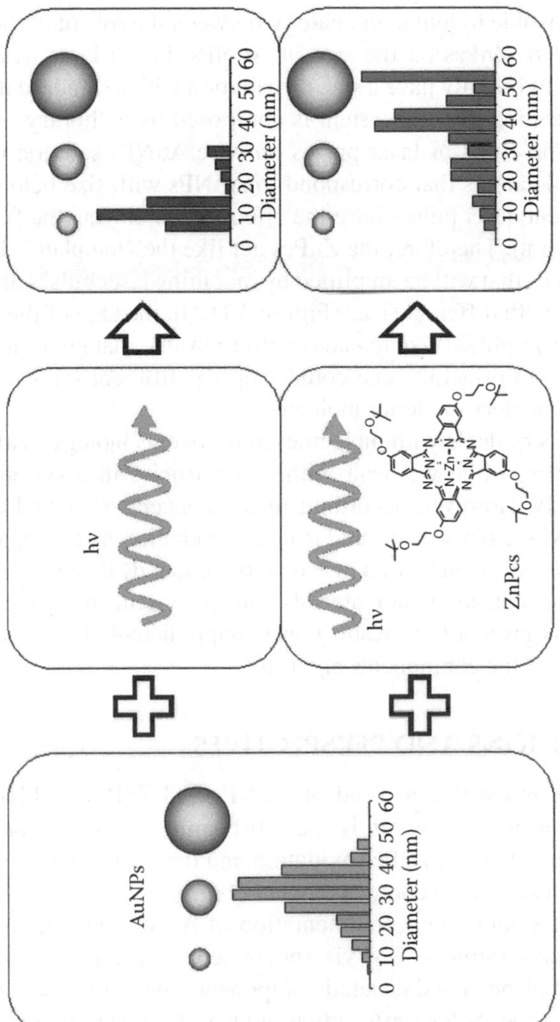

FIGURE 4.11 The relative intensity of the distribution of gold atoms versus AuNPs size (bars) is altered by high-energy 532 nm laser pulses, by amplifying the fraction of smaller AuNPs. When the ZnPcs are also present, the relative intensity of larger AuNPs is amplified.

At the end of the experiment, the peptide distribution in the library was dramatically altered, with the complete destruction of peptides with low binding affinity to carbonic anhydrase and consequently relative amplification of the ratio of the peptide with best binding affinity to carbonic anhydrase. Most important, the ratio of peptides with higher affinity exceeded so far the relative binding affinity ratios of the peptides in the library. Therefore, the experiment showed that it is possible to achieve greater selectivity than the peptide–enzyme affinity ratio by a kinetic mechanism.

Indeed, it is possible to found an analogy between the role of carbonic anhydrase and the influence of ZnPcs on the size of AuNPs during laser irradiation. In our experiment, AuNPs initially have a size distribution with a standard deviation on the order of 50%, meaning that the system is composed by a "library" of AuNPs with different sizes. The effect of laser pulses on bare AuNPs solution is to "amplify" the fraction of gold atoms that correspond to AuNPs with size below 10 nm. When the ZnPcs is present, laser pulses have the effect of amplifying the fraction of larger AuNPs (about 20 nm). Therefore, the ZnPcs act like the "template" determining the size range of AuNPs that will be amplified by incoming laser pulses among the whole library of AuNPs with different sizes (Figure 4.11). In the case of the AuNPs–ZnPcs blend in THF, the amplified size matches with the value that guarantees the optimal OL performances. In principle, one could amplify different sizes in the library by choosing a different electron donor molecule.

Systems chemistry deals with molecules and systems biology deals with biological macromolecules, while we dealt with nanoparticles in a system with a lower level of complexity. However, according to the concepts sketched earlier, we can consider the AuNPs–ZnPcs blend under laser irradiation as a simple case of "systems nanotechnology," namely, of a new discipline that deals with (i) systems where the relevant components are nanomaterials and (ii) systems in which the interaction of the components gives a final result that is unpredictable by the sequence of the interaction of two of the components at a time.

4.5 CONCLUSIONS AND PERSPECTIVES

In conclusion, we found that a blend of AuNPs and ZnPcs in THF shows excellent OL performances. The reason is the self-healing of AuNPs, based on AuNPs aggregation promoted by the ZnPcs oxidation and their subsequent photomelting by incoming laser pulses. It is in contrast to what observed for bare AuNPs, where laser pulses produced the rapid photofragmentation of AuNPs into small nanoparticles, that are poor optical limiters. UV–vis spectroscopy, spectroelectrochemistry, and *Z*-spectroscopy of blended and isolated components suggested that a charge transfer from the ZnPcs to the AuNPs after photoinduced fragmentation is the key of the process. The absence of ligands on the surface of AuNPs obtained by laser ablation also helps the interaction between these two components. Although some of the aspects related to the triggering of the redox process by laser pulses still have to be investigated more, one can predict that other molecules could induce similar effects.

The system composed by AuNPs, ZnPcs, and incoming laser pulses, although at a lower level of complexity, has some interesting analogies with the mixtures studied in systems chemistry and systems biology. Therefore, this three components ensemble can be considered as an embryonic example of systems nanotechnology. Designing self-healing materials at the nanoscale is a hard problem that requires complex solutions and, in the future, systems nanotechnology can provide new opportunities for this field.

ACKNOWLEDGMENTS

The authors acknowledge Danilo Dini, Stefano Polizzi, Jing Sheng, Karl M. Kadish, Mario J.F. Calvete, and Michael Hanack for their contribution to the experimental work.

REFERENCES

Amendola, V., Bakr, O.M., and Stellacci, F. 2010, A study of the surface plasmon resonance of silver nanoparticles by the discrete dipole approximation method: Effect of shape, size, structure, and assembly, *Plasmonics*, 5(1), 85–97.

Amendola, V., Dini, D., Polizzi, S., Shen, J., Kadish, K.M., Calvete, M.J.F., Hanack, M., and Meneghetti, M. 2009, Self-healing of gold nanoparticles in the presence of zinc phthalocyanines and their very efficient nonlinear absorption performances, *J Phys Chem C*, 113(20), 8688–8695.

Amendola, V., Mattei, G., Cusan, C., Prato, M., and Meneghetti, M. 2005, Fullerene non-linear excited state absorption induced by gold nanoparticles light harvesting, *Synth Metals*, 155(2), 283–286.

Amendola, V. and Meneghetti, M. 2009a, Laser ablation synthesis in solution and size manipulation of noble metal nanoparticles, *Phys Chem Chem Phys*, 11, 3805–3821.

Amendola, V. and Meneghetti, M. 2009b, Self-healing at the nanoscale, *Nanoscale*, 1(1), 74–88.

Amendola, V. and Meneghetti, M. 2009c, Size evaluation of gold nanoparticles by UV–vis spectroscopy, *J Phys Chem C*, 113(11), 4277–4285.

Amendola, V., Polizzi, S., and Meneghetti, M. 2006, Laser ablation synthesis of gold nanoparticles in organic solvents, *J Phys Chem B*, 110(14), 7232–7237.

Amendola, V., Polizzi, S., and Meneghetti, M. 2007, Free silver nanoparticles synthesized by laser ablation in organic solvents and their easy functionalization, *Langmuir*, 23(12), 6766–6770.

Boyd, R.W. 2008, *Nonlinear Optics*, Academic Press/Elsevier, Inc., Oxford, U.K.

Bozhevolnyi, S.I., Beermann, J., and Coello, V. 2003, Direct observation of localized second-harmonic enhancement in random metal nanostructures, *Phys Rev Lett*, 90, 197403.

Cheeseman, J.D., Corbett, A.D., Shu, R., Croteau, J., Gleason, J.L., and Kazlauskas, R.J. 2002, Amplification of screening sensitivity through selective destruction: Theory and screening of a library of carbonic anhydrase inhibitors, *J Am Chem Soc*, 124(20), 5692–5701.

Chen, Y., Hanack, M., Araki, Y., and Ito, O. 2005, Axially modified gallium phthalocyanines and naphthalocyanines for optical limiting, *Chem Soc Rev*, 34(6), 517–529.

Chiarabelli, C., Stano, P., and Luisi, P.L. 2009, Chemical approaches to synthetic biology, *Curr Opin Biotechnol*, 20(4), 492–497.

Drexler, K.E. 1992, *Nanosystems: Molecular Machinery, Manufacturing, and Computation*, John Wiley & Sons, Inc., New York.

Elim, H.I., Yang, J., Lee, J., Mi, J., and Jib, W. 2006, Observation of saturable and reverse-saturable absorption at longitudinal surface plasmon resonance in gold nanorods, *Appl Phys Lett*, 88, 083107.

Fang, H., Du, C., Qu, S., Li, Y., Song, Y., Li, H., Liu, H., and Zhu, D. 2002, Self-assembly of the [60] fullerence-substituted oligopyridines on Au nanoparticles and the optical nonlinearities of the nanoparticles, *Chem Phys Lett*, 364, 290–296.

Feynman, R.P. 1959, There's plenty of room at the bottom, *Ann Meet Am Phys Soc*, 23(5), 22–36.

Francois, J., Mostafavi, L.M., Belloni, J., and Delaire, J. 2001, *Phys Chem Chem Phys*, 3, 4965–4971.

Francois, L., Mostafavi, M., Belloni, J., Delouis, J., Delaire, J., and Feneyrou, P. 2000, *J Phys Chem B*, 104, 6133–6137.

Fratzl, P. 2007, Biomimetic materials research: What can we really learn from nature's structural materials? *J Roy Soc Interface,* 4(15), 637.

Goodson, T.I., Varnavsky, O., and Wang, Y. 2004, *Int Rev Phys Chem*, 23, 109–150.

Guarise, C., Manea, F., Zaupa, G., Pasquato, L., Prins, L.J., and Scrimin, P. 2008, Cooperative nanosystems, *J Peptide Sci*, 14(2), 174–183.

Hager, M.D., Greil, P., Leyens, C., van der Zwaag, S., and Schubert, U.S. 2010, Self-healing materials, *Adv Mater*, 22, 5424–5430.

Hasobe, T., Imahori, H., Kamat, P.V., Ahn, T.K., Kim, S.K., Kim, D., Fujimoto, A., Hirakawa, T., and Fukuzumi, S. 2005, Photovoltaic cells using composite nanoclusters of porphyrins and fullerenes with gold nanoparticles, *J Am Chem Soc*, 127, 1216–1228.

Hirsch, T., Shaporenko, A., Mirsky, V.M., and Zharnikov, M. 2007, Monomolecular films of phthalocyanines: Formation, characterization, and expelling by alkanethiols, *Langmuir*, 23, 4373–4377.

Imahori, H. and Fukuzumi, S. 2004, Porphyrin and Fullerene Based Molecular Photovoltaic Devices, *Adv Funct Mater*, 14, 525–536.

Ispasoiu, R.G., Balogh, L., Varnavski, O.P., Tomalia, D.A., and Goodson, III T.J. 2000, Large Optical Limiting from Novel Metal–Dendrimer Nanocomposite Materials, *J Am Chem Soc*, 122, 11005–11006.

Kamat, P.V. 2002, Photophysical, Photochemical and photocatalytic aspects of metal nanoparticles, *J Phys Chem B*, 106, 7729–7744.

Kitano, H. 2002, Systems biology: A brief overview, *Science*, 295(5560), 1662.

Kreibig, U. and Vollmer, M. 1995, *Optical Properties of Metal Clusters*, Springer, Berlin, Germany.

Link, S. and El-Sayed, M.A. 2000, Shape and size dependence of radiative, non-radiative and photothermal properties of gold nanocrystals, *Int Rev Phys Chem*, 19(3), 409–453.

Ludlow, R.F. and Otto, S. 2008, Systems chemistry, *Chem Soc Rev*, 37(1), 101.

Messina, E., Cavallaro, E., Cacciola, A., Iati, M.A., Gucciardi, P.G., Borghese, F., Denti, P. et al. 2011, Plasmon-enhanced optical trapping of gold nanoaggregates with selected optical properties, *ACS Nano*, 5(2), 905–913.

Perry, J.W., Mansour, K., Lee, I.Y.S., Wu, X.L., Bedworth, P.V., Chen, C.T., Ng, D. et al. 1996, Organic optical limiter with a strong nonlinear absorptive response, *Science*, 273, 1533–1536.

Philip, R., Kumar, G.R., Sandhyarani, N., and Pradeep, T. 2000, Picosecond optical nonlinearity in monolayer-protected gold, silver, and gold-silver alloy nanoclusters, *Phys Rev B*, 62, 13160–13166.

Porel, S., Singh, S., Harsha, S.S., Rao, D.N., and Radhakrishnan, T.P. 2005, Nanoparticle-embedded polymer: in situ synthesis, free-standing films with highly monodisperse silver nanoparticles and optical limiting, *Chem Mater*, 17, 9–12.

Qu, S., Du, C., Song, Y., Wang, Y., Gao, Y., Liu, S., Li, Y., and Zhu, D. 2002, Optical nonlinearities and optical limiting properties in gold nanoparticles protected by ligands, *Chem Phys Lett*, 356, 403–408.

Richardson, H.H., Hickman, Z.N., Govorov, A.O., Thomas, A.C., Zhang, W., and Kordesch, M.E. 2006, Thermooptical properties of gold nanoparticles embedded in ice: Characterization of heat generation and melting, *Nano Lett*, 6(4), 783–788.

Sudeep, P.K., Ipe, B.I., Thomas, K.G., George, M.V., Barazzouk, S., Hotchandani, S., and Kamat, P.V. 2002, Fullerene-functionalized gold nanoparticles. A self-assembled photoactive antenna-metal nanocore assembly *Nano Lett*, 2, 29–35.

Sun, W., Dai, Q., Worden, J.G., and Huo, Q. 2005, Optical limiting of a covalently bonded gold nanoparticle/polylysine hybrid material *J Phys Chem B*, 109, 20854–20857.

Sun, Y.P. and Riggs, J.E. 1999, Organic and inorganic optical limiting materials. From fullerenes to nanoparticles *Int Rev Phys Chem*, 18, 43–90.

Thomas, K.G. and Kamat, P.V. 2003, Chromophore-functionalized gold nanoparticles *Acc Chem Res*, 36, 888–898.

Tom, R.T., Nair, A.S., Singh, N., Aslam, M., Nagendra, C.L., Philip, R., Vijayamohanan, K., and Pradeep, T. 2003, Freely Dispersible Au@TiO2, Au@ZrO2, Ag@TiO2, and Ag@ZrO2 Core–Shell Nanoparticles: One-Step Synthesis, Characterization, Spectroscopy, and Optical Limiting Properties *Langmuir*, 19, 3439–3445.

Trave, E., Mattei, G., Mazzoldi, P., Pellegrini, G., Scian, C., Maurizio, C., and Battaglin, G. 2006, Sub-nanometric metallic Au clusters as efficient Er sensitizers in silica *Appl Phys Lett*, 89, 151121.

Wang, G. and Sun, W. 2006, Optical limiting of gold nanoparticle aggregates induced by electrolytes *J Phys Chem B*, 42, 20901–20905.

Wang, Y., Xie, X., and Goodson, T. 2005, Enhanced third-order nonlinear optical properties in dendrimer-metal nanocomposites *Nano Lett*, 5, 2379–2384.

West, R., Wang, Y., and Goodson II, I.T. 2003, Nonlinear absorption properties in novel gold nanostructured topologies *J Phys Chem B*, 107, 3419–3426.

Yakuphanoglu, F., Aydin, M.E., and Kiliçoglu, T. 2006, Photovoltaic properties of Au/β-carotene/n-Si organic solar cells *J Phys Chem B*, 110, 9782–9784.

Yamada, K., Miyajima, K., and Mafuné, F. 2007, Thermionic emission of electrons from gold nanoparticles by nanosecond pulse-laser excitation of interband, *J Phys Chem C*, 111(30), 11246–11251.

Yamada, K., Tokumoto, Y., Nagata, T., and Mafune, F. 2006, Mechanism of laser-induced size-reduction of gold nanoparticles as studied by nanosecond transient absorption spectroscopy, *J Phys Chem B*, 110(24), 11751.

5 Self-Healing in Two-Dimensional Supramolecular Structures
Utilizing Thermodynamic Driving Forces

Markus Lackinger

CONTENTS

5.1 SELF-ASSEMBLY: AN ATTEMPT OF A BRIEF DEFINITION WITH FOCUS ON SUPRAMOLECULAR SYSTEMS

Self-assembly is commonly understood as a process that spontaneously leads to a spatially more ordered arrangement of pre-existent building blocks by virtue of specific interactions. Various definitions can be found in literature,[1] but per se self-assembly is neither limited to a certain length scale, a certain type of building block, nor a certain dimensionality of the self-assembled arrangement. Yet, one important common aspect of various definitions is that self-assembly does not dissipate energy, and can thus proceed without external energy supply. The absence of energy dissipation can be viewed as an important difference between self-assembly and self-organization, where the latter also features a more ordered arrangement, but requires continuous energy supply to maintain this state. Oscillating chemical reactions and schools of fish are prominent examples for self-organization in chemistry and biology. In contrast, self-assembly brings a system toward its thermodynamic equilibrium, consequently promotes the minimization of Gibbs energy. Thus, if kinetically allowed, self-assembly yields the thermodynamically most stable state of a system under given external conditions. This is already the very origin of self-healing properties in any type of self-assembled arrangements, because any perturbation of the equilibrium state will generate a thermodynamic driving force directed to the initial state.

This chapter focuses on particular self-assembled structures, namely, two-dimensional (2D) supramolecular monolayers.[2] Those are commonly surface supported and can be prepared either by vacuum deposition of molecules or through adsorption from solution. Here, we will concentrate on monolayers at the liquid–solid interface, because this environment facilitates processes that are very useful for self-healing. In the first part of this chapter, specific prototypical systems will be introduced, including the environment, suitable building blocks, and predominant intermolecular interactions. Then the quite special experimental situation of monolayers at the liquid–solid interface will be compared with corresponding systems at the vacuum–solid interface, followed by a brief introduction of analytical techniques available for the characterization of such systems. The next section summarizes methods and tools at hand for a thermodynamic description of interfacial monolayers, and discusses possibilities to deliberately tune and engineer their chemical and structural properties. After a description and classification of external influences that may cause healable defects in monolayers, basic requirements for the effectiveness of self-healing mechanisms will be discussed. Although supramolecular interfacial monolayers are up to now primarily a topic of intensifying basic research efforts, conceivable applications of their self-healing properties will be identified and discussed. The final section raises important topics to be addressed and sketches further tasks before these systems can become relevant and useful for any kind of application.

5.2 SUPRAMOLECULAR MONOLAYERS AT THE LIQUID–SOLID INTERFACE

The term "supramolecular" was coined by Jean-Marie Lehn and describes molecular systems beyond single molecules, that is, larger molecular arrangements, which are stabilized by comparatively weak, typically noncovalent intermolecular interactions.[3,4] Prominent examples for supramolecular chemistry include host–guest systems, molecular folding and recognition, and molecular self-assembly. The energetic weakness of intermolecular bonds can also be seen as strength of the supramolecular approach, since intermolecular bonds become reversible at room temperature. This bears important consequences, for instance topological defects that can hardly be avoided during self-assembly of extended structures become correctable, an important requirement for the emergence of long range order systems.

In conclusion, supramolecular monolayers based on weaker, reversible intermolecular bonds are promising candidates for self-healing applications. Various types of weak interactions are suitable and exploited for supramolecular self-assembly, as for instance omnipresent van der Waals interactions, electrostatic interactions, metal-coordination bonds, and hydrogen bonds. Especially, the latter are abundant in natural and artificial systems and owe its importance to their intermediate and widely tunable strength, selectivity, and directionality.[5] The strength of hydrogen bonds can obviously be tuned by virtue of selecting binding partners, that is, the combination of hydrogen bond donors and acceptors, but also by the number and spatial arrangement of "multidentate" hydrogen bonds.[6] A specific spatial arrangement of hydrogen bond donors and acceptors also provides the basis for molecular recognition, since only a binding partner with a fully complementary arrangement of hydrogen bond donors and acceptors will be able to form the ultimately strong interconnect. Under favorable conditions, this principle can already become effective for as little as two or three parallel hydrogen bonds, as for instance in Watson–Crick pairing of DNA bases.

Hydrogen bonds are extremely popular "synthons" for the design of supramolecular systems, because the possibility of forming strong hydrogen bonds very often results in a high formation probability, and thus facilitates a certain degree of predictability of the final structure. Yet, even strong hydrogen bonds may not inevitably form, due to the dominance of competing interactions.[7] On the other hand, even principal nondirectional interactions such as van der Waals forces may be useful to achieve a certain intermolecular arrangement, for example, by the interdigitation of longer alkane chains. The art of "crystal engineering" in supramolecular chemistry is designing molecular building blocks by spatial arrangement of functional groups, such that the intended interactions become effective and are not inhibited or perturbed by competing interactions. Yet, crystal engineering is still a heuristic approach and too often based on trial and error.

The dimensionality of structures targeted by self-assembly is to a large extent encoded in the structure of the building blocks, and mostly depends on the number and topological arrangement of functional groups. Dimensionalities of supramolecular aggregates span the whole range from self-contained OD complexes, over 1D wires and chains,[8,9] over 2D sheet like structures,[10] to 3D arrangements, that is, crystals. Very often but not necessarily 1D and 2D arrangements are surface-supported or confined, where the substrate can exert a pronounced influence on the self-assembly process via templating and registry effects. Self-assembly does not per se imply that the targeted structures have to be fully crystalline, thus exhibit translational symmetry. Yet, crystalline systems are abundant, because of two main advantages—they are in many cases easier to prepare and they are easier to describe and understand. On the other hand, recently demonstrated examples of 2D noncrystalline systems have attracted significant interest.[11,12] Nevertheless, in this chapter we would like to focus on 2D crystalline monolayers prepared by self-assembly at the interface between a solid (typically crystalline) substrate and solution.[13]

At the beginning, a distinction shall be made for surface-supported monolayers between two extreme cases. On one hand, there are strongly surface-anchored systems, where the interaction between the molecule, or more precisely a specific functional (head) group, and the substrate is dominating and intermolecular interactions are of minor importance. On the other hand, there is the reversed case, that is,

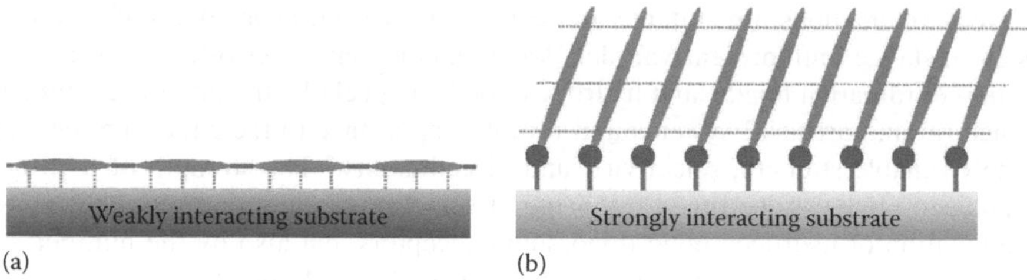

FIGURE 5.1 Principal types of self-assembled monolayers with varying strengths of molecule–molecule vs. molecule–substrate interactions. Strong interactions are indicated by solid lines; weak interactions are indicated by dashed lines. (a) Model of a monolayer with dominating molecule–molecule interactions between typically planar building blocks that adsorb parallel to the surface; molecule–substrate interactions are comparatively weak and do not govern the self-assembled structure. (b) Model of a monolayer with dominating molecule–substrate interaction, typically building blocks are functionalized with anchor groups for covalent surface linkage (indicated as spheres).

the structure is mostly governed by molecule–molecule interactions and molecule–substrate interactions are considerably weaker and do not govern the arrangement. Of course, molecule–substrate interactions still have to be sufficiently strong to confine the monolayer to the surface, but ideally molecule–substrate interactions neither influence the final structure nor do they inhibit the lateral mobility of building blocks during self-assembly. These two borderline cases are schematically sketched in Figure 5.1. In practice, however, even weakly interacting substrates as for instance graphite give raise to a distinct epitaxial relation between adsorbate and substrate lattice, pointing toward a non-negligible influence on structure formation.

While strongly surface-bound monolayers require appropriately functionalized building blocks in combination with a suitable substrate, monolayers that are dominated by lateral molecule–molecule interactions are typically comprised of more or less planar molecules equipped with functional groups for intermolecular interactions. Prototypical covalent anchoring combines thiol head groups (SH) with a Au(111) substrate. Such monolayers are often referred to as self-assembled monolayers (SAMs), where the organic backbone can be either aromatic or aliphatic.[14] SAMs can be prepared by "molecular beaker epitaxy" where the precleaned substrate is just immersed into rather dilute aqueous thiol solutions until complete monolayer coverage is achieved. More elaborate techniques such as nanoimprinting can be used for fabrication of laterally structured samples. Alkane thiols are established as well-studied building blocks for SAMs. It is known that they adsorb almost upright, with a typically somewhat inclined geometry (as sketched in Figure 5.1b) and the resulting monolayer bears similarities with a "molecular English lawn." Since the opposite end of the alkane thiol is pointing away from the surface, additional chemical end groups can be used to functionalize the surface. For instance, termination with carboxylic acid groups renders the surface hydrophilic or particular functional groups can inhibit the nonspecific adsorption of proteins.[15] Evidently, the thickness of alkane-thiol SAMs depends on the length of the alkane tail and its inclination angle with respect to the surface normal.[16] Octane-thiol SAMS have also shown self-healing capacities after exposure to reactive hydrogen atoms through the formation of new densely packed domains and the elimination of domain boundaries.[17]

As already discussed, monolayers where molecule–molecule interactions dominate, mostly planar molecules, very often with aromatic core are used, in combination with chemically inert and weakly interacting substrates. Similarly, this type of monolayers can also easily be prepared by solution techniques. To this end, a droplet of solution is simply applied to a freshly prepared substrate. Although not precisely studied, the self-assembly time scale of molecule–molecule interaction-dominated monolayers is most likely much shorter than for strongly anchored monolayers. The main reason is that the kinetics of rearrangement processes becomes significantly slower with increasing strength of interactions.

As already pointed out, hydrogen bonds are important for self-assembly in general but also for the special case of monolayers. Due to their relatively high bond strength, hydrogen bonds are very often the dominating contribution to the stabilization energy of monolayers and increase their stability and robustness, yet without impairing the self-assembly process. For hydrogen-bonded monolayers, two different types can be roughly distinguished: either fully hydrogen-bonded networks where each molecule is interlinked by hydrogen bonds in 2D or partly hydrogen-bonded networks. In the latter, molecules form 0D hydrogen-bonded complexes or 1D chains. The corresponding monolayer structure consists of a dense packing of these self-contained internally hydrogen-bonded units, in many cases mediated by van der Waals interactions or weak interchain hydrogen bonds.[18,19] Although metal-coordination bonds are very successfully implemented in organic monolayers at the solid–vacuum interface,[20–22] they are hardly represented in organic monolayers at the liquid–solid interface. Most likely, the supply of metal-coordination centers is comparatively easy to realize in vacuum experiments by co-evaporation, but difficult to realize at the liquid–solid interface. Establishing suitable preparation protocols for metal-coordinated monolayers at the liquid–solid interface would be highly rewarding.

Another classification can be made by means of the number of different building blocks constituting the monolayer, where homomeric (one component) and heteromeric (more than one component) systems can be distinguished. Homomeric systems exhibit lower complexity, structural versatility, and tunability, but are significantly easier to prepare. On the other hand, heteromeric systems exhibit more degrees of freedom, for instance the stoichiometric ratio of different building blocks, which leads to increased complexity and structural versatility. For instance, we could show that just two different building blocks are sufficient to obtain six different monolayer structures.[23] The highest degree of complexity reported for ordered monolayers at the liquid–solid interface is a system, which is comprised by four different building blocks.[24] For the successful preparation of heteromeric systems, the building blocks must be compatible, and mixed monolayers are not necessarily a consequence of mixing different building blocks.

As stated earlier, monolayers at the liquid–solid interface can simply be prepared by applying a droplet of solution onto an appropriate substrate. The cartoon in Figure 5.2 illustrates the experimental situation and depicts basic processes of monolayer self-assembly at the liquid–solid interface. Initially, solute molecules bind to the

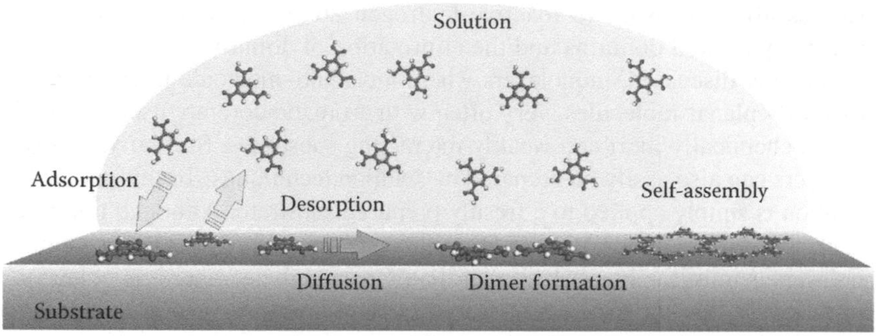

FIGURE 5.2 Sketch of elementary processes of supramolecular monolayer self-assembly at the liquid–solid interface. (Courtesy of Dr. Rico Gutzler, PhD thesis, Ludwig-Maximilians-University, Munich, Germany, 2010.)

surface, because their adsorption is driven by an enthalpic gain, that is, the enthalpy of adsorption is more favorable than the enthalpy of dissolution. After adsorption, molecules can either diffuse across the surface, desorb again into solution, or eventually interact with further adsorbed molecules and associate to larger, less mobile aggregates. Although detailed studies are sparse, by analogy with other growth studies, it is reasonable to assume that monolayer growth similarly requires preceding nucleation, followed by domain growth. Monolayer growth is completed when the surface becomes entirely covered. However, ripening and coalescence of domains in secondary growth processes have also been observed for monolayers at the liquid–solid interface.[25–27] The most interesting and somewhat unique effect for interfacial monolayers is that molecules within the monolayer are in dynamic equilibrium with dissolved molecules in the liquid phase.

In the next paragraph, commonly used building blocks, their molecular structure and functionalization, and different types of monolayers are introduced. Monolayers of long-chain saturated alkanes and fatty acids on a graphite substrate were among the first systems studied at the liquid–solid interface.[28,29] The ease of sample preparation and the high resolution routinely achieved in scanning tunneling microscopy (STM) topographs promoted the field.[13] In the following years, many groups studied various influences of the molecular structure on the monolayer structure. Among those, substitution of side groups helped to understand contrast mechanism in STM imaging,[30,31] but also the influence of aliphatic chain lengths,[32] different head groups like alcohols, and self-assembly from binary solutions that contain two different molecular building blocks[33] were studied. Aromatic moieties were another large group of targeted building blocks with even higher structural versatility than saturated alkane derivates. As a connecting link between both classes hybrid molecules, that is, aromatic cores substituted with alkane or alkoxy chains were also subject of many studies.[34] Yet, it has been shown that purely aromatic building blocks can similarly form stable monolayers even without the assistance of alkane chains, as long as intermolecular interactions are strong enough. An example studied by our group is trimesic acid (1,3,5-benzenetricarboxylic acid), which forms porous monolayers featuring a hexagonal arrangement of ~1 nm wide pores, both at the vacuum–solid[35] and liquid–solid interface.[36] Another common observation in monolayer self-assembly is polymorphism, where the same building block can yield different structures. STM images and corresponding models of two TMA monolayer polymorphs are shown in Figure 5.3. Porous monolayers can serve as host networks for the incorporation of typically geometry matched molecular guests within their pores,[37,38] an effect that has been observed for various systems[39,40] and triggered considerable interest in 2D porous systems.[41] A particularly nice example for a 2D porous hydrogen-bonded network results from heteromeric self-assembly of melamine and perylene tetracarboxylic-dimethlyimine (PTCDI), a system that has similarly first been demonstrated under vacuum conditions,[42] but which could also be demonstrated at the liquid–solid interface and even used for lateral structuring of a SAM.[43] STM topographs and corresponding models of the porous heteromeric melamine and PTCDI system are reproduced in Figure 5.4.

Among the functional groups for hydrogen bonds, carboxylic acid groups are very popular in SAMs.[44] One important feature is that two carboxylic groups can form rather strong cyclic double hydrogen bonds in a self-complementary manner. Nevertheless, alcohol, carboxy, aldehyde, amino, nitrile, and many other functional groups are suitable to promote intermolecular hydrogen bonds in monolayers. Even though alkane chain interdigitation can also yield porous systems, as has impressively been shown for alkoxylated dehydrobenzoannulenes.[45] Another interesting aspect is the scalability of structures and the associated possibility to establish a whole series of isotopological networks. Such networks share the same structural blueprint, that is, feature similar binding motifs and symmetry, but their lattice parameter can be tuned by virtue of varying the length of a spacer group or molecule. The concept of isotopological networks is exemplified by different porous networks, self-assembled from aromatic tricarboxylic acids as shown in Figure 5.5. Another monolayer realization of this concept is the bimolecular system of melamine and fatty acids with different chain lengths.[46] The fatty acid molecules act as spacers between melamine hexamers, and the lattice parameters of otherwise similar networks can be tuned in fine increments of 0.15 nm by combining melamine with the homologous series of fatty acids, as shown in Figure 5.6.

Although self-assembly of monolayers at the vacuum–solid interface may appear rather similar to the liquid–solid interface, important differences and consequences are discussed in the following. Monolayers at the vacuum–solid interface are typically prepared under ultrahigh vacuum (UHV) conditions by *in situ* deposition of molecules onto elaborately prepared samples, mostly single crystals. In order to exclude the influence of contaminations, samples are almost always measured and characterized *in situ*. At the liquid–solid interface, monolayer preparation is much easier—just a droplet of solution needs to be applied to a surface—the presence of the liquid phase has also important consequences not only for self-assembly, but also for self-healing. First, the supernatant liquid phase still contains molecules after monolayer self-assembly, and can thus serve as a reservoir for further solute molecules. In vacuum, the molecular flux onto the surface is only present during monolayer deposition and the resulting coverage is typically adjusted by deposition rate and time.

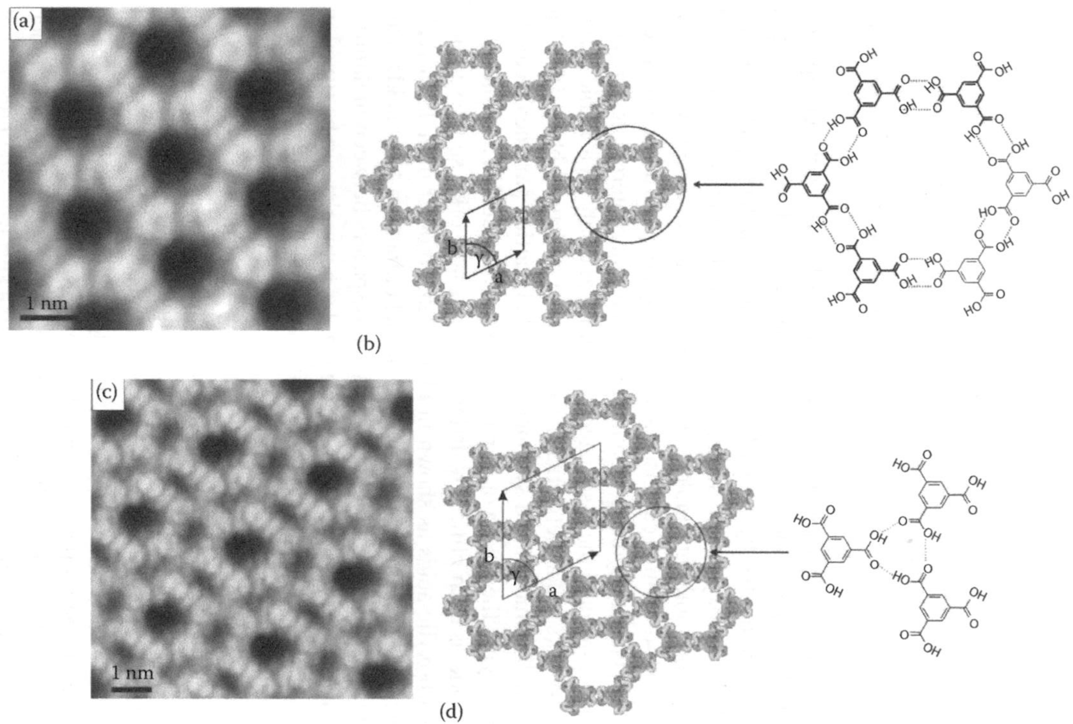

FIGURE 5.3 STM topographs of porous monolayer polymorphs of trimesic acid on graphite and corresponding structural models. (a) Chicken wire structure (lattice parameter 1.7 nm), (b) model of (a) showing the underlying hydrogen bond pattern, which is based on exclusive formation of double hydrogen bonds between carboxylic groups. (c) Flower structure (lattice parameter 2.5 nm), (d) model of (c) showing the underlying hydrogen bond pattern, which also includes cyclic hydrogen bonds between three carboxylic groups. (Adapted and reproduced with permission from Springer Science+Business Media: Self-assembled two-dimensional molecular host-guest architectures from trimesic acid, *Single Mol.*, 3, 2002, 25–31, Griessl, S., Lackinger, M., Edelwirth, M. et al., Copyright Wiley-VCH Verlag GmbH & Co. KGaA.)

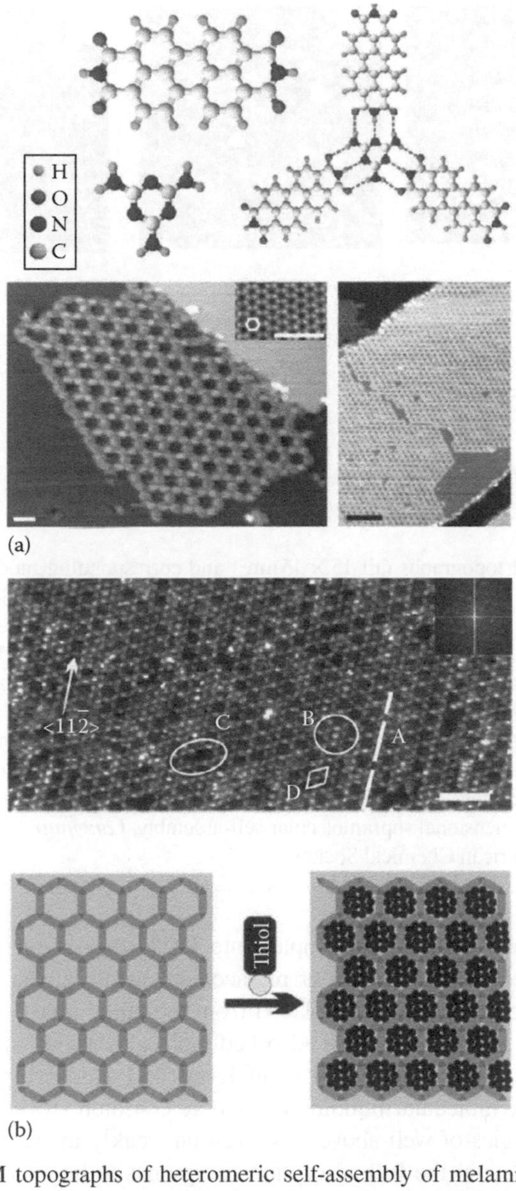

FIGURE 5.4 STM topographs of heteromeric self-assembly of melamine and PTCDI into porous monolayers at (a) the vacuum–solid interface on an Ag-terminated silicon surface and (b) at the liquid–solid interface on Au(111). The upper part of (a) depicts the molecular structures of melamine and PTCDI and the underlying hydrogen bond pattern, where three parallel hydrogen bonds are formed between each PTCDI and melamine molecule. At the liquid–solid interface, the porous monolayers were used as growth templates for thiol-anchored SAMs as shown by the cartoon in the lower part of (b). (Reprinted by permission from Macmillan Publishers Ltd. Theobald, J.A., Oxtoby, N.S., Phillips, M.A. et al., Controlling molecular deposition and layer structure with supramolecular surface assemblies, *Nature*, 424, 1029–1031; Madueno, R., Raisanen, M.T., Silien, C. et al., Functionalizing hydrogen-bonded surface networks with self-assembled monolayers, *Nature*, 454, 618–621, Copyright 2003 and 2008.)

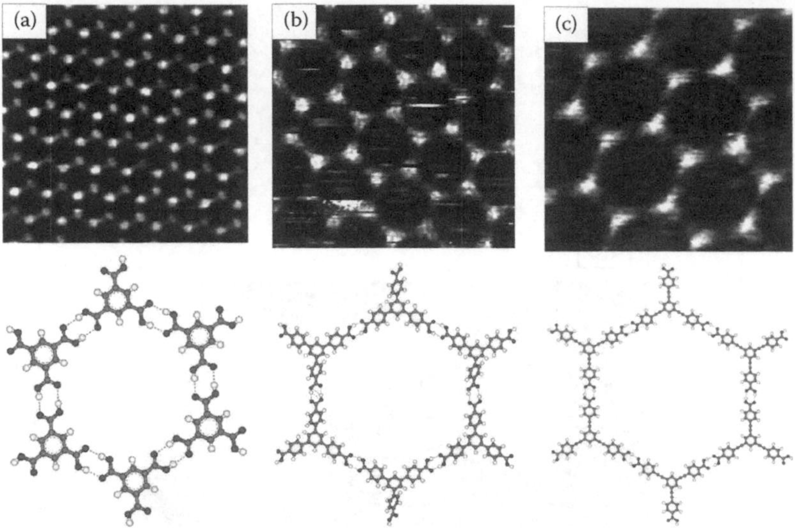

FIGURE 5.5 STM topographs (all 15 × 15 nm²) and corresponding models of isotopological porous networks. (a) benzene-1,3,5-tricarboxylic acid (trimesic acid), (b) 1,3,5-benzenetribenzoic acid (BTB), (c) 1,3,5-tri(4-carboxyphenylethynyl)-2,4,6-trimethylbenzene. For each network, threefold symmetric tricarboxylic acids with aromatic core serve as building blocks. All structures are based on the same hydrogen bond pattern, where all carboxylic groups are interconnected by double hydrogen bonds. Accordingly, lattice parameters and pore sizes scale with the distance of the carboxylic groups to the centers of the molecules. This distance can be altered by virtue of introducing different spacer groups in the molecular structure. (Reprinted and adapted with permission from Lackinger, M. and Heckl, W.M., Carboxylic acids: Versatile building blocks and mediators for two-dimensional supramolecular self-assembly, *Langmuir*, 25, 11307–11321, 2009. Copyright 2009 American Chemical Society.)

When molecules desorb from samples into UHV, they are irretrievably lost and there is no inherent further supply or replacement, while at the liquid–solid phase desorbed molecules can easily be replaced from the liquid phase. Another important difference is the vertical mobility of adsorbed molecules at the liquid–solid interface at room temperature, meaning that they can desorb and re-adsorb. For monolayer self-assembly, mid-size molecular building blocks are common (100–800 amu), resulting in adsorption energies of well above 1 eV even on weakly interacting substrates. For desorption into vacuum the full adsorption energy has to be overcome, which is not possible at room temperature, consequently in UHV larger molecules are confined to the surface. However, the situation at the liquid–solid interface is entirely different; molecules do not desorb into vacuum but into solution, where they are further stabilized by interactions with solvent molecules. Accordingly, the enthalpy of desorption at the liquid–solid interface is substantially reduced by the enthalpy of dissolution. This significantly lower effective desorption enthalpy facilitates desorption of adsorbed molecules into the liquid phase around room temperature. Vertical mobility can result in very different growth mechanisms, but more importantly mediate a dynamic equilibrium between molecules within the monolayer and in the liquid phase.

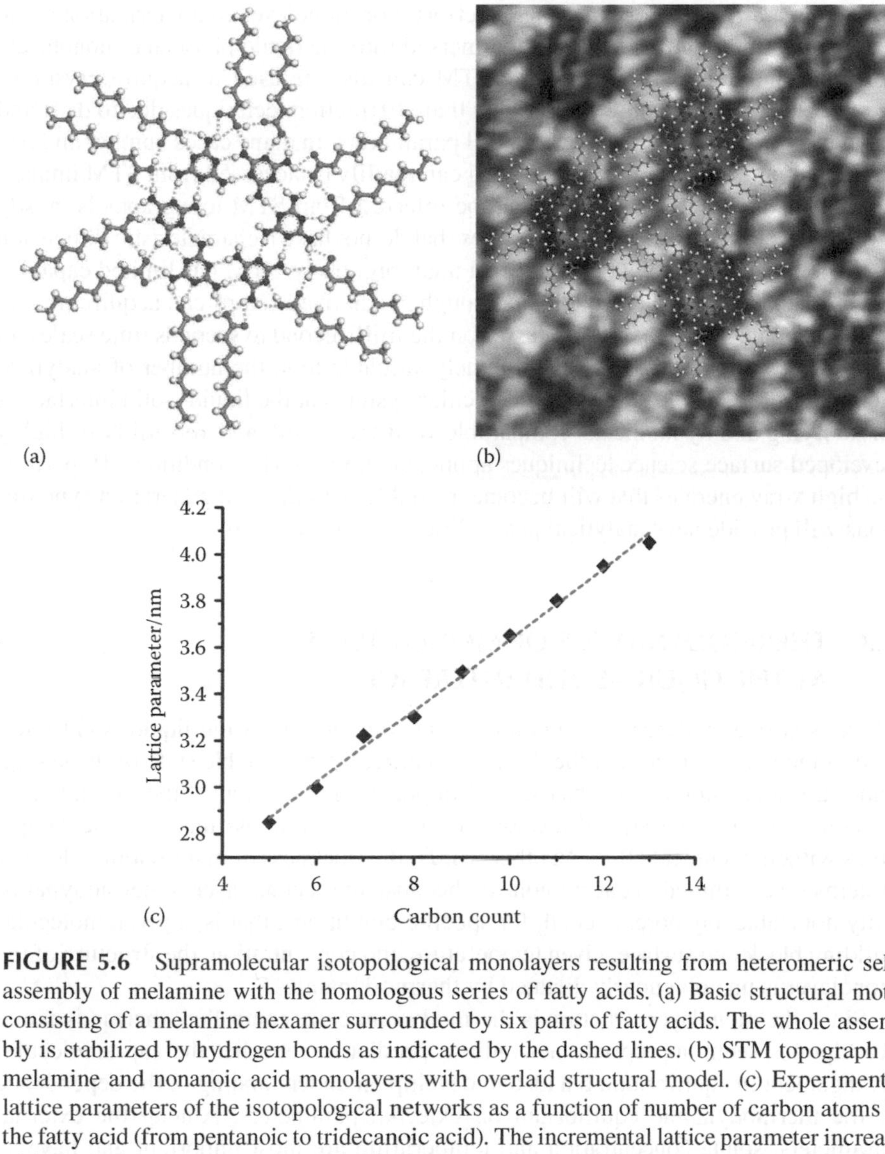

FIGURE 5.6 Supramolecular isotopological monolayers resulting from heteromeric self-assembly of melamine with the homologous series of fatty acids. (a) Basic structural motif consisting of a melamine hexamer surrounded by six pairs of fatty acids. The whole assembly is stabilized by hydrogen bonds as indicated by the dashed lines. (b) STM topograph of melamine and nonanoic acid monolayers with overlaid structural model. (c) Experimental lattice parameters of the isotopological networks as a function of number of carbon atoms in the fatty acid (from pentanoic to tridecanoic acid). The incremental lattice parameter increase of 0.15 nm corresponds to the length difference between subsequent fatty acids. (Reprinted and adapted with permission from Walch, H., Maier, A.K., Heckl, W.M. et al., Isotopological supramolecular networks from melamine and fatty acids, *J. Phys. Chem. C*, 113, 1014–1019, 2009. Copyright 2009 American Chemical Society.)

While for UHV experiments a great variety of different analytical techniques is well-developed and readily available, a major drawback for monolayers at the liquid–solid interface is the lack of complementary analytical techniques for full structural and chemical characterization. Luckily, for many systems *in situ* STM is a suitable real-space technique that can provide submolecularly resolved details of the structure with comparatively small experimental effort. For monolayer characterization at the liquid–solid interface, STM tips are immersed into the liquid phase and monolayers are directly imaged at the interface. STM can also be used to acquire crystallographic data, albeit with lower precision than diffraction techniques due to drift and piezo imperfections. Besides the unit cell parameters, in many cases number and orientation of molecules within the unit cell can readily be accessed from STM images. Yet the chemical information that can be inferred from STM topographs is mostly limited, and especially smaller molecules that do not have a characteristic dimension or shape can very often not be identified unambiguously. STM has limited capabilities concerning the time resolution, although specialized setups can acquire images with video rate, and dynamic processes on the millisecond to seconds time scale can be studied. Although STM is an extremely valuable tool, the number of analytical techniques available to study supramolecular systems at the liquid–solid interface is unsatisfying and by no means comparable with the wealth and versatility of highly developed surface science techniques applicable under UHV conditions. Hopefully, the high x-ray energies that will become available with the next generation synchrotrons will provide new analytical possibilities for such systems.

5.3 THERMODYNAMICS OF MONOLAYERS AT THE LIQUID–SOLID INTERFACE

There is strong evidence that supramolecular monolayers at the liquid–solid interface in many cases represent the thermodynamically most stable state of the system under given conditions. This has several important implications: First of all, homogeneous coverage and layer thickness can be achieved across macroscopic sample areas without technical effort. In other words, the thickness of many supramolecular systems is self-limited to one monolayer, because the second layer is thermodynamically not stable anymore. Second, for specific conditions, that is, a given molecular building block and surface, given temperature and concentration, the structure of the monolayer is unambiguously dictated by thermodynamics.

Since the monolayer structure is thermodynamically controlled, thermodynamic descriptions seem appropriate not only to predict or retrodict the supramolecular arrangement in equilibrium, but also to develop an understanding of the dependence of the thermodynamic equilibrium on external parameters. Among the external parameters, solute concentration and temperature are most important and have so far been studied most intensively.

Different approaches are suitable to describe the thermodynamics of monolayer self-assembly at the liquid–solid interface. An obvious method is to utilize the equality of chemical potentials of all components across the system in thermodynamic equilibrium. Consequently, the chemical potentials of each molecular building block on the surface have to be equal to the chemical potentials of dissolved molecules in solution. While the chemical potential of molecules on the surface is difficult to assess, strongly simplified concepts for the chemical potentials of species in the gas phase or solution are available, most prominently the model of an ideal solution.

The dependence of the chemical potential μ_i of component i on its concentration c_i and temperature T is then given by the following equation:

$$\mu_i = \mu_{0,i} + kT \cdot \ln(c_i)$$

While the standard chemical potential $\mu_{0,i}$ is normally not known, at least changes of the chemical potential as a function of temperature and concentration can be estimated. The ideal solution concept only takes the freedom of translation into account, that is, the translational entropy. Although assuming ideal solutions seems hopelessly oversimplified, this concept has nevertheless been used for an astonishingly accurate description of rather complex systems.[45] Chemical potentials can then be used to evaluate and compare Gibbs free energies per unit area of different monolayer structures,[47] provided that structural data or models are available, which yield molecular area densities. This approach is also feasible for multicomponent systems and allows for instance to simulate phase diagrams in a multidimensional concentration space.[48]

An alternative, more atomistic method for understanding the thermodynamic properties of monolayers at the liquid–solid interface is to evaluate both contributions to Gibbs energy (assuming constant pressure),[7,48] namely, enthalpy and entropy according to

$$\Delta G = \Delta H - T \cdot \Delta S$$

Upon self-assembly, binding enthalpy is typically gained through formation of additional or stronger bonds (negative ΔH), while building blocks loose entropy when they assemble in a more ordered state (negative ΔS). The entropy loss leads to an unfavorable, temperature proportional positive contribution to Gibbs energy, which has to be at least compensated by the enthalpic gain.

In the first approach, the favorable enthalpic contributions in monolayer self-assembly can be subdivided into molecule–substrate and molecule–molecule interactions. Various contributions can also be distinguished for the entropic losses, that is, translational entropy, rotational entropy, conformational entropy, and vibrational entropy.[49] The two main entropic contributions arise from translational and rotational entropy, which are both entirely lost when molecules adsorb on the surface and self-assemble into monolayers. Methods to estimate both contributions are available and for typical monolayer self-assembly conditions, both translational and rotational entropy are in the same order of magnitude.

On the other hand, binding enthalpies can be inferred from a number of theoretical methods at different levels including sophisticated quantum chemical methods. As the simplest approach, molecular mechanics (MM) calculations are becoming increasingly popular because of their simplicity and their favorable scaling behavior rendering them also suitable for larger system sizes. Another advantage of MM is that force fields can easily be parameterized for an optimized description of specific systems.[50]

Density functional theory (DFT) calculations have meanwhile become accessible to a broader community, not at least due to availability of sufficient computational means either on powerful local computers or on increasingly accessible parallel computers. A major shortcoming of DFT for the description of surface-supported supramolecular systems is that standard functionals cannot deal with van der Waals interactions properly. Just in the recent years, however, sophisticated DFT codes based on the so-called hybrid functionals were developed and consider van der Waals contributions by including pairwise interactions between single atoms.[51,52] These methods are extremely promising for a more accurate description, but on the other hand are computationally very expensive, and thus limited to relatively small systems. Nevertheless, for a full evaluation of all thermodynamic contributions to monolayer self-assembly, obtaining robust figures for the entropic loss is much more difficult than for the enthalpic gain.

Instead of a mere understanding of the final state structure, recent theoretical efforts aim at a molecular understanding of nucleation and growth of supramolecular monolayers. Among the suitable theoretical methods, kinetic Monte Carlo (KMC) and molecular dynamics (MD) simulations have gained grounds.[53–55] A main advantage of KMC simulations as opposed to MD is that depending on the level of abstraction, KMC calculations can be computationally efficient and deal with large and more realistic systems sizes and simulate longer time spans. A major disadvantage is their low robustness, that is, the strong influence of simulation parameters on the outcome. For instance, KMC simulations are very sensitive to interaction energies that may be almost considered as fitting parameters. Nevertheless, different growth modes or the consequences of the interplay of various types of interactions can be understood.[53–55] Albeit being able to tackle ever larger and larger system sizes, MD simulations are probably still too limited in system size and simulation times to describe realistic self-assembly scenarios for the complex situation at the liquid–solid interface. Nonetheless, MD can reveal interesting aspects or details of dynamic processes, and maybe in the near future further abstraction and increasing computational power and efficiency will allow to dynamically deal with self-assembly of interfacial monolayers as well.

5.4 LIMITATIONS OF THERMODYNAMIC DESCRIPTIONS

In general, it is very difficult to decide whether a system has adopted the thermodynamically most stable state or whether it is kinetically trapped in a metastable state. The latter situation occurs frequently and gives rise to polymorphism, a common observation for monolayers at the liquid–solid interface.[36,56] For instance in a recent work, Marie et al. observed irreversible structural changes in a monolayer upon annealing, pointing toward conversion of a metastable, kinetically trapped state into the thermodynamically most stable state.[57] Thus, a common way of evaluating whether the system has adopted the thermodynamic equilibrium is tempering the system, in order to provide sufficient thermal energy to overcome kinetic barriers.

Especially, monolayers that are constituted of larger molecular building blocks might be prone to kinetic limitations. For instance, desorption barriers can be too high, thus impairing the vertical mobility and disabling the dynamic equilibrium between monolayer and solution. Also, surface diffusion barriers increase with molecular size and might even limit the lateral mobility of the compound. Another possible source for kinetic limitations is solvent viscosity, since highly viscous solvents hamper diffusion of dissolved molecules in the liquid phase. The bulk diffusivity constant of solute molecules is indirectly proportional to the viscosity, which can change drastically in the experimentally accessible temperature range between ~0°C and ~100°C.[48]

The quantitative thermodynamic description of supramolecular monolayers at the liquid–solid interface is still at its infancy. Although the accuracy of binding enthalpies of molecules—either molecule–molecule or molecule–substrate interactions—is probably satisfying, important enthalpic contributions, for example, the enthalpy of dissolution, have not been evaluated so far. For the similar important entropic contributions, the theoretical situation is less satisfying. As discussed earlier, it seems justified to neglect some entropic contributions (e.g., conformational and vibrational), because they are either very small or hardly change during self-assembly. At least estimates are available for the two most important contributions from translational and rotational entropy. Yet, while the validity of these estimates has only been verified for rather simple model systems, their suitability for more complex systems is on less safe grounds. In summary, much work remains to be done both on the theoretical and experimental side to obtain reliable values for entropy changes upon supramolecular self-assembly.

5.5 STRUCTURAL VERSATILITY BY INTERNAL AND EXTERNAL PARAMETERS

Ideally, the building plan of a supramolecular structure is exclusively encoded in the structure of the building block, that is, the molecule. Yet, several experiments have shown that controlled and less well-controlled external and internal parameters can also affect the resulting structure. One important experimental parameter is the solvent that gives rise to solvent-induced polymorphism.[36,56,58,59] While originally believed to serve merely as a reservoir for molecules, it soon became clear that the solvent can also control the monolayer structure by various effects. Incorporation of solvent molecules into the monolayer structures is only the most obvious effect,[46] and more subtle effects have been postulated.[60] It is impossible to provide a universal model for solvent effects, because its influence might be highly system specific, and in many cases it is even unclear whether the solvent affects the thermodynamics or kinetics of monolayer self-assembly.

Solute concentration is another long time experimentally unattended, albeit in retrospective obviously very important parameter. At least in this case, a common trend was observed for a variety of different systems: the higher the solute concentration in the liquid phase, the higher the packing density of the monolayer polymorphs on the surface and vice versa.[7,45,48] While this observation by itself is very plausible, a thermodynamic explanation is offered by the concentration dependence of translational entropy: The lower the solute concentration, the higher the associated entropic loss of a molecule upon adsorption. Consequently, lower solute concentrations favor entropically cheaper polymorphs with lower packing density.

Albeit hardly studied, temperature is definitely another important parameter for supramolecular self-assembly. Since the magnitude of the entropic contribution is comparable to the enthalpic contribution, the experimental parameter temperature can hold the balance. Many recent experiments have discovered and confirmed the strong influence of temperature on monolayer self-assembly.[48,57,61]

5.6 SELF-HEALING OF MONOLAYERS: HEALABLE DEFECTS AND REQUIREMENTS

Monolayers at the liquid–solid interface can exhibit self-healing properties, because of their tendency to return to the thermodynamically most stable state when this equilibrium is perturbed. In the following section, different kinds of perturbations and defects are distinguished and basic requirements for the effectiveness of thermodynamically driven self-healing are identified.

Concerning the effect of external perturbations, a distinction between local and global influences can be made, where the former impair only a small surface area or even only a single molecule while the rest of the monolayer remains still intact. This could for instance be caused by a local mechanical impact that only removes part of the layer, such as the influence of a scanning tip in scanning probe microscopy experiments. A further local defect might be the deterioration and irreversible damage of single building blocks originating from localized chemical influences, high-energy photons, or particle irradiation.

For highly localized perturbations, the microstructure will not be changed upon self-healing, as long as the damaged area is small compared with the average domain size. In the easiest case, self-healing simply replaces the missing molecules. An example, where molecules that have been removed by a mechanical impact were replaced from the liquid phase will be presented later.

Global perturbations, on the other hand, affect the thermodynamic stability of the monolayer in a sense that they induce desorption of the whole monolayer. This could for instance be caused either by a temperature increase or by a substantial decrease of solute concentration, both of which favor desorption by shifting the equilibrium toward dissolution. In both cases, the enthalpic gain through monolayer formation is not sufficient anymore to compensate the increased entropic loss, resulting in desorption. Since global perturbations shift the thermodynamic equilibrium toward dissolution, self-healing cannot be expected as long as the perturbation persists. Only when the favorable conditions, that is, temperature and/or concentration are restored, the thermodynamic driving force for self-assembly becomes effective and the monolayer re-assembles.

Similar to self-assembly, self-healing can only bring a system back to its thermodynamic equilibrium. In order to take advantage of self-healing, two basic requirements have to be fulfilled. First, the conditions (temperature, solvent, concentration, substrate, etc.) must be adjusted such that the targeted monolayer structure becomes thermodynamically stable again. Second, kinetic traps must not be effective, that is, either the traps have to be eliminated or the available thermal energy must be sufficient to overcome them.

More specifically, this means molecular building blocks must be available and they must be mobile. Their availability is guaranteed as long as the monolayer is in contact with a liquid reservoir for molecules, that is, a solution with high enough solute concentration. Those requirements are definitely fulfilled, when the conditions are even similar to the initial self-assembly situation, where the mobility of dissolved molecules was also guaranteed. For instance, a too low temperature can increase the solvent viscosity to such a degree that self-healing is slowed down to a state where the system becomes kinetically trapped.

Another, maybe not too strict requirement is that molecular building blocks should not be destroyed. Self-assembly, especially of crystalline target structures, only works when all building blocks are similar in structure, conformation, and chemical properties. Self-assembly can be impaired, when building blocks are chemically altered, deteriorated, or even destroyed. What actually happens depends on the affinity of degraded vs. intact molecules to take part in the self-assembly process and the self-assembled structure, hence it is difficult to draw general conclusions.

In the following, two examples are given for self-healing processes in monolayers at the liquid–solid interface after local and global perturbations. The first example is based on a 2D supramolecular host–guest system and demonstrates self-healing of point defects created by a local mechanical impact. The host network is a nanoporous monolayer comprised of trimesic acid (TMA) molecules interconnected through double hydrogen bonds between their carboxylic acid moieties. The hexagonally arranged pores are about ~1.0 nm in diameter and can accommodate size matched coronene molecules as guests,[38] a structural model of a six-membered TMA pore with included coronene guest is reproduced in Figure 5.7a. Such a host–guest system can be prepared by self-assembly from a binary solution, where both TMA host network–forming molecules and coronene guest molecules are dissolved in a fatty acid solvent. Again, the host–guest network can be imaged directly at the liquid–solid interface by STM. As long as the tip–sample interaction is weak during image acquisition, that is, the tunneling current set point is small (<0.1 nA), the host–guest network can be imaged without being perturbed, a representative STM topograph of a coronene-filled TMA monolayer is shown in Figure 5.7b. However, if the tip–sample distance during imaging is decreased by increasing the tunneling current set point, the tip–sample interaction can also increase to such an extent that coronene guest molecules are desorbed through the mechanical impact of the scanning probe, while the highly stable TMA host network still remains unaffected. This situation is exemplified by the STM topograph in Figure 5.7c, where the tunneling current was increased for some scan lines. In this example, three TMA pores were emptied out and appear with dark contrast. Yet, in the presence of the coronene-rich supernatant solution unoccupied pores of the host network are thermodynamically not stable, and become reoccupied on a time scale of minutes. This process can be seen as a self-healing, since the influence of a mechanically induced perturbation of a host–guest network is reversed without the need for external intervention.

The influence of a global thermodynamic perturbation shall be exemplified by the porous network already presented in Figure 5.5b. At room temperature, self-assembly of the tricarboxylic acid 1,3,5-benzenetribenzoic acid (BTB), a larger analog of TMA, from nonanoic acid solutions yields a porous network with larger cavities (2.8 nm diameter as opposed to 1.0 nm for TMA). However, when the temperature of the system is raised above ~53°C, a reversible structural phase transition converts the porous BTB

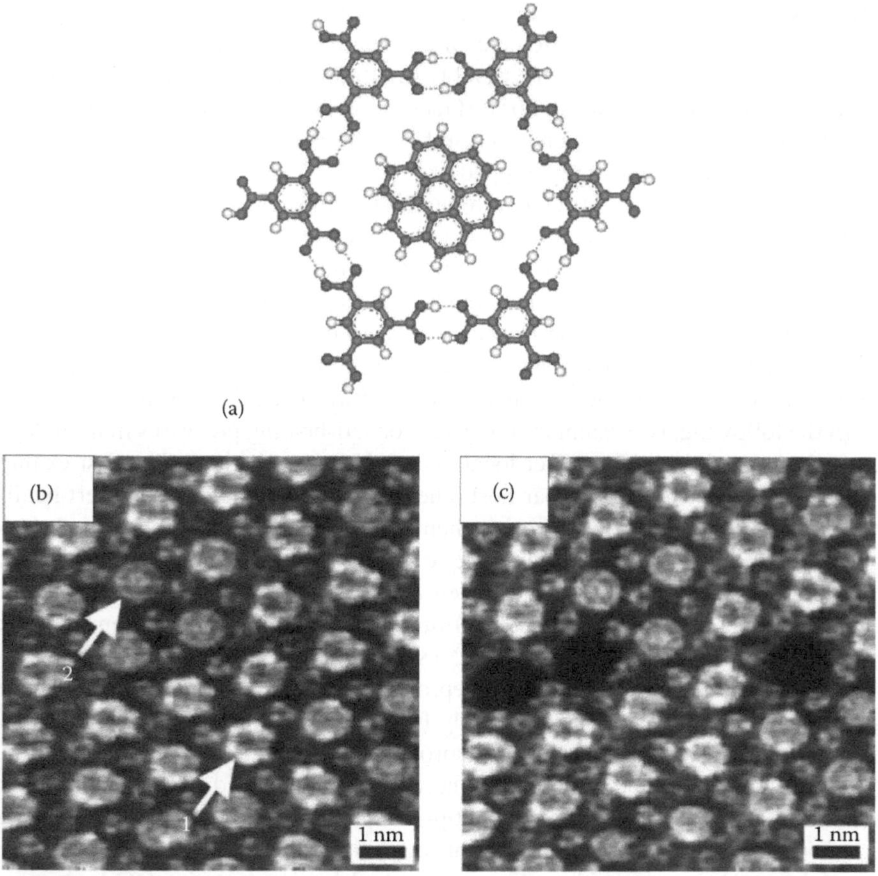

FIGURE 5.7 Porous monolayers of trimesic acid (TMA) can serve as host systems for the incorporation of coronene guest molecules. (a) Structural model of a single six-membered TMA pore with coronene at the center. (b) Corresponding STM topograph of a coronene-filled TMA host network at the liquid–solid interface. Arrows highlight coronene molecules with different STM contrasts, most likely caused by substrate effects. (c) Increasing the STM tunneling current for a few scan lines resulted in the creation of point defects through the increased tip–sample interaction. Three coronene molecules were removed by the scanning probe; the respective empty pores appear with dark contrast. However, self-healing efficiently replaces the missing guest molecules by adsorption from the liquid phase. (Reprinted and adapted with permission from Griessl, S.J.H., Lackinger, M., Jamitzky, F. et al., Incorporation and manipulation of coronene in an organic template structure, *Langmuir*, 20, 9403–9407, 2004. Copyright 2004 American Chemical Society.)

monolayer polymorph into a densely packed row structure.[48] These structural phase transitions are fully reversible. After cooling down again below the transition temperature the original porous network becomes restored. A series of STM topographs demonstrating the phase transition and its reversibility is presented in Figure 5.8: BTB molecules do not desorb from the surface when the temperature is increased, but order in a different arrangement. The high temperature structure consists of rows

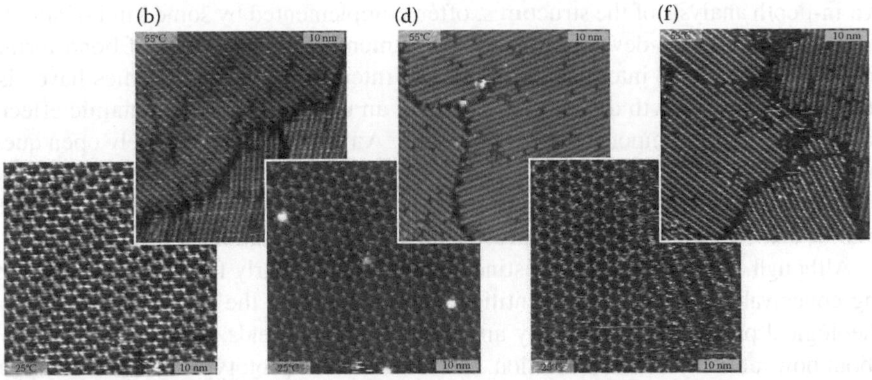

FIGURE 5.8 Self-healing of 1,3,5-benzenetribenzoic acid monolayers at the liquid–solid interface. The system exhibits temperature-induced reversible phase transitions of a porous into a densely packed monolayer polymorph. The STM topographs were acquired at the liquid–solid interface during repeated heat–cool cycles and demonstrate the reversibility of the phase transition. The respective temperatures are stated in the lower (upper) left corner of each image. The cycle starts at the lower left image (a) at room temperature and is continued (b) at 55°C → (c) at 25°C → (d) at 55°C → (e) at 25°C → (f) at 55°C. Increasing the temperature can be seen as a global perturbation of the system, which seriously disturbs the original order. Self-healing sets in when the original conditions, that is, temperature, are restored and the system returns to its thermodynamic equilibrium at room temperature, the porous monolayer structure. (Reprinted and adapted with permission from Gutzler, R., Sirtl, T., Dienstmaier, J.F. et al., Reversible phase transitions in self-assembled mono layers at the liquid-solid interface: Temperature-controlled opening and closing of nanopores, *J. Am. Chem. Soc.*, 132, 5084–5090, 2010. Copyright 2010 American Chemical Society.)

of upright molecules, so the temperature increase destroys the original order of the monolayer with planar adsorbed BTB molecules. Yet, when the system is cooled down again, the original state of the system is restored, because it represents the thermodynamic equilibrium at room temperature. Most likely, the temperature-induced phase transition of the monolayer has a thermodynamic origin, where the densely packed structure becomes thermodynamically more stable at elevated temperature, although a kinetic stabilization of the high temperature phase cannot be fully excluded. Since the phase transition converts an originally nanoporous system with the capability to incorporate guest species into a densely packed structure that cannot incorporate guests anymore, applications with temperature controlled release of guest species are highly promising.

5.7 CONCEIVABLE APPLICATIONS OF SELF-HEALING SUPRAMOLECULAR MONOLAYERS

Up to now, the discussed monolayers at the liquid–solid interface have attracted considerable interest in basic research, and worldwide efforts have led to an impressive amount of diverse high-quality results. Many different systems and properties have been demonstrated, and monolayer structures have been studied in great detail.

An in-depth analysis of the structures, often complemented by some kind of simulation, has aided in the development of a fundamental understanding of bond formation and the complex interplay of competing interactions. Recent studies have also started to tackle growth dynamics and aim at an understanding of dynamic effects, which are similarly important for self-healing. An interesting and widely open question is, however, for what type of applications could these monolayers be suitable and beneficial in a sense that they improve the standard way of accomplishing a defined goal or even open pathways to unprecedented novel applications.

Although, at the moment no distinct application is clearly favored, three promising conceivable fields will be identified and described in the following. Improving rheological properties is definitely among the targeted fields, albeit little is known about how monolayers affect friction. Interestingly, the prototypical model substrate graphite whose usage is very often just motivated by the ease of sample preparation could actually become relevant for rheological applications. Layered materials such as graphite or molybdenite (MoS_2) are already used as lubricants or additives to improve friction properties, by virtue of the small forces required for sliding their sheets against each other. For this type of application, the pivotal question would be the following: How are the rheological properties affected when graphite or molybdenite platelets become covered with supramolecular monolayers? For this purpose, both self-assembly and self-healing are important aspects. While self-assembly will drive initial monolayer formation, self-healing will be responsible for a continuous replenishment. This becomes necessary, because it can be expected that the mechanical forces will constantly damage and remove the monolayer when the graphite platelets are sheered against each other. Provided that the system is adjusted properly, mechanical removal will generate a thermodynamic driving force for renewal of the monolayer and constantly repair the damage. Although rheological applications sound plausible and promising, a first important step would be a rheological characterization of monolayer covered systems. In light of the vast structural versatility of monolayers studied so far, an important goal would be to develop methods for correlating monolayer structures with their rheological properties. Such a systematic approach will at the end facilitate tailored engineering of supramolecular monolayers for different applications.

A second conceivable application of self-healing monolayers lies in the field of surface functionalization. While for the former rheological applications, monolayers with enhanced lateral molecule–molecule interaction seem most promising, strongly surface-anchored SAMs appear better suited for durable surface functionalization.

The initial functionalization of surfaces with molecules can again be achieved by self-assembly or even nanoimprint techniques that facilitate lateral structuring. The concept implies that damages, such as vacancies or extended vacancy islands, can easily be repaired by self-healing techniques. If the application allows, the easiest way to take advantage of self-healing is to constantly supply spare molecules, that is, to have the surface covered with a thin liquid film of solution that serves as reservoir. However, for many applications the functionalized surfaces need to be directly accessible and a liquid film is not tolerable. In this case, self-healing would require an active repairing or regeneration step where new solution is added to the surface or the surface is immersed into solution. An interesting question for this application is whether self-healing mechanisms would be capable of replacing degraded by intact molecules. In any case, a basic requirement is that degraded molecules can still desorb from the surface and take part in the dynamic equilibrium. Then replacement of degraded with intact molecules is statistically favored, as long as the concentration of intact molecules is significantly larger.

More lofty applications for self-healing lie in molecular electronics. The main goal, but at the same time the main challenge of molecular electronics is to realize electronic circuits based on interconnected molecular entities and interfacing them to the real world. Standard lithographic production techniques are definitely not suited for the fabrication of molecular electronics circuits, though self-assembly as a parallel and autonomous technique is in the discussion. A main problem of artificial self-assembled systems, however, is achieving complexity. For instance crystallization, as a subset of self-assembly is an efficient fabrication technique, but results only in periodic systems with little complexity. In molecular electronics, however, demanding tasks require more complex structures. On the other hand, by means of the genetic code and the structure of DNA, nature has impressively demonstrated that self-assembly and self-organization can yield highly complex and information-rich structures. So a primary goal of sophisticated self-assembly is to develop protocols and design structures, which will result in higher complexity. The simplest part of an electronic circuit, a wire has already been realized on the molecular scale by self-assembly techniques. Even for such simple electronic components, self-healing can already be beneficial, maybe in self-healing fuses?

5.8 PERSPECTIVE, OUTLOOK, AND FUTURE TASKS

Although many different types of monolayers—porous vs. densely packed, homomeric vs. heteromeric, aromatic vs. aliphatic, and so on—have already been realized and characterized, the field of supramolecular monolayers has not gone much beyond the stage of basic research. In order to pave the road toward applications, various open questions have to be addressed and problems have to be tackled. A major issue that will definitely have to be addressed is the type of surface. Basic research conducts experiments on extremely well-defined, homogeneous, flat, and atomically clean surfaces, in many cases of single crystals. The choice of substrate in basic research is often mostly motivated by the requirements of the experiment. Technical surfaces, on the other hand, are very often inhomogeneous. Consequently, a logic next step will be to study monolayer self-assembly on "less perfect" substrates. Maybe it is advisable to take small steps in this direction, for instance by using less flat, topographically more pronounced, but still chemically pure surfaces, to identify the influence of substrate imperfections on the self-assembly and self-healing properties.

So far, mostly molecular properties of the monolayers, as for instance their structure, were studied in great detail and little attention has been paid to more macroscopic physical and chemical properties. Rheological properties are a prominent example for uncharacterized monolayer properties. A more complete characterization is highly desirable.

In contrast to covalently anchored SAMs, physisorbed monolayers were in many cases only studied *in situ* at the liquid–solid interface. On the other hand, many applications require direct access to the monolayer, and thus dry samples. So probably the easiest task in the light of applicability is to establish protocols to remove the liquid phase without impairing the monolayer arrangement. To this end, studying systems with superior stabilization through lateral interactions appears most promising. The deliberate design of such systems would require building blocks capable of forming a large number of relatively strong bonds, for example, hydrogen bonds. This step is definitely within reach and once successful methods have been established, it will open the door for new, hopefully insightful experiments. The persistence of such monolayers under ambient conditions can be checked, simply be studying a possible degradation over time. At the liquid–solid interface, complementary analytical techniques have been impaired by the presence of the liquid phase, but could be applied to dry samples in order to gain further knowledge about important monolayer properties.

In my opinion, the most important and promising achievement in this field would be to promote the dialogue between basic and more applied research in different disciplines in order to develop tailored systems for specific applications. Only then SAMs can overcome the not to be underestimated barrier between nanoscience and nanotechnology.

REFERENCES

1. Whitesides, G. M., Mathias, J. P., and Seto, C. T. 1991. Molecular self-assembly and nanochemistry—A chemical strategy for the synthesis of nanostructures. *Science* 254: 1312–1319.
2. Elemans, J. A. A. W., Lei, S. B., and De Feyter, S. 2009. Molecular and supramolecular networks on surfaces: From two-dimensional crystal engineering to reactivity. *Angew. Chem. Int. Ed.* 48: 7298–7332.
3. Lehn, J. M. 1990. Perspectives in supramolecular chemistry—From molecular recognition towards molecular information-processing and self-organization. *Angew. Chem. Int. Ed.* 29: 1304–1319.
4. Lehn, J.-M. 1995. *Supramolecular Chemistry*. VCH, Weinheim, Germany.
5. Steiner, T. 2002. The hydrogen bond in the solid state. *Angew. Chem. Int. Ed.* 41: 48–76.
6. Sherrington, D. C. and Taskinen, K. A. 2001. Self-assembly in synthetic macromolecular systems via multiple hydrogen bonding interactions. *Chem. Soc. Rev.* 30: 83–93.
7. Dienstmaier, J. F., Mahata, K., Walch, H. et al. 2010. On the scalability of supramolecular networks—High packing density vs optimized hydrogen bonds in tricarboxylic acid monolayers. *Langmuir* 26: 10708–10716.
8. Yokoyama, T., Yokoyama, S., Kamikado, T. et al. 2001. Selective assembly on a surface of supramolecular aggregates with controlled size and shape. *Nature* 413: 619–621.
9. Weckesser, J., De Vita, A., Barth, J. V. et al. 2001. Mesoscopic correlation of supramolecular chirality in one-dimensional hydrogen-bonded assemblies. *Phys. Rev. Lett.* 87: 096101.
10. Barth, J. V., Weckesser, J., Cai, C. et al. 2000. Building supramolecular nanostructures at surfaces by hydrogen bonding. *Angew. Chem. Int. Ed.* 39: 1230–1234.
11. Blunt, M. O., Russell, J. C., Champness, N. R. et al. 2010. Templating molecular adsorption using a covalent organic framework. *Chem. Commun.* 46: 7157–7159.
12. Marschall, M., Reichert, J., Weber-Bargioni, A. et al. 2010. Random two-dimensional string networks based on divergent coordination assembly. *Nat. Chem.* 2: 131–137.
13. De Feyter, S. and De Schryver, F. C. 2005. Self-assembly at the liquid/solid interface: STM reveals. *J. Phys. Chem. B* 109: 4290–4302.
14. Schreiber, F. 2000. Structure and growth of self-assembling monolayers. *Prog. Surf. Sci.* 65: 151–256.

15. Ostuni, E., Chapman, R. G., Holmlin, R. E. et al. 2001. A survey of structure–property relationships of surfaces that resist the adsorption of protein. *Langmuir* 17: 5605–5620.
16. Chaki, N. K. and Vijayamohanan, K. 2002. Self-assembled monolayers as a tunable platform for biosensor applications. *Biosens. Bioelectron.* 17: 1–12.
17. Kautz, N. A., Fogarty, D. P., and Kandel, S. A. 2007. Degradation of octanethiol self-assembled monolayers from hydrogen-atom exposure: A molecular-scale study using scanning tunneling microscopy. *Surf. Sci.* 601: L86–L90.
18. Lackinger, M., Griessl, S., Markert, T. et al. 2004. Self-assembly of benzene–dicarboxylic acid isomers at the liquid solid interface: Steric aspects of hydrogen bonding. *J. Phys. Chem. B* 108: 13652–13655.
19. Heininger, C., Kampschulte, L., Heckl, W. M. et al. 2009. Distinct differences in self-assembly of aromatic linear dicarboxylic acids. *Langmuir* 25: 968–972.
20. Barth, J. V. 2007. Molecular architectonic on metal surfaces. *Ann. Rev. Phys. Chem.* 58: 375–407.
21. Barth, J. V. 2009. Fresh perspectives for surface coordination chemistry. *Surf. Sci.* 603: 1533–1541.
22. Li, S. S., Northrop, B. H., Yuan, Q. H. et al. 2009. Surface confined metallosupramolecular architectures: Formation and scanning tunneling microscopy characterization. *Acc. Chem. Res.* 42: 249–259.
23. Kampschulte, L., Werblowsky, T. L., Kishore, R. S. K. et al. 2008. Thermodynamical equilibrium of binary supramolecular networks at the liquid–solid interface. *J. Am. Chem. Soc.* 130: 8502–8507.
24. Adisoejoso, J., Tahara, K., Okuhata, S. et al. 2009. Two-dimensional crystal engineering: A four-component architecture at a liquid–solid interface. *Angew. Chem. Int. Ed.* 48: 7353–7357.
25. Stabel, A., Heinz, R., De Schryver, F. C. et al. 1995. Ostwald ripening of 2-dimensional crystals at the solid–liquid interface. *J. Phys. Chem.* 99: 505–507.
26. Lackinger, M., Griessl, S., Kampschulte, L. et al. 2005. Dynamics of grain boundaries in two-dimensional hydrogen-bonded molecular networks. *Small* 1: 532–539.
27. Kim, K., Plass, K. E., and Matzger, A. J. 2003. Kinetic and thermodynamic forms of a two-dimensional crystal. *Langmuir* 19: 7149–7152.
28. Rabe, J. P. and Buchholz, S. 1991. Commensurability and mobility in 2-dimensional molecular-patterns on graphite. *Science* 253: 424–427.
29. Cyr, D. M., Venkataraman, B., and Flynn, G. W. 1996. STM investigations of organic molecules physisorbed at the liquid–solid interface. *Chem. Mater.* 8: 1600–1615.
30. Giancarlo, L. C. and Flynn, G. W. 2000. Raising flags: Applications of chemical marker groups to study self-assembly, chirality, and orientation of interfacial films by scanning tunneling microscopy. *Acc. Chem. Res.* 33: 491–501.
31. Claypool, C. L., Faglioni, F., Goddard, W. A. et al. 1997. Source of image contrast in STM images of functionalized alkanes on graphite: A systematic functional group approach. *J. Phys. Chem. B* 101: 5978–5995.
32. Florio, G. M., Werblowskyf, T. L., Ilan, B. et al. 2008. Chain-length effects on the self-assembly of short 1-bromoalkane and *n*-alkane monolayers on graphite. *J. Phys. Chem. C* 112: 18067–18075.
33. Venkataraman, B., Breen, J. J., and Flynn, G. W. 1995. Scanning–tunneling-microscopy studies of solvent effects on the adsorption and mobility of triacontane triacontanol molecules adsorbed on graphite. *J. Phys. Chem.* 99: 6608–6619.
34. Gesquiere, A., Abdel-Mottaleb, M. M., De Feyter, S. et al. 2000. Dynamics in physisorbed monolayers of 5-alkoxy-isophthalic acid derivatives at the liquid/solid interface investigated by scanning tunneling microscopy. *Chem. Eur. J.* 6: 3739–3746.

35. Griessl, S., Lackinger, M., Edelwirth, M. et al. 2002. Self-assembled two-dimensional molecular host–guest architectures from trimesic acid. *Single Mol.* 3: 25–31.
36. Lackinger, M., Griessl, S., Heckl, W. M. et al. 2005. Self-assembly of trimesic acid at the liquid–solid interface—A study of solvent-induced polymorphism. *Langmuir* 21: 4984–4988.
37. Griessl, S. J. H., Lackinger, M., Jamitzky, F. et al. 2004. Room-temperature scanning tunneling microscopy manipulation of single C60 molecules at the liquid–solid interface: Playing nanosoccer. *J. Phys. Chem. B* 108: 11556–11560.
38. Griessl, S. J. H., Lackinger, M., Jamitzky, F. et al. 2004. Incorporation and manipulation of coronene in an organic template structure. *Langmuir* 20: 9403–9407.
39. Lu, J., Lei, S. B., Zeng, Q. D. et al. 2004. Template-induced inclusion structures with copper(II) phthalocyanine and coronene as guests in two-dimensional hydrogen-bonded host networks. *J. Phys. Chem. B* 108: 5161–5165.
40. Furukawa, S., Tahara, K., De Schryver, F. C. et al. 2007. Structural transformation of a two-dimensional molecular network in response to selective guest inclusion. *Angew. Chem. Int. Ed.* 46: 2831–2834.
41. Kudernac, T., Lei, S. B., Elemans, J. A. A. W. et al. 2009. Two-dimensional supramolecular self-assembly: Nanoporous networks on surfaces. *Chem. Soc. Rev.* 38: 402–421.
42. Theobald, J. A., Oxtoby, N. S., Phillips, M. A. et al. 2003. Controlling molecular deposition and layer structure with supramolecular surface assemblies. *Nature* 424: 1029–1031.
43. Madueno, R., Raisanen, M. T., Silien, C. et al. 2008. Functionalizing hydrogen-bonded surface networks with self-assembled monolayers. *Nature* 454: 618–621.
44. Lackinger, M. and Heckl, W. M. 2009. Carboxylic acids: Versatile building blocks and mediators for two-dimensional supramolecular self-assembly. *Langmuir* 25: 11307–11321.
45. Lei, S., Tahara, K., De Schryver, F. C. et al. 2008. One building block, two different supramolecular surface-confined patterns: Concentration in control at the solid–liquid interface. *Angew. Chem. Int. Ed.* 47: 2964–2968.
46. Walch, H., Maier, A. K., Heckl, W. M. et al. 2009. Isotopological supramolecular networks from melamine and fatty acids. *J. Phys. Chem. C* 113: 1014–1019.
47. Meier, C., Roos, M., Kunzel, D. et al. 2010. Concentration and coverage dependent adlayer structures: From two-dimensional networks to rotation in a bearing. *J. Phys. Chem. C* 114: 1268–1277.
48. Gutzler, R., Sirtl, T., Dienstmaier, J. F. et al. 2010. Reversible phase transitions in self-assembled mono layers at the liquid–solid interface: Temperature-controlled opening and closing of nanopores. *J. Am. Chem. Soc.* 132: 5084–5090.
49. Mammen, M., Shakhnovich, E. I., Deutch, J. M. et al. 1998. Estimating the entropic cost of self-assembly of multiparticle hydrogen-bonded aggregates based on the cyanuric acid center dot melamine lattice. *J. Org. Chem.* 63: 3821–3830.
50. Gilli, P., Bertolasi, V., Ferretti, V. et al. 1994. Covalent nature of the strong homonuclear hydrogen-bond—Study of the O–H–O system by crystal-structure correlation methods. *J. Am. Chem. Soc.* 116: 909–915.
51. Grimme, S. 2006. Semiempirical GGA-type density functional constructed with a long-range dispersion correction. *J. Comput. Chem.* 27: 1787–1799.
52. Schwabe, T. and Grimme, S. 2007. Double-hybrid density functionals with long-range dispersion corrections: Higher accuracy and extended applicability. *Phys. Chem. Chem. Phys.* 9: 3397–3406.
53. Weber, U. K., Burlakov, V. M., Perdigao, L. M. A. et al. 2008. Role of interaction anisotropy in the formation and stability of molecular templates. *Phys. Rev. Lett.* 100: 156101.

54. Martsinovich, N. and Troisi, A. 2010. Modeling the self-assembly of benzenedicarboxylic acids using Monte Carlo and molecular dynamics simulations. *J. Phys. Chem. C* 114: 4376–4388.
55. Bieri, M., Nguyen, M. T., Groning, O. et al. 2010. Two-dimensional polymer formation on surfaces: Insight into the roles of precursor mobility and reactivity. *J. Am. Chem. Soc.* 132: 16669–16676.
56. Kampschulte, L., Lackinger, M., Maier, A. K. et al. 2006. Solvent induced polymorphism in supramolecular 1,3,5-benzenetribenzoic acid monolayers. *J. Phys. Chem. B* 110: 10829–10836.
57. Marie, C., Silly, F., Tortech, L. et al. 2010. Tuning the packing density of 2D supramolecular self-assemblies at the solid–liquid interface using variable temperature. *ACS Nano*. 4: 1288–1292.
58. Li, Y. B., Ma, Z., Qi, G. C. et al. 2008. Solvent effects on supramolecular networks formed by racemic star-shaped oligofluorene studied by scanning tunneling microscopy. *J. Phys. Chem. C* 112: 8649–8653.
59. Gutzler, R., Lappe, S., Mahata, K. et al. 2009. Aromatic interaction vs. hydrogen bonding in self-assembly at the liquid–solid interface. *Chem. Commun.* 2009: 680–682.
60. Yang, Y. and Wang, C. 2009. Solvent effects on two-dimensional molecular self-assemblies investigated by using scanning tunneling microscopy. *Curr. Opin. Colloid In*. 14: 135–147.
61. English, W. A. and Hipps, K. W. 2008. Stability of a surface adlayer at elevated temperature: Coronene and heptanoic acid on Au(111). *J. Phys. Chem. C* 112: 2026–2031.

6 Pursuit of Long-Lasting Oxygen-Evolving Catalysts for Artificial Photosynthesis

Self-Healing Materials and Molecular-Reinforced Structures

Serena Berardi, Andrea Sartorel, Mauro Carraro, and Marcella Bonchio

CONTENTS

6.1 ARTIFICIAL PHOTOSYNTHESIS

The search for renewable energy sources is becoming day by day more urgent for mankind. The global energy demand is indeed predicted to rise considerably within the next decade, and fossil fuels alone cannot satisfy this energy thirst, due to their limited availability and the Earth pollution caused by their combustion.

Among renewable energy sources, solar light is the most abundant, cheap, and equally distributed. In order to be used, it has to be transformed into other forms that can be subsequently stored and transported.

Artificial photosynthesis converts solar light into chemical energy, by performing the light-driven water splitting into its components: hydrogen (H_2) and oxygen (O_2) (see Scheme 6.1) (Armaroli and Balzani 2007; Balzani et al. 2008). H_2 and O_2 can then be recombined to release their chemical energy (the amount of free energy ΔG associated with the reaction of 2 mol of H_2 with 1 mol of O_2 is 113.38 kcal).

The light-driven splitting of water is actually a complex process, where components dealing with light absorption, energy and electron transfers, and redox catalysis are assembled in a modular approach (Figure 6.1).

Light is first harvested by an antenna, and the energy is transferred to a photosensitizer (P); the latter, in its excited state, promotes an electron transfer from a suitable donor (D) to a suitable acceptor (A); such charge separation is the crucial step of the overall process. Indeed, electrons in A are converged to a hydrogen-evolving catalyst (HEC), able to reduce protons to H_2; at the same time, in the other half-cell, holes in D are used for water oxidation through the intervention of an oxygen-evolving catalyst (OEC).

$$2H_2O \xrightarrow{h\nu} 2H_2 + O_2$$

SCHEME 6.1 Light-driven water splitting into H_2 and O_2.

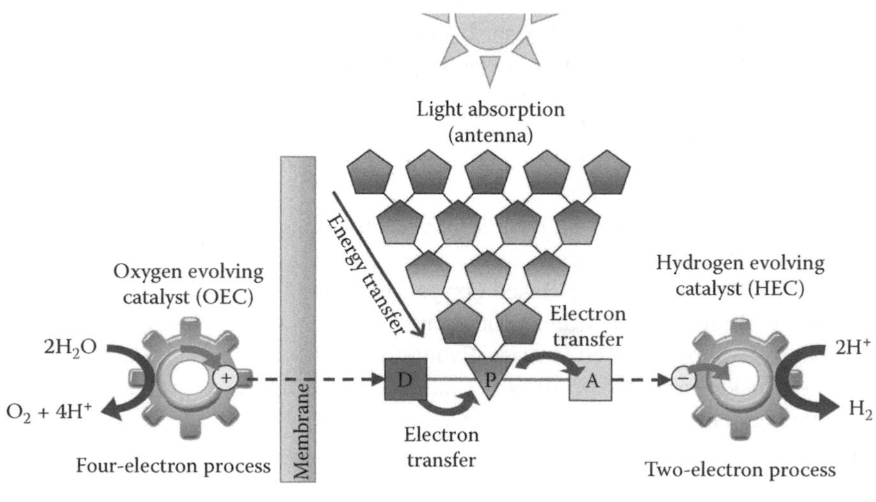

FIGURE 6.1 Schematic representation of the key processes in artificial photosynthesis.

$$\begin{array}{lcl}
\mathrm{OEC} + \mathrm{D(h^+)} & \longrightarrow & \mathrm{OEC^+} + \mathrm{D} \\
\mathrm{OEC^+} + \mathrm{D(h^+)} & \longrightarrow & \mathrm{OEC^{2+}} + \mathrm{D} \\
\mathrm{OEC^{2+}} + \mathrm{D(h^+)} & \longrightarrow & \mathrm{OEC^{3+}} + \mathrm{D} \\
\mathrm{OEC^{3+}} + \mathrm{D(h^+)} & \longrightarrow & \mathrm{OEC^{4+}} + \mathrm{D} \\
\mathrm{OEC^{4+}} + 2\mathrm{H_2O} & \longrightarrow & \mathrm{OEC} + \mathrm{O_2} + 4\mathrm{H^+} \\
\hline
4\mathrm{D(h^+)} + 2\mathrm{H_2O} & \xrightarrow{\mathrm{OEC}} & 4\mathrm{D} + \mathrm{O_2} + 4\mathrm{H^+}
\end{array}$$

SCHEME 6.2 Subsequent 1 e^- oxidations of OEC by photogenerated holes $D(h^+)$ followed by water oxidation by the active form of OEC (i.e. OEC^{4+}).

In order to efficiently exploit solar radiation and to avoid unproductive charge recombination, all these events need to be concerted, and therefore, fast catalysis is needed. In particular, the oxidative half-reaction leading to oxygen is by far more difficult to achieve than the reduction to hydrogen, since it is a $4e^-/4H^+$ process, which also requires the formation of an oxygen–oxygen bond. The OEC is usually constituted by one or several redox active metals, since it should act as a charge pool, by reacting four times with the photogenerated holes in D, prior to transforming water into oxygen in a single step, as described in Scheme 6.2.

Besides the chemical difficulties of these steps, the reactions shown above occur in a very oxidizing environment, since the standard redox potential for the O_2/H_2O couple is 1.23 V vs. normal hydrogen electrode (NHE). These conditions may lead to oxidative damage of the catalyst, with consequent loss of its activity. Therefore, strategies to preserve catalyst activity under turnover conditions must be considered and adopted. In this chapter, we will focus mainly on two approaches:

1. Oxygen-evolving catalysts able to self repair
2. Oxygen-evolving catalysts that are robust and resist under oxidizing conditions

6.2 SELF-HEALING OXYGEN-EVOLVING CATALYSTS

6.2.1 BIOINSPIRED GUIDELINES: THE NATURAL OXYGEN-EVOLVING CENTER, $CaMn_4O_x$

The adoption of a catalytic core, featuring adjacent multitransition metal centers connected by μ-hydroxo/oxo-bridging units, is a winning strategy devised by *Nature* to effect multiple/cascade transformations with minimal energy cost. Indeed, water oxidation occurs with unmatched efficiency at the heart of the Photosystem II (PSII) enzyme, where O_2 evolution is catalyzed by a polynuclear metal-oxo cluster with four manganese and one calcium atom ($CaMn_4O_x$). The structure of the PSII–OEC has been recently addressed by a combined crystallographic, spectroscopic, and computational approach (Figure 6.2a) (Ferreira et al. 2004; Loll et al. 2005; Yano et al. 2006). The more accredited topology involves a $CaMn_3O_4$ fragment, with a trigonal pyramidal arrangement, linked to a fourth "dangling" Mn center via oxo bridges (Figure 6.2a). After submission of this chapter, a new contribution confirmed

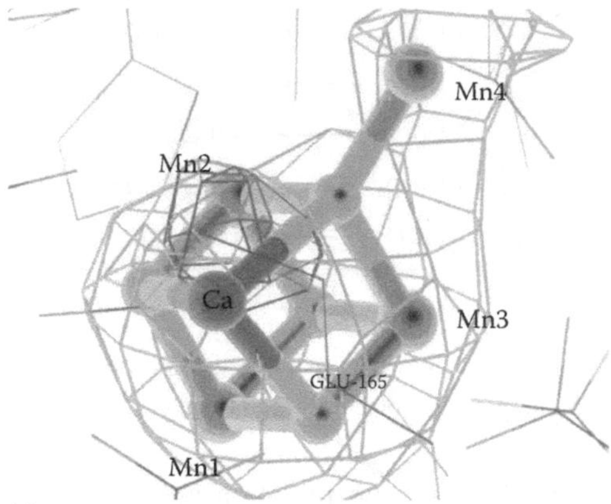

(a)

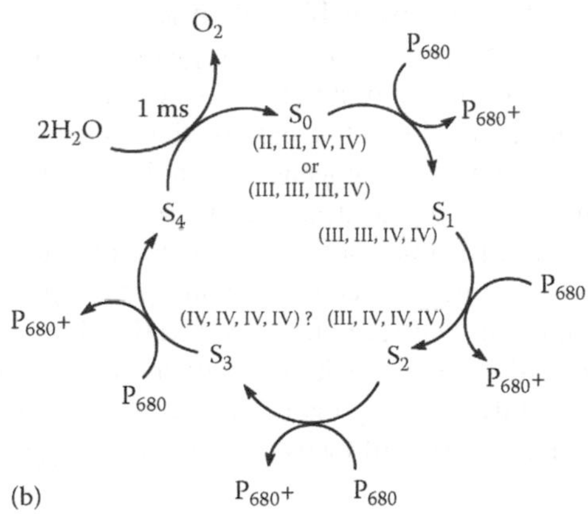

(b)

FIGURE 6.2 (a) Schematic view of the oxygen-evolving center in PSII. The figure was kindly provided by Professor James Barber (Imperial College, London, U.K.) based on the coordinates PDB file 1S5L for the crystal structure of PSII isolated from the cyanobacterium, *Thermosynechococcus elongatus* (Ferreira et al. 2004). (b) The Kok cycle of the oxygen-evolving center in PSII (Kok et al. 1970).

this hypothesis by refining the structure of the $CaMn_4O_x$ cluster at 1.9 Å resolution, revealing the presence of five oxygen bridges and of four water molecules as apical ligands of the metals (Umena et al. 2011).

The OEC in PSII can master the $4e^-/4H^+$ mechanism required for oxygen production through sequential redox steps, thus featuring a unique flexibility of the structural metal core by assistance of the nanostructured proteic environment. The mechanism of water oxidation by the PSII–OEC has been extensively investigated (Haumann et al. 2005; McEvoy and Brudvig 2006; Brudvig 2008; Hammarstrom and Hammes-Schiffer 2009). Although several details still need to be clarified, a general consensus has emerged, indicating that a light-induced four electron oscillation pattern, along five oxidation states S_n ($n = 0$–4), is responsible for the catalytic cycling of the oxygenic $CaMn_4O_x$ core. In this scheme, S_0 is the most reduced and S_4 the four electron oxidized state; when S_4 is generated, it reacts rapidly (1 ms) releasing O_2 and returning to the initial S_0 form of the enzyme (Figure 6.2b; Kok et al. 1970). The photooxidation cycle is driven by a chlorophyll pigment radical cation, called P_{680}^{+}, generated upon illumination (680 nm is the optimal absorption wavelength), which is by far the strongest oxidizing agent known in living systems ($P_{680}/P_{680}^{+} \approx 1.25$ V) (McEvoy and Brudvig 2006).

Within a perfectly merged hybrid nano-environment, oxygenic photosynthesis turns out to effect a solar-activated $4e^-/4H^+$ catalytic routine, with multi-turnover (TON) efficiency yielding up to ca. 400 TON per second, with overall quantum yield close to 10% and operating at a moderate overpotential (ca. 0.3–0.4 V at physiological pH). Herein, the Achilles' heel stems from the intrinsic weakness of the functional components chosen by *Nature* to master such peak performance. Despite some elaborate control/protection strategies, under turnover regime, the PSII enzyme undergoes to fatal damage, thus requiring a perpetual self-healing reconstruction every ca. 30 min (Aro et al. 2005).

6.2.2 Synthetic OEC: Nocera's Co-Pi Catalyst

In 2008, Prof. D. G. Nocera and M. W. Kanan at the Massachusetts Institute of Technology reported the preparation of a novel catalytic electrode, capable of performing water oxidation under benign conditions at low overpotentials (Kanan and Nocera 2008). They observed that dipping an indium–tin oxide (ITO) or a fluorine–tin oxide (FTO) electrode into a neutral (pH 7) aqueous solutions containing Co(II) and phosphate ions, and upon subsequent application of a bias at 1.3 V (vs. NHE), a dark green–black film forms on the electrode surface; the coating was observed by scanning electron microscopy (SEM) to be constituted by coalescence of individual particles of micrometer size, which are visible on the top of the film (Figure 6.3). Deposition of the film is accompanied by increasing vigorous bubbling at the electrode surface, due to water oxidation to oxygen. The thickness of the deposited layer depends on the duration of the bias application and on the initial cobalt concentration; usually prolonged electrolysis produces ca. 3 μm thick films. The microelemental analysis of the film indicated the presence of cobalt and phosphorous in a rough 2:1 ratio; thus, it was possible to argue that the film was plausibly constituted by amorphous Co oxide or hydroxide incorporating a substantial amount of phosphate. Deposition occurs immediately after oxidation of Co(II) to Co(III), observed at 1.14 V vs. NHE under the conditions of preparation of the material, and phosphate plays an important role in this process. Indeed, formation of the film is only observed in protic buffers, such as methylphosphonate and borate (Surendranath et al. 2009). Nevertheless, it is worth noting that partial decomposition of methylphosphonate to phosphate is observed, while the borate media induces a different morphology of the surface, with spherical nodules forming in the early stage of deposition, and merging into larger aggregates upon prolonged electrolysis (Kanan et al. 2009; Surendranath et al. 2009).

6.2.3 Electrocatalytic Oxygen Evolution

The cobalt phosphate film (Co-Pi) displays significant activity in water oxidation. The catalytic tests were performed in neutral water using electrochemical

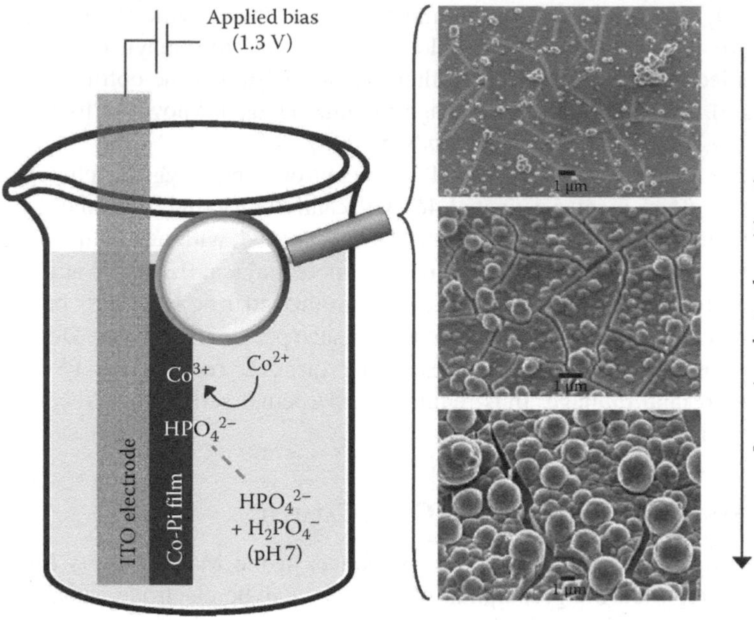

FIGURE 6.3 Schematic procedure for the preparation of the cobalt phosphate catalyst (Co-Pi), and SEM images of the deposited layer. (The SEM images are taken from Kanan, M.W., Surendranath, Y., Nocera, D.G., Cobalt–phosphate oxygen-evolving compound, *Chem. Soc. Rev.*, 38, 109–114, 2009. Reproduced by permission of The Royal Society of Chemistry.)

techniques, in order to evaluate the potential required to observe water oxidation and the amount of oxygen production under prolonged electrolysis. If the Co-Pi-doped electrode is used as the working electrode in a simple cyclic voltammetry experiment, in phosphate buffer at pH 7 under air atmosphere, the onset of the catalytic wave due to water oxidation is observed at 1.1 V vs. NHE (Kanan and Nocera 2008), corresponding to an overpotential of 280 mV (the theoretical potential for the O_2/H_2O couple at pH 7 is 0.82 V, given by the equation $E = 1.23–0.0592 \cdot \text{pH}$). Very high overpotentials are instead observed when bare ITO or FTO electrodes are used as the working electrode.

By performing prolonged electrolysis at 1.29 V (corresponding to 470 mV overpotential), oxygen generation is observed for several hours, until the removal of the bias; the amount of oxygen greatly exceeded the amount of catalyst used: 95 μmol of O_2 were indeed produced after ca. 10 h, with a total amount of catalyst of ca. 0.2 mg. According to the elemental composition of the material (31.1% Co), it is possible to estimate the amount of Co as 1.05 μmol, with a total turnover number per metal center of 90. Noteworthy, all the current passed was used for oxygen production, while the catalytic film does not show any major degradation over the course of the experiment. The doped electrodes can also be stored under ambient atmosphere, and subsequently used in cobalt-free solutions. Notably, oxygen evolution is not limited by the presence of chloride anions in solutions (no Cl^- oxidation to Cl_2 is observed), allowing efficient oxygen evolution also from sea water.

The Co-Pi catalyst has several important features, making it a very attractive and promising candidate for potential applications in a device for artificial photosynthesis: (i) it is constituted by earth-abundant and cheap elements, thus allowing its production on a large scale; (ii) it self-assembles upon application of an electrochemical potential; (iii) it is all-inorganic and therefore robust toward oxidative stress; (iv) it can manage proton exchange events that play an important role in the catalytic cycle of water oxidation; (v) it works in neutral water, at low overpotentials; (vi) it self-repairs under catalytic conditions; and (vii) it can be supported onto semiconductors, in order to achieve light-driven water oxidation. It is worth to highlight that the aforementioned features (i)–(vi) are also observed for the oxygen-evolving center of PSII, thus allowing a comparison of the functional properties of the natural and manmade catalysts.

Another catalyst, similar to Co-Pi, was deposited onto inert electrodes, using nickel instead of cobalt (Dincă et al. 2010). Repetitive cyclic voltammetry experiments on a 1 mM Ni(II) solution in 0.1 M borate buffer at pH 9.2 show an anodic wave at 1.02 V attributed to a $2e^-/3H^+$ oxidation of the nickel centers, coupled with a dimerization process. Immediately following this electrochemical process, the catalytic wave due to water oxidation is initially observed at 1.2 V, and then shifted to 1.15 V under repetitive scans, corresponding to an overpotential of 465 mV (at pH 9.2, the redox potential for the O_2/H_2O couple is 0.685 V).

The catalyst film was prepared by application of a 1.3 V bias in Ni(II) solutions in borate buffer, at pH 9.2; a 3–4 mm thick layer is readily deposited, for which elemental analysis provides a minimal formula of $Ni(III)O(OH)_{2/3}(H_2BO_3)_{1/3} \cdot 1.5H_2O$. The catalytic performance of the Ni–borate-doped electrode was studied in borate buffer at pH 9.2, and revealed an overpotential of ca. 425 mV, with quantitative Faradic efficiency of the current (totally used for oxygen production) and good stability of the catalytic film (no corrosion was observed). Nevertheless, the high overvoltage required to observe catalytic activity and the nonneutral operative conditions of the doped electrode make the nickel-based catalyst less attractive with respect to the cobalt one.

6.2.4 Structure of the Co-Pi Catalyst and Mechanism of Oxygen Production

Elucidation of the chemical structure and valency of the Co-Pi film during catalysis is an essential step toward gaining a mechanistic understanding of water oxidation.

In 2009, Dau and coworkers proposed a structure for the title catalytic layer prepared by cathodic electrodeposition of phosphate buffer (pH 7) containing Cobalt (II) nitrate at 1.4 V vs. NHE, and removed still wet from the solution after 70 min (Risch et al. 2009). X-ray absorption near-edge structure (XANES) spectroscopy of the frozen wet electrode was then performed. The shape of the resulting spectrum reflects the ligand type and coordination of the X-ray-absorbing metal. The results suggest a Co mean oxidation state of 3+ and a near-octahedral coordination of the metal by six oxygen atoms. Extended X-ray absorption fine structure (EXAFS) analysis of the catalyst also confirms the prevalence of $Co(III)O_6$ ligations in the film. Moreover,

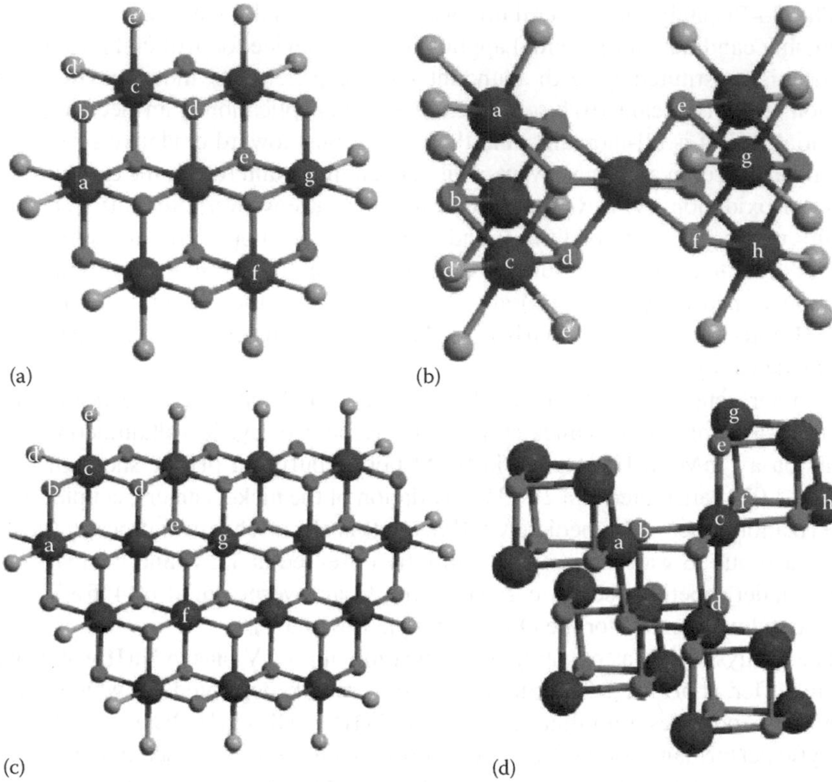

FIGURE 6.4 Structural models for Co-Pi. Bridging oxo/hydroxo ligands are shown in grey; nonbridging oxygen ligands (including water, hydroxide, and phosphate) complete the octahedral coordination geometry of each peripheral Co ion (dark grey) and are shown in light grey. (a) Edge-sharing MCC model for surface Co-Pi. (b) Corner-sharing cubane model for surface Co-Pi. (c) Edge-sharing MCC model for bulk Co-Pi. (d) Corner-sharing cubane model for bulk Co-Pi. (Reprinted with permission from Kanan, M.W., Yano, J., Surendranath, Y., Dincă, M., Yachandra, V.K., Nocera, D.G., Structure and valency of a cobalt-phosphate water oxidation catalyst determined by *in situ* X-ray spectroscopy, *J. Am. Chem. Soc.*, 132, 13692–13701, 2010. Copyright 2010 American Chemical Society.)

the observed Co–Co distance (2.81 Å) is typical for Co atoms connected by di-μ-oxo bridges, like in Co cubanes (Dimitrou et al. 1993; Ama et al. 2000). Considering all these data, Dau proposed the structural motif reported in Figure 6.4b for the bulk catalyst that consist of clusters of complete or incomplete Co-oxo cubanes, sharing Co corners.

Moreover, the X-ray absorption spectroscopy, XAS (EXAFS + XANES) data exclude direct Co-P bonds, and also phosphate oxygens as bridging ligands for cobalt atoms. Terminal phosphate ligation is instead conceivable, as well as potassium ligation to three Co-bridging oxygen, leading to a $Co_3K(\mu\text{-}O)_4$ cubane moiety, similar to the $Mn_3Ca(\mu\text{-}O)_4$ catalytic motif of PSII described earlier (Ferreira et al. 2004; Dau et al. 2008; Sproviero et al. 2008).

In 2010, the group of Nocera reported an *in situ* X-ray study on the title cobalt catalyst that led to a different structural model of that reported by Dau and coworkers (Kanan et al. 2010). In particular, the measurements were made for two different samples, a thin one and a bulk one, both deposited onto ITO electrodes. The former was prepared by electrodepositions of 0.1 M phosphate buffer solutions containing 0.5 mM of Co(II), carried out at 1.1 V (vs. NHE). At this potential, H_2O oxidation activity of the deposited catalyst is minimal, and essentially all of the current can be attributed to Co(II)/Co(III) oxidation events. In this case, few monolayers of Co-Pi film are deposited, so nearly all of the Co-Pi material is exposed to the electrolyte during catalysis. This extremely thin sample can thus mimic the surface of a thicker sample.

The bulk Co-Pi sample was instead prepared by performing electrodepositions at a higher potential (1.25 V vs. 1.1 V used for the preparation of the thin sample). At this potential, deposition occurs in parallel with water oxidation, so the amount of charge passed does not directly indicate the amount of Co(III) ions deposited.

After the depositions, a potential of 1.25 V was applied to both the electrodes (the thin and the bulk) and XAS (EXAFS and XANES) spectra were collected *in situ*, since the electrolysis cell is modified in order to have an X-ray transparent window, facing the ITO/Co-Pi coating. The cell was then switched to open circuit, and then switched back to 1.25 V.

Both the XANES edge position and the EXAFS spectra of the Co-Pi catalysts show similarities to those of the Co(III) oxide model, that is, CoO(OH), suggesting that a viable model for Co-Pi may consist of Co-oxo/hydroxo clusters composed of edge-sharing CoO_6 octahedra, the basic structural components of cobaltates (Ama et al. 2000). In particular, the observed values of Co–O vector (1.89 Å), Co–Co vector (2.82 Å), and the number of neighbors in each shell, *N* value (ca. 3.4 in surface and ca. 4.5 in bulk Co-Pi) are consistent with a molecular cobaltate cluster (MCC, i.e., a non-extended structure) containing seven Co ions, in which the Co_2O_2 units result planar (Figure 6.4a).

It is worth noting that the XAS data for the thin Co-Pi film are consistent with both Nocera's and Dau's models (corner-sharing cubane). Nevertheless, significant differences between the surface and bulk Co-Pi EXAFS spectra are better accounted by an MCC model, for which the *N* values for the nearest-neighbor Co–Co fit better. In Figure 6.4c and d, both bulk MCC and bulk corner-sharing cubane Co-Pi are represented.

The MCC model over a corner-sharing cubane model can also be argued on the basis of the Co/H_2O Pourbaix diagram (Chivot et al. 2008) and of structural data for other synthetic Co-oxo compounds. For example, the same edge-sharing CoO_6 octahedral structural motif of the MCC model is found in the reduced Co_4O_{16} core of a Co polyoxometalate (POM) recently reported as water oxidation catalyst, see next paragraph (Yin et al. 2010).

As already stated, the absence of a Co-P vector is also pointed out by means of EXAFS analysis. This observation reflects the absence of direct phosphate-Co binding in Co-Pi, but it is possible, however, to hypothesize coordination of exchangeable HPO_4^{2-} to the terminal sites of the outer ring of the cobalt ions.

In situ XAS analysis also allow obtaining information on the Co valency during catalysis and valency changes when the potential is removed. In particular, for the title catalyst, EXAFS data are consistent with a Co valency ≥3. Moreover, by means of cyclic voltammetry, the formal oxidation state of cobalt in Co-Pi film is reported to be 3+ prior to catalysis. A catalytic wave is obtained upon further oxidation of the Co(III) film, as confirmed by the detection of Co(IV) by means of electronic paramagnetic resonance (EPR) studies (vide infra) (McAlpin et al. 2010). The analysis of the XANES spectra also indicate a continuous reduction of the Co-Pi catalyst upon successive switching from a potential sufficient for sustained water oxidation (1.25 V) to open circuit. This kind of activity is in contrast with the properties of solid-state cobaltate materials with high Co valencies (Takada et al. 2004; Sakurai et al. 2006, 2007; Ren et al. 2007) and suggests that the molecular dimensions of the clusters found in Co-Pi may be essential for catalysis.

As mentioned earlier, Nocera and coworkers also reported an EPR study on the Co-Pi catalyst films (McAlpin et al. 2010). The registered EPR signals correspond to populations of both paramagnetic Co(II) and Co(IV) species. By increasing the deposition voltage (in order to support also water oxidation), the population of low-spin Co(IV) containing species rises, concurrently with a decrease in the population of Co(II) species. This potential dependence suggests that Co(IV) species are predominantly generated during electrocatalytic water oxidation at neutral pH.

Further studies on the speciation of the Co-Pi catalyst film will be needed in order to better understand the mechanism of the oxygen evolution reaction (OER). In particular, Nocera and coworkers reported the electrochemical kinetics and ^{18}O isotope experiments of the OER in neutral water catalyzed by ultrathin (<100 nm) Co-Pi films, prepared by low-potential electrodeposition in order to decrease both mass and charge transport limitations to the OER (Surendranath et al. 2010).

The Co-Pi catalysts exhibit a Tafel slope approximately equal to 2.3 × RT/F (59 mV/decade) for the O_2 evolution from water in neutral solutions, supporting a mechanism involving a reversible one-electron transfer prior to the chemical turnover-limiting step (Gileadi 1993).

The rate law for the OER catalysis is expected to be pH dependent, since the reaction itself involves the removal of four protons per equivalent of O_2 generated. In particular, both potentiostatic and galvanostatic experiments indicates an inverse first-order dependence of the current density on proton activity, consistent with the loss of a single proton in an equilibrium step, prior to the turnover-limiting process.

Moreover, the electrochemical rate law exhibits also a zeroth-order dependence of the current density on phosphate, that is, on the proton-accepting electrolyte.

Hence, all these electrokinetic studies suggest a mechanism involving a rapid and reversible $1e^-/1H^+$ equilibrium followed by the chemical turnover-limiting step, involving oxygen–oxygen bond coupling.

In particular, considering that the deposition process entails the oxidation of the initial Co(II) to Co(III), the pre-equilibrium may involve the redox transition between Co(III)-OH and Co(IV)-O in the Co(III)–Co(IV) mixed valence clusters (Figure 6.5).

Another interesting aspect of the OER concerns the nature of the O–O bond formation, that is, the rate-limiting step that has been studied by means of ^{18}O-labeling experiments.

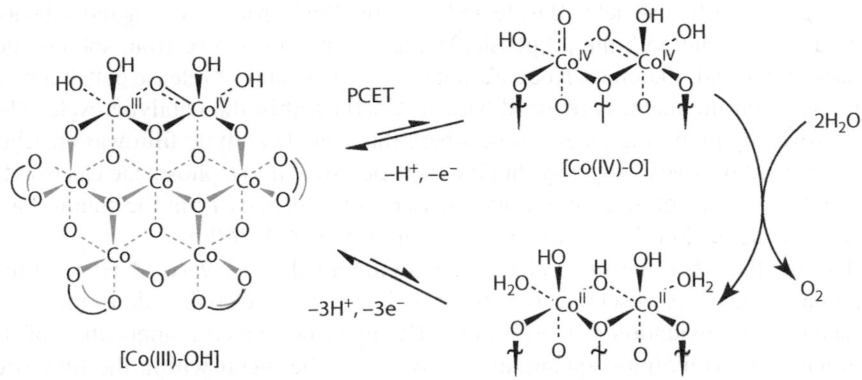

FIGURE 6.5 Proposed pathway for OER catalyzed by Co-Pi. A proton-coupled electron transfer (PCET) pre-equilibrium is followed by the turnover-limiting O–O bond forming step. Curved lines denote phosphate, or OH_x terminal or bridging ligands. (Reprinted with permission from Surendranath, Y., Kanan, M.W., Nocera, D.G., Mechanistic studies of the oxygen evolution reaction by a cobalt-phosphate catalyst at neutral pH, *J. Am. Chem. Soc.*, 132, 16501–16509, 2010. Copyright 2010 American Chemical Society.)

^{18}O-enriched catalyst films were prepared by electrodepositions at 1.10 V of Co(II) in a phosphate buffer enriched with 87% $^{18}OH_2$. The electrodes, covered with the thin film, were then subjected to electrolysis at 1.3 V in the presence of Co(II)-free phosphate buffer, enriched with 87% $^{18}OH_2$. Upon conclusion of electrolysis, mass spectrometry analyses were then performed, and $^{32}O_2$, $^{34}O_2$, and $^{36}O_2$ signals detected.

Isotopic distributions of the O_2 evolved from these $^{18}OH_2$-enriched films indicate that a significant percentage of the estimated total ^{18}O contained in the films was progressively extruded in the form of $^{34}O_2$ and, to a lesser extent, $^{36}O_2$, concomitant with the catalytic production of a large excess of $^{32}O_2$. These data are consistent with a slow exchange of the label from catalyst films to the bulk solvent, and the participation of μ-oxo/μ-hydroxide moieties in O_2 production. However, mechanistic details on the formation of the O–O bond still remain to be determined.

6.2.5 Self-Healing Mechanism

Metal-based catalysts involved in redox transformations are often subject to ligand exchange and partial structural reorganization, both factors that may preclude the stability and long-term performance of the catalyst itself. Water oxidation in particular poses severe challenges to the integrity of catalysts, since it is a $4e^-/4H^+$ process taking place under highly oxidizing conditions, and therefore the possibility of self-repairing the catalyst is highly desirable. In particular, the necessity to repair the cobalt-based oxygen-evolving film is due to the different nature of the oxidation states of the metal involved in the catalytic cycle, namely, Co(II), Co(III), and Co(IV) (Shafirovich et al. 1980; Brunschwig et al. 1983). Indeed, Co(II) is usually a high spin ion, whereas Co(III) and superior state are low spin; as a consequence, Co(II) is labile to ligand substitution, whereas Co(III) and Co(IV) are quite inert in an oxygen atom ligand field (Basolo and Pearson 1967). Moreover, ligand substitution rates correlate with the propensity of metal ions to dissolve from solid oxides (Casey 1991), and, therefore, Nocera's water oxidation catalyst release cobalt ions in solution when the metal is reduced back to Co(II) within the catalytic cycle. This was proved by preparing an electrode where the coated catalytic film was enriched with ^{57}Co radioactive isotope; by dipping such electrode into a phosphate electrolyte, and in the absence of an external bias, increase of radioactivity in the solution was observed, due to slow dissolution of ^{57}Co (Lutterman et al. 2009).

Differently, when the same electrode is maintained at 1.3 V (vs. NHE), no dissolution of cobalt is observed until the bias is removed. Cobalt redeposition may be achieved by oxidation of Co(II) to Co(III) upon prolonged reapplication of the external bias, with almost quantitative recovery of the metal within the film after 14 h (0.002% of cobalt still present in solution).

By performing similar isotope-labeling experiments, the authors clarified the role of phosphate, which is actually responsible for the self-repairing of the catalyst. Indeed, by enriching the catalytic film with ^{32}P radioactive isotope, it was possible to monitor phosphorous exchange in the presence and in the absence of the applied bias. The rate of ^{32}P leaching in solution by application of 1.3 V potential was two times lower than in the absence of the external bias. Accordingly, dipping a non-isotopically enriched coated electrode into a phosphate solution containing ^{32}P resulted in a superior enrichment of ^{32}P in the catalytic film in the absence of applied potential rather than in the presence of 1.3 V bias. In summary, higher phosphate exchange between the catalytic film and the solution is observed in the absence of an external potential, with respect to electrodes where a 1.3 V bias is applied. The reason why phosphate exchange is higher than cobalt one is probably ascribed to the involvement of the latter in robust metal–oxygen frameworks.

Phosphate electrolyte ensures then the stability of the catalytic film, by promoting redeposition of Co(III), generated from electrochemical oxidation of Co(II) ions leached in solution.

Cobalt leaching is indeed irreversible in the absence of phosphate or other proton-accepting electrolytes; in this case, cobalt dissolution increases with increasing the applied potential, as a consequence of Co-Pi film corrosion by the protons produced together with water oxidation. Furthermore, adding phosphate to the corroding films leads to rapid redeposition of Cobalt onto the catalyst film.

6.2.6 Applications of the Co-Pi Catalyst in Light-Driven Water Oxidation

An important step forward for the application of Nocera's catalyst in a device for artificial photosynthesis is represented by its assembling in a photoanode, in order to achieve light-driven water oxidation. This can be accomplished by depositing the Co-Pi catalytic film onto suitable semiconductors (see scheme below, Figure 6.6); upon light absorption, an electron is promoted from the valence to the conduction band of the semiconductor, and the hole in the valence band is filled by the electrons of the cobalt catalyst. As a result, the latter is transformed into its oxidized

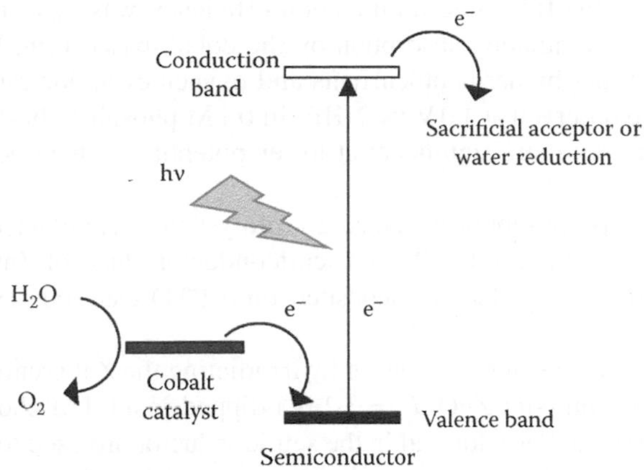

FIGURE 6.6 Schematic representation of light-driven oxygen production by coupling the Co-Pi catalyst with a semiconductor.

form, capable of extracting electrons from water, thus producing oxygen. The electron promoted in the conduction band of the semiconductor is generally transferred to a sacrificial acceptor, but in principle it could also be used for water reduction to generate hydrogen (Figure 6.6). The energy level of the orbitals and the band gap of the semiconductor play a major role in the efficiency of the overall process.

In the recent literature, a couple of examples dealing with anchoring the Co-Pi film onto semiconductors were reported. Gamelin et al. prepared a composite photoanode by electrodeposition of Nocera's catalyst onto mesostructured hematite (α-Fe_2O_3)-based electrodes (Zhong et al. 2009; Sun et al. 2010; Zhong and Gamelin 2010). Hematite is indeed inexpensive and resistant to oxidative stress, while it absorbs visible light ($E_g = 2.1$ eV) (Kay et al. 2006); the major drawback is that the potential of the conduction band is not sufficient to drive water reduction to hydrogen, so the system needs to be coupled with a second photoelectrochemical electrode. The electrodeposition of Nocera's catalyst onto hematite was performed under similar conditions with respect to the original work, by application of a 1.29 V bias (vs. NHE) for 1 h to a hematite electrode dipped in a phosphate buffer containing Co(II) ions. The catalytic film obtained under these conditions is ca. 200 nm thick and conforms to the topology of the hematite surface, as revealed by SEM images. By measuring photocurrents as a function of the applied voltage under simulated 1 sun AM1.5 solar irradiation, the resulting photoanode shows a >350 mV reduction of the bias voltage required for promoting water oxidation, with respect to simple hematite in 1 M aqueous NaOH (pH 13.6). Cobalt leaching in solution was excluded by several converging tests, including continuous photocatalysis for 10 h (Zhong et al. 2009). Subsequent improvements of this system include: (i) reduction of nonproductive photon absorption by the catalyst, (ii) use of the photoanode at pH 8, and (iii) enhancement of oxygen evolution (Zhong and Gamelin 2010). All these goals were achieved by reducing the time of Co-Pi electrodeposition onto hematite, from the original 1 h to 15–30 min. These conditions lead to the deposition of smaller patches of the catalyst (<<100 nm thick), rather than obtainment of the 200 nm thick film with 1 h electrodeposition; the front side illumination efficiency was then improved due to minor unproductive radiation absorption by the cobalt-based film. With such photoanodes, ca. 5 times higher photocurrents and oxygen evolution rates with respect to hematite were observed at 1.0 V vs. NHE, in 0.1 M phosphate buffer at pH 8. The improvement is even more significant at lower potentials, where hematite alone is almost inactive.

A second example of supporting Nocera's catalyst onto semiconductors deals with photochemical deposition of Co-Pi on a semiconductor phase of ZnO rods (100 nm diameter, 600 nm average length) deposited onto FTO electrode (Steinmiller and Choi 2009).

Photochemical deposition is achieved by irradiating the ZnO semiconductor with UV light (λ = 302 nm—for ZnO, E_g = 3.2 eV) dipped in a 0.1 M phosphate solution containing Co(II); the holes formed in the semiconductor are used to oxidize Co(II) to Co(III), causing its deposition onto the electrode surface. Concurrently, the electrons promoted in the conducting band of ZnO are consumed by the reduction of water or dissolved O_2. Interestingly, after 30 min of irradiation, the catalyst is deposited as 10–30 nm nanoparticles, as revealed by SEM images.

Indeed, this method overcomes electrochemical deposition because it avoids the use of an external bias to produce the catalytic phase, and it places the cobalt catalyst where the holes are most readily available, thus reducing the total amount of catalyst deposited and therefore its interference with photon absorption from the semiconductor. Also, photodeposition could be an alternative to electrodeposition when the semiconductor is subject to dissolution or decomposition upon application of an anodic bias; moreover, unnecessary deposition of the catalyst directly onto the inert FTO electrode is avoided (catalyst deposited on the FTO electrode cannot participate in light-driven O_2 evolution). The performance of the ZnO photoanodes in photocurrent generation was investigated in 0.1 M phosphate buffer solution at pH 11.5, upon UV illumination. Both original ZnO electrode and cobalt-doped one generated anodic photocurrents; however, the onset potential of the photocurrent was reduced by 0.23 V in the presence of the Co-Pi; moreover, a general enhancement of the photocurrent in a wide potential range is associated with the presence of the Co-based catalyst.

6.3 ROBUST, ALL-INORGANIC MOLECULAR OXYGEN-EVOLVING CATALYSTS

6.3.1 Polyoxometalates

Polyoxometalates (POMs) are molecular, polyanionic, multi-metal oxygen cluster anions, generally assembling from aqueous solutions of early transition metals (V, Nb, Ta, Mo, W) in their highest oxidation states, depending on specific conditions such as pH and temperature (Pope 1983; Pope and Muller 2001). In Figure 6.7, two examples of POMs structures are reported. These species display remarkable functional properties that depend on their elemental composition, structure, and associated counterion, and that make them of interest for different disciplines, such as catalysis (Bonchio et al. 2003, 2004, 2005, 2007; Mizuno et al. 2005; Carraro et al. 2006a; Berardi et al. 2007; Sartorel et al. 2007, 2008),

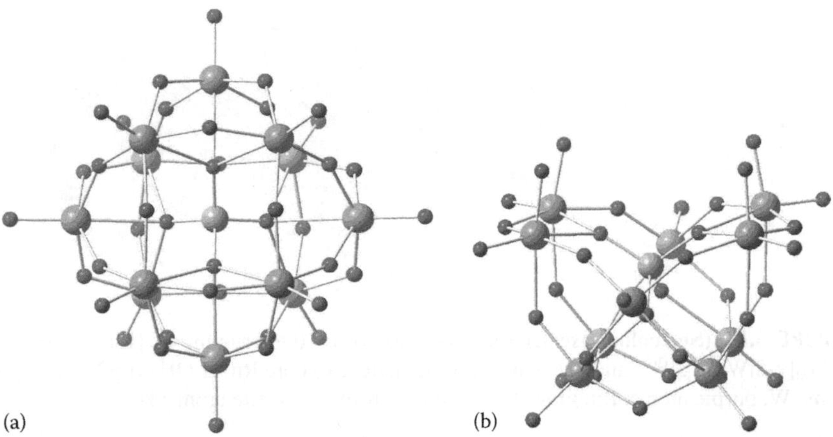

(a) (b)

FIGURE 6.7 **(See color insert.)** (a) Representation of the structure of the $[\alpha\text{-SiW}_{12}O_{40}]^{4-}$ polyoxometalate. (b) Representation of the structure of $[\gamma\text{-SiW}_{10}O_{36}]^{8-}$ polyoxometalate. Blue atoms: W; gray atoms: Si; red atoms: O.

materials sciences (Müller and Kögerler 1999; Zeng et al. 2000; Carraro et al. 2008), and medicine (Judd et al. 2001).

In particular, the widespread use of POMs in catalysis stems from two important features of these all-inorganic molecular oxides: (i) their ability of acting as ligands for catalytically active transition metals and (ii) their extreme robustness under thermal and oxidative stress.

Indeed, some POMs are characterized by defects in the structure (the so-called "vacant" or "lacunary" POMs, see Figure 6.7b), displaying nucleophilic oxygens, that constitute a polydentate-binding site for transition metals. The resulting complexes show remarkable stability, especially toward oxidizing conditions, since the POM framework is constituted by metals in their highest oxidation states, and therefore it is inert to oxidative degradation. Moreover, POMs can be conveniently embedded in suitable materials, such as polymeric membranes (Carraro et al. 2006b) and chitosan nanocomposites (Geisberger et al. 2011), that can further protect them under reaction conditions, while driving their reactivity. Thus, all these features make POMs ideal candidates to design structurally stable, molecular OEC.

6.3.2 Polyoxometalate-Based Oxygen-Evolving Catalysts

The first POM-based species proposed as an OEC was a ruthenium derivative, $Na_{14}[Ru^{III}{}_2Zn_2(H_2O)_2(ZnW_9O_{34})_2]$ (Howells et al. 2004). The oxygen-evolving activity of this species was studied by pulsed voltammetry in aqueous phosphate buffer (pH 8), with electrochemical oxygen generation observed at ca. 0.75 V (vs. NHE), close to the thermodynamic potential. However, some structural disorders indicated that the actual catalyst could be composed by a mixture of isomeric cluster; indeed, the mechanism of this species was not further investigated.

The first structurally characterized POM-based species acting as an OEC was reported in 2008 completely independently by two groups (Geletii et al. 2008;

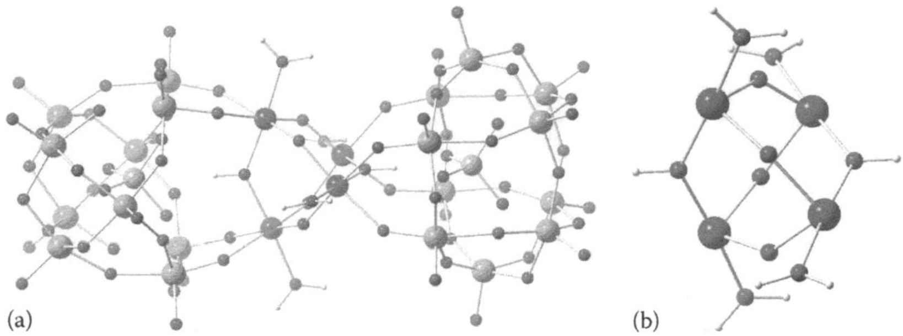

FIGURE 6.8 (See color insert.) Representation of (a) the structure of $\{Ru_4(\mu\text{-}OH)_2(\mu\text{-}O)_4(H_2O)_4[\gamma\text{-}SiW_{10}O_{36}]\}^{10-}$, and of (b) the tetraruthenate-oxo core $Ru_4(\mu\text{-}OH)_2(\mu\text{-}O)_4(H_2O)_4$. Blue atoms: W; purple atoms: Ru; gray atoms: Si; red atoms: O; white atoms: H.

Sartorel et al. 2008). $\{Ru_4(\mu\text{-}OH)_2(\mu\text{-}O)_4(H_2O)_4[\gamma\text{-}SiW_{10}O_{36}]\}^{10-}$ is readily synthesized by reacting the POM $[\gamma\text{-}SiW_{10}O_{36}]^{8-}$ (shown in Figure 6.7b) with ruthenium precursors in aqueous solution, and crystals of the product are obtained in good yields. The crystallographic analysis reveals the embedding of an adamantane like tetraruthenium core by the two POM units (Figure 6.8). The tetraruthenate core shows actually some structural analogies with the OEC in PSII (see Figure 6.2), as both are constituted by four redox active transition metals connected through oxygen bridges. The Ru-O connectivities in the tetraruthenate core resemble also the surface defects of the rutile structural motif of RuO_2, which was one of the first OEC reported to date (Kiwi and Grätzel 1978). Thus, the tetraruthenium core in $\{Ru_4(\mu\text{-}OH)_2(\mu\text{-}O)_4(H_2O)_4[\gamma\text{-}SiW_{10}O_{36}]\}^{10-}$ can be considered as an active fragment of an inorganic oxide, embedded in a totally inorganic molecular structure.

The activity of $\{Ru_4(\mu\text{-}OH)_2(\mu\text{-}O)_4(H_2O)_4[\gamma\text{-}SiW_{10}O_{36}]\}^{10-}$ as a water oxidation catalyst was first studied in homogeneous solution, using sacrificial electron acceptors such as Ce(IV) (Sartorel et al. 2008) or $Ru(bpy)_3^{3+}$ (bpy = 2,2′-bipyridine; Geletii et al. 2008). Adding an excess of Ce(IV), vigorous oxygen bubbling is observed from aqueous solutions of $\{Ru_4(\mu\text{-}OH)_2(\mu\text{-}O)_4(H_2O)_4[\gamma\text{-}SiW_{10}O_{36}]\}^{10-}$; the catalyst cycles up to 500 turnovers (defined as the number of moles of oxygen per mole of catalyst), with an initial turnover frequency of 0.125 $cycles \cdot s^{-1}$ (Sartorel et al. 2008). With $Ru(bpy)_3^{3+}$ as the oxidant, up to 18 turnovers are achieved, with an initial turnover frequency of 0.45–0.60 s^{-1} (Geletii et al. 2008).

The mechanism of oxygen production by $\{Ru_4(\mu\text{-}OH)_2(\mu\text{-}O)_4(H_2O)_4[\gamma\text{-}SiW_{10}O_{36}]\}^{10-}$ is still debated (Geletii et al. 2009a; Sartorel et al. 2009; Quiñonero et al. 2010; Sartorel et al. 2011). This is mainly due to the difficulties in characterizing the competent intermediates, involved in the catalytic cycle. A general consensus has however emerged on some relevant points:

1. $\{Ru_4(\mu\text{-}OH)_2(\mu\text{-}O)_4(H_2O)_4[\gamma\text{-}SiW_{10}O_{36}]\}^{10-}$ undergoes several consecutive 1 e^- oxidations that finally lead to an active form of the catalyst, capable of oxidizing water and restoring the initial state of the catalyst itself.
2. The consecutive 1 e^- oxidations seem to be coupled with proton losses from the tetraruthenate core; this is confirmed by a shift of the catalyst redox

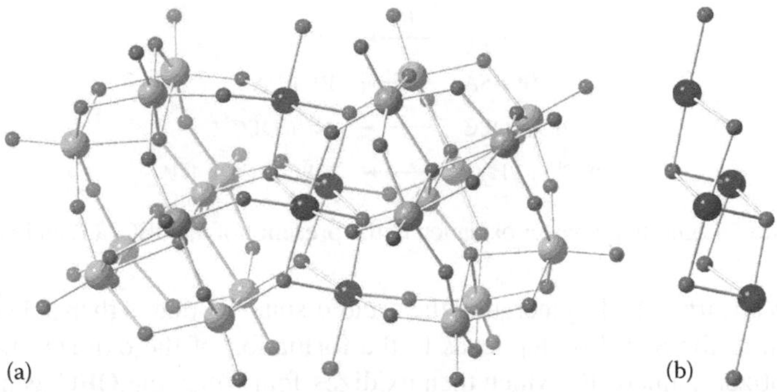

FIGURE 6.9 **(See color insert.)** Representation of (a) the structure of $[Co_4(H_2O)_2(PW_9O_{34})_2]^{10-}$, and of (b) the tetracobalt-oxo core $Co_4(\mu\text{-}O)_4(H_2O)_2$. Blue atoms: W; dark blue atoms: Co; orange atoms: P; red atoms: O.

waves upon changing of the pH (Geletii et al. 2009a), together with the appearance of bands attributed to Ru(V)-OH vibrations in the resonant Raman spectra of high valent intermediates of $\{Ru_4(\mu\text{-}OH)_2(\mu\text{-}O)_4(H_2O)_4[\gamma\text{-}SiW_{10}O_{36}]\}^{10-}$ (Sartorel et al. 2009).

3. Water oxidation by $\{Ru_4(\mu\text{-}OH)_2(\mu\text{-}O)_4(H_2O)_4[\gamma\text{-}SiW_{10}O_{36}]\}^{10-}$ is an actual $4e^-$ process, since the formation of hydrogen peroxide as an intermediate oxidation product was excluded on the basis of a kinetic argumentation (Sartorel et al. 2011).

Notably, these features evidence further strict analogies of the tetraruthenate catalyst with the OEC in PSII (see discussion on the Kok cycle described earlier), thus providing an outstanding proof of biomimetic catalysis.

Two years after the contemporary reports on the tetraruthenate catalyst, a cobalt-based POM was reported to be an actual OEC (Yin et al. 2010). The interest in cobalt stems from its higher abundance on the Earth's crust with respect to ruthenium (25–30 ppm for Co, 10^{-3} ppm for Ru), and its consequently lower cost. The structure of the species $[Co_4(H_2O)_2(PW_9O_{34})_2]^{10-}$ is reported in Figure 6.9, and is constituted by a Co_4O_4 active core, stabilized by two POM units. The activity of this species as an OEC was studied in aqueous buffer media (pH 8), using $Ru(bpy)_3^{3+}$ as the sacrificial oxidant. A remarkable turnover frequency of $5\,s^{-1}$ was observed under very dilute conditions, while the stability of the catalyst was confirmed by convergent spectroscopic techniques (Yin et al. 2010).

6.3.3 Light-Driven Water Oxidation

The promising results achieved by using POM-based OEC in the presence of sacrificial oxidants under dark conditions paved the way to explore their activity in light-driven water oxidation. Light-driven water oxidation is generally achieved in the presence of a P, a sacrificial electron acceptor SA, and the OEC, according to Scheme 6.3.

$$\mathrm{P} \xrightarrow{h\nu} {}^{*}\mathrm{P}$$

$$ {}^{*}\mathrm{P} + \mathrm{SA} \longrightarrow \mathrm{P}^{+} + \mathrm{SA}^{-}$$

$$4\mathrm{P}^{+} + \mathrm{OEC} \longrightarrow 4\mathrm{P} + \mathrm{OEC}^{4+}$$

$$\mathrm{OEC}^{4+} + 2\mathrm{H_2O} \longrightarrow \mathrm{OEC} + \mathrm{O_2} + 4\mathrm{H}^{+}$$

SCHEME 6.3 Light-driven water oxidation in the presence of an OEC, a P, and SA.

Light is absorbed by P, generating the excited state *P, that is then able to transfer an electron to the SA; this step leads to the formation of the oxidized form of the photosensitizer, namely, P^{+}, which then oxidizes four times the OEC, generating its active form, OEC^{4+}, which is able to oxidize water and restore the initial state of the catalyst.

Together with the aforementioned reactions, several competing unproductive pathways may occur, leading to low efficiency of the light-driven oxygen production. The quantum efficiency of the process may be defined as the number of photons that are actually used to produce oxygen, divided by the total number of photons, and therefore it comprises between 0 and 1. In the case of POM-based OECs, ruthenium polypyridine complexes have been used as P, in combination with the persulfate anion, $S_2O_8^{2-}$, as the sacrificial electron acceptor (Geletii et al. 2009b; Besson et al. 2010; Puntoriero et al. 2010; Huang et al. 2011). The results are summarized in Table 6.1.

First, the tetraruthenium-based OEC, $\{Ru_4(\mu\text{-}OH)_2(\mu\text{-}O)_4(H_2O)_4[\gamma\text{-}SiW_{10}O_{36}]\}^{10-}$, was coupled with $Ru(bpy)_3^{2+}$, which is the most commonly used photosensitizer (λ_{MAX} = 454 nm). Under these conditions, the OEC cycles up to 350 turnovers, achieving quantum efficiencies in the range 0.09–0.14, depending on the reaction medium (#1 and #2 in Table 6.1). However, an important limit of $Ru(bpy)_3^{2+}$ as the P is the limited absorption in the visible region; to overcome this issue, polynuclear P with bridging ligands (such as 2,3-bis(2′-pyridyl)pyrazine, dpp) have been employed (Larsen et al. 2007). Indeed, coupling $\{Ru_4(\mu\text{-}OH)_2(\mu\text{-}O)_4(H_2O)_4[\gamma\text{-}SiW_{10}O_{36}]\}^{10-}$ with

TABLE 6.1
Light-Driven Water Oxidation with POM-Based OEC and Ruthenium Polypyridine as Photosensitizers

#	OEC	P	λ/Medium	Quantum Efficiency	Reference
1	$\{Ru_4(\mu\text{-}OH)_2(\mu\text{-}O)_4(H_2O)_4[\gamma\text{-}SiW_{10}O_{36}]\}^{10-}$	$Ru(bpy)_3^{2+}$	420–520 nm/phosphate buffer, pH 7.2	0.09	Geletii et al. (2009)
2	$\{Ru_4(\mu\text{-}OH)_2(\mu\text{-}O)_4(H_2O)_4[\gamma\text{-}SiW_{10}O_{36}]\}^{10-}$	$Ru(bpy)_3^{2+}$	420–520 nm/Na_2SiF_6/$NaHCO_3$, pH 5.8	0.14	Besson et al. (2010)
3	$\{Ru_4(\mu\text{-}OH)_2(\mu\text{-}O)_4(H_2O)_4[\gamma\text{-}SiW_{10}O_{36}]\}^{10-}$	$Ru\{(\mu\text{-}dpp)Ru(bpy)_2\}_3^{8+}$	550 nm/phosphate buffer, pH 7.2	0.60	Puntoriero et al. (2010)
4	$[Co_4(H_2O)_2(PW_9O_{34})_2]^{10-}$	$Ru(bpy)_3^{2+}$	420–470 nm/borate buffer, pH 8	0.30	Huang et al. (2011)

the tetranuclear P $[Ru\{(\mu\text{-dpp})Ru(bpy)_2\}_3]^{8+}$ allows efficient oxygen production by an adoption of an extended fraction of visible light, with $\lambda > 550\,nm$ (#3 in Table 6.1); in this system, an astounding quantum efficiency of 0.60 was observed (Puntoriero et al. 2010).

Recently, also the cobalt-based $[Co_4(H_2O)_2(PW_9O_{34})_2]^{10-}$ was efficiently used in light-driven oxygen production, with $Ru(bpy)_3^{2+}$ as the P (Huang et al. 2011); the system is indeed promising, since quantum efficiencies up to 0.30 are observed, even though it does not operate at neutral pH (#4 in Table 6.1).

Such very good performance of POM-based catalyst with ruthenium-based polypyridine P is very promising; one of the key factors that enable the achievement of these high quantum efficiencies is the rate of the electron transfer from the OEC to the oxidized form of the P. Indeed, the reaction rate between photogenerated $Ru(bpy)_3^{3+}$ and the $\{Ru_4(\mu\text{-OH})_2(\mu\text{-O})_4(H_2O)_4[\gamma\text{-SiW}_{10}O_{36}]\}^{10-}$ OEC was investigated by nanosecond laser flash photolysis, which revealed a bimolecular process with a rate constant of $2.1 \cdot 10^9\ M^{-1}\ s^{-1}$, close to the diffusion-controlled limiting rate; for the sake of comparison, when iridium oxide nanoparticles are employed as OEC this rate is three orders of magnitude lower (Orlandi et al. 2010). It is then reasonable to conclude that such an efficient electron transfer is ascribable to the complementary charge of the cationic $Ru(bpy)_3^{2+}$ and the anionic POM-based OEC.

6.3.4 Nanostructured Oxygen-Evolving Anodes

In the perspective of adopting the POM-based OECs in a device for artificial photosynthesis, it is necessary to support them onto electrodes, in order to develop an oxygen evolving anode, operating in heterogeneous conditions. In such an electrode, the choice of the support material for the OEC has a primary importance on its efficiency, since it has to satisfy several requirements: (i) it has to be a good electron conductor; (ii) it has to be robust enough, in order to be stable under the highly oxidizing catalytic conditions; (iii) it must display a high surface area, to maximize the active surface upon support of a catalyst; (iv) the electrical contact between the catalyst and the electrode surface must be optimized, in order to have an efficient electron transport. In this sense, a proof-of-principle example of such an electrode was recently reported, by using a conductive bed of multi-wall carbon nanotubes as the support material for the $\{Ru_4(\mu\text{-OH})_2(\mu\text{-O})_4(H_2O)_4[\gamma\text{-SiW}_{10}O_{36}]\}^{10-}$ OEC (Toma et al. 2010). To anchor the anionic, POM-based OEC onto the surface of the tubes, electrostatic interactions were exploited, by functionalizing the nanotubes with positively charged, dendrimeric ammonium moieties (Figure 6.10). Several spectroscopic and microscopic techniques revealed that the structure of the OEC was maintained also in the hybrid material, and that its deposition on the tubes occurred mainly as single molecules, which is crucial in order to exploit all the redox centers and access single-site catalysis.

The composite hybrid material was then drop cast onto ITO electrodes, and the activity of such an electrode in oxygen evolution was evaluated by electrochemical techniques. Upon application of an external bias, when the electrode is dipped in phosphate buffer at pH 7, the catalytic wave due to water oxidation occurs at overpotentials of ca. 0.35 V. Bare ITO alone and ITO doped with multi-wall carbon nanotubes display instead very high overpotentials, confirming the role of the

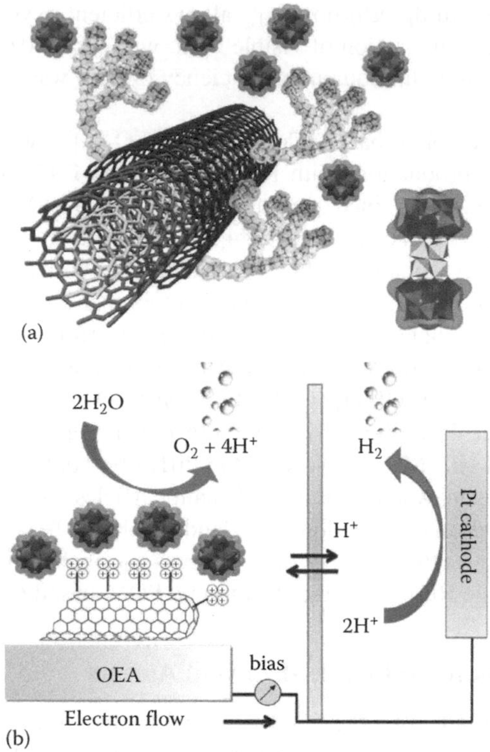

FIGURE 6.10 (a) Representation of the hybrid material where the $\{Ru_4(\mu\text{-}OH)_2(\mu\text{-}O)_4(H_2O)_4[\gamma\text{-}SiW_{10}O_{36}]\}^{10-}$ OEC is anchored onto dendron-functionalized multi-wall carbon nanotubes. (b) Scheme of a complete electrochemical cell for water splitting. (Adapted from Toma, F.M. et al., *Nat. Chem.*, 2, 826, 2010 [Toma et al. 2010].)

POM-based OEC in the catalytic activity of the electrode. Significantly, the conducting features of the nanotubes are essential to achieve efficient catalysis, since their substitution with amorphous carbon leads to a significant abatement of the electrode performance, in terms of catalytic current due to oxygen production.

6.4 CONCLUSIONS AND PERSPECTIVES

In the last few years, there have been considerable efforts to develop new molecules and materials exhibiting activity as water oxidation catalysts. In order to propose a valuable system for large-scale applications, one of the main requests for an oxygen-evolving catalyst is its long durability under reaction conditions. In this chapter, two complementary approaches to solve this problem have been presented, the first based on a self-healing cobalt-based material, the second exploiting the robustness of totally inorganic molecular oxo clusters. Recently, Nocera's Co-Pi catalyst has been supported on an "artificial leaf," a device with dimension of a playing card that splits water using sunlight, with 5.5% efficiency in converting it into hydrogen fuel (Service 2011). This device is based on a silicon wafer, with a

HEC and the OEC Co-Pi catalyst being spread onto the opposite sides of the wafer; this device will be probably an enormous step toward efficient conversion of solar light into chemical energy.

Nevertheless, there is still probably a long way to go to improve the overall efficiency of the process, maximizing H_2 production as a function of irradiating light. In this perspective, the study of molecular catalysts that drive the redox processes may help in the comprehension of the mechanistic details, allowing improvement of the performance and control of the reactivity by tuning competent features of the catalyst.

ACKNOWLEDGMENTS

Financial support from the University of Padova (Progetto Strategico 2008 HELIOS prot. STPD08RCX, PRAT CPDA104105/10), MIUR (PRIN n. 20085M27SS) and Fondazione Cariparo (Nano-Mode, progetti di eccellenza 2010) is gratefully acknowledged.

REFERENCES

Ama, T., Rashid, M. M., Yonemura, T., Kawaguchi, H., Yasui, T. 2000. Cobalt(III) complexes containing incomplete C_3O_4 or complete Co_4O_4 cubane core. *Coord. Chem. Rev.* 198: 101–116.

Armaroli, N., Balzani, V. 2007. The future of energy supply: Challenges and opportunities. *Angew. Chem. Int. Ed.* 46: 52–66.

Aro, E.-M., Suorsa, M., Rokka, A., Allahverdiyeva, Y., Paakkarinen, V., Saleem, A. et al. 2005. Dynamics of photosystem II: A proteomic approach to thylakoid protein complexes. *J. Exp. Bot.* 56: 347–356.

Balzani, V., Credi, A., Venturi, M. 2008. Photochemical conversion of solar energy. *Chem. Sus. Chem.* 1: 26–58.

Basolo, F., Pearson, R. G. 1967. *Mechanism of Inorganic Reactions*. John Wiley & Sons, Inc., New York.

Berardi, S., Bonchio, M., Carraro, M., Conte, V., Sartorel, A., Scorrano, G. 2007. Fast catalytic epoxidation with H_2O_2 and $[\gamma\text{-SiW}_{10}O_{36}(PhPO)_2]^{4-}$ in ionic liquids under microwave irradiation. *J. Org. Chem.* 72: 8954–8957.

Besson, C., Huang, Z., Geletii, Y. V., Lense, S., Hardcastle, K. I., Musaev, D. G. et al. 2010. $Cs_9[(\gamma\text{-PW}_{10}O_{36})_2Ru_4O_5(OH)(H_2O)_4]$, a new all-inorganic, soluble catalyst for the efficient visible-light-driven oxidation of water. *Chem. Commun.* 46: 2784–2786.

Bonchio, M., Carraro, M., Farinazzo, A., Sartorel, A., Scorrano, G., Kortz, U. 2007. Aerobic epoxidation by iron-substituted polyoxotungstates: Evidence for a metal initiated autooxidation mechanism. *J. Mol. Catal. A Chem.* 262: 36–40.

Bonchio, M., Carraro, M., Scorrano, G., Bagno, A. 2004. Photooxidation in water by new hybrid molecular photocatalysts integrating an organic sensitizer with a polyoxometalate core. *Adv. Synth. Catal.* 346: 648–654.

Bonchio, M., Carraro, M., Scorrano, G., Fontananova, E., Drioli, E. 2003. Heterogeneous photooxidation of alcohols in water by photocatalytic membranes incorporating decatungstate. *Adv. Synth. Catal.* 345: 1119–1126.

Bonchio, M., Carraro, M., Scorrano, G., Kortz, U. 2005. Microwave-assisted fast cyclohexane oxygenation catalyzed by iron-substituted polyoxotungstates. *Adv. Synth. Catal.* 347: 1909–1912.

Brudvig, G., Ed. 2008. The role of manganese in Photosystem II. *Coord. Chem. Rev.* 252: 231–468.

Brunschwig, B. S., Chou, M. H., Creutz, Q., Ghosh, P., Sutin, N. 1983. Mechanisms of water oxidation to oxygen: Cobalt(IV) as an intermediate in the aquocobalt(II)-catalyzed reaction. *J. Am. Chem. Soc.* 105: 4832–4833.

Carraro, M., Gardan, M., Scorrano, G., Drioli, E., Fontananova, E., Bonchio, M. 2006b. Solvent-free, heterogeneous photooxidation of hydrocarbons by Hyflon® membranes embedding a fluorous-tagged decatungstate. *Chem. Commun.* 43: 4533–4535.

Carraro, M., Sandei, L., Sartorel, A., Scorrano, G., Bonchio, M. 2006a. Hybrid polyoxotungstates as 2nd-generation POM-based catalysts for MW-assisted H_2O_2 activation. *Org. Lett.* 8: 3671–3674.

Carraro, M., Sartorel, A., Scorrano, G. et al. 2008. Chiral strandberg-type molybdates $[(RPO_3)_2Mo_5O_{15}]^{2-}$ as molecular gelators: Self-assembled fibrillar nanostructures with enhanced optical activity. *Angew. Chem. Int. Ed.* 47: 7275–7279.

Casey, W. H. 1991. On the relative dissolution rates of some oxide and orthosilicate minerals. *J. Colloid Interface Sci.* 146: 586–589.

Chivot, J., Mendoza, L., Mansour, C., Pauporté, T., Cassir, M. 2008. New insight in the behaviour of Co–H_2O system at 25–150°C, based on revised Pourbaix diagrams. *Corros. Sci.* 50: 62–69.

Dau, H., Grundmeier, A., Loja, P., Haumann, M. 2008. On the structure of the manganese complex of photosystem II: Extended-range EXAFS data and specific atomic-resolution models for four S-states. *Philos. Trans. R. Soc. B* 363: 1237–1244.

Dimitrou, K., Folting, K., Streib, W. E., Christou, G. 1993. "Dimerization" of the $[Co^{III}_2(OH)_2]$ Core to the first example of a $[Co^{III}_4O_4]$ cubane: Potential insights into photosynthetic water oxidation. *J. Am. Chem. Soc.* 115: 6432–6433.

Dincă, M., Surendranath, Y., Nocera, D. G. 2010. Nickel–borate oxygen-evolving catalyst that functions under benign conditions. *Proc. Natl. Acad. Sci. USA.* 107: 10337–10341.

Ferreira, K. N., Iverson, T. M., Maghlaoui, K., Barber, J., Iwata, S. 2004. Architecture of the photosynthetic oxygen-evolving centre. *Science* 303: 1831–1838.

Geisberger, G., Paulus, S., Carraro, M., Bonchio, M., Patzke, G. R. 2011. Synthesis, characterization and cytotoxicity of polyoxometalate/carboxymethyl chitosan nanocomposites. *Chem. Eur. J.* 17: 4619–4625.

Geletii, Y. V., Besson, C., Hou, Y. et al. 2009a. Structural, physicochemical, and reactivity properties of an all-inorganic, highly active tetraruthenium homogeneous catalyst for water oxidation. *J. Am. Chem. Soc.* 131: 17361–17370.

Geletii, Y. V., Botar, B., Kögerler, P. et al. 2008. An all-inorganic, stable, and highly active tetraruthenium homogeneous catalyst for water oxidation. *Angew. Chem. Int. Ed.* 47: 3898–3902.

Geletii, Y. V., Huang, Z., Hou, Y., Musaev, D. G., Lian, T., Hill, C. L. 2009b. Homogeneous light-driven water oxidation catalyzed by a tetraruthenium complex with all inorganic ligands. *J. Am. Chem. Soc.* 131: 7522–7523.

Gileadi, E. 1993. *Electrode Kinetics for Chemists, Chemical Engineers, and Materials Scientist.* Wiley-VCH, New York, pp. 127–184.

Hammarstrom, L., Hammes-Schiffer, S., Eds. 2009. Artificial photosynthesis and solar fuels. *Acc. Chem. Res.* 42: 1859–2029.

Haumann, M., Liebisch, P., Müller, C., Barra, M., Grabolle, M., Dau, H. 2005. Photosynthetic O_2 formation tracked by time-resolved X-ray experiments. *Science* 310: 1019–1021.

Howells, A. R., Sankarraj, A., Shannon, C. 2004. A diruthenium-substituted polyoxometalate as an electrocatalyst for oxygen generation. *J. Am. Chem. Soc.* 126: 12258–12259.

Huang, Z., Luo, Z., Geletii, Y. V. et al. 2011. Efficient light-driven carbon-free cobalt-based molecular catalyst for water oxidation. *J. Am. Chem. Soc.* 133: 2068–2071.

Judd, D. A., Nettles, J. H., Nevins, N. et al. 2001. Polyoxometalate HIV-1 protease inhibitors. A new mode of protease inhibition. *J. Am. Chem. Soc.* 123: 886–897.

Kanan, M. W., Nocera, D. G. 2008. In situ formation of an oxygen-evolving catalyst in neutral water containing phosphate and Co^{2+}. *Science* 321: 1072–1075.

Kanan, M. W., Surendranath, Y., Nocera, D. G. 2009. Cobalt–phosphate oxygen-evolving compound. *Chem. Soc. Rev*. 38: 109–114.

Kanan, M. W., Yano, J., Surendranath, Y., Dincă, M., Yachandra, V. K., Nocera, D. G. 2010. Structure and valency of a cobalt-phosphate water oxidation catalyst determined by in situ X-ray spectroscopy. *J. Am. Chem. Soc*. 132: 13692–13701.

Kay, A., Cesar, I., Grätzel, M. 2006. New benchmark for water photooxidation by nanostructured α-Fe_2O_3 films. *J. Am. Chem. Soc*. 128: 15714–15721.

Kiwi, J., Grätzel, M. 1978. Oxygen evolution from water via redox catalysis. *Angew. Chem. Int. Ed*. 17: 860–861.

Kok, B., Forbush, B., McGloin, M. 1970. Cooperation of charges in photosynthetic O_2 evolution-I. A linear four step mechanism. *Photochem. Photobiol*. 11: 457–475.

Larsen, J., Puntoriero, F., Pascher, T., McClenaghan, N., Campagna, S., Åkesson, E., Sundström, V. 2007. Extending the light-harvesting properties of transition-metal dendrimers. *ChemPhysChem* 8: 2643–2651.

Loll, B., Kern, J., Saenger, W., Zouni, A., Biesiadka, J. 2005. Towards complete cofactor arrangement in the 3.0Å resolution structure of photosystem II. *Nature* 438: 1040–1044.

Lutterman, D. A., Surendranath, Y., Nocera, D. G. 2009. A self-healing oxygen-evolving catalyst. *J. Am. Chem. Soc*. 131: 3838–3839.

McAlpin, J. G., Surendranath, Y., Dincă, M., Stich, T. A., Stoian, S. A., Casey, W. H., Nocera, D. G., Britt, R. D. 2010. EPR evidence for Co(IV) species produced during water oxidation at neutral pH. *J. Am. Chem. Soc*. 132: 6882–6883.

McEvoy, J. P., Brudvig, G. W. 2006. Water-splitting chemistry of Photosystem II. *Chem. Rev*. 106: 4455–4483.

Mizuno, N., Yamaguchi, K., Kamata, K. 2005. Epoxidation of olefins with hydrogen peroxide catalyzed by polyoxometalates. *Coord. Chem. Rev*. 249: 1944–1956.

Müller, A., Kögerler, P. 1999. From simple building blocks to structures with increasing size and complexity. *Coord. Chem. Rev*. 182: 3–17.

Orlandi, M., Argazzi, R., Sartorel, A., Carraro, M., Scorrano, G., Bonchio, M. et al. 2010. Ruthenium polyoxometalate water splitting catalyst: Very fast hole scavenging from photogenerated oxidants. *Chem. Commun*. 46: 3152–3154.

Pope, M. T. 1983. *Heteropoly and Isopoly Oxometalates*. Springer-Verlag, New York.

Pope, M. T., Muller, A. 2001. *Polyoxometalate Chemistry from Topology via Self-Assembly to Applications*. Kluwer, Dordrecht, the Netherlands.

Puntoriero, F., La Ganga, G., Sartorel, A., Carraro, M., Scorrano, G., Bonchio, M. et al. 2010. Photo-induced water oxidation with tetra-nuclear ruthenium sensitizer and catalyst: A unique 4 × 4 ruthenium interplay triggering high efficiency with low-energy visible light. *Chem. Commun*. 46: 4725–4727.

Quiñonero, D., Kaledin, A. L., Kuznetsov, E., Geletii, Y. V., Besson, C., Hill, C. L., Musaev, D. G. 2010. Computational studies of the geometry and electronic structure of an all-inorganic and homogeneous tetra-Ru-polyoxotungstate catalyst for water oxidation and its four subsequent one-electron oxidized forms. *J. Phys. Chem. A* 114: 535–542.

Ren, Z., Luo, J., Xu, Z., Cao, G. 2007. Tuning co valence state in cobalt oxyhydrate superconductor by postreduction. *Chem. Mater*. 19: 4432–4435.

Risch, M., Khare, V., Zaharieva, I., Gerencser, L., Chernev, P., Dau, H. 2009. Cobalt-oxo core of a water-oxidizing catalyst film. *J. Am. Chem. Soc*. 131: 6936–6937.

Sakurai, H., Osada, M., Takayama-Muromachi, E. 2007. Hydration of sodium cobalt oxide. *Chem. Mater*. 19: 6073–6076.

Sakurai, H., Tsujii, N., Suzuki, O. et al. 2006. Valence and Na content dependences of superconductivity in $Na_xCoO_2 \cdot yH_2O$. *Phys. Rev. B: Condens. Matter. Phys*. 74(9): 092502/1–4.

Sartorel, A., Carraro, M., Bagno, A., Scorrano, G., Bonchio, M. 2007. Asymmetric tetra-protonation triggers catalytic epoxidation by γ-$[(SiO_4)W_{10}O_{32}]^{8-}$. A new perspective in the assignment of the competent catalyst. *Angew. Chem. Int. Ed.* 46: 3255–3258.

Sartorel, A., Carraro, M., Scorrano, G., De Zorzi, R., Geremia, S., McDaniel, N. D., Bernhard, S., Bonchio, M. 2008. Polyoxometalate embedding of a tetraruthenium(IV)-oxo-core by template-directed metalation of $[\gamma\text{-}SiW_{10}O_{36}]^{8-}$: A totally inorganic oxygen-evolving catalyst. *J. Am. Chem. Soc.* 130: 5006–5007.

Sartorel, A., Miró, P., Salvadori, E., Romain, S., Carraro, M., Scorrano, G., Di Valentin, M., Llobet, A., Bo, C., Bonchio, M. 2009. Water oxidation at a tetraruthenate core stabilized by polyoxometalate ligands: Experimental and computational evidence to trace the competent intermediates. *J. Am. Chem. Soc.* 131: 16051–16053.

Sartorel, A., Truccolo, M., Berardi, S., Gardan, M., Carraro, M., Toma, F. M., Scorrano, G., Prato, M., Bonchio, M. 2011. Oxygenic polyoxometalates: A new class of molecular propellers. *Chem. Commun.* 47: 1716–1718.

Service, R. F. 2011. Artificial leaf turns sunlight into a cheap energy source. *Science* 332: 25.

Shafirovich, V. Y., Khannanov, N. K., Strelets, V. V. 1980. Chemical and light-induced catalytic water oxidation. *Nouv. J. Chim.* 4: 81–84.

Sproviero, E. M., Gascon, J. A., McEvoy, J. P., Brudvig, G. W., Batista, V. S. 2008. A model of the oxygen-evolving centre of photosystem II predicted by structural refinement based on EXAFS simulations. *J. Am. Chem. Soc.* 130: 6728–6730.

Steinmiller, E. M. P., Choi, K.-S. 2009. Photochemical deposition of cobalt-based oxygen evolving catalyst on a semiconductor photoanode for solar oxygen production. *Proc. Natl. Acad. Sci. USA* 106: 20633–20636.

Sun, J. W., Zhong, D. K., Gamelin, D. R. 2010. Composite photoanodes for photoelectrochemical solar water splitting. *Energy Environ. Sci.* 3: 1252–1261.

Surendranath, Y., Dincă, M., Nocera, D. G. 2009. Electrolyte-dependent electrosynthesis and activity of cobalt-based water oxidation catalysts. *J. Am. Chem. Soc.* 131: 2615–2620.

Surendranath, Y., Kanan, M. W., Nocera, D. G. 2010. Mechanistic studies of the oxygen evolution reaction by a cobalt-phosphate catalyst at neutral pH. *J. Am. Chem. Soc.* 132: 16501–16509.

Takada, K., Fukuda, K., Osada, M. et al. 2004. Chemical composition and crystal structure of superconducting sodium cobalt oxide bilayer-hydrate. *J. Mater. Chem.* 14: 1448–1453.

Toma, F. M., Sartorel, A., Iurlo, M. et al. 2010. Efficient water oxidation at carbon nanotube–polyoxometalate electrocatalytic interfaces. *Nat. Chem.* 2: 826–831.

Umena, Y., Kawakami, K., Shen, J.-R., Kamiya, N. 2011. Crystal structure of oxygen-evolving photosystem II at a resolution of 1.9 Å. *Nature* 473: 55–61.

Yano, J., Kern, J., Sauer, K. et al. 2006. Where water is oxidized to dioxygen: Structure of the photosynthetic Mn_4Ca cluster. *Science* 314: 821–825.

Yin, Q., Tan, J. M., Besson, C. et al. 2010. A fast soluble carbon-free molecular water oxidation catalyst based on abundant metals. *Science* 328: 342–345.

Zeng, H., Newkome, G. R., Hill, C. L. 2000. Poly(polyoxometalate) dendrimers: Molecular prototypes of new catalytic materials. *Angew. Chem. Int. Ed.* 39: 1771–1774.

Zhong, D. K., Gamelin, D. R. 2010. Photoelectrochemical water oxidation by cobalt catalyst ("Co-Pi")/α-Fe_2O_3 composite photoanodes: Oxygen evolution and resolution of a kinetic bottleneck. *J. Am. Chem. Soc.* 132: 4202–4207.

Zhong, D. K., Sun, J., Inumaru, H., Gamelin, D. R. 2009. Solar water oxidation by composite catalyst/α-Fe_2O_3 photoanodes. *J. Am. Chem. Soc.* 131: 6086–6087.

7 Dynamic Self-Assembly of Nanoscale Components for Solar Energy Conversion

Ardemis A. Boghossian, Moon-Ho Ham, Jong Hyun Choi, and Michael S. Strano

CONTENTS

7.1 INTRODUCTION

For centuries, plants have utilized dynamic mechanisms of self-repair to develop fault-tolerant energy conversion schemes essential for withstanding climatic variability and fluctuation of solar flux [1–5]. These mechanisms ensure the continual regeneration of plants responsible for enhanced photoefficiency and the indefinite prolongation of plant lifetime. On the other hand, most synthetic light conversion devices to date have remained largely static, lacking robustness in fault tolerance, photostability [6], material abundance [7], photoefficiency [8–10], and cost effectiveness [11,12]. Recent efforts have focused on the incorporation of biomimetic mechanisms into synthetic photovoltaic devices. These studies often rely on the direct implementation of naturally derived photosynthetic complexes into synthetic photoelectrochemical cells, relying on synthesis procedures on tethering of the extracted proteins to various substrates [13–20]. In a recent study [21], scientists have expanded into using photosynthetic, extracted protein to synthesize the first photoelectrochemical cell capable of mimicking key aspects of the self-repair cycle used in the chloroplasts of plants. The remainder of this chapter focuses on chloroplast structure, function, and self-repair mechanisms, followed by a thorough discussion on its application to synthetic devices and the kinetic and thermodynamics quantification of the self-assembly process integral to the self-repair process.

7.2 BACKGROUND

7.2.1 Chloroplast Structure and Function

The chloroplast is the smallest unit of energy conversion within the plant cell (Figure 7.1). Similar to most organelles, it consists of an outer and an inner surrounding membrane. These surrounding membranes enclose a thick fluid known as the stroma, which occupies the area in between the stacks of thylakoid membranes, or grana. The lumen, which is the compartmental area bound by thylakoid membranes, is an acidic environment that aids in proton transfer in the photophosphorylation reactions that occur during photosynthesis. The primary site for the light-dependent reactions that occur during photosynthesis, however, actually occurs within the thylakoid membranes themselves [22–25].

The thylakoid membrane contains a series of integral proteins that are responsible for the sequential steps that occur during photosynthesis (Figure 7.1). The four

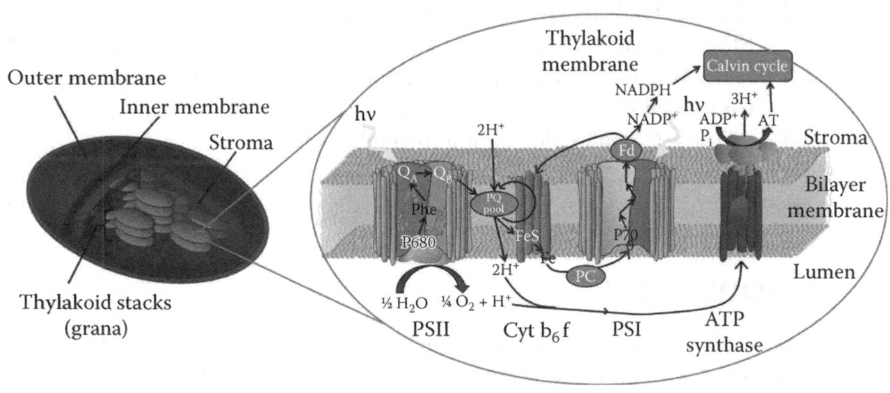

FIGURE 7.1 Chloroplast structure. The chloroplast (left) consists of both outer and inner membranes that house stacks of thylakoid membranes, or grana. Surrounded by the fluid-filled stroma, the thylakoid membranes enclose an inner lumen. The light reactions of photosynthesis are primarily conducted by a series of thylakoid membrane-embedded protein complexes (right). The key complexes of interest are PSII, cytochrome b (cyt b_6f), PSI, and ATP synthase.

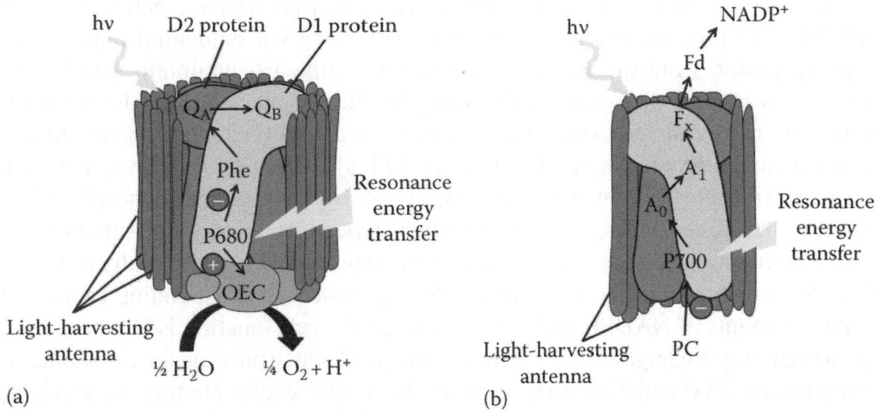

FIGURE 7.2 Structure of photosystems II and I. (a) PSII is the first protein complex in the series of complexes responsible for the light reactions. Energy is absorbed by the P680 site, generating an electron–hole pair. The electron is shuttled to pheophytin (Phe) and the quinine sites, Q_A and Q_B, prior to being expelled into the plastoquinone pool. The hole, which remains at the P680, is used in the oxidation of water by the OEC. Both direct light absorption and energy transfer from the surrounding light-harvesting antenna contribute to the energy absorption by the P680 site. (b) PSI ultimately acquires the electron emitted by PSII after it has been degraded by cytochrome b (cyt b_6f). Absorption of energy at the P700 is used to excite the degraded electron to ultimately contribute to NAPH production for the Calvin–Benson cycle. As with PSII, both direct absorption and energy transfer from surrounded antenna contribute to electron excitation at the P700 site.

primary protein complexes involved in photosynthesis is photosystem II (PSII), cytochrome b (cyt b_6f), photosystem I (PSI), and adenosine triphosphate (ATP) synthase. PSII (Figure 7.2a) comprises the D1 and D2 proteins surrounded by an array of light-harvesting antennas. Upon illumination, light is absorbed at ~680 nm at the P680 site on the D1 protein. A broader range of the solar spectrum is additionally absorbed by the surrounding light-harvesting antenna. Light absorbed by the antenna is transferred to the P680 site via electron energy transfer (EET), and, in combination with light directly absorbed by P680, is used in electron excitation and electron–hole separation. Separation of the electron–hole pair is marked by electron transfer to the pheophytin (Phe) site on the D1 protein. The retention of the hole at the P680 site, which is subsequently acquired by the manganese cluster via a redox-active tyrosine, is used to catalyze the oxidation of water in the production of oxygen via oxygen-evolving complexes (OEC) according to the reaction

$$2H_2O \xrightarrow{hv} O_2 + 4e^- + 4H^+$$

The electron is transferred to the static Q_A site in the D2 protein and subsequently acquired by the Q_B site in the D1 protein. After two reductions by Q_A, acquisition of the electron by the mobile Q_B site is marked by the reduction of the Q_B plastoquinone to plastoquinol. From the Q_B site, the electron is ejected into the platoquinone pool outside of PSII, where it is brought to the next major protein complex, cyt b_6f.

The cyt b_6f complex, which nominally serves to transfer the electron from PSII and PSI, also provides the proton gradient necessary via sequential redox centers to pump proton from the stroma to the internal lumen to maintain acidic conditions necessary for the phosphorylation cycle. Hence, electron transfer across the photosystems is coupled to the proton movement across the membrane to create an electrochemical proton potential that drives ATP synthesis. ATP behaves as a means of energy transport within the plant cell, cyclically undergoing dephosphorylation to form adenosine diphosphate (ADP) when expending energy. In a parallel process, nicotinamide adenine dinucleotide phosphate ($NADP^+$) is cyclically reduced to NADPH, which ultimately serves as a reducing power when expending energy. The relative amounts of NADPH and ATP necessary for cell function is balanced by cyt b_6f, which may undergo either cyclic or noncyclic electron transfer to balance the formation of ATP and NADPH, respectively. Under cyclic photophosphorylation, electrons extracted from both PSI and PSII are used directly for ATP synthesis and the production of NAPH is halted. Under noncyclic conditions, electron degradation from PSII is used to pump protons into the lumen to power ATP synthase.

Once the electron is ultimately expelled from cyt b_6f, it is received by plastocyanin (PC), which serves as a mediator for electron transfer to PSI (Figure 7.2b). Similar to the PSII, PSI consists of two primary proteins, surface adhesin proteins A (psaA) and B (psaB), that contain the primary sites for electron transfer. The P700 site, which is located on psaA, behaves much similar to the P680 sire in the PSII. It intercepts the electron from the PC. Under illumination, light is absorbed at ~700 nm to re-excite the electron after it has undergone degradation in cyt b_6f. The surrounding light-harvesting antennae behave analogously to those surrounding PSII, which serve to expand the absorption spectrum of the photosynthetic dyes. Direct absorption by the P700 site, along with EET from the light-harvesting antenna, is used to excite the electron to the modified chlorophyll (A_0) and the phylloquinone (A_1) sites. The electron is ultimately transferred to the iron–sulfur complex consisting of the F_x, F_a and F_b sites that direct electron transfer toward ferredoxin, which facilitates the reduction of NAP^+ to NADPH [26,27].

The fourth, and final, major protein of complex of interest in the thylakoid membrane is ATP synthase. As alluded to by its name, ATP synthase is primarily responsible for the chemical synthesis and storage of energy in the form of ATP. Hydrogen ions extracted from the acidic, luminal side of the membrane drive the conversion of ADP to ATP. Energy extraction from ATP subsequently occurs within the stroma of the chloroplast via the Calvin–Benson Cycle, which is responsible for glucose production within the plant cell according to the reaction [28]

$$6CO_2 + 12NADPH + 12H^+ + 6H_2O + 18ATP \rightarrow C_6H_{12}O_6 + 6H_2O + 12NADP^+ + 18ADP + 18P_i$$

7.2.2 D1 Protein Self-Repair Cycle

While under continuous illumination, several of the photosynthetic proteins, namely the D1 protein, become photodamaged. As discussed in Section 7.2.1, the D1 protein contains the P680 site, the site primarily responsible for initial electron–hole

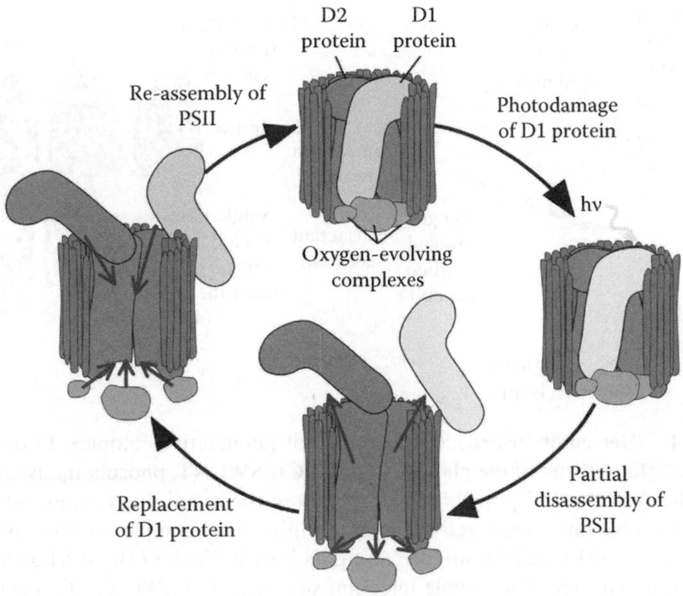

FIGURE 7.3 PSII self-repair cycle. A functional PSII (top) is under continuous illumination, resulting in photodamage of the D1 protein over time (right). Denaturing of the D1 protein induces the spontaneous, partial disassembly of PSII, expelling the photodamaged protein (bottom). The photodamaged protein is replaced by a newly synthesized protein (left). Introduction of the new D1 protein induces the spontaneous reassembly of the complex to create a functional PSII (top).

separation. To address this issue, plants have evolved highly sophisticated mechanisms for autonomous repair of the damaged protein (Figure 7.3). Photodamage of the D1 protein induces the partial disassembly of PSII. Upon release of the photodamaged protein from the complex, the protein diffuses toward the outer stroma for degradation. Meanwhile, the new, functional D1 protein, which is synthesized alongside the stroma-exposed membrane, diffuses toward the disassembled PSII complex within the appressed region of the thylakoid membrane. Introduction of the newly synthesized protein triggers the spontaneous incorporation of the D1 protein in the reassembly of the complex into a functional PSII. The reversible self-assembly and disassembly of PSII and the formation of kinetically trapped metastable thermodynamic states in the interim are key to this self-repair cycle [29–31].

7.3 BIOMIMETIC REGENERATION OF A PHOTOELECTROCHEMICAL COMPLEX

In a recent study [21], researchers developed the first synthetic photoelectrochemical complex capable of mimicking key aspects of this self-repair cycle (Figure 7.4). An aqueous solution consisting of the phospholipid 1,2-dimyristoyl-*sn*-glycero-3-phosphocholine (DMPC), membrane scaffold proteins (MSPs), single-walled carbon nanotubes (SWCNTs), and sodium cholate surfactant (SC) is dialyzed to selectively

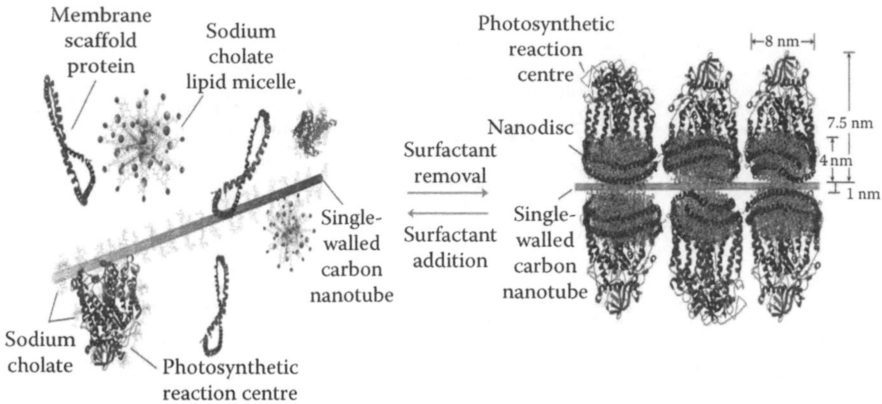

FIGURE 7.4 (See color insert.) Self-assembly of photoactive complex. In the disassembled state, a solution containing photosynthetic RCs, SWCNT, phospholipids, and MSPs is dispersed using SC (left). Upon dialysis of the latter, the remaining components spontaneously self-assemble into a photoactive complex (right). This complex consists of a SWCNT with lipid bilayer disks, or NDs aligned along its length. Each of these NDs can house up one RC, orientated such that its hole injection site faces the SWCNT. This self-assembly process is completely reversible in that the readdition of surfactant once more disassembles the complex into its initial micellar state (left). (Reprinted from Macmillan Publishers Ltd. *Nat. Chem.*, Ham, M.H. et al., Photoelectrochemical complexes for solar energy conversion that chemically and autonomously regenerate, 2(11), 929–936, Copyright 2010.)

remove surfactant from the mixture. Upon surfactant removal, the remaining components spontaneously self-assemble into the complex shown in Figure 7.4 (right). The assembled complex consists of an individual SWCNT with lipid bilayers, or nanodisks (NDs) on either side of the nanotube. As shown in Figure 7.5, NDs contain DMPC lipids arranged in a bilayer wherein the hydrophilic heads are facing outward toward the aqueous solution and the hydrophobic tails are sandwiched in between. Two strands of MSP wrap along the circumference of the 4 nm high disks.

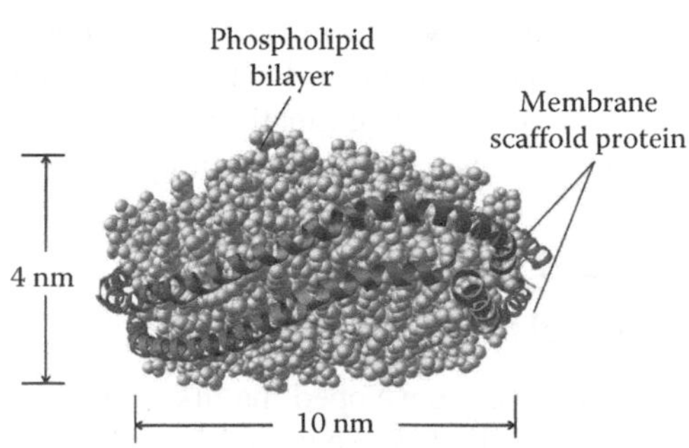

FIGURE 7.5 ND structure. The ND consists of a phospholipid (1,2-dimyristoyl-*sn*-glycero-3-phosphocholine; DMPC) bilayer surrounded by two strands of MSP. The particular NDs used in this study are ~4 nm high and 10 nm wide.

These NDs align along the length of the nanotube to form a one-dimensional array of disks. These disks contain photosynthetic RCs that are hypothesized to be specifically orientated such that hydrophobic end containing the P680 site is facing the nanotube and the hydrophilic end containing the Q_A and Q_B sites is facing outward toward the solution. This configuration, which is reminiscent of that of the protein within its native bilayer membrane, utilizes the relative variation in hydrophobicity throughout the protein to interface with the other nanocomponents of the system at a specific orientation. Reintroduction of the surfactant to a solution containing the assembled complex disassembles the complex into its initial micellar state. Structural characterization of the complex was verified using a atomic force microscopy (AFM), small-angle neutron scattering (SANS), and spectroscopic measurements described in Section 7.3.2 [21].

7.3.1 Purification of the Assembled Complex

Removal of surfactant from the micellar solution forms not only these RC-ND-SWCNT complexes, but also RC-ND, ND-SWCNT, and ND complexes. Therefore, characterization of the RC-ND-SWCNT complex requires additional steps for the purification of the complex from the reaction mixture. Purification in this study was achieved via density gradient centrifugation. In density gradient centrifugation, separation is obtained based on size and density differences in the reaction products. After centrifugation, samples were fractionated into 250 μL aliquots and subject to both fluorescence and absorbance measurements. Fluorescence emissions of a lipid-soluble Laurdan dye were used to track ND formation in the hydrophobic phase. Photoluminescence (PL) measurements were used to monitor the formation of RC-containing complexes. These measurements, along with nanotube fluorescence emissions (discussed in Section 7.3.3) were used to track and isolate the RC-ND-SWCNT fraction within the centrifuged sample [21].

7.3.2 SWCNT Fluorescence Measurements

In addition to emission from Laurdan and RCs, fluorescence emissions from SWCNTs were also used to track ND formation. Unlike the Laurdan and RC emissions, however, SWCNT emissions demonstrate a unique chirality dependency. A discussion on this chirality-based behavior requires a fundamental understanding of SWCNT photophysics.

SWCNTs are essentially rolled up sheets of graphene demonstrating unidirectional confinement of exciton and electron mobility. The direction in which the graphene sheet is rolled determines both the electrical and dimensional properties of the SWCNT (Figure 7.6). SWCNTs demonstrate conductivities varying from metallic to semiconducting and diameters ranging from 0.4 nm to several nanometers. The nomenclature used to label these different nanotube chiralities is illustrated in Figure 7.6a, where a two-dimensional graphene lattice is overlain with a pair of axes located 30° apart from one another. Any carbon atom on the coordinate system can labeled with pair of (n,m) coordinates that represent the number of carbon atoms along each of the axes. A nanotube that is rolled from the origin of the graphene lattice to

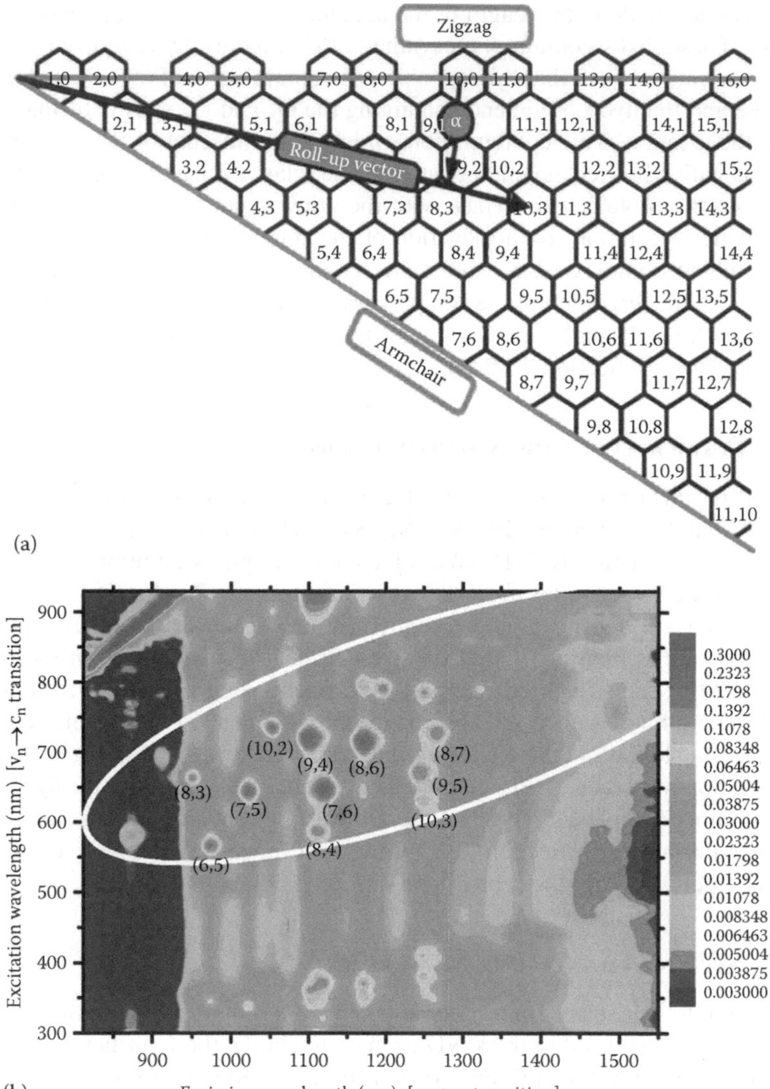

FIGURE 7.6 **(See color insert.)** Nanotube chirality. (a) Nanotube nomenclature relies on the utilization of coordinate system wherein the two axes ("zigzag" and "armchair") are 30° apart overlaying a graphene sheet. A specified carbon atom in the sheet can be located by counting the number of carbon atoms along each of the axes direction to generate a pair of coordinates or indices representative of the atom location. By rolling the graphene from the origin to *q*-specific carbon atom in the lattice, the nanotube that is formed is named according to the pair of indices that designate the location of the specified carbon atom. (b) A PL plot examines the fluorescence emission intensity of a nanotube solution over a range of excitation wavelengths. A solution consisting of individually suspended nanotubes contains several peaks, with each peak corresponding to a specific type of nanotube. (From Bachilo, S.M. et al., Structure-assigned optical spectra of single-walled carbon nanotubes, *Science*, 298(5602), 2361–2366, 2002. Reprinted with permission of AAAS.)

a specified carbon atom within the coordinate system is named according to the coordinates of that atom. For instance, a graphene lattice that is rolled up from the origin to the carbon atom located six carbon atoms along the *x*-axis and five carbon atoms along the *y*-axis is the (6,5) nanotube. Nanotubes labeled as (*n*,0) are called "zigzag" nanotubes, whereas those labeled as (*n*,*n*) are called "armchair" nanotubes. Similarly, "near-armchair" nanotubes are those labeled as (*n*,*n*−1) [32].

As a consequence of the variation of bandgap with nanotube chirality, each nanotube chirality demonstrates a unique excitation–emission fluorescence profile [33]. The excitation–emissions fluorescence map of a solution of nanotubes suspended in sodium cholate is shown in Figure 7.6b. As shown in the plot, each fluorescence peak is labeled by a pair of coordinates corresponding to the particular nanotube responsible for the fluorescence at the specific wavelength [34]. When sodium cholate is dialyzed from this system in the presence of DMPC and MSP, the near-armchair nanotubes selectively demonstrate a redshift in their emission wavelengths. This redshifting is completely reversible in that the readdition of surfactant to the solution returns the fluorescence emission peaks to their initial, blue-shifted location. Hence, SWCNT fluorescence can be used a means of monitoring ND formation [21]. As will be discussed in Section 7.4, this fluorescence shifting serves as a tool for monitoring ND formation.

7.3.3 Photoelectrochemical Measurements

The photoelectrochemical properties of the assembled RC-ND-SWCNT complex were characterized to evaluate the practical implications of complex self-assembly. The setup for the photoelectrochemical cell is shown in Figure 7.7. A 700 nM solution containing the RC-ND-SWCNT was placed within a polydimethylsiloxane (PDMS) mold and illuminated below using a 785 nm laser. A SWNT-cast film was used as a working electrode, with a Pt counter electrode and Ag/AgCl reference. Under illumination, the cell produces a photocurrent of 22 nA, yielding a 40% per complex efficiency. Overall cell efficiency, which is <1% for a 700 nM cell, has been shown to increase with increasing concentration to approach the 40% per nanotube efficiency [21].

Utilization of the dynamic, self-assembling capabilities of the RC-ND-SWCNT complex was subsequently used to demonstrate regeneration of the photoelectrochemical cell. As shown in Figure 7.7a, in addition to the aforementioned setup consisting of the mould, source of illumination, and pertinent electrodes, the setup also contained two dialyzers. The first dialyzer (Figure 7.7a, left) is used for surfactant removal to initiate complex self-assembly. The second dialyzer (Figure 7.7a, right) is used for surfactant addition for the disassembly of the complex. Photocurrent is monitored under continuous illumination as the RCs become photodamaged, and the corresponding photocurrent undergoes photodecay (Figure 7.7b). When ~20% of the initial photocurrent is achieved, surfactant is introduced into the system to initiate complex disassembly. The photodamaged components are removed from the system. Introduction of newly synthesized RCs is followed by the removal of surfactant from the solution and autonomous reassembly of the photoactive complex. Incorporation of the newly synthesized protein into the assembled complex recovers

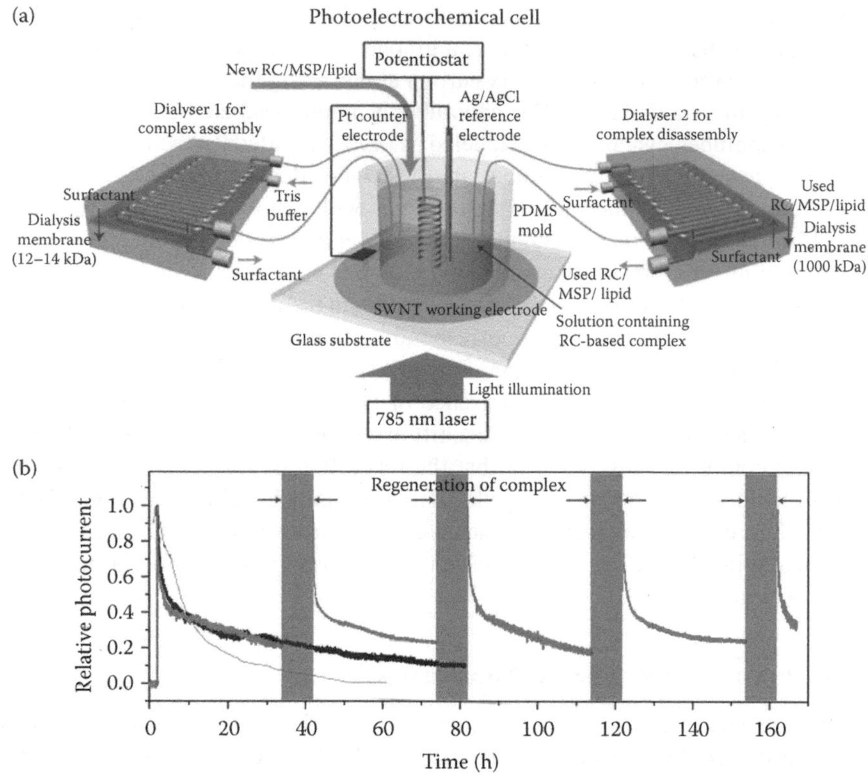

FIGURE 7.7 Photoelectrochemical measurements. (a) The setup for the photoelectrochemical measurements consists of a PDMS mold containing 700 nM RC-ND-SWCNT solution. The sample is illuminated below using a 786 nm laser, and photocurrent is measured using a SWCNT-cast film working electrode, a Pt counter electrode, and an Ag/AgCl reference electrode. Utilization of a regeneration cycle requires the addition of two dialyzers. The first dialyzer is responsible for the removal of surfactant from the solution for complex assembly (left). The second dialyzer is responsible for the addition of surfactant to the solution for complex disassembly and the removal of photodamaged components. The damaged components are replaced by new components, which are added directly to the PDMS solution. (b) The RC-ND-SWCNT solution demonstrates a decay in photocurrent (black) when under continuous illumination. This decay rate is comparable to those found in the literature for a solid-state dye-sensitized solar cell (DSSC) [36]. Utilization of the regeneration cycle allows for the recovery of photocurrent upon replacement of the damaged components. Operation of the regeneration cycle for 8 h after 32 h of illumination results in a 300% increase in overall efficiency over 168 h. (Reprinted from Macmillan Publishers Ltd. *Nat. Chem.*, Ham, M.H. et al., Photoelectrochemical complexes for solar energy conversion that chemically and autonomously regenerate, 2(11), 929–936, Copyright 2010.)

the initial photocurrent of the photoelectrochemical cell. Upon photodecay of this newly synthesized complex, the complex is once more subjected to a regeneration cycle wherein photodamaged components are disassembled and replaced with newly isolated RCs. Utilization of this regeneration cycle over 168 h increases overall cell efficiency by over 300% and prolongs cell lifetime indefinitely [21].

7.4 KINETIC MODEL FOR THE SELF-ASSEMBLY OF PHOTOACTIVE COMPLEXES

Key to the regeneration cycle of the photoelectrochemical cell is the ability of the photoactive to autonomously and reversibly self-assemble via chemical signaling alone. A fundamental understanding of the kinetics behind the self-assembly process can only be achieved with the development of a quantitative model of the dialysis [36]. A starting reaction mixture consists of 12.86 mM DMPC, 0.13 mM MSP, 0.36 mM SWCNT, and 60 mM SC surfactant. Prior to dialysis, the complex is in the disassembled state, and the micellar solution contains an initial concentration of surfactant. After the reaction mixture concentrations have equilibrated, it is placed within the dialysis setup shown in Figure 7.8. As shown in the setup, the reaction solution is contained within a 10-kDa MWCO dialysis membrane, and the solution-containing membrane is suspended within a constantly stirred beaker of 10 mM Tris buffer (pH 7.4).

Solution dialysis was modeled as three-stage process: equilibration prior to dialysis, dialysis above the critical micelle concentration (CMC), and dialysis below the CMC. Each of these stages is characterized by a different physical state and corresponding system of equations. The first stage, equilibration, occurs spontaneously upon addition of the various reaction components to the aqueous reaction mixture. In particular, introduction of proteins, lipids, surfactant, and nanotubes to an aqueous solution results in the spontaneous formation of various micelles (surfactant micelles, surfactant–lipid micelles) as well as suspended nanotubes (surfactant-suspended nanotubes, lipid-suspended nanotubes, protein-suspended nanotubes).

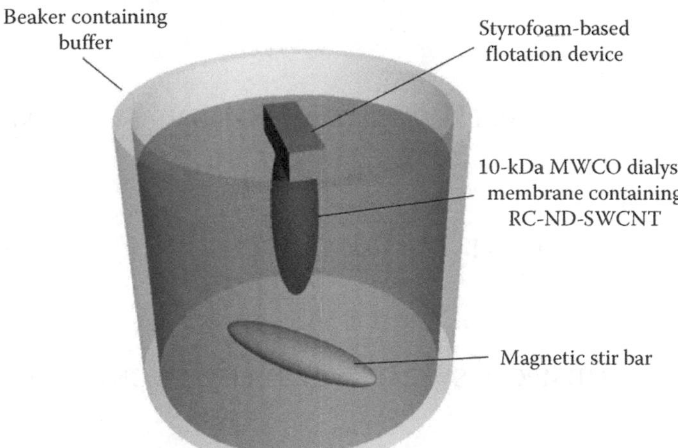

FIGURE 7.8 Dialysis setup. A beaker containing Tris buffer suspends a 10-kDa MWCO dialysis membrane filled with RC-ND-SWCNT solution. The membrane remains suspended using stryofoam-based flotation. A magnetic stir bar is placed within the beaker to ensure constant stirring in order to enhance convective mass transfer of the surfactant from within the membrane solution to the surrounding buffered solution. The dialysis rate is controlled by varying the concentration of surfactant in the surrounding buffer.

These agglomerates form spontaneously prior to solution dialysis. The second stage of the model is the dialysis of the solution above the CMC. Solution dialysis is initially marked by the removal of surfactant from the reaction mixture. Although the concentration of the predialysis agglomerates may vary according to the concentration of surfactant, no new species or agglomerates are formed until surfactant concentration approaches the CMC. Dialysis below the CMC, which is modeled in the third stage, introduces the formation of new agglomerates, including ND, ND-SWCNT, vesicle, and lipid bilayer formation. The details behind the kinetics of each stage are described in greater detail in the following sections [36].

7.4.1 Stage 1: Equilibration prior to Dialysis

Prior to dialysis, when individual components of the reaction mixture are added to the aqueous solution, they spontaneously self-assemble into thermodynamically stable agglomerates. Such agglomerates include the formation of micelles and various nanotube suspensions. In order to model the dialysis of the mixture, one must determine the starting, equilibrated concentrations of these various agglomerates. The formation of these agglomerates can be modeled by the following set of reactions:

$$A[\mathrm{SC}] \underset{k_{1r}}{\overset{k_{1f}}{\rightleftarrows}} [\mathrm{SC}]_{\mathrm{micelle}} \qquad \text{(Reaction 7.1)}$$

$$B[\mathrm{DMPC}] + C[\mathrm{SC}] \underset{k_{2r}}{\overset{k_{2f}}{\rightleftarrows}} [\mathrm{SC\text{-}DMPC}]_{\mathrm{micelle}} \qquad \text{(Reaction 7.2)}$$

$$[\mathrm{MSP}] + D[\mathrm{SC}] \underset{k_{3r}}{\overset{k_{3f}}{\rightleftarrows}} [\mathrm{SC\text{-}MSP}] \qquad \text{(Reaction 7.3)}$$

$$[\mathrm{SWCNT}] + E[\mathrm{SC}] \underset{k_{4r}}{\overset{k_{4f}}{\rightleftarrows}} [\mathrm{SC\text{-}SWCNT}] \qquad \text{(Reaction 7.4)}$$

$$F[\mathrm{DMPC}] + [\mathrm{SWCNT}] \underset{k_{5r}}{\overset{k_{5f}}{\rightleftarrows}} [\mathrm{DMPC\text{-}SWCNT}] \qquad \text{(Reaction 7.5)}$$

$$G[\mathrm{MSP}] + [\mathrm{SWCNT}] \underset{k_{6r}}{\overset{k_{6f}}{\rightleftarrows}} [\mathrm{MSP\text{-}SWCNT}] \qquad \text{(Reaction 7.6)}$$

$$2[\mathrm{SWCNT}] \xrightarrow{k_{7f}} [\mathrm{SWCNT}_2] \qquad \text{(Reaction 7.7)}$$

where [SC] is the concentration of free SC monomers, $[SC]_{micelle}$ is the concentration of SC micelles, [DMPC] is the concentration of free DMPC lipids, $[SC\text{-}DMPC]_{micelle}$ is the concentration of SC-DMPC mixed micelles, [MSP] is the concentration of MSP, [SC-MSP] is the concentration of SC-suspended MSP, [SWCNT] is the concentration of free SWCNTs, [SC-SWCNT] is the concentration of SC-suspended SWCNTs, [DMPC-SWCNT] is the concentration of DMPC-suspended SWCNTs, [MSP-SWCNT] is the concentration of MSP-suspended SWCNTs, and $[SWCNT_2]$ represents the concentration of bundled nanotubes. The coefficients A-G represent the stoichiometric quantities of their respective reactants, and k_f and k_r are the forward and reverse rate constants for each reaction. The concentration of each of the species over time, t, in the reaction mixture can be modeled according to mass-action kinetics:

$$\begin{aligned}\frac{d[\text{SC}]}{dt} &= -C * k_{2f}[\text{DMPC}][\text{SC}][\text{SC-DMPC}] + C * k_{2r}[\text{SC-DMPC}] \\ &\quad - D * k_{3f}[\text{MSP}][\text{SC}] + D * k_{3r}[\text{SC-MSP}] - E * k_{4f}[\text{SWNT}][\text{SC}] \\ &\quad + E * k_{4r}[\text{SC-SWNT}]\end{aligned} \tag{7.1}$$

$$\begin{aligned}\frac{d[\text{DMPC}]}{dt} &= -B * k_{2f}[\text{DMPC}][\text{SC}][\text{SC-DMPC}] + B * k_{2r}[\text{SC-DMPC}] \\ &\quad - F * k_{5f}[\text{DMPC}][\text{SWNT}] + F * k_{5r}[\text{DMPC-SWNT}]\end{aligned} \tag{7.2}$$

$$\begin{aligned}\frac{d[\text{MSP}]}{dt} &= -k_{3f}[\text{MSP}][\text{SC}] + k_{3r}[\text{SC-MSP}] - G * k_{6f}[\text{MSP}][\text{SWNT}] \\ &\quad + G * k_{6r}[\text{MSP-SWNT}]\end{aligned} \tag{7.3}$$

$$\begin{aligned}\frac{d[\text{SWCNT}]}{dt} &= -k_{4f}[\text{SWCNT}][\text{SC}] + k_{4r}[\text{SC-SWCNT}] - k_{5f}[\text{DMPC}][\text{SWCNT}] \\ &\quad + k_{5r}[\text{DMPC-SWCNT}] - k_{6f}[\text{MSP}][\text{SWCNT}] + k_{6r}[\text{MSP-SWCNT}] \\ &\quad - 2 * k_{7f}[\text{SWCNT}]^2 + 2 * k_{7r}[\text{SWCNT}_2]\end{aligned} \tag{7.4}$$

$$\frac{D[\text{SC-MSP}]}{dt} = k_{3f}[\text{MSP}][\text{SC}] - k_{3r}[\text{SC-MSP}] \tag{7.5}$$

$$\frac{d[\text{SC-SWNT}]}{dt} = k_{4f}[\text{SC}][\text{SWNT}] - k_{4r}[\text{SC-SWNT}] \tag{7.6}$$

$$\frac{d[\text{DMPC-SWNT}]}{dt} = k_{5f}[\text{DMPC}][\text{SWNT}] - k_{5r}[\text{DMPC-SWNT}] \tag{7.7}$$

$$\frac{d[\text{MSP-SWNT}]}{dt} = k_{6f}[\text{MSP}][\text{SWNT}] - k_{6r}[\text{MSP-SWNT}] \tag{7.8}$$

$$\frac{d[\text{SWNT}_2]}{dt} = k_{7f}[\text{SWNT}]^2 - k_{7r}[\text{SWNT}_2] \tag{7.9}$$

To mimic experimental conditions, equilibration of this reaction mixture was modeled by monitoring species' concentrations for approximately an hour, after which pseudo-equilibration is established as species' concentrations remain constant. The final concentrations, which serve as starting concentration in the next stage of the model, are only metastable provided the irreversible formation of SWCNT bundles, which would tend the system toward complete bundling as time increases indefinitely [36].

7.4.2 Stage 2: Dialysis above the CMC

The metastable concentrations obtained after system equilibration serves as the starting concentrations of the components upon solution dialysis. In addition to the aforementioned set of reactions, dialysis introduces another reaction demonstrative of the removal of surfactant from the reaction mixture

$$[\text{SC}] \rightarrow [\text{SC}]_{\text{removed}} \tag{Reaction 7.8}$$

where $[\text{SC}]_{\text{removed}}$ is the concentration of SC removed from the reaction solution. The surfactant removal rate can be modeled as an exponential decay reminiscent of kidney dialysis rates:

$$\frac{d[\text{SC}]_{\text{removed}}}{dt} = r[\text{SC}]_{\text{initial}}\, e^{-rt} \tag{7.10}$$

where

r is the dialysis rate

$[\text{SC}]_{\text{initial}}$ is the initial SC concentration

To account for the removal of SC from the system, an additional term must be added to the SC balance (Equation 7.1):

$$\begin{aligned}\frac{d[\text{SC}]}{dt} &= -C * k_{2f}[\text{DMPC}][\text{SC}][\text{SC-DMPC}] + C * k_{2r}[\text{SC-DMPC}] \\ &\quad - D * k_{3f}[\text{MSP}][\text{SC}] + D * k_{3r}[\text{SC-MSP}] - E * k_{4f}[\text{SWNT}][\text{SC}] \\ &\quad + E * k_{4r}[\text{SC-SWNT}] - r[\text{SC}]_{\text{initial}}\, e^{-rt}\end{aligned} \tag{7.11}$$

Equations 7.2 through 7.11 characterize the second stage of the model, dialysis above the CMC.

7.4.3 Stage 3: Dialysis below the CMC

As SC is continually removed from the system, the reaction solution approaches the third stage of the model, which is dialysis below the CMC. Dialysis below the CMC is characterized by the introduction of new, supramolecular agglomerates consisting of surfactant-based vesicles, bilayers, and, most importantly, ND and ND-SWCNT formation. The vesicle and lipid bilayer formation is marked by a variation in the lipid-to-surfactant ratio shown in Reaction 7.2 and a variation in the corresponding rate constants. A more thorough discussion on surfactant- and lipid-based agglomeration is provided in Section 7.4.4. Formation of ND and ND-SWCNT complexes is summarized by the reactions shown as follows:

$$H[\text{DMPC}] + I[\text{MSP}] \underset{k_{9r}}{\overset{k_{9f}}{\rightleftarrows}} [\text{ND}] \quad \text{(Reaction 7.9)}$$

$$J[\text{ND}] + [\text{SWCNT}] \underset{k_{10r}}{\overset{k_{10f}}{\rightleftarrows}} [\text{ND-SWCNT}] \quad \text{(Reaction 7.10)}$$

where

[ND] designates ND concentration

[ND-SWCNT] is ND-SWCNT formation

The corresponding rates of formation for ND and ND-SWCNT are described as follows:

$$\frac{d[\text{ND}]}{dt} = k_{8f}[\text{DMPC}][\text{MSP}] - k_{8r}[\text{ND}] - H * k_{9f}[\text{ND}][\text{SWNT}] + H * k_{9r}[\text{ND-SWNT}] \quad (7.12)$$

$$\frac{d[\text{ND-SWNT}]}{dt} = k_{9f}[\text{ND}][\text{SWNT}] - k_{9r}[\text{ND-SWNT}] \quad (7.13)$$

Additional terms must be added to Equations 7.2 through 7.4 to account for DMPC, MSP, and SWCNT participation in ND and ND-SWCNT formation:

$$\frac{d[\text{DMPC}]}{dt} = -B * k_{2f}[\text{DMPC}][\text{SC}] + B * k_{2r}[\text{SC-DMPC}] - F * k_{5f}[\text{DMPC}][\text{SWCNT}]$$
$$+ F * k_{5r}[\text{DMPC-SWCNT}] - 150 * k_{8f}[\text{DMPC}][\text{MSP}] + 150 * k_{8r}[\text{ND}] \quad (7.14)$$

$$\frac{d[\text{MSP}]}{dt} = -k_{3f}[\text{MSP}][\text{SC}] + k_{3r}[\text{SC-MSP}] - G * k_{6f}[\text{MSP}][\text{SWCNT}]$$
$$+ G * k_{6r}[\text{MSP-SWCNT}] - 2 * k_{8f}[\text{DMPC}][\text{MSP}] + 2 * k_{8r}[\text{ND}] \quad (7.15)$$

$$\begin{aligned}\frac{d[\text{SWCNT}]}{dt} &= -k_{4f}[\text{SWCNT}][\text{SC}] + k_{4r}[\text{SC-SWCNT}] - k_{5f}[\text{DMPC}][\text{SWCNT}]\\ &\quad + k_{5r}[\text{DMPC-SWCNT}] - k_{6f}[\text{MSP}][\text{SWCNT}] + k_{6r}[\text{MSP-SWCNT}]\\ &\quad - 2*k_{7f}[\text{SWCNT}]^2 + 2*k_{7r}[\text{SWCNT}_2] - k_{9f}[\text{ND}][\text{SWCNT}]\\ &\quad + k_{9r}[\text{ND-SWCNT}]\end{aligned} \tag{7.16}$$

Dialysis below the CMC is thus modeled using Equations 7.5 through 7.16.

7.4.4 Determination of Kinetic Rate Constants and Stoichiometric Coefficients

Introduction of surfactant molecules to an aqueous solution results in the spontaneous formation of thermodynamically favored agglomerates. The size and composition of these agglomerates vary based on the relative concentration of surfactant in solution as well as temperature. In particular, high concentrations of surfactant exceeding the CMC yield spherical micelles. In the presence of surfactant, these micelles suspend less soluble, individual lipid molecules in solution. As surfactant concentration decreases, the solution transitions from a micellar to a transitional, and finally, to a lamellar phase yielding planar, lipid bilayers.

Though the agglomerates undergo various structural changes throughout dialysis, kinetically, these changes can be captured by variation of the surfactant-to-lipid ratio, Re, the agglomerate size, N, and the forward and reverse rate constants. Specifically, with decreasing surfactant concentration, Re decreases and N increases. Mathematically, these two parameters are reflected in the stoichiometric coefficients B and C:

$$B = \frac{N}{1+\text{Re}} \tag{7.17}$$

$$C = \frac{\text{Re}*N}{1+\text{Re}} \tag{7.18}$$

Values for Re and N were varied throughout dialysis, with dependency on surfactant concentration and the corresponding phase (micellar, transitional, and lamellar). Similarly, forward and reverse rate constants were also varied according to the surfactant phase of the solution. Corresponding values for Re, N, k_f, and k_r (for Reactions 7.1 and 7.2) were determined on the onset of each phase according to literature values. Intermittent values varied linearly within each of the phases.

The kinetic rate constants and stoichiometric coefficients for the remaining reactions (Reactions 7.3 through 7.7, 7.9, and 7.10) were determined using a combination of empirical approximations and experimental data. The kinetic rate constants for Reactions 7.3 through 7.7 were calculated according to diffusion-controlled kinetics. The stoichiometric coefficients D, F, and G were calculated based on surface area coverage of the larger molecule. Coefficients *E*, *H*, and *I* were calculated according to the literature, and coefficient *J* was determined based on AFM results [36].

7.4.5 Results: Complex Formation throughout Dialysis

The aforementioned model was implemented to determine the forward and reverse rate kinetic constants for ND and ND-SWCNT formation (Reactions 7.9 and 7.10). The model was fit to the normalized ND-SWCNT concentration over time, and the best-fit rate constants are $k_{9f} = 70\,\text{mM}^{-1}\,\text{s}^{-1}$ and $k_{10f} = 5.4 \times 10^2\,\text{mM}^{-1}\,\text{s}^{-1}$ (Figure 7.9a).

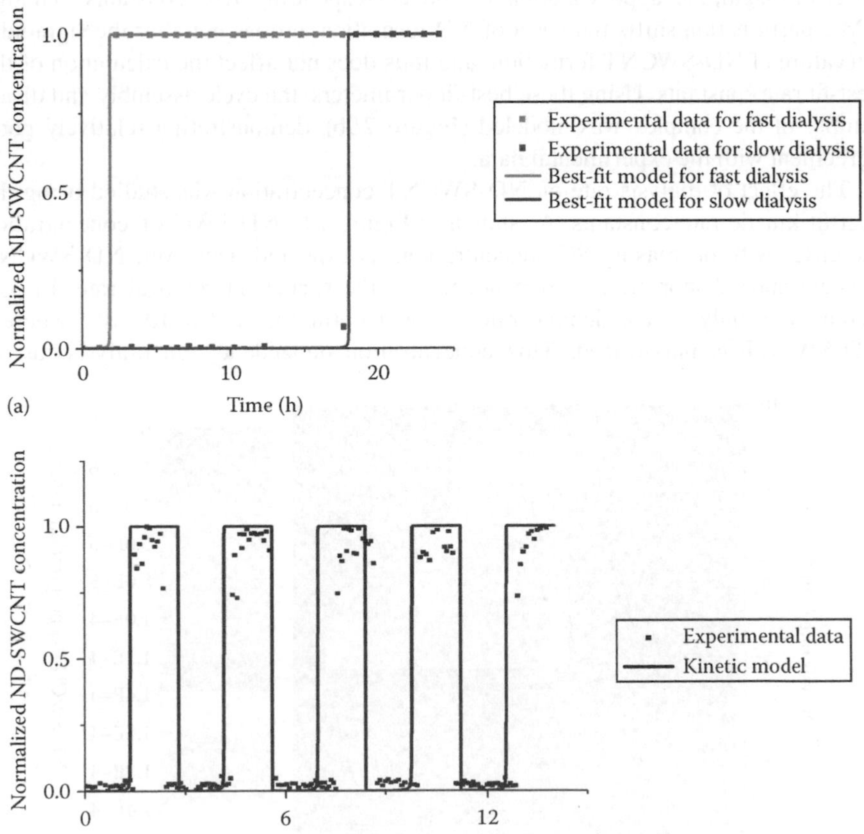

FIGURE 7.9 Comparison of self-assembly kinetic model and experimental data. (a) Experimental data (dots) of normalized ND-SWCNT concentration over time were fitted by the kinetic model for both slow (black) and fast (grey) dialysis rates. The best-fit traces (lines) yield a forward rate constant of $79\,\text{mM}^{-1}\,\text{s}^{-1}$ for ND formation and $5.4 \times 10^2\,\text{mM}^{-1}\,\text{s}^{-1}$ for ND-SWCNT formation. (b) The best-fit rate parameters were used to simulate the cyclic assembly and disassembly of the ND-SWCNT complex, which is in relatively good agreement with the experimental data (dots). (Reprinted with permission from Boghossian, A.A. et al., Dynamic and reversible self-assembly of photoelectrochemical complexes based on lipid bilayer disks, photosynthetic reaction centers, and single-walled carbon nanotubes, *Langmuir*, 27(5), 1599–1609, 2011. Copyright 2011 American Chemical Society.)

In a diffusion-controlled solution, the relatively bulky ND and SWCNT components are expected to demonstrate a smaller rate constant of formation compared with that of the more mobile lipid molecules during ND formation. However, in this case, the opposite scenario holds: the ND-SWCNT rate constant is larger than the ND rate constant, which is indicative of a nondiffusion-controlled condition.

The sensitivity of the best-fit trace with respect to variations in calculated ND and ND-SWCNT rate constants and perturbations in the predicted CMC was also examined. The results of this study reveals that the best-fit traces are relatively sensitive to perturbations to ND and ND-SWCNT rate constants, and thus can offer a valid, order-of-magnitude approximation of the corresponding rate constants. Though CMC perturbation shifts the onset of ND formation, it does not alter the sigmoidal curvature of ND-SWCNT formation, and thus does not affect the calculation of the best-fit rate constants. Using these best-fit parameters, the cycle assembly and disassembly of the complex was modeled (Figure 7.9b), demonstrating relatively good agreement with the experimental data.

The effect of dialysis rate on ND-SWCNT concentration was studied using the best-fit kinetic rate constants. As shown in Figure 7.10 ND-SWCNT concentration increases with decreasing SC concentration, as expected. However, ND-SWCNT concentration demonstrates a dependence on the surfactant removal rate. In particular, the dialysis rate demonstrates a local optimum of 8×10^{-4} s^{-1}, wherein ND-SWCNT is maximized. This concentration dependency on dialysis rate is

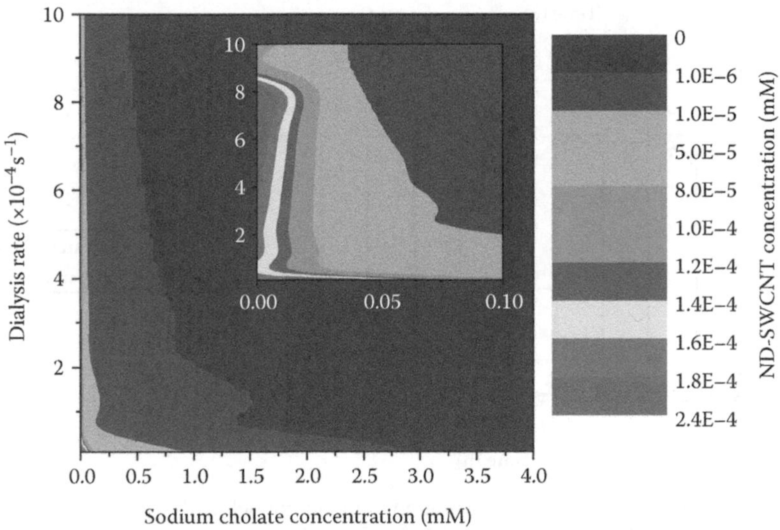

FIGURE 7.10 **(See color insert.)** Effect of dialysis rate on complex formation. The kinetic model was used to predict ND-SWCNT concentration as a function of surfactant concentration and dialysis rate. Complex concentration increases with decreasing surfactant concentration. Complex concentration is also affected by the rate of surfactant dialysis, with a local dialysis rate optimum of $\sim 8 \times 10^{-4}$ s^{-1}. (Reprinted with permission from Boghossian, A.A. et al., Dynamic and reversible self-assembly of photoelectrochemical complexes based on lipid bilayer disks, photosynthetic reaction centers, and single-walled carbon nanotubes, *Langmuir*, 27(5), 1599–1609, 2011. Copyright 2011 American Chemical Society.)

attributed to opposing contributions from increased ND-SWCNT formation at faster rates, the competitive, irreversible formation of SWCNT bundles, and the relatively slower formation of NDs [36].

7.5 CONCLUSIONS

Plants have evolved highly sophisticated mechanisms of self-repair wherein photodamaged proteins trigger the autonomous disassembly of a photoactive protein complex. Upon disassembly, the photodamaged protein is replaced with a newly biosynthesized protein, and the remaining components spontaneously self-assemble to form the functional photoactive protein complex. This self-repair cycle contributes to both enhanced photosynthetic efficiency and prolonged cell lifetime.

The advantages of this self-repair cycle have recently been realized in photovoltaics with the development of the first synthetic photoelectrochemical complex capable of mimicking key aspects of this self-repair cycle. In this platform, the removal and addition of surfactant triggers the spontaneous assembly and disassembly, respectively, of the photoactive RC-ND-SWCNT complex. The structural configuration and composition of this complex was verified using AFM, SANS, and PL measurements, and the complex demonstrates 40% external quantum efficiency. Utilization of a regeneration cycle wherein photodamaged components of the photoactive complex are autonomously replaces demonstrates a 300% increase in efficiency over 168 h of illumination and the indefinite prolongation of cell lifetime.

The kinetics of this self-assembly process was modeled and fit to experimental data capturing the formation of the ND-SWCNT complex. According to best-fit parameters, the predicted forward rate constants for ND and ND-SWCNT formation are $k_{9f} = 70\,mM^{-1}\,s^{-1}$ and $k_{10f} = 5.4 \times 10^2\,mM^{-1}\,s^{-1}$. These parameters were used to model the effect of surfactant removal rate on ND-SWCNT concentration. The results of this study predict an optimal dialysis rate of approximating $8 \times 10^{-4}\,s^{-1}$, wherein ND-SWCNT concentration is maximized.

REFERENCES

1. Krause, G.H., in *Causes of Photooxidative Stress and Amelioration of Defense Systems in Plants*, C.H. Foyer and P.M. Mullineaux, Eds., 1994, Boca Raton, FL: CRC Press, pp. 43–76.
2. Singh, M., Turnover of D1 protein encoded by psbA; gene in higher plants and cyanobacteria sustains photosynthetic efficiency to maintain plant productivity under photoinhibitory irradiance. *Photosynthetica*, 2000, 38(2): 161–169.
3. Iwai, M. et al., Live-cell imaging of photosystem II antenna dissociation during state transitions. *Proceedings of the National Academy of Sciences*, 2010, 107(5): 2337–2342.
4. Dannehl, H. et al., Changes in D1-protein turnover and recovery of photosystem II activity precede accumulation of chlorophyll in plants after release from mineral stress. *Planta*, 1996, 199(1): 34–42.
5. Drepper, F. et al., Lateral diffusion of an integral membrane protein: Monte Carlo analysis of the migration of phosphorylated light-harvesting complex II in the thylakoid membrane. *Biochemistry*, 1993, 32(44): 11915–11922.
6. Jorgensen, M., K. Norrman, and F.C. Krebs, Stability/degradation of polymer solar cells. *Solar Energy Materials and Solar Cells*, 2008, 92(7): 686–714.

7. Andersson, B.A., Materials availability for large-scale thin-film photovoltaics. *Progress in Photovoltaics*, 2000, 8(1): 61–76.
8. Birkmire, R.W. and B.E. McCandless, CdTe thin film technology: Leading thin film PV into the future. *Current Opinion in Solid State & Materials Science*, 2010, 14(6): 139–142.
9. Fahr, S., C. Rockstuhl, and F. Lederer, Improving the efficiency of thin film tandem solar cells by plasmonic intermediate reflectors. *Photonics and Nanostructures—Fundamentals and Applications*, 2010, 8(4): 291–296.
10. Kosyachenko, L.A. and E.V. Grushko, Open-circuit voltage, fill factor, and efficiency of a CdS/CdTe solar cell. *Semiconductors*, 2010, 44(10): 1375–1382.
11. Goetzberger, A., C. Hebling, and H.W. Schock, Photovoltaic materials, history, status and outlook. *Materials Science & Engineering R—Reports*, 2003, 40(1): 1–46.
12. Pizzini, S., Towards solar grade silicon: Challenges and benefits for low cost photovoltaics. *Solar Energy Materials and Solar Cells*, 2010, 94(9): 1528–1533.
13. Faulkner, C.J. et al., Rapid assembly of photosystem I monolayers on gold electrodes. *Langmuir*, 2008, 24(16): 8409–8412.
14. Frolov, L. et al., Fabrication of a photoelectronic device by direct chemical binding of the photosynthetic reaction center protein to metal surfaces. *Advanced Materials*, 2005, 17(20): 2434–2437.
15. Ciesielski, P.N. et al., Functionalized nanoporous gold leaf electrode films for the immobilization of photosystem I. *ACS Nano*, 2008, 2(12): 2465–2472.
16. Terasaki, N. et al., Bio-photo sensor: Cyanobacterial photosystem I coupled with transistor via molecular wire. *Biochimica et Biophysica Acta—Bioenergetics*, 2007, 1767(6): 653–659.
17. Terasaki, N. et al., Plugging a molecular wire into photosystem I: Reconstitution of the photoelectric conversion system on a gold electrode. *Angewandte Chemie—International Edition*, 2009, 48(9): 1585–1587.
18. Trammell, S.A. et al., Orientated binding of photosynthetic reaction centers on gold using Ni-NTA self-assembled monolayers. *Biosensors & Bioelectronics*, 2004, 19(12): 1649–1655.
19. Lebedev, N. et al., Conductive wiring of immobilized photosynthetic reaction center to electrode by cytochrome c. *Journal of the American Chemical Society*, 2006, 128(37): 12044–12045.
20. Carmeli, I. et al., A photosynthetic reaction center covalently bound to carbon nanotubes. *Advanced Materials*, 2007, 19(22): 3901–3904.
21. Ham, M.H. et al., Photoelectrochemical complexes for solar energy conversion that chemically and autonomously regenerate. *Nature Chemistry*, 2010, 2(11): 929–936.
22. Bidlack, J.E.S. and S.H. Jansky, *Introductory Plant Biology*, 2003, New York: McGraw-Hill.
23. Asimov, I., *Photosynthesis*, 1968, New York: Basic Books, Inc.
24. Rabinowitch, E.G., *Photosynthesis*, 1969, London, U.K.: John Wiley.
25. Reece, J.C., *Biology*, 2005, San Francisco, CA: Pearson, Benjamin Cummings.
26. Blankenship, R.E., *Molecular Mechanisms of Photosynthesis*, 2nd edn., 2008, New York: John Wiley & Sons Inc.
27. Blankenship, R.E. and S. Govindjee, Photosynthesis, in *McGraw Hill Encyclopedia of Science and Technology*, 2007, New York: McGraw-Hill Professional.
28. Berg, J.M.T., L. John, and L. Stryer, *The Calvin Cycle and the Pentose Phosphate Pathway. Biochemistry*, 2002, New York: W.H. Freeman and Company.
29. Melis, A., Photosystem-II damage and repair cycle in chloroplasts: What modulates the rate of photodamage in vivo? *Trends in Plant Science*, 1999, 4(4): 130–135.
30. Prazil, O.A. and I. Ohad, *The Photosystems: Structure, Function and Molecular Biology*, J. Barber, Ed., 1992, Amsterdam, the Netherlands: Elsevier Science Publishers.

31. Constant, S. et al., Expression of the psbA gene during photoinhibition and recovery in *Synechocystis* PCC 6714: Inhibition and damage of transcriptional and translational machinery prevent the restoration of photosystem II activity. *Plant Molecular Biology*, 1997, 34(1): 1–13.
32. Saito, R., M.S. Dresselhaus, and G. Dresselhaus, *Physical Properties of Carbon Nanotubes*, 1998, London, U.K.: Imperial College Press.
33. Dresselhaus, M.S. et al., Exciton photophysics of carbon nanotubes. *Annual Review of Physical Chemistry*, 2007, 58: 719–747.
34. Bachilo, S.M. et al., Structure-assigned optical spectra of single-walled carbon nanotubes. *Science*, 2002, 298(5602): 2361–2366.
35. Boghossian, A.A. et al., Dynamic and reversible self-assembly of photoelectrochemical complexes based on lipid bilayer disks, photosynthetic reaction centers, and single-walled carbon nanotubes. *Langmuir*, 2011, 27(5): 1599–1609.
36. Biancardo, M., K. West, and F.C. Krebs, Quasi-solid-state dye-sensitized solar cells: Pt and PEDOT:PSS counter-electrodes applied to gel electrolyte assemblies. *Journal of Photochemistry and Photobiology A: Chemistry*, 2007, 187(2–3): 395–401.

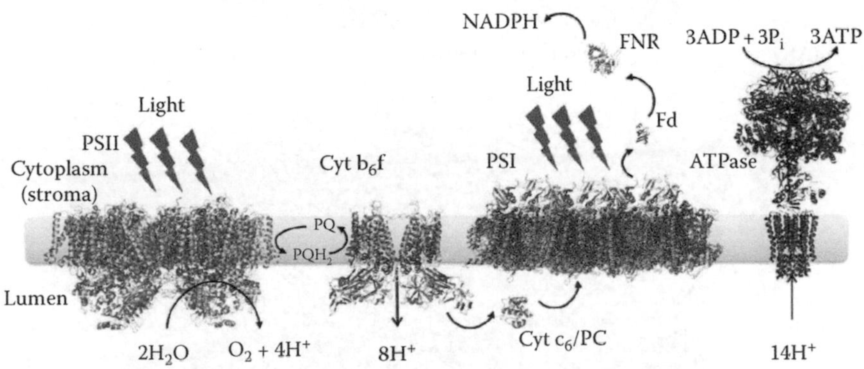

FIGURE 1.1 Protein components of the photosynthetic electron transport chain. Electron transport from water to $NADP^+$ is driven by PSII extracting electrons from water and transferring them to plastoquinone (PQ) to produce plastoquinol (PQH_2). PSI catalyzes the reduction of ferredoxin and, subsequently, formation of NADPH via ferredoxin-$NADP^+$ reductase (FNR). Oxidized PSI is re-reduced by electrons from the cytochrome b_6f complex (Cyt b_6f) via cytochrome c_6 (Cyt c_6) or plastocyanin (PC), whereas PQH_2 produced by PSII reduces oxidized Cyt b_6f. Electron transport is coupled to the translocation of protons from the cytoplasm in cyanobacteria (or stroma in chloroplasts) to the lumen to produce a proton-motive force. In the case of Cyt b_6f, the Q-cycle operates (not shown in the figure to aid clarity) to give a stoichiometry of eight H^+ deposited in the lumen per four electrons transported to PSI. Additional translocation of protons across the membrane is mediated by cyclic electron flow around PSI (data not shown). The proton-motive force coupled to movement of protons back across the membrane through the ATP synthase (ATPase) drives synthesis of ATP from ADP and inorganic phosphate (P_i). The protein complexes shown are the crystal structures derived from various cyanobacteria.

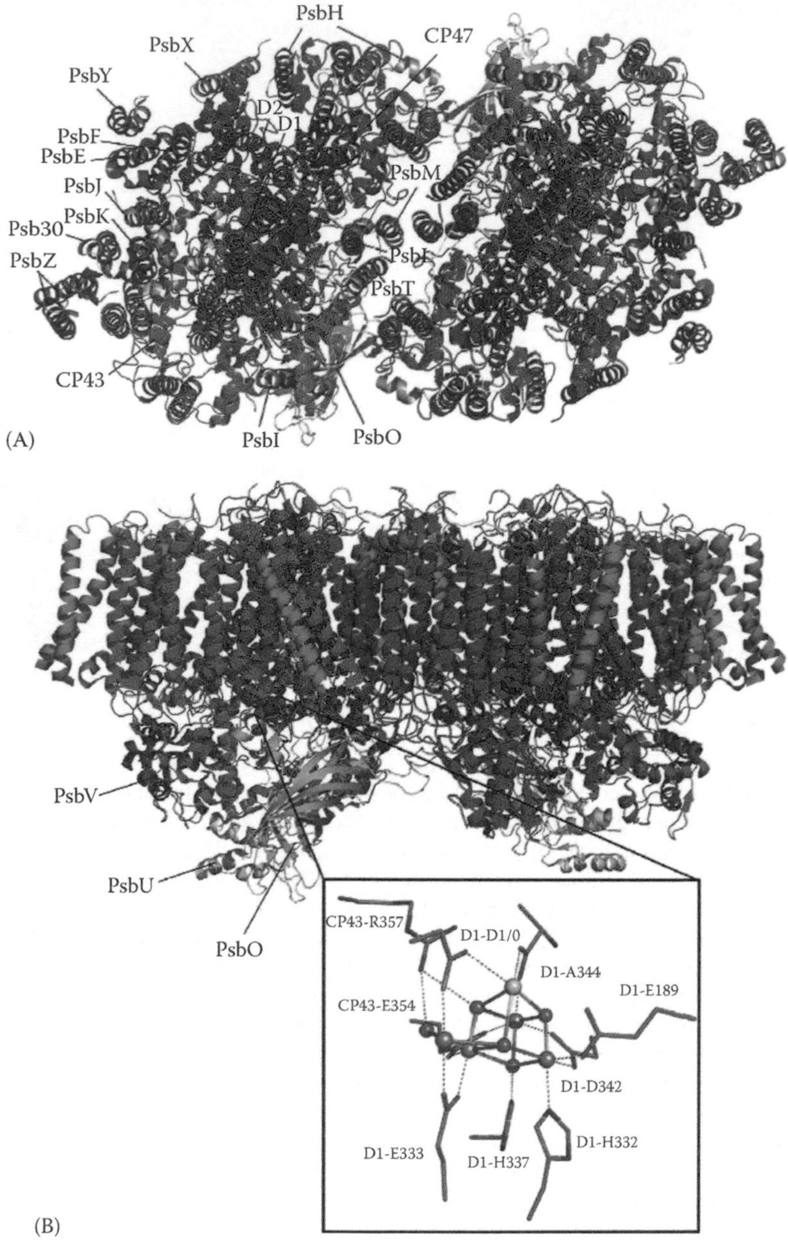

FIGURE 1.2 Crystal structure of the PSII complex isolated from thermophilic cyanobacteria. View of the homodimeric complex from *Thermosynechococcus elongatus* from the cytoplasmic side of the thylakoid membrane (panel A) and perpendicular to the membrane normal (panel B). The 20 subunits have been annotated in panel A and color-coded: D1 (yellow), D2 (orange), CP43 (green), CP47 (red), cytochrome b-559 (purple), PsbO (violet), PsbV (dark blue), and PsbU (light blue). The remaining 11 small transmembrane subunits are shown in gray. For clarity, pigments have been omitted. The inset in panel B shows the structure of the Mn_4CaO_5 cluster involved in water oxidation, with coordinating amino acid residues, determined for *Thermosynechococcus vulcanus* (Umena et al. 2011). The Ca ion is shown in gray, the Mn ions in violet, and the oxo bridges in red. For clarity, bound waters have been omitted. The figure was created with the software Pymol (http://pymol.sourceforge.net, version 0.99) and the PDB files 3BZ1 and 3BZ2 (Guskov et al. 2009) and 3ARC (Umena et al. 2011).

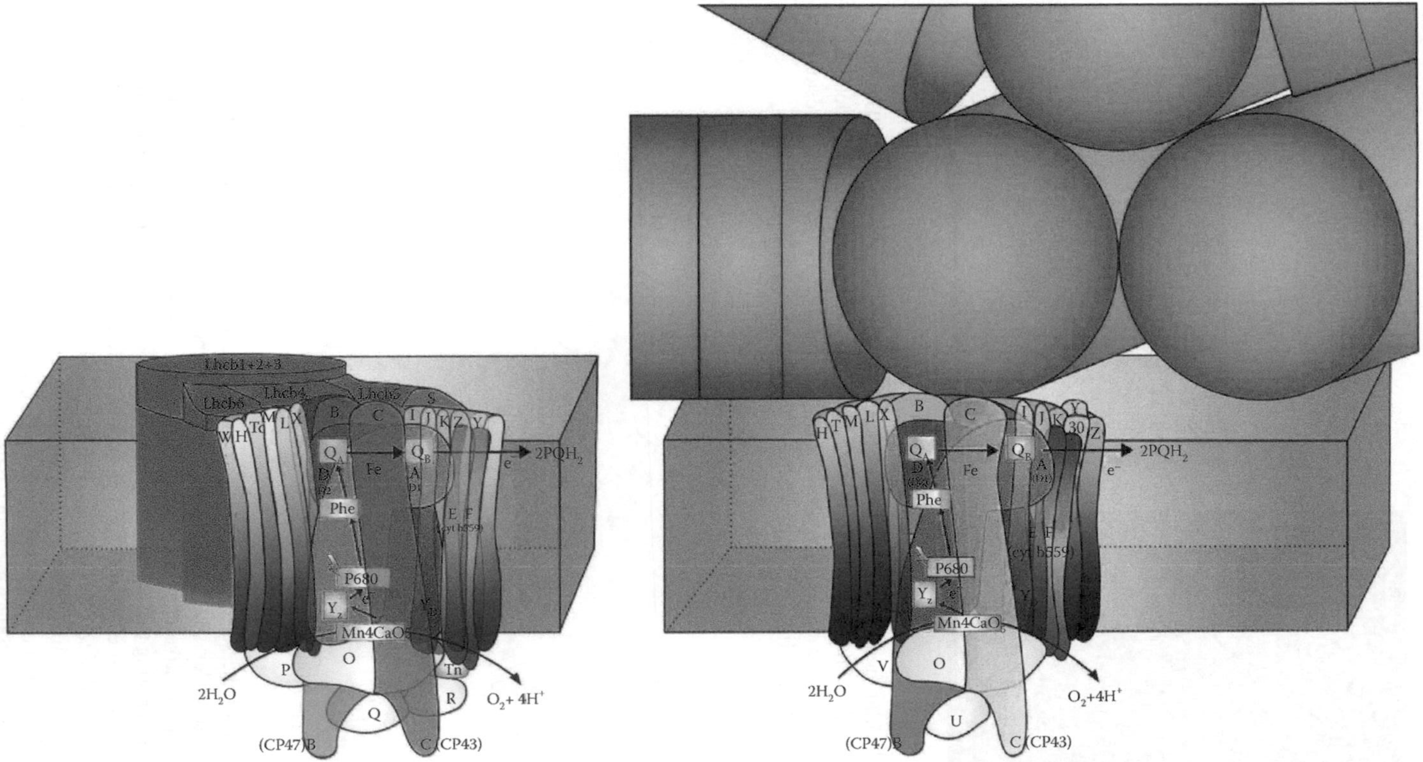

FIGURE 1.3 Comparison of PSII in chloroplasts (left panel) and cyanobacteria (right panel). Excitation of the primary electron donor, P680, leads to stepwise reduction of the pheophytin electron acceptor (Phe) and the plastoquinones, Q_A and Q_B, located close to a non-heme iron (Fe). PQH_2 is produced after two photoacts. On the donor side, $P680^+$ oxidizes tyrosine, Y_z, which in turn oxidizes the Mn_4CaO_5 oxygen-evolving center. Other redox-active components not involved directly in water oxidation include a second redox-active tyrosine, Y_D, located within the D2 subunit and the heterodimeric Cyt b-559 complex. Subunits are annotated so that PsbA is labeled A, and so on. Chloroplasts contain an integral light-harvesting system (Lhcb subunits), whereas cyanobacteria contain the extremely large phycobilisome (large blue–green cylinders) that docks on to the cytoplasmic surface of the thylakoid membrane. The figure was kindly provided by Dr. Jon Nield (http://www.queenmaryphotosynthesis.org//nield//).

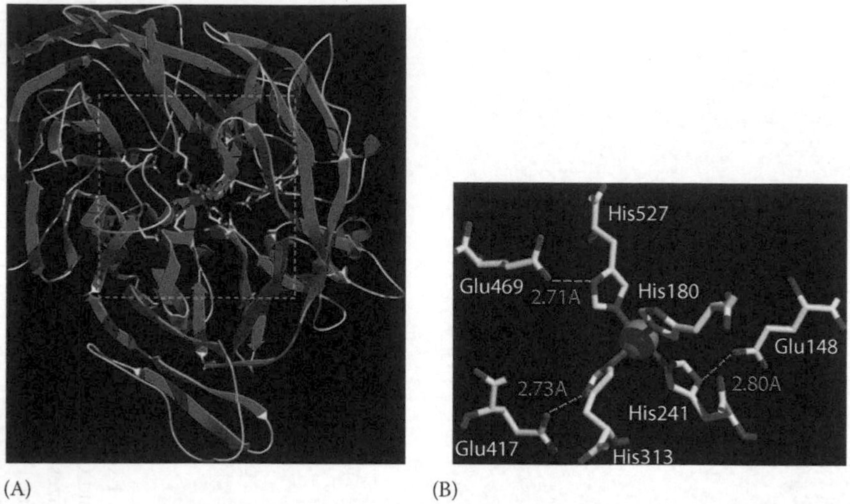

FIGURE 2.4 Molecular model of the bovine RPE65 iron-binding site. (A) Overall 3D structure of RPE65, where the conserved Fe^{2+} site marked with purple dashed line. (B) A detailed view of the RPE65 structure in the vicinity of iron-binding site. The conserved residues His180, His241, His313, and His527 coordinate the catalytically essential iron ion.

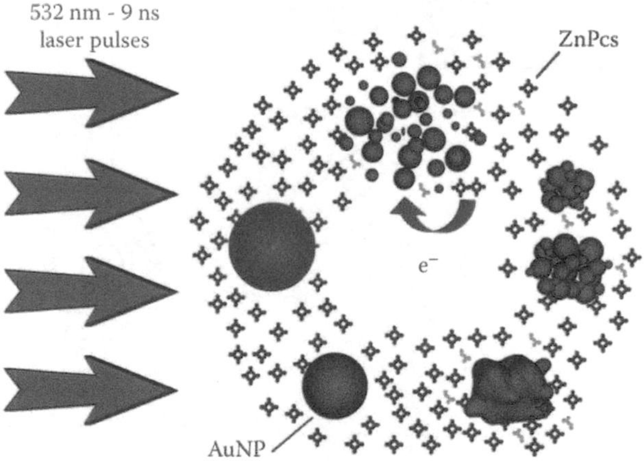

FIGURE 4.9 Sketch of the self-healing mechanism for AuNPs in the presence of ZnPcs. From left in clockwise order: AuNPs absorb photons and heat up over the fragmentation threshold; some of the positive charges of photofragmented AuNPs are neutralized by the oxidation of ZnPcs, promoting their aggregation; aggregates are photomelted into new spherical AuNPs, that are ready to efficiently limit light again. (Reprinted with permission from Amendola, V., Dini, D., Polizzi, S., Shen, J., Kadish, K.M., Calvete, M.J.F., Hanack, M., and Meneghetti, M., Self-healing of gold nanoparticles in the presence of zinc phthalocyanines and their very efficient nonlinear absorption performances, *J. Phys. Chem. C*, 113(20), 8688–8695, 2009. Copyright 2009 American Chemical Society.)

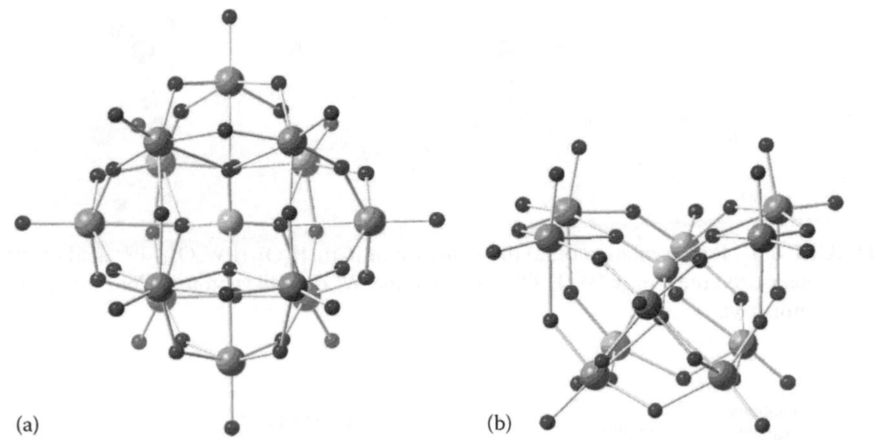

FIGURE 6.7 (a) Representation of the structure of the $[\alpha\text{-}SiW_{12}O_{40}]^{4-}$ polyoxometalate. (b) Representation of the structure of $[\gamma\text{-}SiW_{10}O_{36}]^{8-}$ polyoxometalate. Blue atoms: W; gray atoms: Si; red atoms: O.

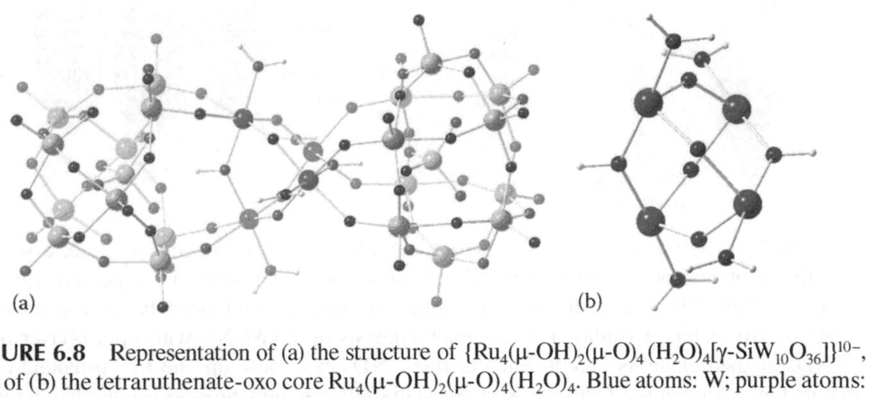

FIGURE 6.8 Representation of (a) the structure of $\{Ru_4(\mu\text{-}OH)_2(\mu\text{-}O)_4(H_2O)_4[\gamma\text{-}SiW_{10}O_{36}]\}^{10-}$, and of (b) the tetraruthenate-oxo core $Ru_4(\mu\text{-}OH)_2(\mu\text{-}O)_4(H_2O)_4$. Blue atoms: W; purple atoms: Ru; gray atoms: Si; red atoms: O; white atoms: H.

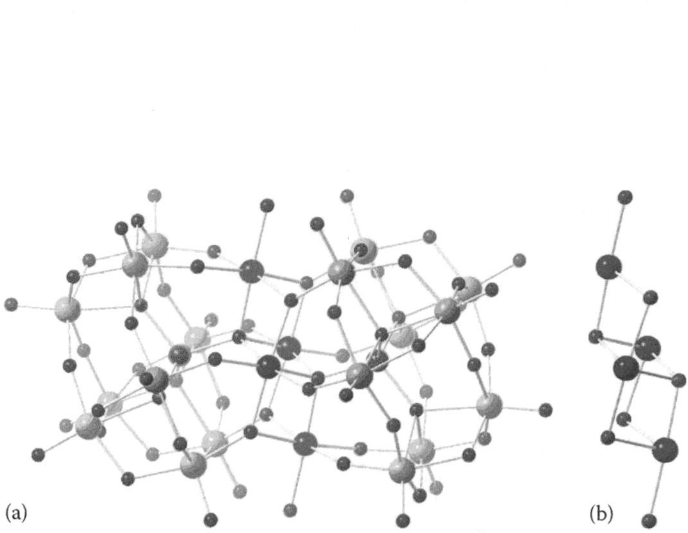

FIGURE 6.9 Representation of (a) the structure of $[Co_4(H_2O)_2(PW_9O_{34})_2]^{10-}$, and of (b) the tetracobalt-oxo core $Co_4(\mu\text{-}O)_4(H_2O)_2$. Blue atoms: W; dark blue atoms: Co; orange atoms: P; red atoms: O.

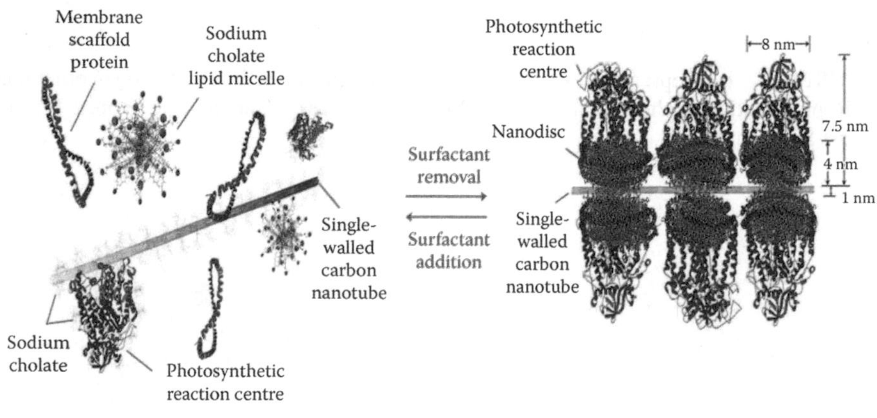

FIGURE 7.4 Self-assembly of photoactive complex. In the disassembled state, a solution containing photosynthetic RCs, SWCNT, phospholipids, and MSPs is dispersed using SC (left). Upon dialysis of the latter, the remaining components spontaneously self-assemble into a photoactive complex (right). This complex consists of a SWCNT with lipid bilayer disks, or NDs aligned along its length. Each of these NDs can house up one RC, orientated such that its hole injection site faces the SWCNT. This self-assembly process is completely reversible in that the readdition of surfactant once more disassembles the complex into its initial micellar state (left). (Reprinted from Macmillan Publishers Ltd. *Nat. Chem.*, Ham, M.H. et al., Photoelectrochemical complexes for solar energy conversion that chemically and autonomously regenerate, 2(11), 929–936, Copyright 2010.)

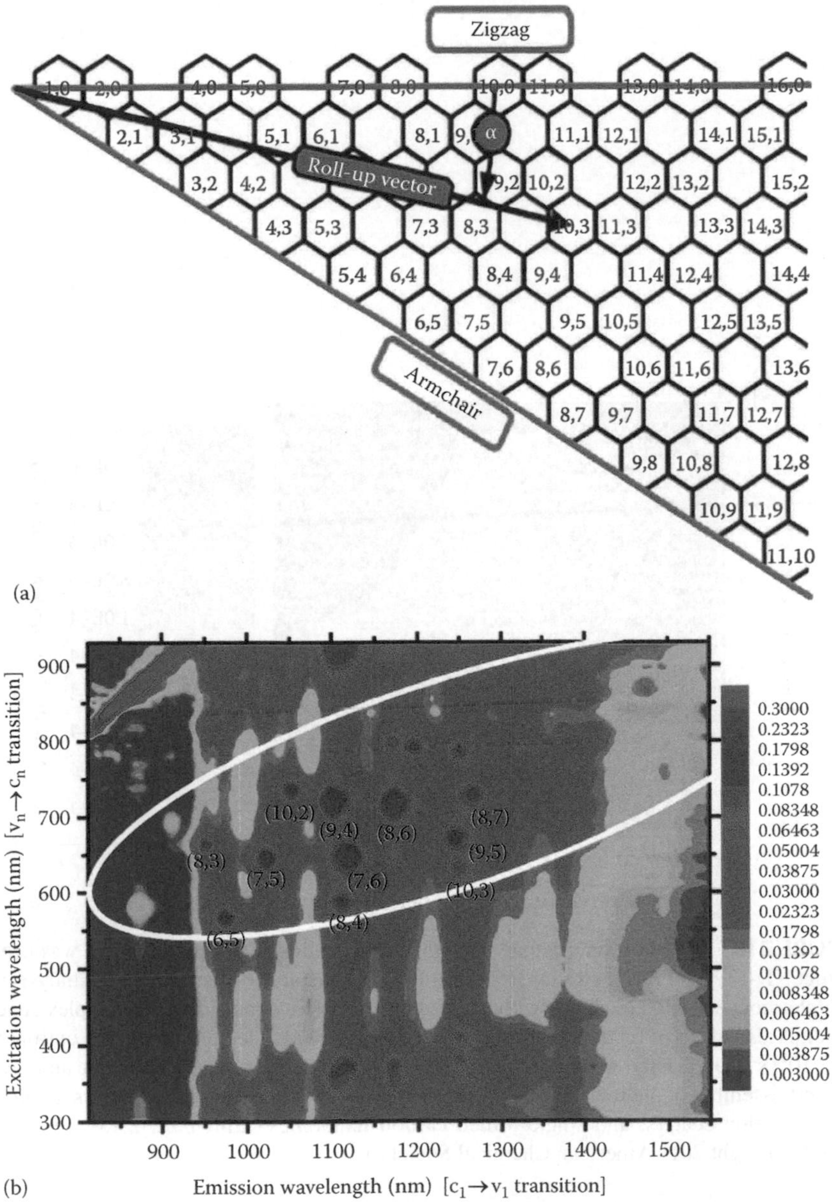

FIGURE 7.6 Nanotube chirality. (a) Nanotube nomenclature relies on the utilization of coordinate system wherein the two axes ("zigzag" and "armchair") are 30° apart overlaying a graphene sheet. A specified carbon atom in the sheet can be located by counting the number of carbon atoms along each of the axes direction to generate a pair of coordinates or indices representative of the atom location. By rolling the graphene from the origin to *q*-specific carbon atom in the lattice, the nanotube that is formed is named according to the pair of indices that designate the location of the specified carbon atom. (b) A PL plot examines the fluorescence emission intensity of a nanotube solution over a range of excitation wavelengths. A solution consisting of individually suspended nanotubes contains several peaks, with each peak corresponding to a specific type of nanotube. (From Bachilo, S.M. et al., Structure-assigned optical spectra of single-walled carbon nanotubes, *Science*, 298(5602), 2361–2366, 2002. Reprinted with permission of AAAS.)

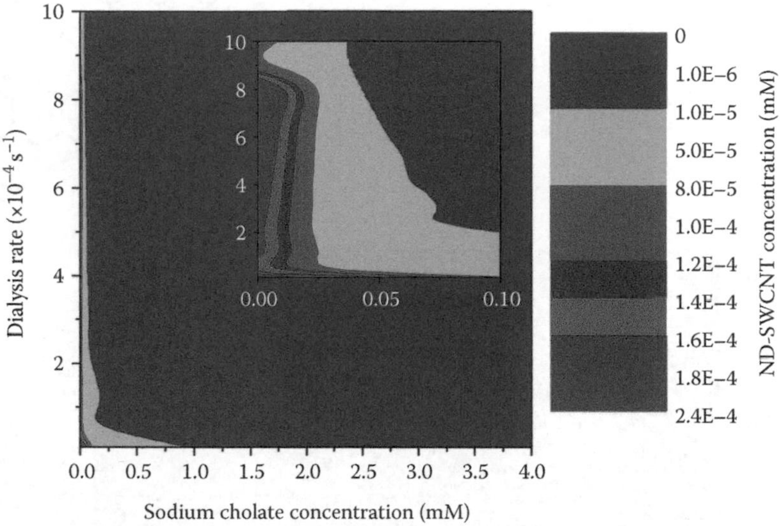

FIGURE 7.10 Effect of dialysis rate on complex formation. The kinetic model was used to predict ND-SWCNT concentration as a function of surfactant concentration and dialysis rate. Complex concentration increases with decreasing surfactant concentration. Complex concentration is also affected by the rate of surfactant dialysis, with a local dialysis rate optimum of $\sim 8 \times 10^{-4}$ s^{-1}. (Reprinted with permission from Boghossian, A.A. et al., Dynamic and reversible self-assembly of photoelectrochemical complexes based on lipid bilayer disks, photosynthetic reaction centers, and single-walled carbon nanotubes, *Langmuir*, 27(5), 1599–1609, 2011. Copyright 2011 American Chemical Society.)

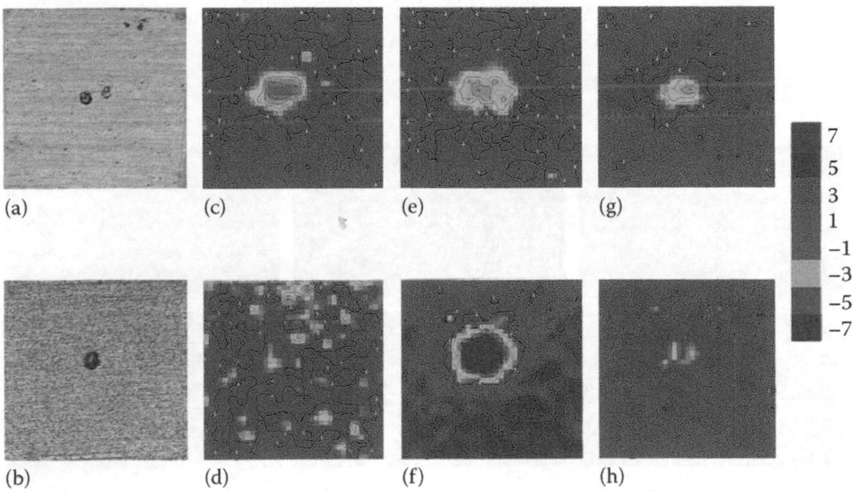

FIGURE 10.6 SVET maps of the ionic currents measured above the surface of the defected (a, b) AA2024 coated with undoped sol–gel pretreatment (c, e, g) and with that impregnated by nanoreservoirs (d, f, h). The maps were obtained in 5 h (c, d), 24 h (e, f), and 26 h (g, h) after defects formation. Scale units: $\mu A/cm^2$. Scanned area: 2 mm × 2 mm. (From Shchukin, D.G., Zheludkevich, M.L., Yasakau, K.A., Lamaka, S.V., Ferreira, M.G.S., Möhwald, H. Layer-by-layer assembled nanocontainers for self-healing corrosion protection. *Adv. Mater.* 2006. 18. 1672–1678. Copyright Wiley-VCH Verlag GmbH & Co. KGaA. Reprinted with permission.)

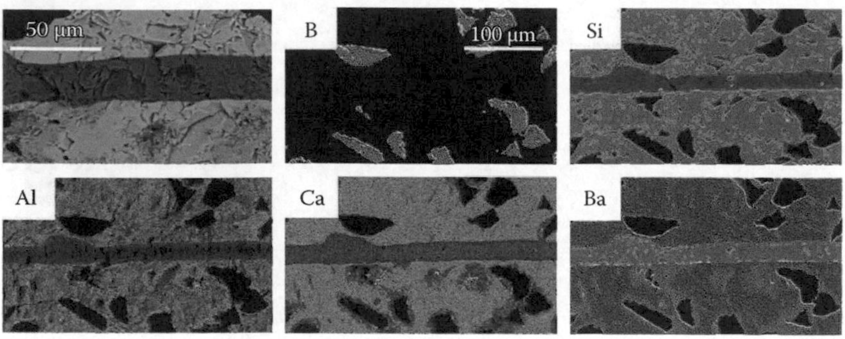

FIGURE 11.17 *Ex situ* Castaing microprobe analysis of a fractured glass-ceramic/B_4C sample healed under air at 700°C for 1 h: SEM picture and x-ray element maps after heat treatment. (Reprinted from Coillot, D. et al., *Adv. Eng. Mater.*, 13, 430, 2011.)

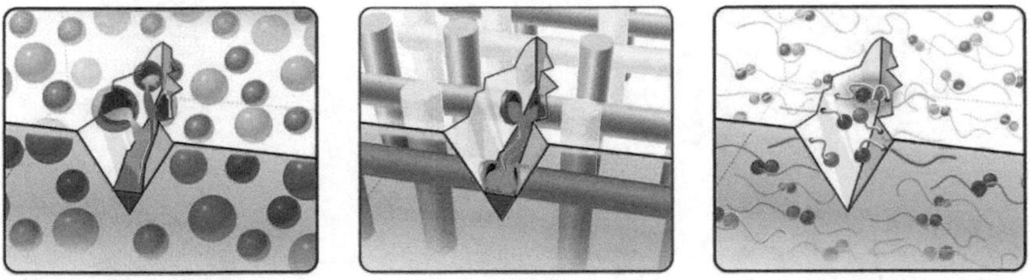

FIGURE 12.2 Concepts for self-healing polymers based on microcapsules (left), vascular system (middle), and reversible interactions (right). (Reproduced from Blaiszik, B.J. et al., *Annu. Rev. Mater. Res.*, 40, 211, 2010. With permission from Annual Reviews.)

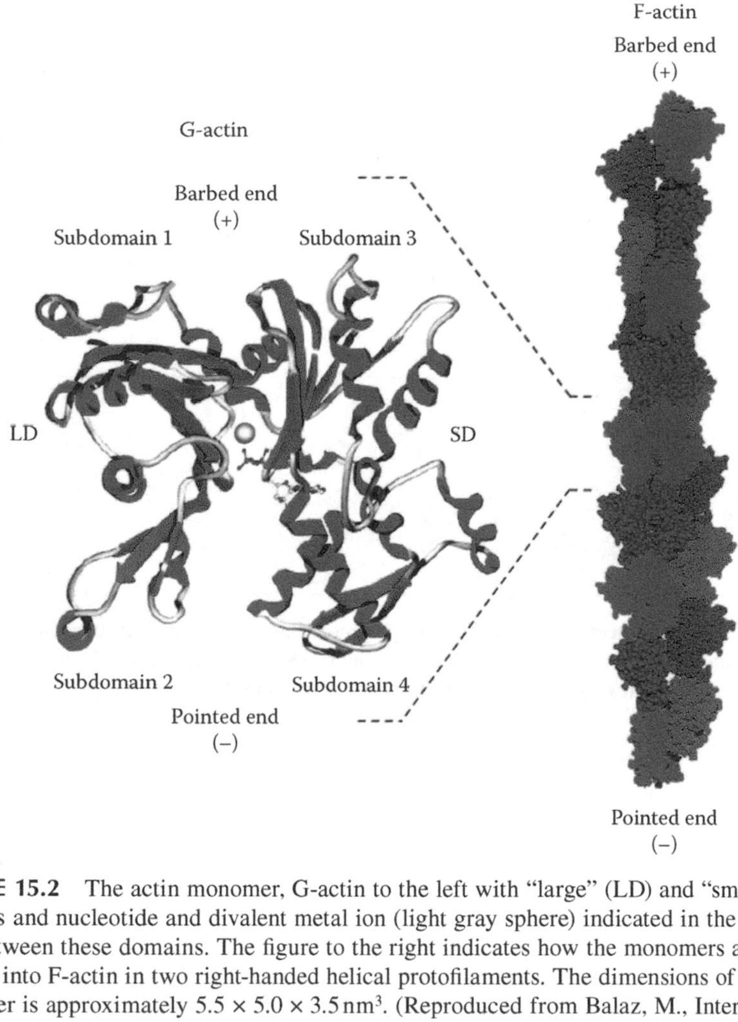

FIGURE 15.2 The actin monomer, G-actin to the left with "large" (LD) and "small" (SD) domains and nucleotide and divalent metal ion (light gray sphere) indicated in the catalytic cleft between these domains. The figure to the right indicates how the monomers are incorporated into F-actin in two right-handed helical protofilaments. The dimensions of the actin monomer is approximately $5.5 \times 5.0 \times 3.5\,nm^3$. (Reproduced from Balaz, M., Interaction of actomyosin with synthetic materials—Effects on motor function and potential for exploitation, PhD thesis, University of Kalmar, Kalmar, Sweden, p. 202, 2008.)

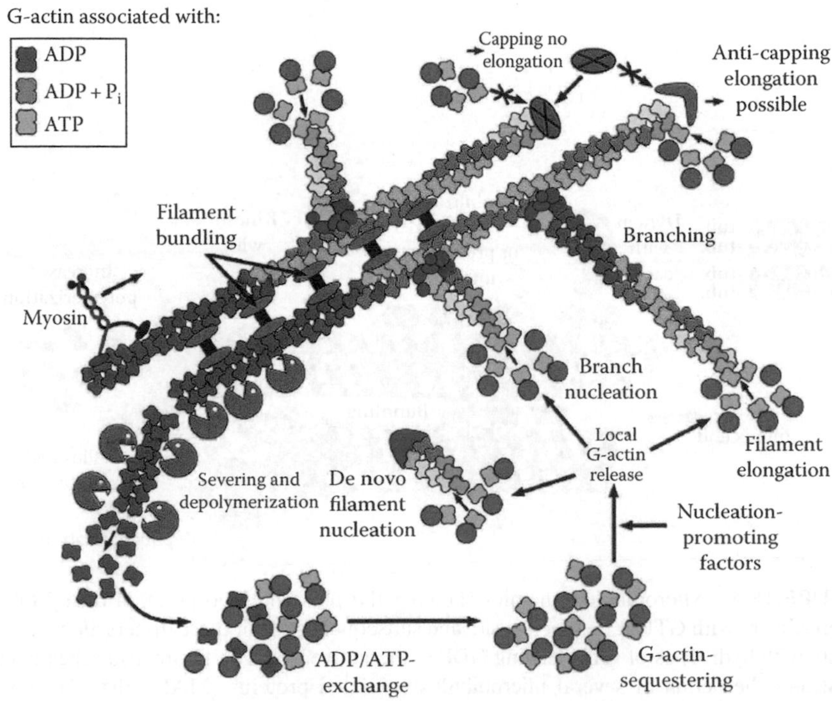

FIGURE 15.3 Schematic illustration of actin dynamics showing the incorporation of G-actin–ATP at the plus end and subsequent shift of these monomers while they are first hydrolyzing ATP followed by release of inorganic phosphate (P_i). The figure also schematically illustrates the action of several actin-binding proteins that (1) nucleate formation of new filaments (e.g., formins), (2) nucleate branching of filaments (e.g., the ARP 2/3 complex), (3) sever (cut, e.g., gelsolin) and depolymerize (e.g., cofilin) filaments, (4) sequester G-actin monomers (e.g., profilin), (5) cap filaments to block further monomer incorporation (e.g., Cap Z and gelsolin), and (6) bundle filaments in different geometric patterns (e.g., filamin, fascin, α-actinin). (Reprinted from *Curr. Opin. Neurobiol.*, 18, Witte, H. and Bradke, F., The role of the cytoskeleton during neuronal polarization, 479–487, Copyright 2008, with permission from Elsevier.)

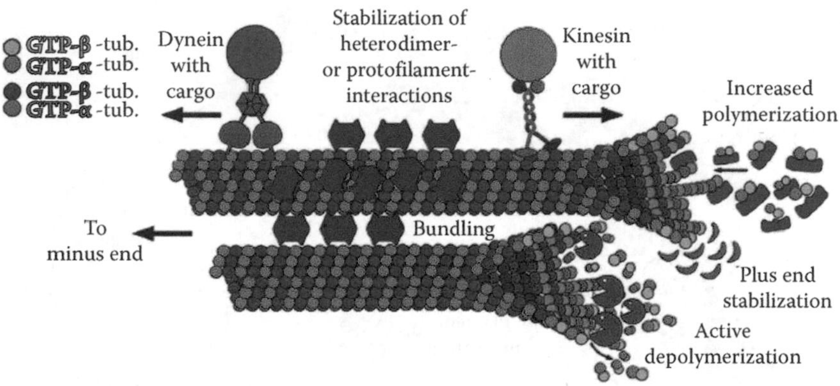

FIGURE 15.4 Microtubule dynamics showing the plus end incorporation of α/β-tubulin heterodimers with GTP at their active site and subsequent shift of these dimers along the filaments with hydrolysis of GTP, leaving GDP at the active site. The figure also schematically illustrates the action of several microtubule-associated proteins (MAPs) that (1) stabilize dimer interactions within the filament (e.g., tau), (2) link neighboring filaments or facilitate bundling (e.g., MAP2), (3) facilitate polymerization by binding to α/β-tubulin heterodimers (e.g., CRMP-2), and (4) stabilize the dynamic plus end (e.g., EB1). (Reprinted from *Curr. Opin. Neurobiol.*, 18, Witte, H. and Bradke, F., The role of the cytoskeleton during neuronal polarization, 479–487, Copyright 2008, with permission from Elsevier.)

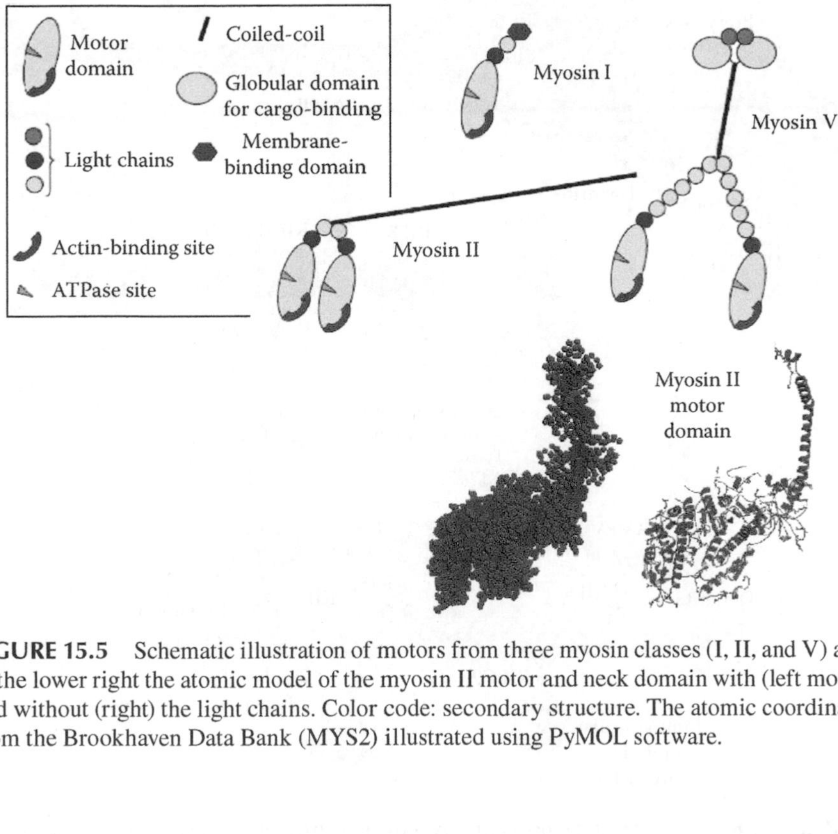

FIGURE 15.5 Schematic illustration of motors from three myosin classes (I, II, and V) and, to the lower right the atomic model of the myosin II motor and neck domain with (left model) and without (right) the light chains. Color code: secondary structure. The atomic coordinates from the Brookhaven Data Bank (MYS2) illustrated using PyMOL software.

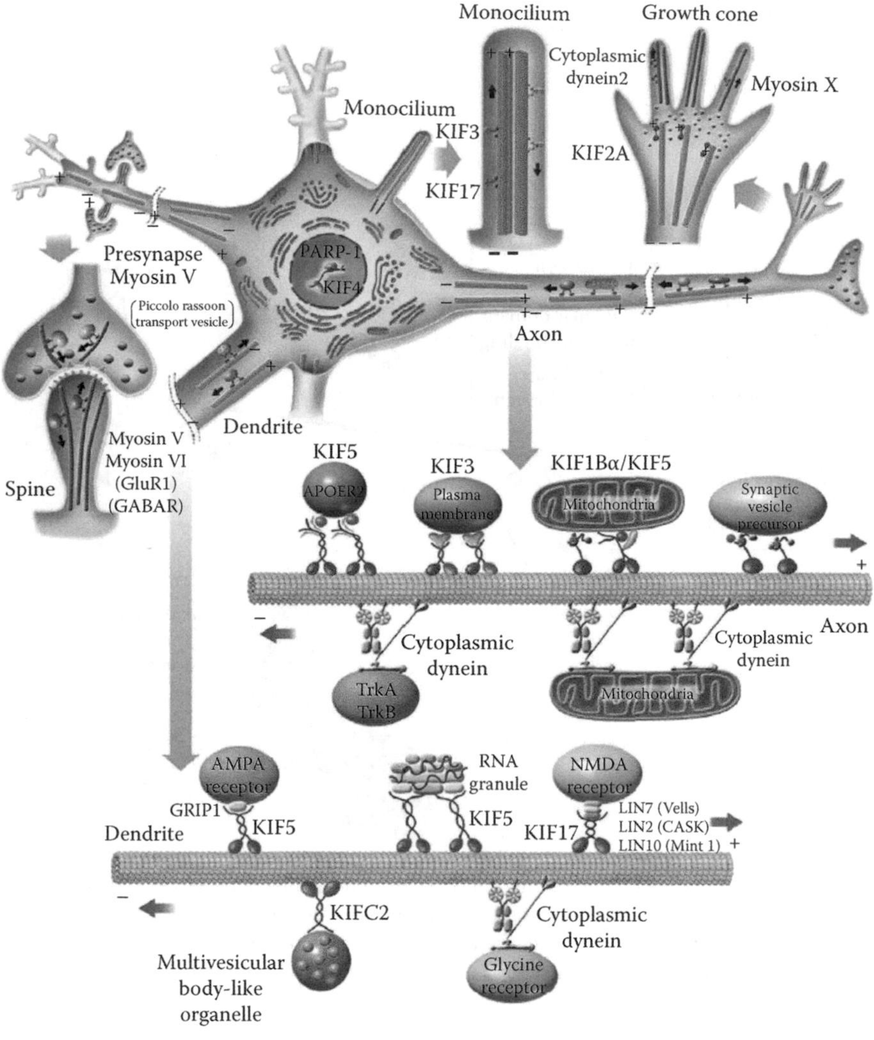

FIGURE 15.6 The role of cytoplasmic dynein and different myosin and kinesin (KIF) classes of motors in a neuron (e.g., in axonal/dendritic transport, synaptic plasticity/receptor recycling, and growth-cone motility). Actin filaments (red) and microtubules (yellow). (Reprinted from *Neuron*, 68, Hirokawa, N., Niwa, S., and Tanaka, Y., Molecular motors in neurons: Transport mechanisms and roles in brain function, development, and disease, 610–638, Copyright 2010, with permission from Elsevier.)

8 Self-Healing Nanocomposites
Role and Activation of Inorganic Moieties and Hybrid Nanophases

Ivano Alessandri

CONTENTS

8.1 INTRODUCTION

Nanocomposites, that is, multiphase materials resulting from combination of a bulk matrix and one or more nanodimensional phases, play a key role in the development of new materials and technologies (Ajayan et al. 2003, and references therein). In general, nanocomposites can be classified in three different categories, on the basis of their main bulk phase, the so-called "matrix," which can be made of metals, ceramics, or polymers. In particular, polymer matrix nanocomposites are showing an ever-increasing extension toward manifold areas of materials science and technology, with applications in materials for transportation, energy, health, packaging, information and communication, etc. In many cases, the interest in these materials is due not only to their structural but also to their functional properties. In nanocomposites, the synergetic combination of bulk phase and nanophase can result in significant improvements of both chemical and physical properties. According to the definitions given by Vaia and Wagner (2004), nanophases can mainly operate in two ways. In the simplest and most frequently observed cases, nanophases play just as nanoscale fillers for polymer matrices, yielding a direct enhancement of some properties in comparison with the corresponding unfilled matrices or macrocomposites. Moreover, nanophases can provide novel, unique properties, which can give rise to entirely new materials, just like supramolecular assemblies may exhibit new functionalities with respect to the single molecular moieties which they are made of. In the latter case, nanocomposites are indicated as "polymer/inorganic hybrids" or "molecular composites." Whatever the type of nanocomposite is, developing strategies to prevent damages or repair them should be considered a mandatory task to accomplish in order to take full advantage of their potential. In this context, the term "damage" is not limited to indicate a physical alteration of the structure, but is more generally extended to all the changes that can irreversibly jeopardize either functionalities or performances of these materials. In general, prevention of detrimental effects can rely on different strategies, which can be classified, on the basis of the mechanisms involved in the protection process, as passive or active (Fischer 2010). The former strategy makes use of protecting agents or coating layers as passive barriers against chemical, photo-, thermal and mechanical degradation. This route is on/off type, since damages can be only prevented, but not repaired, once they have occurred. Moreover, passive protection can fail when damage-triggering defects are already present in the original materials. This limitation is due to the absence of any feedback systems that can detect damage (which is often much localized) before it reaches a critical extent, and intervene to repair it. Nowadays, passive protection is still the most widely used strategy, due to their universal applicability. On the other hand, systems including mechanisms of self-repair are indicated as "active." In this case, structural and functional damages can be detected and repaired thanks to different active agents that can trigger self-healing in response either to external stimuli or directly to damage itself. Therefore, self-healing materials are generally distinguished between nonautonomous (or stimuli-assisted) and autonomous systems (Hager et al. 2010). Nonautonomous self-healing can be induced and controlled by various parameters, such as heat, light, mechanical forces, chemical reactions, pH, etc. On the other hand, autonomous self-healing does not need any external intervention, as the damage itself triggers repair processes. Autonomous systems behave as smart, adaptive materials. However, up to now only few examples of autonomic self-healing materials have been successfully reported, and the self-repairing mechanisms have been addressed only to restoring of mechanical properties. Autonomous self-healing processes can be further distinguished between intrinsic and extrinsic. Intrinsic autonomous self-healing is the ultimate goal for most applications, in particular for mending mechanical failures, since it is based on the formation of either covalent or noncovalent chemical bonds between cracked interfaces. Unfortunately, engineering intrinsic processes is quite hard for most of the materials. Extrinsic autonomous self-healing requires the presence of externally loaded healing agents, which can be triggered by a mechanical damage (White et al. 2001).

Nanocomposites and, in particular, polymer matrix nanocomposites can be suitably engineered in order to create both autonomous and nonautonomous self-healing materials. Self-repairing in polymer matrix nanocomposites encompasses several key factors and can be dealt with from different standpoints. The reader is referred to other excellent reviews for a comprehensive description of self-healing composites and nanocomposites (see, e.g., Wu et al. 2008; Yuan et al. 2008; Gosh 2009; Liu and Urban 2010; Mauldin and Kessler 2010; Murphy and Wudl 2010).

Healing processes in nanocomposites can proceed through different basic strategies, such as crack filling by healing agents, diffusion, bond reformation, and improvement of the original properties.

This chapter will be mainly focused on the role played by inorganic (metal or metal oxides) and organic–inorganic hybrid moieties in most of these processes. It is known that inorganic moieties allow for enhanced performances of nanocomposites and provide specific functionalities to polymeric matrices. In the case of self-healing nanocomposites, they can operate over different levels, often tightly intertwined with each other. The scheme reported in Figure 8.1 roughly summarizes the main types of inorganic and hybrid organic/inorganic moieties in self-healing nanocomposites, which will be reviewed in this chapter.

We can distinguish among the following cases:

1. Inorganic nanofillers (metal or metal oxide nanoparticles and nanostructures), which increase the mechanical properties of self-healing nanocomposites and, sometimes, can contribute to the self-healing mechanism itself (Section 8.2).
2. Metal and metal oxide nanoparticles and molecular species, taking part to the self-healing process with a functional role (Section 8.3). In this case, further subsets are given by
   a. Metal centers, inorganic complexes catalysts, and precursors, which trigger self-healing processes (Sections 8.3.1 and 8.3.4).
   b. Magnetic nanoparticles, which activate self-healing in response to external magnetic fields (Section 8.3.2).
   c. Nanostructures and metallorganic polymers, which can undergo resistive heating (Section 8.3.3).
   d. Metal nanoparticles, which can promote light-driven and/or optothermal-stimulated self-healing (Section 8.3.4).

The functional roles of the inorganic moieties can be further detailed on the basis of the stimuli exploited for their activation. In this regard, activation strategies based on mechanical, chemical, and electromagnetic stimuli and their effects on nanophases and molecular species will be reviewed. Figure 8.2 shows a synopsis of the examples discussed in this chapter, classified on the basis of their activating stimuli and their specific contributions to self-healing processes.

Though mainly focused on inorganic moieties, in few specific cases, this review will briefly touch other types of nanophases, such as carbon nanotubes or photoresponsive organic molecules, in order to give wider perspectives on specific topics. Similarly, in some other cases, the discussion will be extended to macrocomposites, since most of the strategies utilized for the latter could be conveniently adapted to the nanocounterparts. This treatment will follow two levels of discussion. The most investigated examples of inorganic or inorganic/organic molecular species will be

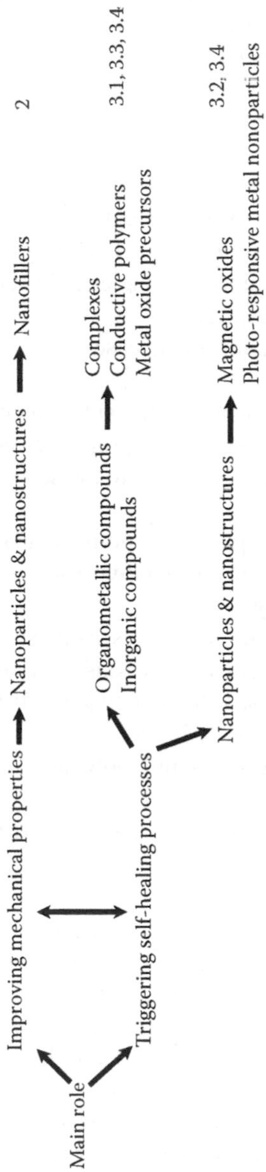

FIGURE 8.1 Scheme of the main types of inorganic and organic/inorganic hybrid nanophases and molecular species presented in this chapter.

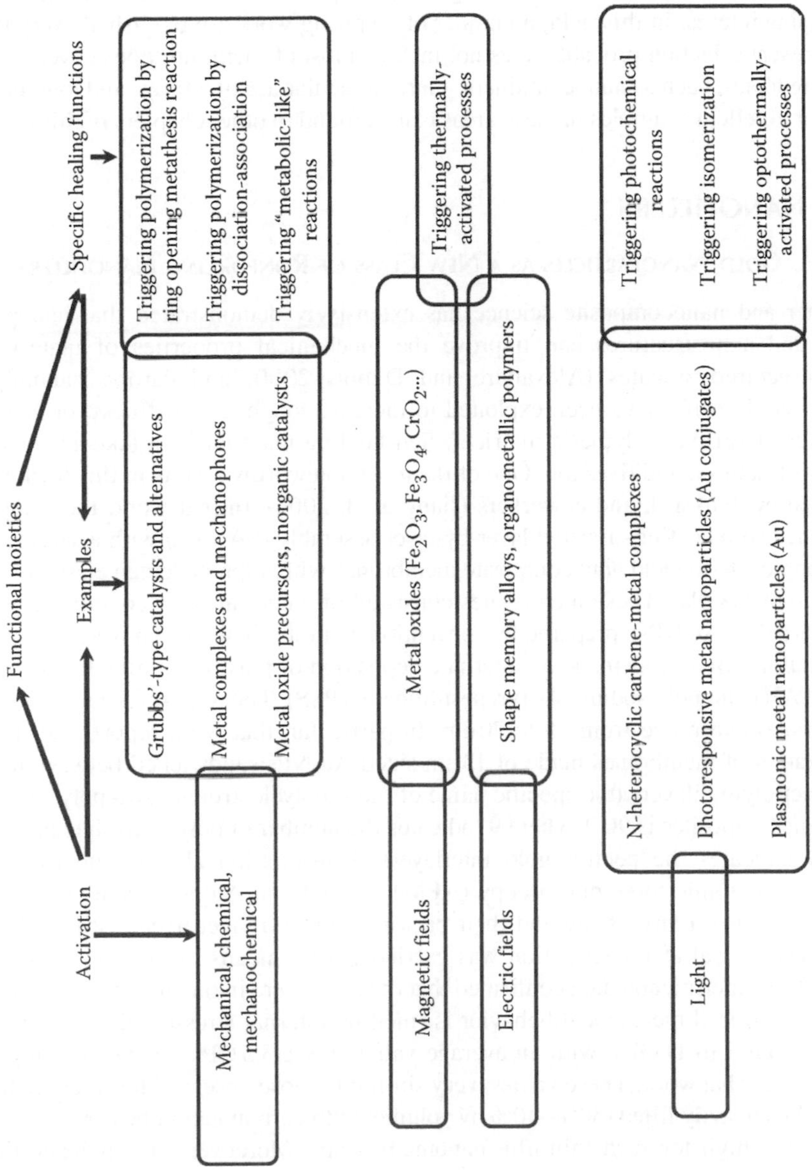

FIGURE 8.2 Synopsis of the functional moieties discussed in this review, classified on the basis of their activation and role in self-healing processes.

the object of a general overview, whereas more recent approaches, such the use of gold or magnetic metal oxide nanoparticles, will be analyzed in more detail. Instead of providing a report on well-established solutions, most of the selected examples aim to give an outlook on some exciting research lines, which could stimulate new advances in fabrication of self-healing nanocomposites. Finally, due to the ever-increasing interest in this field, a number of inspiring work are published everyday. The present selection probably does not include most of them, nor does cover some exciting topics, such as nanocontainers, supramolecular architectures, and ionomers. Further excellent examples in these areas can be found in other chapters of this book.

8.2 NANOFILLERS

8.2.1 Gold Nanoparticles as a New Class of Reinforcing Nanofillers

Polymer and nanocomposite science has extensively demonstrated that nanoparticles and nanostructures can improve the mechanical properties of materials. Nanostructured silicates (Alexandre and Dubois 2000) and carbon nanotubes (Sahoo et al. 2010) have been exploited to increase toughness, stiffness, or elastic modulus of various polymeric matrices. Self-healing materials can take advantage of nanostructured additives too. One of the most noteworthy results in this field was reported by Tsuksruk and coworkers (Jiang et al. 2004). In that work, the authors combined spin coating–assisted layer-by-layer assembly (SA-LbL) with a sacrificial layer approach, to yield nanocomposite membranes with unprecedented elastic modulus (up to 11 GPa). These membranes consisted of a central layer containing gold nanoparticles (Au NPs), prepared by conventional citrate route, which were wrapped with multilayered films made by alternate deposition of poly(allylamine hydrochloride) (PAH) and poly(sodium 4-styrenesulfonate) (PSS). The total thickness of these membranes can range from 25 to 70 nm. In particular, that work reports the characterization of membranes made of 13 nm sized Au NPs sandwiched between nine polyelectrolyte bilayers (the specific name of these polyelectrolyte–Au–polyelectrolyte nanocomposites is 9G9, where 9 indicates the number of polyelectrolyte bilayer, and G indicates the central gold interlayer). Transmission electron microscopy (TEM) and atomic force microscopy (AFM) showed that Au NPs were uniformly distributed over a large scale and their packing density ranged from <2% to 25%. Micromechanical characterization was carried out by means of bulging tests and AFM. The experimental data confirmed that the films were in the membrane regime, where the overall mechanical behavior is ruled by internal stresses. Elastic moduli ranged from 3 to 11 GPa, with an average value of 8 ± 3.5 GPa for the 20 samples examined in that work. These values, very similar to those observed for micrometer-sized films heavily filled (30%–40% by volume) with carbon nanotubes or clays, are remarkably high for such thin-film nanomembranes. Moreover, they indicate that defect-free, very thin membranes with comparatively large areas (0.4 mm^2) can be fabricated through this route. Further investigations revealed that the elastic modulus is directly related to the concentration of Au NPs. As already observed for other layer-by-layer membranes, the elastic properties were affected by moisture and temperature. In particular, under normal variation of humidity (20%–50%) the values of elastic modulus oscillate 30%–40% from its average, whereas they only slightly decreased upon thermal treatment at 120°C for several hours. However, these nanomembranes were much less sensitive to external factors in comparison with analogous gold-free materials. In addition, they were easily manipulated and transferred onto a solid substrate without being damaged. AFM measurements of the mechanical properties at the nanoscale found out that the bending stiffness of 9G9 freely suspended nanomembranes was about $2\,Nm^{-1}$, a value that surpasses the bending parameters reported for polymer multilayers of microcapsules (Dubreuil et al. 2003). All of these data showed that these nanomembranes could be used to sense deflection at both nano- and microscale, with a dynamic range of detectable pressures of about 10^8. The bulging tests also revealed an unexpected property of these nanocomposites: autonomic self-healing. In fact, upon application of an external pressure (4 kPa), these membranes were visibly damaged. However, the original shape was gradually restored after few seconds and the original micromechanical properties were fully recovered after some hours. The authors proposed different factors to explain this remarkable viscoelastic autorecovery, which is undoubtedly related to the particular structure of these materials. As often observed for nanocomposites, the overall properties could take advantage of the mutual synergetic interactions among different components. In the present case, the high level of spreading of polymer chains in the plane of the film, which was already observed for simple layer-by-layer grown materials, could be incremented by the spinning-assisted procedure, which is also expected to give rise to densely packed polymer polyelectrolyte chains. An additional contribution to autorecovery was sought in prestretching, originated by membrane shrinking upon drying. Moreover, Au NPs allowed establishing a dense network of weak, sacrificial bonds between polyelectrolyte layers with oppositely charged groups, which can quickly reorganize upon removal of an external load. This mechanism was also suggested to explain the mechanical properties of other nanocomposites and might be extended to recently developing nanohybrid materials based on biopolymers (Kharlampieva et al. 2010).

Recent literature also reported further outstanding examples of improved mechanical properties achieved thanks to the presence of Au NPs. One case is represented by thin films based on microgels. South and Lyon (2010) reported the fabrication of self-healing polymer films that can be mechanically damaged and quickly repaired by simple water exposure.

These materials were made of anionic hydrogel microparticles (microgels) and linear polycations. Microgels were composed of *N*-isopropylacrylamide (NIPAm: 71 mol%) and acrylic acid (AA: 26 mol%), cross-linked by poly(ethylene glycol) diacrylate (PEGDA-575, Mw = 575, 3.5 mol%), whereas poly(diallyldimethylammonium chloride) (PDADMAC) was used as a polycation moiety. The thin films were prepared by layer-by-layer deposition, alternating anionic and cationic moieties. These films were easily indented and damaged by mechanical stress. However, these films underwent complete self-repair by adding water. This process was very fast (on a time scale of seconds) and did not show apparent desorption of microgels from the film. Moreover, neither high salt solution concentration (1 M) nor moderate temperature (50°C) could prevent the self-healing process. Though still fare from being fully understood, this behavior was related to the Coulombic reversible interactions that held together the hydrogel films. Upon cracking of the film, these polyanion–polycation interactions could be interrupted, leaving an excess of positive or negative charges at well-separated spatial regions. Hydration allows for a general redistribution of microgels, which results in restoring of polyanion–polycation interactions. The mechanical properties of these films were strongly enhanced by addition of Au NPs (Park et al. 2011). Au NPs, with an average size of 13 nm, were prepared by conventional citrate methods and the concentration of the final nanoparticle solution was 1.4×10^{-8} M. The swollen hydrogels were soaked for 1 h into Au NP solutions with three different concentrations, corresponding to 100%, 50%, and 10% of the original one. Optical data suggested that Au NPs are uniformly distributed all over the film and no significant aggregation occurred. A general scheme of the overall process is shown in Figure 8.3.

The evaluation of self-healing capabilities of these composites was given by stretching tests carried out by means of homemade, micrometer-actuated apparatus and compared with those exhibited by gold-free samples. One end of each specimen was fixed and the other end was pulled by the micrometer-controlled translational stage. Three different uniaxial strains $\varepsilon = 0.1$, 0.2, and 0.3 ($\varepsilon = (L - L_0/L_0)$ (where L is the length of stretched material and L_0 is its original, nonstretched length) were applied to the specimens. The gold-free reference sample was visibly damaged upon application of the smallest strain, whereas gold-loaded samples exhibited a remarkable high resistance. In particular, a straightforward correlation between gold concentration and mechanical resistance was observed. Samples with the highest Au concentration (those obtained from the original, nondiluted Au NP solution) were not damaged at $\varepsilon = 0.1$ and showed the best resistance to the subsequent strains. All the samples were rapidly healed by simple immersion in water for a few seconds, and self-healing was repeatedly observed over several cycles. Improved mechanical properties were qualitatively explained by accounting for different effects.

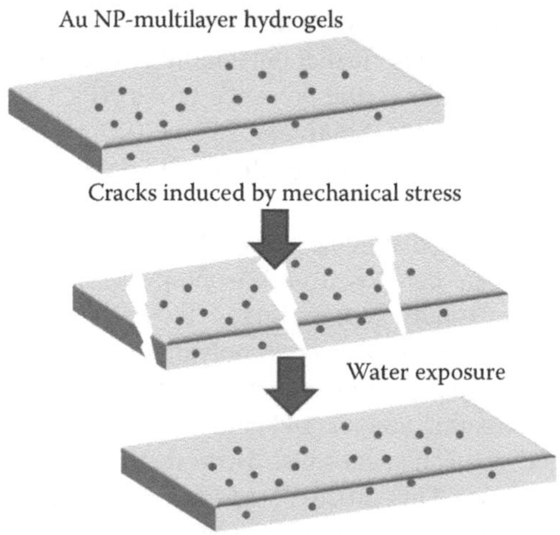

FIGURE 8.3 Scheme showing self-healing multilayer hydrogel films incorporating Au NPs. As a result of Au NPs incorporation, the mechanical properties of the films are strongly improved. Self-healing is activated by water exposure.

The first was the filling effect of Au NPs. Gold-free hydrogels have an elastic modulus of ~100 kPa, whereas the elastic modulus of Au is much higher (~7 GPa), so that Au NPs could be able to shield cracks as hard fillers usually do with polymeric materials (Faber and Evans 1983; Kinloch and Taylor 2002; Jiang et al. 2005).

Moreover, since microgels are characterized by high density of voids and crack initiation is usually triggered by their presence, Au NPs fillers can significantly contribute to inhibiting crack generation and propagation. In addition, due to the excess of positive charge resulting from polycations adsorbed on the microgel surface, the Au NPs, which are expected to be negatively charged because of the citrate-capping layers, could be preferentially localized close to the polycationic moieties, giving rise to a densely packed structure that prevents cracks from nucleation. The remarkable elasticity exhibited by gold-loaded nanocomposites was considered an additional key effect to explain mechanical properties. The Coulombic interactions among Au NPs, polycations, and microgel could absorb energy during stretching, making the film more elastic (Kausch and Plummer 1994). However, a more detailed understanding of the mechanisms needs further experiments.

In summary, Au NPs may provide self-healing nanocomposites with new, unique assets. While their reinforcing function has been demonstrated, yet not fully understood, there is still plenty of room for research aimed to their exploitation as stimuli-responsive media for generating a new class of self-healing materials. In particular, light-triggered self-healing nanocomposites could take advantage of both mechanical reinforcement and optothermal efficiency associated to these nanoparticles. Moreover, manifold extensions to other metal nanoparticles and nanostructures can be envisaged in the next future.

8.2.2 New Perspectives for "Traditional" Reinforcing Fillers

Reviewing of reinforcing fillers in nanocomposites cannot get out of a quick outlook on traditional approaches that have been successfully employed in macrocomposites, in order to seek possible sources of inspiration for nanophases. An approach that deserves to be mentioned makes use of hollow glass fibers as reinforcing fillers that can be loaded with healing agents. This allows reaching a twofold goal. Hollow glass fibers enhance mechanical and structural properties of the composites (Hucker et al. 1999, 2002, 2003; Trask et al. 2007). At the same time, they can be exploited as reservoirs for healing agents (cyanoacrylates, epoxy resins, etc.) (Dry 1996). Although these systems cannot be considered as nanocomposites (the diameters of the hollow glass fibers are in the order of tens of micrometers), the concept of hollow-fiber reservoir could be adapted to release nanoparticles or reactive precursors in order to trigger self-healing. This releasing system could further improve their functionalities by adding dyes or fluorescent tags that allow detecting damage and following the flow of the healing agents, as already done for studying self-repairing mechanisms in macrocomposites (Bleay et al. 2001).

Relying on similar considerations, Lanzara et al. proposed to utilize carbon nanotubes as reinforcing fillers capable of loading healing agents (Lanzara et al. 2009). In their investigation, carried out through molecular dynamics (MD) simulations, they demonstrated that single-walled carbon nanotubes can release healing agents stored inside them upon cracking. Healing agents can be released in few picoseconds, depending on both temperature and size of the cracks. Literature reports a number of examples in which carbon nanotubes have been used for storing either molecules, from H_2 (Dillon et al. 1997) to DNA (Gao et al. 2003), or metal compounds (Seraphin et al. 1993; Guerret-Plécourt et al. 1994). In principle, molecular or nanostructured catalysts could be embedded into carbon nanotubes and selectively released upon cracking. In addition, carbon nanotubes accomplish the function of a valuable reinforcement, as widely documented in literature (see, e.g., Mamedov et al. 2002) and are, in general, more efficient than macrocontainers. In perspective, analogous approaches could be envisaged, in which metal oxide nanotubes could be employed as both reinforcing fillers and nanostructured carriers for healing agents.

For a specific review on nanocontainers, the reader is referred to another chapter of this book (see Chapter 10).

8.3 FUNCTIONAL NANOPHASE AND MOLECULAR MOIETIES

8.3.1 Inorganic Catalysts for Self-Healing Reactions

Metal centers can be conveniently exploited to catalyze self-healing processes. In this regard, one of the most successful examples of autonomous extrinsic self-healing is represented by works of White and coworkers (White et al. 2001; Kessler et al. 2003). These systems are based on a "crack filling healing" approach, in which fluid healing agents are able to flow and fill damaged regions. Repairing is then obtained through chemical reactions or physical processes. Liquid healing agents usually consists of monomeric moieties that are stored by microcapsules (in general, urea–formaldehyde) embedded into the polymer matrix. Upon mechanical damage, the capsules break up, releasing liquid monomers that are driven to the cracks by capillarity. Polymerization and subsequent healing are promoted by inorganic catalysts dispersed into the matrix. Catalysts should meet some specific requirements. In particular, they should be well dispersed throughout the matrix, in order to maximize their efficiency, and easily dissolved in the liquid healing agent. In addition, they should be stable over a wide range of temperature, chemically compatible with the polymer matrix and highly reactive with healing agents. In general, the best healing results have been achieved through ring opening metathesis polymerization (ROMP) of dicyclopentadiene (DCPD) liquid monomers (Figure 8.4).

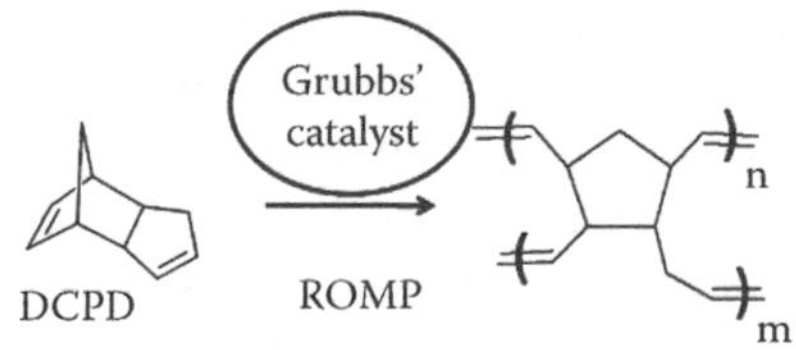

FIGURE 8.4 General scheme of a ROMP used for self-healing. DCPD monomers yield cross-linked networks, through a ROMP reaction promoted by ruthenium-based Grubbs' catalysts.

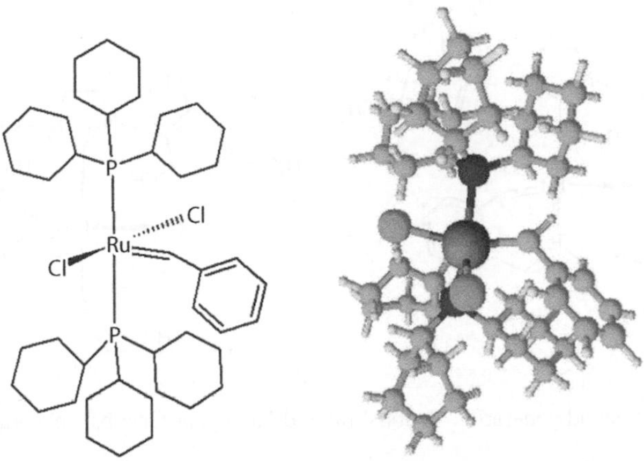

FIGURE 8.5 First generation Grubbs' catalyst used for promoting self-healing reactions.

Catalysts are the key factors that allow self-healing to occur efficiently. In particular, ROMP reactions utilize Grubbs' catalysts, that is, ruthenium(IV) catalysts that take their name from Robert H. Grubbs who was awarded the Nobel prize in Chemistry in 2005 (Figure 8.5).

Initially, self-healing reactions made use of first generation Grubbs' catalyst, that is, bis(tricyclohexylphosphine)benzylidene ruthenium(IV) dichloride. In ROMP reactions, the ruthenium(IV) center coordinates the highly strained C–C double bond of DCPD. Upon cycloaddition with the ruthenium–carbine, a metallocyclobutane is formed, which promotes opening of the DCPD strained ring. This forms a new unit that is added to the growing polymer chain (Sanford et al. 2001; Grubbs 2006).

Although very useful as proof-of-concept demonstrators, the first generation Grubbs' catalysts undergo deactivation upon long exposures to air and moisture. In addition, they can be deactivated upon interaction with some curing agents and, most important, they can agglomerate, provoking detrimental delamination. Other critical factors are the rate of dissolution of the catalyst in healing agents and the amount of the catalyst that can be found on the crack region. Unfortunately dissolution rate, which depends on the size of catalyst particles, goes along with deactivation: smaller particles undergo faster dissolution, but equally fast deactivation and vice versa. Two different approaches were proposed to tackle these issues. Taber and Frankowski first observed that paraffin can strongly stabilize Grubbs' catalysts (Taber and Frankowski 2003). White and Sottos' groups took advantage of this achievement by encapsulating the catalysts in wax microspheres, which can be easily dissolved upon interaction with the healing agent. This method allowed solving the deactivation problems at the expense of a modest (9%) decrease of reactivity due to wax embedding. Moreover, encapsulation improved catalysts dispersion, resulting in more uniform healing processes and remarkable reduction of the catalyst concentration. Due to this approach 93% of healing efficiency, defined as the % ratio between fracture toughnesses of healed and virgin material, was obtained using only 0.75 wt% of catalyst, that is, 90% lower than the amount required for nonencapsulated materials

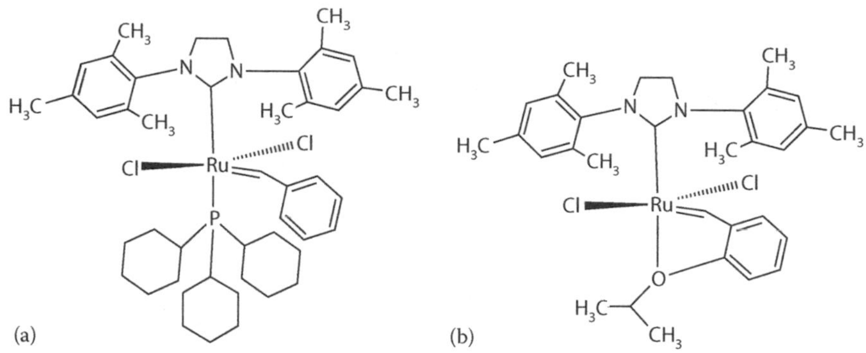

FIGURE 8.6 Second generation Grubbs' (a) and Hoveyda–Grubbs' (b) catalysts.

(Rule et al. 2005). The same groups successfully generalized the application of this method by extending it to more challenging matrices, as those requiring curing procedures that are expected to strongly deactivate the catalysts. Also in these cases, the catalyst activity was almost completely maintained (Wilson et al. 2008a).

The second approach explored newer Grubbs' catalysts as possible alternative to the first generation ones. White and Sottos' groups compared the performances of first generation Grubbs' with second generation Grubbs' and second generation Hoveyda–Grubbs' catalysts (Wilson et al. 2008b). As shown by their molecular structures (Figure 8.6), the three catalysts have different chemical groups coordinated around the ruthenium(IV) center.

The Hoveyda–Grubbs' catalysts exhibited the best performances in terms of reaction rates and matrix-induced deactivation. No significant improvement of healing efficiency was reported under normal test conditions, but second generation Grubbs' catalysts exhibited the best thermal stability, which made them more efficient in healing tests at 125°C.

Although other healing agents, such as those based on norbornene, can further improve the self-healing performances, the use of very expensive ruthenium catalysts still remain a major limitations for a wider diffusion of these nanocomposites.

Kamphaus et al. (2008) tested WCl_6 as a possible alternative to the Grubbs' catalysts. WCl_6 is actually expected to be a catalyst precursor, which can be activated by either alkylation with phenylacetylene or oxidation in air. Nonylphenol was added to increase dissolution in DCPD. Tungsten(VI) precursors exhibited several major drawbacks. They yielded a remarkable decrease of toughness, which could be partially recovered by adding silane coupling agents. Moreover, they tended to aggregate and underwent strong deactivation. Although the wax-encapsulation method allows keeping their activity, the latter vanishes after air exposure for 24 h. Finally, healing is observed only for quite high concentrations of the catalyst precursor, and several deactivating side processes can occur in preparation stages. However, using 12 wt% WCl_6 the authors were able to obtain the maximum healing efficiency, which was only 20%.

Since a major drawback of Grubbs' catalysts is given by their tendency to deactivation upon air exposure, Cho et al. (2006) investigated catalysts that could be activated by direct exposure to air moisture. One of these systems is represented by organotin compounds, which have been known to be activated by water since pioneering work by van der Weij, which dates back to 1980 (van der Weij 1980). In their work, which was carried out in collaboration with Sottos and White, the authors utilized di-*n*-butyltin dilaurate as an organotin catalyst to promote polycondensation of phase-separated droplets containing hydroxy end–functionalized polydimethylsiloxane (HOPDMS) and polydiethoxysiloxane (PDES). HOPDMS, which serves as liquid healing agent, undergoes phase separation in the polymer matrix, so that it can be directly mixed without any further encapsulation. Moreover, the organotin catalyst is encapsulated within polyurethane beads embedded in a vinyl ester matrix and can be easily released when the capsules are broken by mechanical damage. Interaction between catalyst and liquid healing triggers polycondensation, which allows the damage to be repaired. These catalysts offer manifold advantages since they are stable under either wet or humid environments. A maximum 24% healing efficiency was achieved using 12 wt% PDMS, 3.6 wt% microcapsules containing *n*-butyltin dilaurate, and 4 wt% methylacryloxypropyl triethoxysilane used as an adhesion promoter. Although the healing efficiency is far from that achieved by conventional systems, these materials are able to strongly reduce the crack stress and prevent crack propagation, which are the most important tasks required to self-healing materials working in corrosive environments. Further significant improvements were achieved by Keller et al. (2007) using platinum catalyst complexes entrapped within poly(urea–formaldehyde) capsules together with a high-molecular-weight vinyl-functionalized PDMS. This capsule triggers self-healing by reacting with the active sites of a second PDMS copolymer, yielding a "healing phase," which is the same as the matrix. These systems allowed achieving 75% healing efficiency and excellent recovery of the mechanical properties, yet the issues due to the cost of the catalyst remain still open.

Another interesting use of metal complexes relies on the comparatively high lability of their metal–ligand bonds. Metal complexes embedded into polymer matrices can allow for self-healing through bond reformation. Indeed, the dissociation of these weak metal–ligand bonds upon mechanical stress may prevent degradation of the polymer matrix. This is due to the fact that the covalent bonds of polymer matrices are much more stable than those of the embedded metal complexes. In principle, an appropriate selection of metal centers and ligands on the basis of their reciprocal affinity should allow exploiting the easy dissociation–association reactions for undoing the effect of mechanical stresses before irreversible break of the matrix covalent bonds could occur. Kersey et al. (2006) investigated the metal–ligand dissociation of bispalladium–pyridine complexes induced by the movement of an AFM probe. One year later, the same group reported the first application of these complexes in self-healing materials. In particular, palladium and platinum biscomplexes were cross-linked to pyridine moieties attached to methacrylate chains (Figure 8.7).

When subjected to shear stress, the resulting hybrid polymer gels showed metal–ligand dissociation followed by rapid reassociation (Kersey et al. 2007). Thus, metal complexes acts as stress-bearing groups, which can be broken instead of the polymer covalent bonds upon application of a comparatively low stress and reformed upon stress relief. Moreover, the mechanical properties of these materials and in particular, their strength, increased upon addition of the metal complexes.

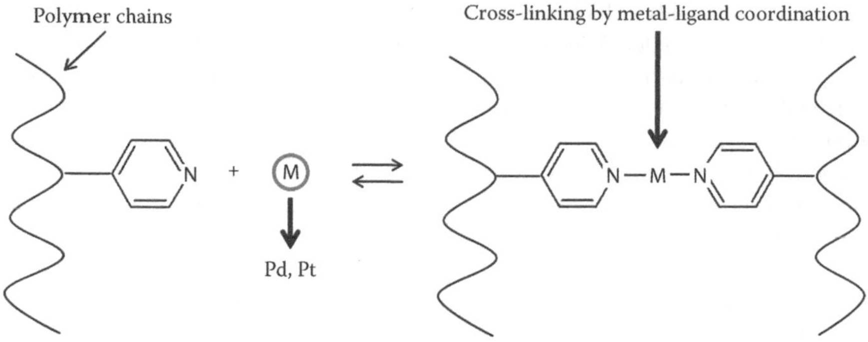

FIGURE 8.7 General scheme showing cross-linking mediated by reversible coordination of two pyridine to Pt or Pd metal centers. Metal–ligand dissociation and reassociation equilibria have been investigated by Kersey et al. (2007) as possible mechanisms for self-healing processes.

Using another approach, Varghese et al. (2006) synthesized hydrophilic polymer gels that can be nonautonomously repaired by soaking into $CuCl_2$ solutions. This procedure is allowed by the presence of a number of dangling carboxyl groups in the polymer matrix. Upon mechanical fracture, these groups are exposed to external environment and can coordinate to the copper chloride solution. As a result, a significant recovery of the mechanical properties was reported.

Metal complexes have been also investigated as possible mechanophores, that is, chemical units that can be activated upon mechanical stress to promote reinforcement of the polymer matrix before its failure. Mechanophores can be considered the ultimate evolution of the aforementioned metal complexes, which trigger self-healing through metal–ligand dissociation–association mechanisms. In fact, stress-induced dissociation of metal–ligand bonds could be exploited to activate latent catalysts, which, in turn, can promote self-healing reactions. The Sijbesma's group demonstrated that diphosphine telechelic poly(tetrahydrofuran) polymers complexed to palladium or platinum halides preferentially dissociate through the metal–phosphine bond upon application of ultrasound (Paulusse and Sijbesma 2004). In 2009, the same group reported that both silver and ruthenium coordinated to N-heterocyclic carbene (NHC) ligands can undergo fast dissociation of the metal–ligand bonds upon sonication (Karthikeyan et al. 2008; Piermattei et al. 2009). The key step for yielding latent catalysts that can be activated by mechanical stress is their functionalization with long polymeric chains, since the shear forces originating from cavitation bubbles produced by ultrasound induce stretching and dissociation in polymer chains but not in small molecules. Ag(I) complexes with polymer-functionalized NHCs were used to catalyze transesterification between benzyl alcohol and vinyl acetate. Sonication was carried out at a frequency of 20 kHz, under inert argon atmosphere, at about 6°C. This proof-of-concept was extended to Ru(IV) biscomplexes with polymer-functionalized NHCs, which successfully catalyzed both ring closing and opening metathesis polymerization. Although very interesting in their basic mechanism, up to now metal catalyst mechanophores have not found any reliable application in self-healing nanocomposites. The main issue is related to transferring the ultrasound-based shear stress generation from solution to solid-state materials. To the best of our knowledge, an actual mechanochemical triggering of metal complex mechanophores in polymer matrices upon mechanical stress has not yet been reported.

The few cases of active mechanophores were represented by gem-dichlorocylopropane (Lenhardt et al. 2009) and poly(methyl acrylate)/poly(methyl methacrylate)-functionalized spiropyrans (Davis et al. 2009).

In the first case, the mechanophores were attached along the backbone of a single polymer chain. Sonication led to formation of 2,3-dichloroalkenes by electrocyclic ring opening, but no self-healing processes have been activated. Moreover, spiropyran mechanophores were able to activate a reversible electrocyclic ring opening reaction to give red-colored merocyanine. This reaction does not give rise to polymerization, cross-linking, or other self-repairing processes, but could be used for mapping mechanochemical transduction through a polymer. However, due to the great potential associated to metal complex mechanophores in view of their use for triggering local self-repair, this field deserves further investigations.

Finally, metal centers and, more generally, inorganic moieties, can lead self-repairing reactions by mimicking other natural self-healing strategies, such as continuous reshaping and metabolic cycles. Takeda et al. (2003) adapted these concepts to different polymers such as polycarbonate (PC), polyether–ketone (PEK), polybutylene–terephthalate (PBT), and polyphenylene–ether (PPE). For example, sodium carbonate is a powerful healing agent for PC which, being the result of polymerization between bisphenol-A and phosgene, exhibits phenoxy and phenyl as terminal groups. PC can be damaged upon exposure to heat and hydrolysis in acid environments. In particular, the carbonate bonds are preferentially damaged by these factors, leading to an overall increase of the phenoxy groups. Addition of sodium carbonate catalyzes the reaction between phenoxy and phenyl groups, thus promoting the reformation of PC chains. Analogous processes were reported for PBT and PEK. PPE can be also self-repaired by an inverse reaction mechanism catalyzed by Cu(II). Also in this case, the polymer matrix undergoes structural damages upon exposure to heat, light, and/or mechanical stress. As a result, the polymer chains are broken into radical species, which are stabilized by a hydrogen donor. Cu(II) can complex the radical moieties, oxidizing them and reducing itself to Cu(I). Oxidation restores polymer chains by eliminating two protons. Cu(I) species diffuse through the polymer network and are reoxidized to Cu(II) by reaction with atmospheric oxygen. Finally, the two protons react with the remaining oxygen to yield water, which is eliminated by diffusion out of the polymer matrix. However, these "metabolic" self-healing processes can be applied only to a few thermoplastic matrices, capable of recombining chain ends through specific reaction mechanisms. Developing new strategies to extend this approach to other matrices is a challenging goal for upcoming research in this field.

Another example showing conversion of potentially detrimental agents into self-healing promoters was given by Balkus and coworkers (Liu et al. 2008). In this case, the authors tackled the problem of extending lifetime of thin-film organic electronic devices. These materials are particularly sensitive to air and moisture, so that composites made by alternating polymer and metal oxide layers are commonly used to reduce oxygen and water adsorption/diffusion. However, the presence of metal oxide layers makes these composites quite brittle. Cracks or scratches can be easily introduced by flexing or wearing and, as a consequence, oxygen and water can enter the polymer networks, reducing the material lifetime. To overcome this limit, the authors designed nanocomposites that can be healed when environmental moisture penetrates the polymer networks through defects or cracks. In these systems, $TiCl_4$ was used as a precursor for generating TiO_2 by hydrolysis upon interaction with environmental moisture. $TiCl_4$ was encapsulated into the pores of poly(lactic) acid (PLA) fibers prepared by electrospinning. PLA fibers undergo degradation upon exposure to water, so that the healing agent can be delivered and converted into nanocrystalline, anatase-type, titania (Figure 8.8).

On the basis of UV–vis characterization, the authors estimated that the loading of $TiCl_4$ was 24.6 mmol/g of PLA fibers. Upon infiltration of $TiCl_4$, the porous PLA fibers were sealed by toluene. After first proof-of-concept experiments showing the successful hydrolysis of $TiCl_4$ under controlled humidity, this approach was extended to planar *p*-methyl methacrylate (*p*-MMA) films embedded with $TiCl_4$-filled PLA fibers through spin coating and coated with aluminum oxide (80 nm) by atomic layer deposition. Upon cracking by application of a flexing load, the films were exposed to 80% relative humidity (RH) for 45 days. SEM analysis revealed that anatase nanoparticles had been selectively grown along the cracks and at few other specific regions corresponding to pristine defects of the alumina layer. The latter observation is particularly important since it means that the self-healing delivery systems can allow repairing not only mechanical-induced damages, but also preexisting defects that can be preferential sites for future failure. This site-specific delivery system shows an example of multiple-level protection, in which polymer, metal

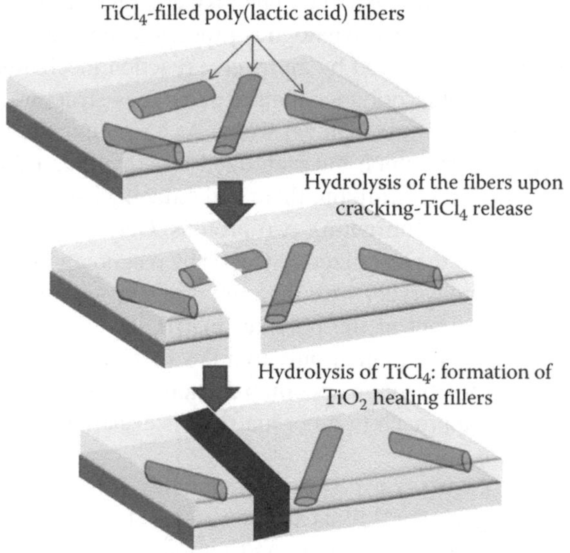

FIGURE 8.8 Scheme showing self-healing of multilayered composites by metal oxide precursors infiltrated into water-degradable poly(lactic acid) fibers. (See Liu, H.A. et al., *Adv. Funct. Mater.*, 18, 3620, 2008.)

oxide layers, self-healing precursors and material architecture cooperate together for preventing and/or restoring the functionality of thin-film devices. Here environmental water is no more a detrimental factor, but a necessary component of the self-healing process. Moreover, this approach could also be integrated into microvascular networks, similar to those originally developed by White and Sottos' groups for achieving an efficient, bioinspired delivery of both encapsulated Grubbs' catalysts and self-healing agents (Toohey et al. 2007). $TiCl_4$ is particularly suited to play as self-healing precursor, due to its high reactivity with moisture, high volatility and capability to flow through the cracks. However, the authors are aware that though excellent for proof-of-concept demonstration, this precursor suffers from some major drawbacks in view of practical applications. For example, the hydrolysis reaction of $TiCl_4$ yields gaseous HCl as a by-product, which strongly limits the general use of this precursor. Trimethylaluminum (TMA), an alumina precursor that yields methanol as a by-product of the hydrolysis reaction, has been suggested as a possible alternative to $TiCl_4$. However, to date and to the best of our knowledge, examples of TMA-based self-healing nanocomposites have not been reported.

In this regard, the main challenges of future research will be related to design, synthesis, and integration of new metal oxide precursors, which can be rapidly hydrolyzed without generation of detrimental by-products.

Further improvements can be expected from development of solid-state healing agents. The latter could greatly simplify most of the critical fabrication steps, such as encapsulation and its related issues. Computational works on self-assembled nanoparticles, which can directly restore functionalities of polymer composites, were reported by Balazs and coworkers (Lee et al. 2004). A more detailed review on this topic can be found in another chapter of this book (see Chapter 13).

8.3.2 Nanoparticles for Magnetic Field–Triggered Self-Healing

Promising examples of self-healing materials that feature inorganic moieties are given by magnetic nanocomposites. Here, magnetic compounds or nanoparticles are dispersed into the polymer matrix. Mechanical damages can be repaired by simple electromagnetic induction. Indeed, upon activation by an external magnetic field, the nanoparticle magnetic moment can oscillate. In particular, two types of oscillations can occur. The first is a Brownian-type oscillation, which is due to the whole rotation of a magnetic particle suspended in a fluid. The second type is due to rotation of the magnetic moment within the magnetic core (Néel relaxation). Embedding nanoparticles into a polymer matrix can strongly limit the Brownian motion, so that Néel relaxation mechanism is preferentially activated by magnetic induction. Oscillations dissipate energy as heat, so that the magnetic nanoparticles act as nanoheaters. Heating effects can be harnessed for mending the polymer matrix (Figure 8.9).

Inductive heating offers a number of advantages in comparison with other thermally activated self-healing strategies, since it allows remote control of the process with no need of either contacts or internal circuitry. Important advantages can also be found in comparison with light-triggered self-healing systems, because utilization of magnetic field allows getting rid of the still unsolved issues related to light attenuation in thick nanocomposites.

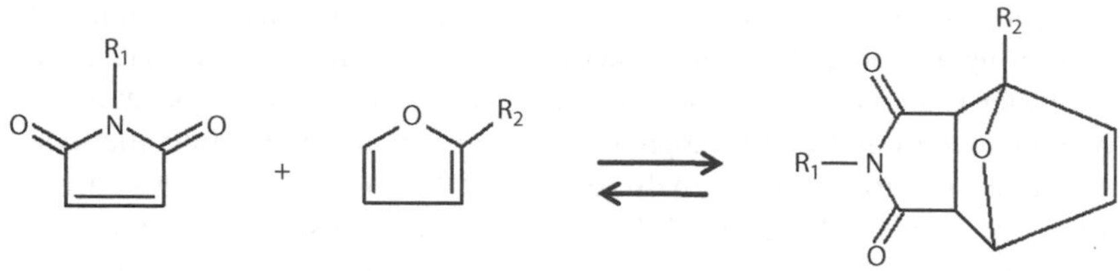

FIGURE 8.9 Diels–Alder reaction between maleimide- and furan-based monomers.

Different approaches have been investigated so far. Schmide (2006) pioneered the use of superparamagnetic magnetite nanoparticles for biodegradable shape memory polymers. These materials consisted of oligo(ε-caprolactone)-grafted Fe_3O_4 nanoparticles (Fe_3O_4 40 wt%, size: 11 nm) embedded into shape memory polymers obtained from oligo(ε-caprolactone)dimethacrylate and butylacrylate monomers. In these experiments, the permanent shape of the nanocomposites is transformed into a temporary shape, which is stabilized by a crystalline phase of oligo(ε-caprolactone) segments. The nanocomposites were then heated (using a commercial generator, 300 kHz, 5 kW) above the transition temperature, which represents the boundary between temporary and permanent shape. As a result, the original shape was fully recorded through selective heating stimulated by magnetic induction. Other proof-of-concept experiments demonstrated that magnetic ferrite particles can be exploited for healing nanocomposites by magnetic induction (Duenas et al. 2006).

More recently, Corten and Urban (2009) demonstrated that thermoplastic polymethacrylates loaded with superparamagnetic γ-Fe_2O_3 undergo complete self-healing upon application of an external oscillating magnetic field. In that work, γ-Fe_2O_3 nanoparticles were dispersed into polymeric films made of *p*-methyl methacrylate/*n*-butylacrylate/heptadecafluorodecyl methacrylate (*p*-MMA/*n*BA/HDFMA). The iron oxide nanoparticles were synthesized by mixing of $FeCl_2$ and $FeCl_3$ solutions (1:2 molar ratio) followed by addition of NH_4OH at pH 11. Sodium oleate was added to the γ-Fe_2O_3 colloidal suspension to prevent nanoparticle aggregation. The resulting stabilized γ-Fe_2O_3 nanoparticles were then added to an aqueous phase containing the three monomers: *p*-MMA, *n*BA, and HDFMA. TEM analysis revealed that iron oxide nanoparticles tend to be attracted by HDFMA moieties and to accumulate around them, so preventing their further aggregation. Upon coalescing the colloidal dispersion for 72 h at 23°C and 55% RH, the iron oxide nanoparticles resulted uniformly dispersed over the polymer matrix. Two different concentrations of iron oxide, 4.3% and 14% w/w, were tested. Both the specimens were superparamagnetic, as expected from uniformly dispersed single magnetic domain nanoparticles. Single magnetic domains are usually observed in γ-Fe_2O_3 nanoparticles smaller than 30 nm. In the present case, 12 nm–sized iron oxide nanoparticles were employed.

The self-healing proof-of-concept experiment was carried out by cutting the films into two physically separated parts, which were then brought into contact. A magnetic field oscillating at 278 kHz was applied through a three-loop copper coil for 2 h, allowing the cut regions to be fully mended. Self-healing was repeated several times, giving analogous results. Moreover, the mechanical properties (Young's modulus and strain) of those films were the same before and after repairing. A more detailed structural characterization based on combination of scanning electron microscopy (SEM) and internal reflection infrared imaging (IRIRI) data revealed that though the mechanical properties remained unchanged, morphology was significantly modified. In particular, upon application of an oscillating magnetic field, the *p*-HDFMA phase underwent a remarkable stratification in the same area where the iron oxide nanoparticles were detected. The Fe_2O_3 nanoparticles oscillate under magnetic stimulation, so generating a flow of the amorphous phases of interfacial regions. This flow allows the film to be repaired. The energy associated to both Néel and Brownian relaxations are converted to heat, giving rise to hyperthermic effects. The mechanism of magnetic field–induced self-healing was explained by considering that the Néel relaxation time, τ_N (defined as $\tau_0 \exp(K_{eff} V/kT)$, where K_{eff} is the anisotropy constant = 1.6×10^4 J/m^3 for Fe_2O_3, k is the Boltzmann constant, T is the temperature, and V is the volume of the nanoparticle) is ~10^{-7} s. Thus, the magnetic moment can return to its original state with efficient heat dissipation through Néel relaxation. At the nanoparticle–polymer interface, the Brownian relaxation leads to significantly longer relaxation times in the order of 10^{-2} s. Although a reliable evaluation of the local temperature reached upon application of the magnetic field was not possible (the authors measured 43°C, but independently observed a pure thermally induced amorphous flow only at $T \geq 120$°C), the main benefits of a magnetic field–triggered self-healing compared with a simple thermal activation were the complete retention of mechanical properties and the absence of visible scars on the mended surface.

In 2010, Bowman and coworkers (Adzima et al. 2010) reported the magnetic field–induced self-healing of a Diels–Alder cross-linked network embedded with chromium dioxide (CrO_2). Using Diels–Alder reactions for self-healing purposes was pioneered by Wudl and coworkers (Chen et al. 2002).

Bowman and coworkers exploited Diels–Alder reaction between furan and maleimide functionalities (Figure 8.10).

In this specific case, a trisfuran (pentaerythritol propoxylate tris(3-(furfurylthiol) propionate), PPTF and a bismaleimide (1,1′-(methylene-di-4,1-phenylene)bismaleimide), DPBM) were used as monomers. At low temperature, the forward reaction is favored, yielding bond formation and gelation. Above 92°C the opposite reaction predominates, allowing reversible bond breaking. The authors aimed to take advantage of hysteresis heating developed by application of an alternating magnetic field to polymer matrices containing ferromagnetic metals or compounds. Above the Curie temperature (T_C) of this material, the magnetic susceptibility vanishes and heating ceases. Thus, unlike superparamagnetic effects or induced currents, hysteresis heating provides a self-limiting mechanism that is precisely controlled by the Curie temperature of the susceptible material. Here the main problem was related to the choice of appropriate magnetic moieties, which should have T_C between 92°C (activation of retro-Diels–Alder reaction) and the decomposition temperature of the polymer matrices (in general below 300°C). The authors opted for CrO_2 since it exhibits T_C at 113°C, whereas common ferromagnetic materials, such as Ni, Fe, and Co, are unsuitable for organic materials as their T_C are well above those of polymer decomposition (358°C, 770°C, and 1130°C, respectively). In general, the application of an external magnetic field results in increased steady-state temperatures. Above a critical

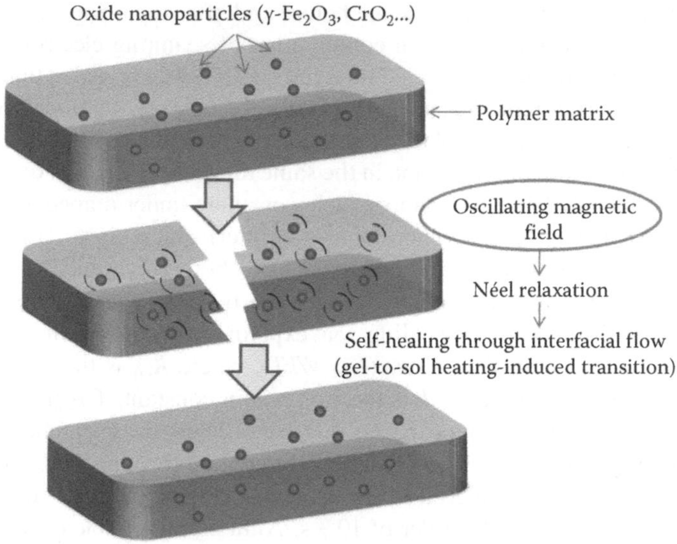

FIGURE 8.10 General scheme showing polymer self-repairing upon application of an external oscillating magnetic field. Efficient heat dissipation through Néel relaxation is exploited for triggering self-healing processes.

magnetic field, the steady-state temperature is nearly independent of the applied field (a very weak rise is observed, due to the magnetic coil resistive heating). The authors tested three different concentrations of CrO_2: 0.1, 1, and 10 wt%. In all of the cases, the observed maximum steady-state temperatures were about 120°C, 135°C, and 146°C, respectively, thus well above the Curie temperature, suggesting the presence of impurities and possible particle agglomeration. The critical magnetic field is in the order of mT and is inversely dependent on the CrO_2 content (about 13 mT for 0.1 wt% samples and 8 mT for the 10 wt% ones). More importantly, before reaching the maximum steady-state temperature, both 1 and 10 wt% samples exhibited well-defined hysteresis loops, demonstrating the nonlinear nature of magnetic-induced heating in chromium oxides. The authors showed that above 110°C the composite material behaves as a viscous liquid and could be easily reshaped using simple teflon molds. This cycle was repeated several times, exhibiting the complete recovery of both flexural modulus and ultimate strength. These experiments demonstrated that hybrid nanocomposites based on thermoreversible gels containing magnetic susceptible nanoparticles can be a promising route for nonautonomic self-healing. The local generation of hysteresis heating can be a suitable trigger for gel-to-sol transitions, which allows the self-healing material flow. This strategy could be particularly useful, since it does not require a perfect realignment of the fractured parts. As long as the induced temperature can exceed the gel-to-sol transition without reaching thermal decomposition, side reactions can be negligible.

Due to the number of advantages described earlier, magnetic activation is expected to be one of the main focuses for most of the forthcoming research on stimuli-responsive self-healing nanocomposites. In this regard, synthesis of new metal and metal oxide nanophases that can trigger self-healing processes by remote magnetic field activation is witnessing an ever-increasing interest.

In particular, future challenges will be related to the possible activation of self-healing mechanisms no more limited to either Diels–Alder or pure thermal effects, but capable of exploiting new reactions and phenomena.

8.3.3 Nanophases for Electrically Triggered Self-Healing

Incorporation of conductive moieties can be a major breakthrough for developing new classes of self-healing nanocomposites. Fractures are expected to cause significant increases of electrical resistance in these materials. If the material is integrated into a circuit containing a voltmeter, this drop in conductivity could be detected and offset by application of an external electric field. The latter induces cracks to act as local sources of heat. Localized heating can be directly exploited to heal damages and restore functional properties of the material. Another very interesting opportunity is related to the introduction of an electronic feedback, which could allow getting a real-time diagnostics of the structural status of materials. As a result, mechanical failures and microcracks can be early detected and repaired. Moreover, materials could record stress events at the microscale. To date, the most advanced contributions to self-healing conducting composites come from aerospace research. Carbon fiber–reinforced polymer composites have been the first examples in these sense. For example, Abry et al. (1999) and the Hayes and coworker (Hou and Hayes 2002) made use of measurements of electrical resistance for monitoring damages in carbon fiber–reinforced polymers. Later on, further studies on resistive heating (Yarlagadda et al. 2002) opened the door for investigating carbon fiber–reinforced compounds that can be self-healed through Diels–Alder reaction triggered by application of external electrical fields (Wang et al. 2007). Park et al. (2008) successfully demonstrated that DCPF-based polymers can be repaired by applying electrical currents through a graphite/epoxy substrate. Resistive heating at 70°C–100°C resulted in full mending of the mechanically induced microcracks, and the structural quality of the repaired materials was better than that obtained by convective heating. In another study, Kirkby et al. (2008) explored the use of Ni/Ti/Cu shape memory alloys to reach a twofold goal: the metal alloy can be used as an active medium for conducting resistive heating and, at the same time, the thermally induced austenite-to-martensite phase transition can be exploited to exert a mechanical force that can further drive crack mending. The shape memory alloy was added to an epoxy composite containing wax-encapsulated Grubbs' catalysts, which have been previously described. Upon activation by resistive heating at 80°C for 30 min, the self-healing DCPD operated very efficiently and the thermally driven contraction of the alloy strongly contribute to close the fractures. Optimized materials exhibited a maximum healing of 98%. All of these examples and concepts could be extended, in principle, to other classes of composites, in which the conductive moieties could be given by nanostructured units or organometallic polymers as well. For example, replacement of carbon fibers with carbon nanotubes offers exciting benefits in monitoring and self-healing of mechanical damages. Studies on conducting carbon nanotubes networks dispersed into an epoxy polymer matrix and their use in failure sensing were firstly reported by Thostenson and Chou (2006). The authors used direct-current measurements for detecting damages induced in the polymer matrix. Clear-cut shifts in the sensing curve were revealed upon irreversible damage induced by static loads. However, fatigue-induced damages were not considered in this study. One year later, Koraktar and coworkers (Zhang et al. 2007) demonstrate the full potential of both single and multi-walled carbon nanotubes in revealing and quantifying extent and propagation of fatigue-induced defects. The carbon nanotubes were highly dispersed into a commercial epoxy resin. This work features the key role of through-thickness conductivity measurements for detecting delamination. This is a major advantage in comparison with carbon fiber–reinforced composites. Due to their uniform dispersion all over the polymer, carbon nanotubes allow any region of the material to be analyzed by simple resistance measurements, without sophisticated circuitry and electronics. Moreover, very-low-weight fractions (~0.5%–1%) of carbon nanotubes are enough for making these composites conductive. The volume electrical resistivity of the composites reduces by two orders of magnitude by increasing the weight fraction of the nanotubes from 0.1% to 1%. In addition, multi-walled carbon nanotubes are about one order of magnitude more conductive than single-walled ones over these weight fractions, so that the authors limited their investigation to multi-walled species. They found that electrical resistance linearly changes as a function of the interface size, and is also very sensitive to delamination length. Due to such a large resistance change variations, material failures can be precisely localized with micrometer resolution and repaired by the cracked interfaces at 150°C. The self-healing efficiency of optimized samples was 70%. In perspective, the authors envisage using of higher electrical currents to induce the selective heating up of nanotubes, with consequent fast mending of the cracks as soon as they appear. This process could be driven automatically by remote computers, which makes these composites particularly suited for technology transfer. More recently, similar strategies employed graphene instead of carbon nanotubes. For example, Cheng and coworkers (Xiao et al. 2010) fabricated graphene-based self-healable shape memory polymer composites. In this case nanolayered graphene, synthesized by microwave plasma enhanced chemical vapor deposition, was incorporated into epoxy shape memory polymer composites. The shape memory properties of the polymer matrix provided thermal healing capability into the composites, which was strongly enhanced at ultralow filler contents of 0.0025 and 0.0125 vol%. In general, both nanotubes and graphene-based nanocomposites hold great promise as potential electrically driven self-healing materials. Moreover, integration of these nanostructures in microcapsule-based autonomous self-healing systems is one of the future goals of the Moore's group, which has already synthesized microcapsules containing suspensions of carbon nanotubes (Caruso et al. 2009).

"Bottom-up" approaches can also be explored, in order to fabricate conductive composites from a molecular level. Obvious advantages in terms of sensitivity, efficiency, and selectivity may be envisioned. However, although dynamic polymerizations can be performed in several ways, in general the reversible reactions involved in these processes are carried out by nonconductive moieties, so that resistive heating cannot operate. A possible solution, which has been proposed by Bielawski and coworkers (Williams et al. 2007), might be given by complexes formed between

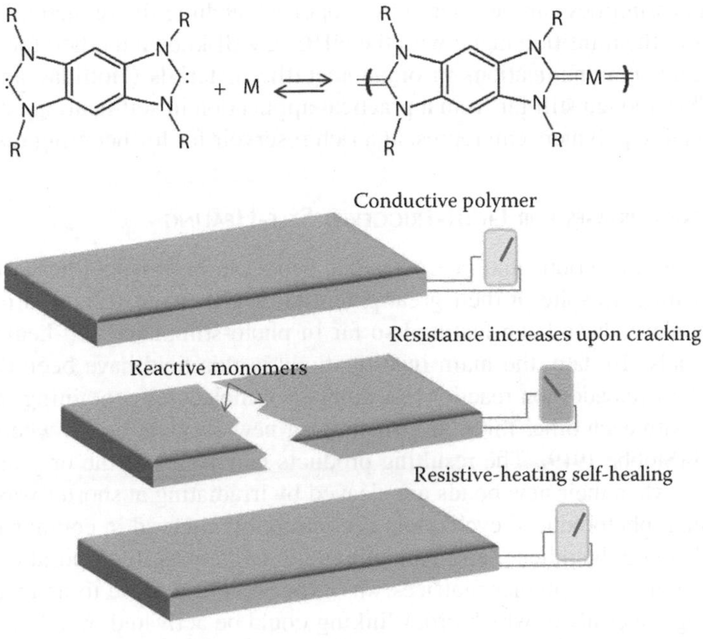

FIGURE 8.11 Scheme showing a possible exploitation of organometallic conductive polymers as self-healing materials. The equilibrium between monomer species and polymers can be controlled by resistive heating addressed through an external feedback circuit. Upon cracking, the resistance increases and the resulting heat promotes polymerization of the monomer species. (See Williams, K.A. et al., *J. Roy. Soc. Interface*, 4, 359, 2007.)

NHCs and transition metals. The authors showed that NHCs can react with Ni, Pd, or Pt salts to yield organometallic polymers (see Figure 8.11). The electrical conductivity of these polymers, measured through a four-point setup, is 10^{-3} S/cm. The polymerization is fully reversible and thermally activated, so that resistive heating could be exploited as an external stimulus to induce reconstruction of cracked chains. In their proof-of-concept, 800 nm thick polymer films were scored with a razor blade. Upon heating at 200°C for 25 min, the surface damages were smoothened. Better results were obtained when dimethyl sulfoxide (DMSO) vapor is introduced in a sealed vessel containing the specimen. In this case an actual refilling of the crack was observed, because the organic solvent facilitates reformation of the NHC–metal broken bonds. As pointed out by the authors, many challenges must be still overcome in order to use organometallic polymers in resistive heating–stimulated self-healing materials. In particular, their dependence on solvent, which must be added as vapor or liquid in a second step, should be eliminated. As a possible solution, the authors suggested incorporation of bulky N-alkyl groups, such as 2,3-dimethylbutyl, into the carbene moieties. These modifications should frustrate crystallization, reduce viscosities of the respective polymers, and promote the material flow into neighboring microcracks resulting from depolymerization. Another mandatory step forward is required by the conductivity of these materials, which must be enhanced to more than or equal to 1 S/cm, in order to make them of practical usefulness. However, manifold possibilities can be offered by properly matching the reduction–oxidation potentials of the transition metal with the NHC, a well-known method for maximizing electronic communications in organometallic materials (Holliday and Swager 2005). Thus, though still far from a practical application in self-healing composites, organometallic polymers can represent a rich reservoir for forthcoming research.

8.3.4 Nanophases for Light-Triggered Self-Healing

Light-triggered reactions and processes can represent promising alternative routes for self-healing. In spite of their great potential, neither nanostructured phases nor metal complexes have been featured so far in photo-stimulated self-healable polymer materials. To date, the main investigations in this field have been devoted to Diels–Alder cycloaddition reactions. A number of molecules containing olefins can be linked with each other through formation of new covalent bonds when properly irradiated (Stobbe 1919). The resulting products may revert to the original state of single units when their new bonds are cleaved by irradiating at shorter wavelengths. In particular, photoinduced cyclization is commonly observed in coumarin, anthracene, maleimide, butadiene, and cinnamic acid derivatives. Incorporation of these photoactive units in polymer matrices was expected to give rise to nonautonomous self-healing materials in which cross-linking could be activated by UV irradiation. For example, coumarin units undergo [2+2] cycloaddition upon irradiation in the 310–355 nm range, resulting in coumarin dimers linked through a newly formed cyclobutane bridge (Hasegawa et al. 1972). This reaction can be reverted by irradiation at 277 nm and is characterized by high yields also in the solid state (Ramasubbu et al. 1982). Studies on photogelation of coumarin moieties showed successful cross-linking of the polymer matrix (Chujo et al. 1990). In spite of those encouraging preliminary results, coumarin derivatives have not been extensively investigated for self-healing applications. Moreover, one of the most important works in this field has been carried out using cinnamic acid derivatives (Chung et al. 2004). In this work, the authors demonstrated an example of photochemically induced self-healing of a polymer matrix. This strategy relies on the hypothesis that the highly strained cyclobutane rings, which bridge two cinnamoyl units, are preferentially broken upon crack propagation. Appropriate irradiation induce photocyclization, thus restoring the pristine cyclobutane units and mechanical strength of the material. The major limitation of this system is low healing efficiency (14%), which can be slightly improved (up to 26%) by combination with thermal treatment. Again, very few alternatives to this approach have been proposed so far. One of these makes use of allyl sulfide moieties that can be cross-linked through addition–fragmentation chain transfer reactions (Scott et al. 2005). In this case, upon irradiation the polymer backbone undergoes cleavage and subsequent rearrangement in a new configuration, characterized by a lower stress state. As already mentioned earlier, a striking feature of this overview on light-driven self-healing is the absence of work reporting direct utilization of metal complex catalysts as those employed for ring opening metathesis or analogous processes. This is in contrast with the results of a number of studies that highlight the key role of transition metal catalysts and photosensitizers in polymerization reactions (Stoll and Hecht 2010, and references therein). For example, Mühlebach and coworkers showed the positive effects of light on polymerization reactions based on olefin metathesis, which are promoted by ruthenium catalysts analogous to those previously reported (Section 8.3.1) (Karlen et al. 1995).

The authors demonstrated the high efficiency of ruthenium–arene sandwich complexes in photochemical activation of ROMP of several strained bicyclic olefins. Unfortunately, these catalysts exhibited a certain activity also before irradiation. In a following work, the same group discovered that activity can be significantly reduced by replacing ruthenium with osmium (Hafner et al. 1997). These new catalysts were able to trigger ROMP of norbornene (one of the monomers that have been employed in self-healing systems) only upon UV irradiation with a mercury lamp, allowing controlling the reaction through an external photo-stimulus. Further control of light-driven ROMP reaction can be offered by NHCs that have shown excellent capabilities of tailoring both electronic and steric properties of the metal center (Wanzlick 1962). For example, Noels and coworkers demonstrated that [Ru(*p*-cymene)Cl_2NHC] complexes can promote polymerization of cyclooctene by simple irradiation with a commercial 40 W fluorescent light bulb (Delaude et al. 2001). Many further examples of photoactive ruthenium–carbene complexes have been reported in the last years (Stoll and Hecht 2010, and references therein), yet their potential exploitation to self-healing have been overlooked. Although manifold challenging issues could arise in this regard, it should be recalled that analogous catalysts have been already employed in microcapsule-based autonomous self-healing composites activated by mechanical stress. Thus, most of the skills coming from those studies could be suitably adapted to photoactive catalysts. For example, Sriram (2002) explored photoinduced ROMP of either norbornene or DCPD, as a complementary process to the free radical ROMP. This work confirmed that ROMP reactions can be successfully carried out with monomers commonly used as self-healing agents. However, fabrication of self-healing nanocomposites has not yet been reported. In case of success, smart materials capable of both autonomous and stimulated self-healing could be achieved.

New, exciting perspectives for light-triggered self-healing nanocomposites can be also given by metal nanoparticles. In particular, we will focus our attention on a couple of examples that bring out a new concept of self-healing and could open exciting perspectives for fabrication of new materials.

The first example was given by Grzybowski and coworkers who realized hybrid materials based on photoresponsive metal nanoparticles (gold or silver) embedded into flexible organogels (Klajn et al. 2009). The 5 nm–sized metal nanoparticles coated with dodecylamine and azobenzene-terminated thiols ((11-mercaptoundecanoxy)azobenzene, MUA) were added to a poly(methyl methacrylate) (PMMA) toluene solution, which was cast into flexible preheated polymeric molds. The resulting gel photopapers were characterized by bright colors originating from metal nanoparticles dispersed into the flexible matrix. These materials can be impressed by any graphical images by simple exposure to UV light (365 nm, 0.7–10 mW/cm^2) irradiated through proper masks. UV irradiation induced a visible change of color in the composite polymer, whose intensity depended on irradiation time. When UV light was switched off, the written images underwent progressive self-erasing, which turned out into complete restoration of the original, virgin surface. The overall process is due to the following mechanism: first, UV irradiation induces *trans*-to-*cis* isomerization of the azobenzene groups of MUA. As a result, the dipole moment increases, stimulating attractive interactions between metal nanoparticles. When the degree of coverage, χ, of NPs by MUA was $0.23 < \chi < 0.34$, the NPs clustered reversibly into metastable supraspherical aggregates, whose average size was ~150 nm. These aggregates undergo spontaneous disassembly when UV light is switched off. It should be noted that erasing is not caused by diffusion of the aggregates, but just by their disintegration, which is ultimately ruled by the *cis*-to-*trans* retro-isomerization rate. Depending on the degree of MUA coverage, the self-erasing (or, from another standpoint, the self-healing) time ranged from hours to days. Heat or visible light exposure shortened the erasing time to few seconds, whereas dark conditions increased stability over a week. The writing/erasing process fully worked for several hundred cycles.

In this case gold is used just as a "color maker," since the aggregation–disintegration is governed by isomerization of the azobenzene ligands, and self-healing is meant as a merely functional process, because it does not involve any structural change of the polymer matrix.

Other promising strategies for promoting light-driven self-healing phenomena in nanocomposites can rely on thermoplasmonics, a neologism recently introduced to indicate the generation of heat upon interaction of a light source (in general a laser) with metal nanoparticles. For example, the very highly efficient optothermal conversion exhibited by gold nanoshells and aggregates has been extensively exploited for introducing selective modifications and patterns in polymer matrices (Alessandri and Depero 2008; Maity et al. 2011). Moreover, plasmonic heating can also be conveniently harnessed to induce phase transition in metal oxides with high spatial resolution (Alessandri et al. 2009, 2011). Different examples of functional self-healing based on plasmonic heating can be found in a work on pressure-sensitive adhesive (PSA) nanocomposites (Alessandri 2010). PSAs, a class of materials including acrylics, polyurethanes, polyesters, or silicones, are widely used for a variety of applications in everyday life (Satas 1989). In particular, upon modification with electrically conductive fillers, such as carbon or metals, they can be applied as antistatic self-adhesive tapes for electromagnetic shielding purposes in various contexts. In this work commercial carbon-filled PSAs, obtained by blending butyl acrylate (BA; $T_g = -54°C$) 37 wt%, 2-ethylhexyl acrylate (2-EHA; $T_g = 70°C$) 60 wt%, acrylic acid (AA; $T_g = 106°C$) 3 wt%, monomers and carbon black (with a maximum nominal content of 15 wt% with reference to the polymer), were adapted to work as laser-writable and laser-rewritable adhesive substrates. This system is based on cooperative interplay between viscoelastic properties of PSAs and enhanced thermal conductivity provided by a thin overlayer of gold. Depending on final applications, the stored information was either preserved or erased through proper surface modifications (e.g., by adding protecting coatings). From another standpoint, this system can be seen as an example of multiple self-healing processes working on the same material at different stages. In fact, the written patterns result from surface damages induced by laser. The overall interaction represents a sort of "ballistic stimulus" for these nanocomposites. Upon laser stimulation (the "writing" stage), the material is able to restore either its structural integrity or its functionality through different self-healing mechanisms. The writing stage was carried out by focusing a continuous wave (cw) He–Ne laser (λ = 632.8 nm, instrumental output power: 4.5 mW) onto the surface of gold-coated PSAs by means of microscope objectives with different numerical apertures. Point-by-point static irradiation results in the formation of nearly spherical rings on the PSA surface. These rings are delimited by regular rims consisting of gold aggregates tethered into the polymer matrix. Under the same operating conditions (power of the laser, exposure time, thickness of the Au coating, depth of focus, etc.), the reproducibility of these features is quite good in terms of size and wall thickness and the outcomes of writing processes can be tailored by a direct approach. More complex patterns and figures can be obtained by combining this point-by-point approach with a line scanning operating mode, as well as by driving irradiation onto selected zones. Moreover, more spatially limited features can be introduced by direct laser irradiation of the rings through optical filters that attenuate the power of the incoming radiation. There are several parameters influencing the writing process. Most of them, in particular those coming from the optical setup of this writing system (such as power of laser, depth of focus, etc.), are common with other laser writing processes (Alessandri and Depero 2008). Herein, the most interesting issues are related to the role of the gold layer. Au-PSAs exhibited a strong plasmon resonance band at about 500 nm. As the plasmonic band is far from being overlapped with the wavelength of the exciting laser, possible enhanced absorption effects due to the gold coating do not play a major role in writing process. On the contrary, for Au-PSAs the contribution of scattering and reflection increases in red and near-infrared, so that the overall absorption in this region is lower than that of uncoated PSAs. This is a key factor, since PSAs are good light absorbers and moderate conductors, so that full matching of laser excitation and plasmon resonance would generate too strong heating effects, which could not be harnessed to get predictable, reproducible and, definitely, useful patterns. Moreover, since both thermal and electrical surface conductivities are strongly dependent on the thickness of the gold layer, the latter is a critical parameter to control the effects of the laser interaction. Optimal thicknesses are in the range of 12–20 nm. If this thickness is too low, heat accumulation gives rise to explosive processes, which are difficult to be controlled. Moreover, thicker Au coatings (30 nm or more) are too reflecting and cannot be exploited for inducing surface modifications. The mechanism of ring formation was qualitatively described as follows. Upon laser irradiation, the local temperature of Au-PSAs increases. Due to the Gaussian shape of the laser beam, the temperature distribution reaches its maximum at the center of the focus and decreases as a radial function. The power of the laser at the center of the focused beam is high enough to yield substrate ablation. As a consequence, small pierced spots appear in the central part of the rings. Unlike the holes generated upon irradiation of uncoated PSAs, in the case of Au-PSAs the size of the ablated area did not significantly exceed the diameter of laser beam waist, even after prolonged exposure. Due to the high thermal conductivity of Au (318 W/m/K at 25°C for bulk samples; in thin films with thicknesses of 15–20 nm this value can range from 300 to <100 W/m/K, depending on both electron surface and grain boundary scattering) (Feng et al. 2009), heat cannot be accumulated as for uncoated samples but it can efficiently propagate to the surrounding area, leading the underlying polymer to soften and melt. Thermal gradients drive the melting material from the hot center to the cold peripheral regions (thermocapillary Marangoni's flow). When the laser is switched off, the fluid is instantaneously supercooled leaving spherical rings as "snapshots" of the original hydrodynamic flow. Au coating is necessary to ensure good conductivity and to uniformly propagate heat through the surface. Using less conductive random aggregates or ordered networks of Au nanoparticles did not give rise to micrometer-sized rings but only to local dimples limited in size by nanoparticle/aggregate dimensions.

Interestingly, the written patterns undergo spontaneous self-erasing in air. As previously discussed for the case of MUA-coated Au nanoparticles described by Klajn et al. (2009), self-erasing is another form of functional self-healing, in which alterations of the material surface can be progressively reduced or fully eliminated. In the present case, all of the polymeric parts were etched and only weak footprints of the original rings were detected after 24 h. The remaining footprints were constituted by coalesced gold and carbonaceous species resulting from polymer degradation. Self-degradation can be due to various factors, probably in combination with each other. Au aggregates and nanoparticles embedded into the ring walls can catalyze the polymer degradation very efficiently. At the same time, we should recall that the composite walls resulted from partial degradation of PSAs. This process is known to yield peroxides and highly reactive species that can trigger subsequent self-degradation. Moreover, simple photodegradation can be ruled out as the process occurred also under dark conditions. The most important fact is that written tracks self-erase within a defined period of time, which depend on their size and is on the timescale of hours. More efficient erasing can be obtained by dipping the freshly written Au-PSA into water. These substrates can be usually dissolved in ethyl acetate or isopropanol and are known to be insensitive to water. Upon laser irradiation this holds true, with the important exception of the written tracks, which can be easily removed. This can be regarded as a further proof of alteration of the chemical nature of the rims. Thus, although difficult to be demonstrated with quantitative data (*in situ* Raman and infrared microspectroscopy analyses were ineffective for extracting any useful chemical information, and other characterization techniques can significantly alter the substrate or do not have sufficient spatial resolution or sensitivity), the generation of a large number of hydrophilic, polar moieties in PSA substrates might be considered among the effects of laser irradiation. In the case of water removal the PSA surfaces came back fully clean, as the aggregated gold was mechanically washed away together with the organic slurry. Moreover, it has also been demonstrated that self-erasing can be fully prevented by preliminary deposition of a protective oxide-coating layer.

These systems also display another type of self-healing process, even more interesting for our discussion. This process was observed, for example, on the walls of rings generated by laser writing. When laser is focused onto a selected area of the wall at low power (<1 mW) small holes, with diameter range $\sim 2-4\ \mu m$, are generated. These holes undergo spontaneous self-healing and are fully mended within few minutes. Specific examples show that $2\ \mu m$ holes were refilled in <3 min. The self-healing rate, estimated on the basis of optical measurements of refilling time, was $0.92\ \mu m/min$. Similar self-healing tests, carried out on other rims, confirmed that the healing rate was around $1\ \mu m/min$. When a new hole is generated in the healed region, it undergoes identical self-healing.

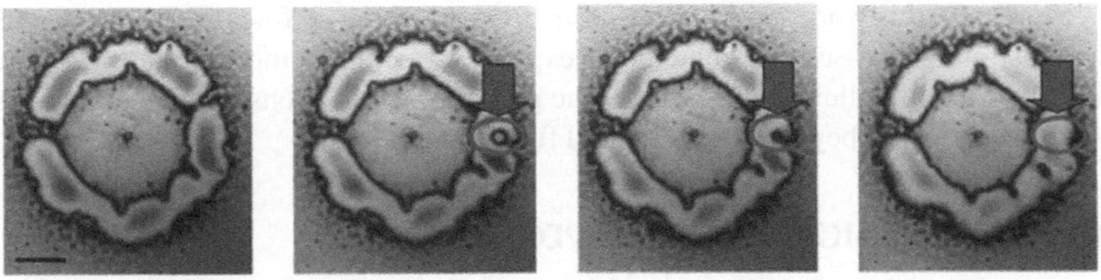

FIGURE 8.12 Example of self-healing in gold-coated pressure-sensitive adhesives. Laser irradiation generates a circular hole (indicated by an arrow in snapshot 2) that undergoes rapid self-healing within few minutes. Scale bar: 5 μm. (See Alessandri, I., *Small*, 6,1679, 2010.)

These cycles were repeated several times, showing high reproducibility of the process. An example of this kind of self-healing is given in Figure 8.12.

After the first cycle, the new self-healing processes were slightly slower, each one occurring at the same rate (about 0.27 μm/min). Moreover, under the same exposure time, the initial sizes of the holes achieved upon subsequent rewriting cycles were progressively smaller than that observed in the case of first irradiation. Thus, although rewriting and self-healing processes are reproducible, slight, yet significant, differences in terms of spatial resolution and self-healing rate can be observed. In this case, self-healing relies on the viscoelastic response of the composite to heat generated by interaction between laser and Au NPs. A possible explanation of this mechanism may be given by the analysis of the dynamic competition between two different Marangoni effects occurring upon laser irradiation. As we observed before for ring formation, spherical holes are generated through thermocapillary forces that push the irradiated material toward cooler peripheral regions. Once the laser has been turned off, the slushy composite (which is constituted by Au NP aggregates embedded into a blend of melted adhesive, olefins, and alcohols by-products resulting from thermal degradation of the substrate) (Czech and Pelech 2009) reverts its radial expansion and fill back the empty space left behind. At this stage, the rate of self-healing is primarily driven by chemo-capillary flow, induced by surface compositional gradients. Repeated laser irradiation can introduce many sources of compositional differences that can directly affect the viscoelastic properties of the ring walls. For example, most of the components responsible for PSA tackiness undergo thermal degradation, so that sequential laser irradiation can be detrimental for viscoelasticity of these regions. Moreover, the writing tests previously demonstrated that the walls of rings are gold–polymer composites in which the most efficient heat generation occurs where the gold aggregates are more concentrated and larger in size. As sequential irradiation of the same region gave rise to smaller and smaller holes, one may infer that the gold concentration undergo progressive decrease upon laser exposures. Thus, heating effects and thermal gradients are expected to be less pronounced as the number of laser cycle increases. This can account for leveling of viscoelastic behavior, which is reflected by stabilization of self-healing rates.

These investigations could be extended to other adhesives and polymer-based viscoelastic systems, opening new perspectives for light-driven stimuli-responsive materials.

Moreover, this work shows another way to think about self-healing processes. By demonstrating that self-healing can be exploited to control the lifetime of a given information, it implicitly suggests that the potential applications of self-healing processes can go well beyond structural and functional repair.

8.4 CONCLUSIONS AND PERSPECTIVES

The examples presented in this review feature the key role of inorganic nanophases and hybrid molecular species in self-healing composites. Metal centers rule self-healing processes based on reversible metal–ligand coordination and polymerization reactions. Inorganic catalysts and oxide precursors are involved in "metabolic-like" self-healing cycles, which take advantage of environmental components, such as air or water, to restore both structural and functional properties of polymeric matrices. Metal and metal oxide nanoparticles can be exploited to trigger and control self-healing through electromagnetic fields and, in many cases, they can also provide a significant improvement of the material mechanical properties. Due to the number of materials and applications, nanocomposites must adopt different strategies for making self-healing efficient and advantageous in comparison with conventional passive protection. For example, Au NP-reinforced hydrogels could be used as biomaterials that can autonomic self-repair from damages introduced by either wear and tear or surgical handling, metal oxide precursors could be crucial for fabricating thin-film organic electronic devices based on multilayer composites, and so on.

In a few cases, such as microcapsule-based self-healing elastomers, thermosets, and powder coatings, the research outcomes have been already transferred to industrial production. More sophisticated strategies, such as those requiring a real-time electrical feedback or light-driven activation, are still far from scaling-up. However, in view of a real technological transfer over a large scale all of the self-healing nanocomposites must face two main issues: economic feasibility and, partially related to this, long-term repeatability of the self-healing processes. In this context, many aspects should be optimized and nanophases could provide decisive breakthroughs. New strategies, based on still unexplored reactions and catalysts, could be expected to extend the range of applications and reduce the costs of these materials. Similarly, due to their unique functional properties, metal and metal oxide nanoparticles could be exploited in a variety of self-healing processes, which could occur under cheap, safe, and environmental-friendly conditions. Moreover, advances in synthesis and characterization of hybrid nanostructures and interfaces could make nanotechnology a leading discipline for self-healing systems. Several tightly intertwined goals can be envisaged for future research. Among those, the development of advanced nanophases that can contemporary reinforce the host matrix and trigger self-healing processes might be one of the most viable on the midterm. This review reported some examples that can be taken as useful proof-of-concept for future investigations. Further major advances can be expected from integration of nanoalloys in polymer matrices, in the attempt to achieve shape memory nanocomposites. At the same time, electric field–responsive gels and, more generally, stimuli-responsive interfaces could be intensively investigated and integrated in complex, multifunctional materials. In this context, thermo- and magnetoplasmonic nanostructures represent powerful tools that are still largely unexplored. Future nanocomposites should be able to activate different self-healing mechanisms (autonomic and/or nonautonomic) in the same matrix. This redundancy would be very useful, not only for enhancing the self-healing capabilities of these materials, but also to make them responsive to different stimuli. This would allow mimicking and (at least conceptually) approaching the adaptive self-healing strategies observed in natural systems. Finally, results and by-products of self-healing processes could be exploited to introduce new functionalities in these materials, as observed for pressure-sensitive composites. The latter, more subtle strategy, overcomes the merely restoring function of self-healing and opens new exciting perspectives for smart materials.

REFERENCES

Abry, J. C., Bochard, S., Chateauminois, A., Salvia, M., and Giraud, G. 1999. In situ detection of damage in CFRP laminates by electrical resistance measurements. *Composite Science and Technology* 59: 925–935.

Adzima, B. J., Kloxin, C. J., and Bowman, C. N. 2010. Externally triggered healing of a thermoreversible covalent network via self-limited hysteresis heating. *Advanced Materials* 22: 2784–2787.

Ajayan, P. M., Schadler, L. S., and Braun, P. V. 2003. *Nanocomposite Science and Technology*. Weinheim, Germany: Wiley-VCH.

Alessandri, I. 2010. Writing, self-healing and self erasing of pressure-sensitive adhesives. *Small* 6: 1679–1685.

Alessandri, I. and Depero, L. E. 2008. Laser-induced modification of polymeric beads coated with gold nanoparticles. *Nanotechnology* 19. Article No: 305301.

Alessandri, I. and Depero, L. E. 2009. Using plasmonic heating of gold nanoparticles to generate local SER(R)S-active TiO_2 spots. *Chemical Communications* 17: 2359–2361.

Alessandri, I., Ferroni, M., and Depero, L. E. 2009. In situ plasmon-heating-induced generation of Au/TiO_2 "hot spots" on colloidal crystals. *ChemPhysChem* 10: 1017–1022.

Alessandri, I., Ferroni, M., and Depero, L. E. 2011. Plasmon-assisted, spatially resolved laser generation of transition metal oxides from liquid precursors. *Journal of Physical Chemistry C* 115: 5174–5180.

Alexandre, M. and Dubois, P. 2000. Polymer-layered silicate nanocomposites: Preparation, properties and uses of a new class of materials. *Materials Science and Engineering R: Reports* 28: 1–63.

Bleay, S. M., Loader, C. B., Hawyes, V. J., Humberstone, L., and Curtis, P. T. 2001. A smart repair system for polymer matrix composites. *Composites Part A* 32: 1767–1776.

Caruso, M. M., Schelkopf, S. R., Jackson, A. C., Landry, A. M., Braun, P. V., and Moore, J. S. 2009. Microcapsules containing suspensions of carbon nanotubes. *Journal of Materials Chemistry* 19: 6093–6096.

Chen, X., Dam, M. A., Ono, K., Mal, A., Shen, H., Nutt, S. R., Sheran, K., and Wudl, F. 2002. A thermally re-mendable cross-linked polymeric material. *Science* 295: 1698–1702.

Cho, S. H., Andersson, M., White, S. R., Sottos, N. R., and Braun, P. V. 2006. Polydimethylsiloxane-based self-healing materials. *Advanced Materials* 18: 997–1000.

Chujo, Y., Sada, K., and Saegusa, T. 1990. Polyoxazoline having a coumarin moiety as a pendant group. Synthesis and photogelation. *Macromolecules* 23: 2693–2697.

Chung, C. M., Roh, Y. S., Cho, S. Y., and Kim, J. G. 2004. Crack healing in polymeric materials via photochemical [2+2] cycloaddition. *Chemistry of Materials* 16: 3982–3984.

Corten, C. C. and Urban, M. W. 2009. Repairing polymers using an oscillating magnetic field. *Advanced Materials* 21: 5011–5015.

Czech, Z. and Pelech, R. 2009. The thermal degradation of acrylic pressure-sensitive adhesives based on butyl acrylate and acrylic acid. *Progress in Organic Coatings* 65: 84–87.

Davis, D. A., Hamilton, A., Yang, J., Cremar, L. D., van Gough, D., Potisek, S. L., Ong, M. T. et al. 2009. Force-induced activation of covalent bonds in mechanoresponsive polymeric materials. *Nature* 459: 68–72.

Delaude, L., Demonceau, A. F., and Noels, A. F. 2001. Visible light induced ring-opening metathesis polymerization of cyclooctene. *Chemical Communications* 11: 986–987.

Dillon, A. C., Jones, K. M., Bekkedhal, T. A., Kiang, C. H., Bethune, D. S., and Heben, J. 1997. Storage of hydrogen in single-walled carbon nanotubes. *Nature* 386: 377–379.

Dry, C. 1996. Procedures developed for self-repair of polymer matrix composite materials. *Composite Structures* 35: 263–269.

Dubreuil, F., Elsner, N., and Fery, A. 2003. Elastic properties of polyelectrolyte capsules studied by atomic force microscopy and RICM. *European Physics Journal E* 12: 215–221.

Duenas, T., Bolanos, E., Murphy, E., Mal, A., Wudl, F., and Schaffner, C. 2006. Multifunctional self-healing morphing composites. In *Proceedings for the 25th Army Science Conference*, Orlando, FL, November 27–30, Paper No: GP-03.

Faber, K. T. and Evans, A. G. 1983. Crack deflection processes. 1. Theory. *Acta Metallurgica* 31: 565–576.

Feng, B., Li, Z., and Zhang, Y. 2009. Prediction of size effect on thermal conductivity of nanoscale metallic films. *Thin Solid Films* 517: 2083–2807.

Fischer, H. 2010. Self-repairing material systems—A dream or a reality? *Natural Science* 8: 873–901.

Gao, H., Kong, Y., Cui, D., and Ozkan, C. S. 2003. Spontaneous insertion of DNA oligonucleotides into carbon nanotubes. *Nano Letters* 3: 471–473.

Ghosh, S. K. 2009. *Self-Healing Materials: Fundamentals, Design Strategies and Applications*. Weinheim, Germany: Wiley-VCH.

Grubbs, R. H. 2006. Olefin-metathesis catalysts for the preparation of molecules and materials. *Angewandte Chemie International Edition* 45: 3760–3765.

Guerret-Plécourt, C., Le Bouar, Y., Loiseau, A., and Pascard, H. 1994. Relation between metal electronic structure and morphology of metal compounds inside carbon nanotubes. *Nature* 372: 761–765.

Hafner, A., Mühlebach, A., and van der Schaaf, P. A. 1997. One-component catalysts for thermal and photoinduced ring opening metathesis polymerization. *Angewandte Chemie International Edition* 36: 2121–2124.

Hager, M. D., Greil, P., Leyens, C., van der Zwaag, S., and Schubert, U. S. 2010. Self-healing materials. *Advanced Materials* 22: 5424–5430.

Hasegawa, M., Suzuki, Y., and Kita, N. 1972. Photocleavage of coumarin dimers. *Chemistry Letters* 1: 317–320.

Holliday, B. J. and Swager, T. M. 2005. Conducting metallopolymers: The roles of molecular architecture and redox matching. *Chemical Communications* 1: 23–26.

Hou, L. and Hayes, S. A. 2002. A resistance-based damage location sensor for carbon–fibre composites. *Smart Materials and Structures* 11: 966–999.

Hucker, M., Bond, I. P., Bleay, S., and Haq, S. 2003. Experimental evaluation of unidirectional hollow glass fibre/epoxy composites under compressive loading. *Composites Part A* 34: 927–932.

Hucker, M. J., Bond, I. P., Foreman, A., and Hudd, J. 1999. Optimisation of hollow glass fibres and their composites. *Advanced Compounds Letters* 8: 181–189.

Hucker, M. J., Bond, I. P., Haq, S., Bleay, S., and Foreman, A. 2002. Influence of manufacturing parameters on the tensile strength of hollow and solid glass fibres. *Journal of Materials Science* 38:7: 309–315.

Jiang, C. Y., Markutsya, S., Pikus, Y., and Tsukruk, V. V. 2004. Freely suspended nanocomposite membranes as highly sensitive sensors. *Nature Materials* 3: 721–728.

Jiang, C. Y., Markutsya, S., Shulha, H., and Tsukruk, V. V. 2005. Freely suspended gold nanoparticle arrays. *Advanced Materials* 17: 1669–1673.

Kamphaus, J. M., Rule, J. D., Moore, J. S., Sottos, N. R., and White, S. R. 2008. A new self-healing epoxy with tungsten(IV) chloride catalyst. *Journal of the Royal Society Interface* 5: 95–103.

Karlen, T., Ludi, A., Mühlebach, P., Bernhard, C., and Pharisa, C. 1995. Photoinduced ring-opening metathesis polymerization (PROMP) of strained bicyclic olefins with ruthenium complexes of the type [(eta(6)-arene (1)Ru(eta(6)-arene(2))](2+) and [Ru(NC-R)(6)](2+). *Journal of Polymer Science Part A* 33: 1665–1674.

Karthikeyan, S., Potisek, S. L., Piermattei, A., and Sijbesma, R. P. 2008. Highly efficient mechanochemical scission of silver–carbene coordination polymers. *Journal of the American Chemical Society* 130: 14968–14969.

Kausch, H. H. and Plummer, C. J. G. 1994. The role of individual chains in polymer deformation. *Polymer* 35: 3848–3857.

Keller, M. W., White, S. R., and Sottos, N. R. 2007. A self-healing polydimethylsiloxane elastomer. *Advanced Functional Materials* 17: 2399–2404.

Kersey, F. R., Loveless, D. M., and Craig, S. I. 2007. A hybrid polymer gel with controlled rates of cross-link rupture and self-repair. *Journal of the Royal Society Interface* 4: 373–380.

Kersey, F. R., Yount, W. C., and Craig, S. I. 2006. Single-molecule force spectroscopy of bimolecular reactions: System homology in the mechanical activation of ligand substitution reactions. *Journal of the American Chemical Society* 128: 3886–3887.

Kessler, M. R., Sottos, N. R., and White, S. R. 2003. Self-healing structural composite materials. *Composites Part A—Applied Science and Manufacturing* 34: 743–753.

Kharlampieva, E., Kozlovskaya, V., Gunawidjaja, R., Shevchenko, V. V., Vaia, R., Naik, R. R., Kaplan, D. L., and Tsukruk, V. V. 2010. Flexible silk-inorganic nanocomposites: From transparent to highly reflective. *Advanced Functional Materials* 20: 840–846.

Kinloch, A. J. and Taylor, A. C. 2002. The toughening of cyanate–ester polymers—Part I—Physical modification using particles, fibres and woven-mats. *Journal of Materials Science* 37: 433–460.

Kirkby, E., Rule, J., Michaud, V., Sottos, N. R., White, S. R., and Manson, J. A. 2008. Embedded shape-memory alloy wires for improved performance of self-healing polymers. *Advanced Functional Materials* 18: 2253–2260.

Klajn, J., Wesson, P. J., Bishop, K. J. M., and Grzybowski, B. A. 2009. Writing self-erasing images using metastable nanoparticle "inks." *Angewandte Chemie-International Edition* 48: 7035–7039.

Lanzara, G., Yoon, Y., Liu, H., Peng, S., and Lee, W.-I. 2009. Carbon nanotube reservoirs for self-healing materials. *Nanotechnology* 20. Article No. 335704.

Lee, J. Y., Buxton, G. A., and Balazs, A. C. 2004. Using nanoparticles to create self-healing composites. *Journal of Chemical Physics* 124: 5531–5540.

Lenhardt, J. M., Black, A. L., and Craig, S. L. 2009. Gem–dichlorocyclopropanes as abundant and efficient mechanophores in polybutadiene copolymers under mechanical stress. *Journal of the American Chemical Society* 131: 10818–10819.

Liu, H. A., Gnade, B. E., and Balkus, Jr., K. J. 2008. A delivery system for self-healing inorganic films. *Advanced Functional Materials* 18: 3620–3629.

Liu, F. and Urban, M. W. 2010. Recent advances and challenges in designing stimuli-responsive polymers. *Progress in Polymer Science* 35: 3–23.

Maity, S., Downen, L. N., Bochinski, J. R., and Clarke, L. I. 2011. Embedded metal nanoparticles as localized heat sources: An alternative processing approach for complex polymeric materials. *Polymer* 52: 1674–1685.

Mamedov, A. A., Kotov, N. A., Prato, M., Guldi, D. M., Wicksted, J. P., and Hirsch, A. 2002. Molecular design of strong single-wall carbon nanotube/polyelectrolyte multilayer composites. *Nature Materials* 1: 190–194.

Mauldin, T. C. and Kessler, M. R. 2010. Self-healing polymers and composites. *International Materials Reviews* 55: 317–346.

Murphy, E. B. and Wudl, F. 2010. The world of smart healable materials. *Progress in Polymer Science* 35: 223–251.

Park, C. W., South, A. B., Hu, X., Verdes, C., Kim, J.-D., and Lyon, A. 2011. Gold nanoparticles reinforce self-healing microgel multilayers. *Colloid Polymer Science* 289: 583–590.

Park, J. S., Takahashi, K., Guo, Z., Wang, Y., Bolanos, E., Hamann-Schaffner, C., Murphy, E., Wudl, F., and Hahn, H. T. 2008. Towards development of a self-healing composite using a mendable polymer and resistive heating. *Journal of Composite Materials* 42: 2869–2881.

Paulusse, J. M. J. and Sijbesma, R. P. 2004. Reversible mechanochemistry of a Pd(II) coordination polymer. *Angewandte Chemie International Edition* 43: 4460–4462.

Piermattei, A., Karthikeyan, S., and Sijbesma, R. P. 2009. Activating catalysts with mechanical force. *Nature Chemistry* 1: 133–137.

Ramasubbu, N., Row, T. N. G., Venkatesan, K., Ramamurthy, V., and Rao, C. N. R. 1982. Photodimerization of coumarins in the solid state. *Journal of the Chemical Society Chemical Communications* 3: 178–179.

Rule, J. D., Brown, E. N., Sottos, N. R., White, S. R., and Moore, J. S. 2005. Wax-protected catalyst microspheres for efficient self-healing materials. *Advanced Materials* 17: 205–208.

Sahoo, N. G., Rana, S., Cho, J. W., Li, L., and Chan, S. H. 2010. Polymer nanocomposites based on functionalized carbon nanotubes. *Progress in Polymer Science* 35: 837–867.

Sanford, M. S., Love, J. A., and Grubbs, R. H. 2001. Mechanism and activity of ruthenium olefin metathesis catalysis. *Journal of the American Chemical Society* 123: 6543–6554.

Satas, D. 1989. *Handbook of Pressure Sensitive Adhesives*, 3rd edn. New York: Van Nostrand Reinhold.

Schmide, A. M. 2006. Electromagnetic activation of shape memory polymer networks containing magnetic nanoparticles. *Macromolecular Rapid Communications* 27: 1168–1172.

Scott, T. F., Schneider, A. D., Cook, W. D., and Bowman, C. N. 2005. Photoinduced plasticity in cross-linked polymers. *Science* 308: 1615–1617.

Seraphin, S., Zhou, D., Jiao, J., Withers, J. C., and Loutfy, Y. 1993. Yttrium carbide in nanotubes. *Nature* 362: 503.

South, A. B. and Lyon, L. A. 2010. Autonomic self-healing of hydrogel thin films. *Angewandte Chemie-International Edition* 49: 767–771.

Sriram, S. R. 2002. Development of self-healing polymer composites and photoinduced ring opening metathesis polymerisation, PhD dissertation. University of Illinois at Urbana-Champaign, Urbana, IL.

Stobbe, H. 1919. Light reactions of the allo- and iso-cinnamic acids. *Chemische Berichte* 52B: 666–672.

Stoll, R. S. and Hecht, S. 2010. Light-gated catalysts. *Angewandte Chemie International Edition* 49: 5054–5075.

Taber, D. F. and Frankowski, K. J. 2003. Grubbs' catalyst in paraffin: An air-stable preparation for alkene metathesis. *Journal of Organic Chemistry* 68: 6047–6048.

Takeda, K., Tanahashi, M., and Unno, H. 2003. Self-repairing mechanism of plastics. *Science and Technology of Advanced Materials* 4: 435–444.

Thostenson, E. T. and Chou, T. W. 2006. Carbon nanotube networks: Sensing of distributed strain and damage for life prediction and self healing. *Advanced Materials* 18: 2837–2841.

Toohey, K. S., Sottos, N. R., Lewis, J. A., Moore, J. S., and White, S. R. 2007. Self-healing materials with microvascular networks. *Nature Materials* 6: 581–585.

Trask, R. S., Williams, G. J., and Bond, I. P. 2007. Bioinspired self-healing of advanced composite structures using hollow glass fibres. *Journal of the Royal Society Interface* 4: 363–371.

Vaia, R. A. and Wagner, H. D. 2004. Framework for nanocomposites. *Materials Today* 7: 32–37.

van der Weij, F. W. 1980. The action of tin compounds in condensation-type RTV silicon rubbers. *Macromolecular Chemistry* 180: 2541–2548.

Varghese, S., Lele, A., and Mashelkar, R. 2006. Metal-ion mediated healing of gels. *Journal of Polymer Science A* 44A: 666–670.

Wang, Y., Bolanos, E., Wudl, F., Hahn, T., and Kwok, N. 2007. Self-healing polymers and composites based on thermal activation. *Proceedings SPIE* 6526: 65261I.

Wanzlick, H. W. 1962. Aspects of nucleophilic carbene chemistry. *Angewandte Chemie International Edition* 1: 75–80.

White, S. R., Sottos, N. R., Geubelle, P. H., Moore, J. S., Kessler, M. R., Sriram, S. R., Brown, E. N., and Viswanathan, S. 2001. Autonomic healing of polymer composites. *Nature* 409: 794–797.

Williams, K. A., Boydston, A. J., and Bielawski, C. W. 2007. Towards electrically conductive, self-healing materials. *Journal of the Royal Society Interface* 4: 359–362.

Wilson, G. O., Caruso, M. M., Reimer, N. T., White, S. R., Sottos, N. R., and Moore, J. S. 2008b. Evaluation of ruthenium catalysts for ring-opening metathesis polymerization-based self-healing applications. *Chemistry of Materials* 18: 44–52.

Wilson, G. O., Moore, J. S., White, S. R., Sottos, N. R., and Andersson, H. 2008a. Autonomic healing of epoxy vinyl esters via ring opening metathesis polymerization. *Advanced Functional Materials* 18: 44–52.

Wu, D. Y., Meure, S., and Solomon, D. 2008. Self-healing polymeric materials: A review of recent developments. *Progress in Polymer Science* 33: 479–522.

Xiao, X. C., Xie, T., and Cheng, Y. T. 2010. Self-healable graphene polymer composites. *Journal of Materials Chemistry* 20: 3508–3514.

Yarlagadda, S., Kim, H. J., Gillespie, J. W., Shevchenko, N. B., and Fink, B. K. 2002. A study on the induction heating of conductive fiber reinforced composites. *Journal of Composite Materials* 36: 401–421.

Yuan, Y. C., Yin, T., Rong, M. Z., and Zhang, M. Q. 2008. Self healing in polymers and polymer composites. Concepts, realization and outlook: A review. *Express Polymer Letters* 2: 238–250.

Zhang, W., Sakalkar, V., and Koraktar, N. 2007. In situ health monitoring and repair in composites using carbon nanotube additives. *Applied Physics Letters* 91. Article No: 133102.

9 Thermoreversibility in Polymeric Systems
Chemical and Physical Aspects

C. Toncelli, D. De Reus, A.A. Broekhuis, and F. Picchioni

CONTENTS

9.1 INTRODUCTION

Thermoreversible cross-linking represents one of the most promising and studied fields in polymer chemistry, as testified by the progressively increasing number of publications submitted in the last 20 years. The reasons for such interest can be properly testified by the advantages that a selective bond cleavage (i.e., of the cross-linking points vs. the one along the polymeric backbone) brings in applications such as self-healing, shape-memory materials, and recyclable and removable thermosets. Indeed, thermosets are generally susceptible to the formation of (micro) cracks, scratches, or deformation. The cross-linked nature of these materials makes it very difficult to repair or reshape them, and usually the material has to be replaced. Reversible cross-linking provides a way to heal these cracks (Tian et al. 2009). This method has the potential to increase the lifetime of many cross-linked materials while the corresponding technology can be potentially simplified to the level at which application, for example, of a blow-dryer might potentially be applied to repair a scratched or eroded material. Moreover, in many applications (e.g., electronics), expensive equipment is embedded in thermosets for various purposes. The use of a reworkable thermoset would allow the recovery of such equipment without causing any damage (Liu et al. 2004). Moreover, the polymeric material might be reused, leading to both environmental and cost advantages. If thermosets are used in environments with varying temperatures or fluctuating loads, cracks might form as well. If the thermoset is reinforced, the difference in the thermal expansion coefficient between the matrix and the reinforcing material can even speed up this process. An example could be the use of a self-healing thermoset in asphalt (Saxegaard 2003), so that cracks that form during cold periods are automatically repaired in warmer ones. This has the potential to extend the useful life of the asphalt, thereby decreasing maintenance costs. Recycling of cross-linked materials is usually combined with a severe degradation in properties and functionality. Car tires, for example, are often shredded and end up as cheap fillers in new tires, plastics, or even landfills. While the use of thermoplastic materials is not possible due to the required properties, the use of reversible cross-links allows thermosets to be reshaped or re-extruded (Ladd 1967). This results in a more functional recycling path, which reduces the amount of virgin material needed in the recycling process.

Thermally reversible cross-links broadly fit into two categories: physical and chemical cross-links (Figure 9.1). Examples of physically reversible cross-links are hydrogen bonds (Peng and Abetz 2005; Nair et al. 2008), ionic (Hennink and Van Nostrum 2002), and van der Waals interactions (e.g., SBS rubber) (Chen et al. 2002; Peng and Abetz 2005). By far, the most studied systems are based on ionic interactions most probably due to the relative easiness of the preparation procedures. Chemically reversible cross-links are based on the covalent bonds that can be broken by external influences. Although this can be achieved by several methods (e.g., shear stress and light), the focus of this chapter is thermal reversibility. The most studied chemical systems in this context are the Diels–Alder (DA) reaction, ionene formation, sulfur bridges, and special cases of thermally reversible ester and amide bonds. From the earlier considerations, it appears clear how this basic concept (thermal reversibility of cross-linking points) constitutes the conceptual basis of several different research fields, including but not limited to the design of self-healing polymeric materials. Objective of this chapter is to provide a critical overview of the most commonly studied physical interactions and chemical bonds used for the preparation of thermoreversible materials. In particular, by looking at the nature of the interaction in connection with the chemical environment in which the latter takes place, one might be able to define general concepts relating the chemistry of the system to the end-properties of the formed network. For every kind of chemical

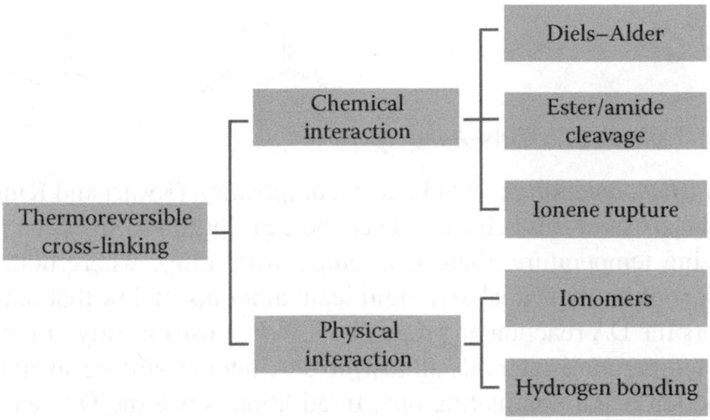

FIGURE 9.1 Overview of the main thermoreversible cross-linking systems known by literature.

bond/physical interaction, an overview of the most studied system will be provided in Sections 9.2 and 9.3. Every section is closed by a discussion part in which the basic concepts pertinent to the kind of interaction are reviewed and framed in a general context helpful for the (molecular) design of new thermoreversible materials. Finally, general remarks will be provided in Section 9.4.

9.2 CHEMICAL INTERACTIONS

9.2.1 Diels–Alder Reaction

Especially the DA reaction is often applied to create thermoreversible cross-links. With these systems, multiple methods to make chemically thermoreversible polymers have been explored. For example, small molecules with multiple dienophile and diene functionalities have been cross-linked to yield thermoreversible polymer networks (Chen et al. 2002, 2003). Dienes or dienophiles have been attached to polymeric backbones, after which the obtained polymers were cross-linked with either a cross-linker such as (methylene-di-*p*-phenylene)-bismaleimide, or another polymer containing a corresponding functionality (Chujo et al. 1990; Canary and Stevens 1992; Goiti et al. 2003). Moreover, epoxy resins have been synthesized, which contain reacted DA adducts even before the reaction that makes them soluble upon heating, facilitating easy removal of the resin (McElhanon et al. 2002). The use of DA chemistry for thermoreversible systems is not only limited to the cross-linking reaction but it has also been used for improving the interaction of the base polymer with inorganic fillers (Imai et al. 2000; Adachi et al. 2004). The DA reaction is probably one of the best-known organic reactions. It is a six-electron [4+2] cycloaddition (Figure 9.2). During the reaction, a 1,3-diene reacts with the π-bond of a dienophile to form a new ring. The kinetics increase with an electron-rich diene and an electron-poor dienophile (Grossman 2003) (or the other way around). Already in the article by Diels and Alder (1929) where the discovery of the DA reaction was described, the authors noticed that the adduct of furan and maleic anhydride decomposes into the starting materials at a temperature of 125°C. This reaction is known as the retro

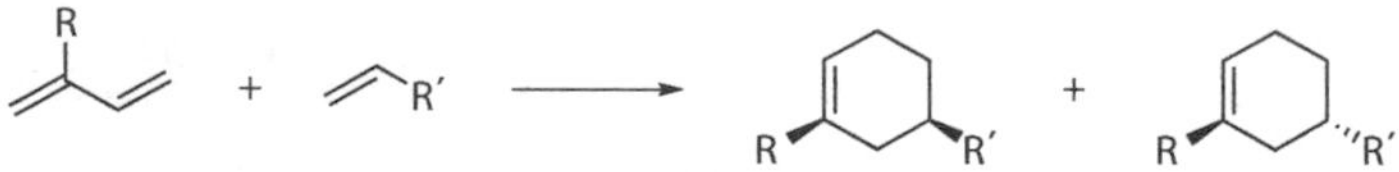

FIGURE 9.2 Example of a Diels–Alder reaction.

Diels–Alder (rDA) reaction or retrodiene decomposition (Kwart and King 1968), and many DA adducts show this behavior. Since the equilibrium shifts to the rDA reaction with increasing temperature, there is a temperature range where both the adduct, diene and dienophile are available in significant amounts. Below that range, the equilibrium favors the DA reaction and the system will consist mostly of the DA adduct. Above that temperature interval, almost no DA adduct will be available because the rDA reaction is the dominating one. In addition, since the DA reaction follows second-order kinetics and the rDA reaction follows first-order ones, the equilibrium between the two states does not only depend on the temperature, but it is also a function of the concentrations of the diene and dienophile. This implies that the de-cross-linking temperature will increase with increasing concentrations of the diene and dienophile. Since the DA reaction is a self-contained reaction (all atoms in the initial molecules are also present in the adduct), and since it can be used in a wide temperature range, it is a very suitable system to prepare thermoreversible polymers. Different combinations diene–dienophile are described in the following sections.

9.2.1.1 Furan–Maleimide

The system furan–maleimide is without any doubt the most popularly studied for thermoreversible cross-linking (an overview of the most commonly used systems is provided in Figure 9.3). A 1969 patent by Craven (1969) is the first literature recording of a thermally reversible cross-linked polymer based on the furan–maleimide system (Figure 9.3a and b). The patent describes the synthesis, cross-linking, and remolding of different polyesters, polyureas, and polyamides with pending furan groups. The obtained polymers were cross-linked with several bis- and trismaleimides by mixing them at 140°C, after which the product was molded and cooled to room temperature, where it cross-linked. Upon heating to 120°C–140°C, the polymer could be remolded, after which it cross-linked again upon cooling.

On the basis of this general idea, many studies followed this initial invention in trying to prepare polymeric materials cross-linked through the use of furan and bismaleimide groups. For example, a number of thermoreversible networks have been prepared using modified polystyrene or copolymers of styrene and furfurylmethacrylate (Figure 9.3a through c). Stevens and Jenkins, for example, have grafted maleimide groups onto polystyrene using a mild Friedel–Crafts reaction in 1979 (Figure 9.3a). Concentrations of up to 20 mol% have been grafted onto the polymer, which was then cross-linked by difurfuryl adipate. Since the polymer is cross-linked by using the furan–maleimide system, thermoreversibility is expected; however, this is not mentioned by the authors (Stevens and Jenkins 1979).

Many studies rely actually on simple cycles of gel formation–solubilization to show the desired thermal reversibiity. Indeed, Canary and Stevens (1992) studied the same kind of polymer (Figure 9.3a) also with 32 mol% substitution for thermoreversibility. They found that a solution of the polymer and the cross-linker formed a gel

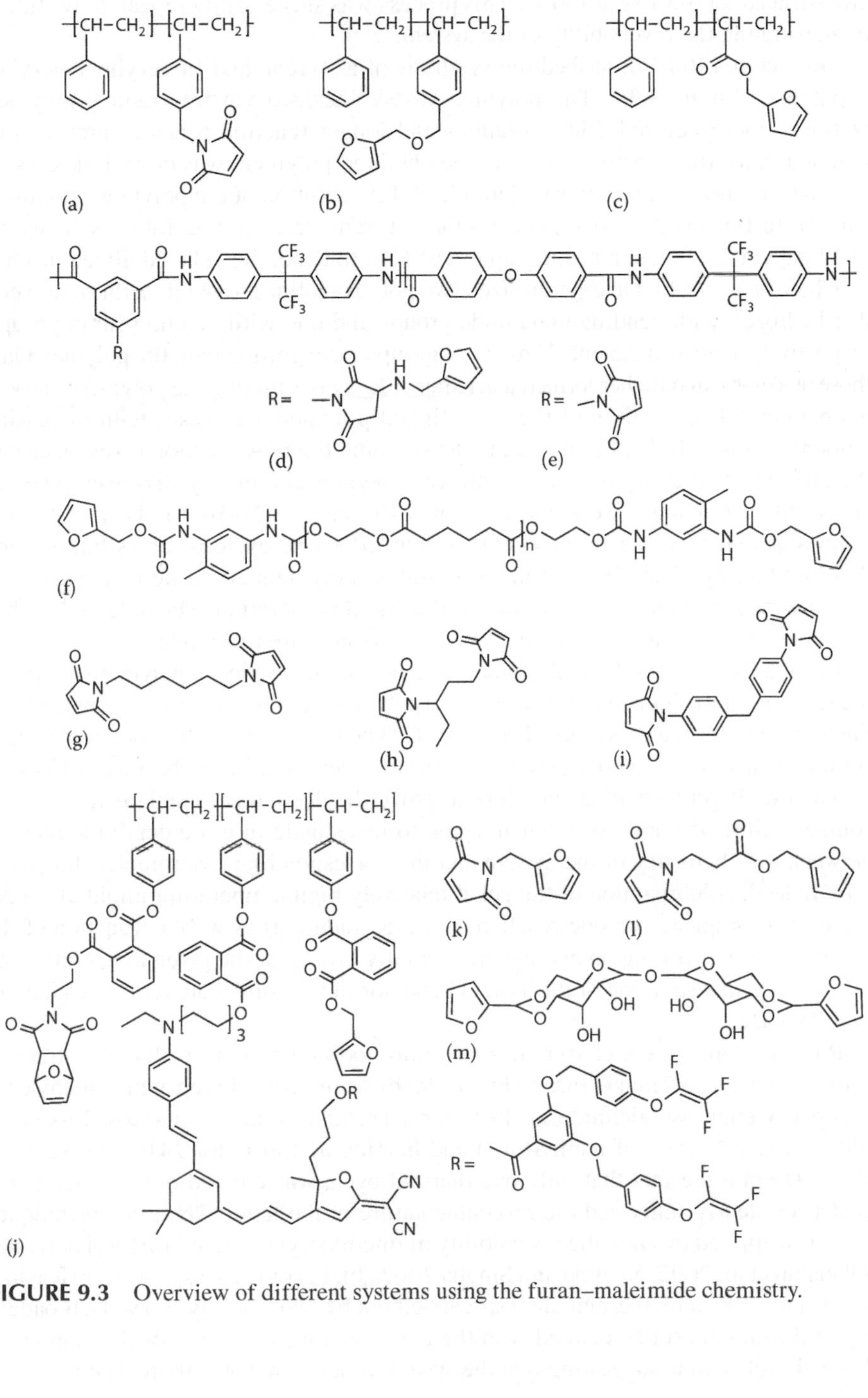

FIGURE 9.3 Overview of different systems using the furan–maleimide chemistry.

within 75 min at 64°C and became a liquid within 2.5 min at 150°C. A solution of the polymer with a furan-substituted polystyrene instead of difurfuryl adipate de-cross-linked within 15 s at 150°C. This process was successfully repeated five times, demonstrating the reversibility of the system.

Goiti et al. (2001) described the synthesis of a styrene and furfurylmethacrylate copolymer (Figure 9.3c). The polymer was synthesized via free radical polymerization, which required dilute solutions and longer reaction times to prevent side reactions with furan π-bonds. One of the obtained polymers was cross-linked using (methylene-di-*p*-phenylene)bismaleimide. A 12% solution of the polymer containing 5 mol% furfurylmethacrylate and 95 mol% styrene cross-linked into a viscous gel in 12 days at 25°C. Upon heating up to 110°C in toluene, the gel solubilized in 1 h.

Chujo et al. (1990) have synthesized two polyoxazolines to obtain a thermoreversible hydrogel, with pending maleimide groups and one with pending furan groups, respectively. Between 2.5 and 22 mol% of groups were grafted onto the polymer. Only those at 4 mol% and higher formed a gel after 7 days after mixing the polymers at room temperature. The swelling of the cross-linked polymers decreases with increasing amounts of cross-links. Upon heating the solution, the rDA reaction takes place and the amount of swelling increases, until the polymer completely dissolves. The de-cross-linked polymers were analyzed by GPC and ^{1}H-NMR to show that there is no apparent increase in molecular weight after one cycle of cross-linking and de-cross-linking. The obtained de-cross-linked polymers could be re-cross-linked by the same procedure that was used for the initial preparation. The obtained gel has the same properties as the first gel. Thus, the systems are reversible.

Liu and Chen (2007) have also modified a polyamide to yield a polymer with pending furan groups (Figure 9.3d). The obtained polymer was mixed with the polymer from Figure 9.3e to form a cross-linked gel at 50°C. The reaction proved to be reversible, as confirmed also by FTIR, since the gel returns to a clear solution by heating to 150°C.

The use of gel formation/solubility to prove the desired reversible behavior is of course of limited importance when trying to investigate in more depth the relation between the chemistry of the system and the corresponding macromolecular structure. Indeed, solubilization of the gel at relatively high temperature might also take place as consequence of side reactions (e.g., degradation) as well as function of the amount of cross-linking points that are factually broken at the given temperature. In this respect, investigation of the system behavior in the solid state represents a more useful choice.

Random copolymers of styrene and furan-modified styrene with concentrations between 5.5 and 97 mol% furan (Figure 9.3b) were cross-linked with (methylene-di-*p*-phenylene)-bismaleimide at 40°C for 24 h, and could be de-cross-linked by adding a large excess of methylfuran and heating at 130°C for 24 h (Goussé et al. 1998). Despite the fact that only one thermal cycle was carried out, spectroscopic evidences clearly suggested the reversible nature of the latter. The same techniques were also applied to study the reversibility of (meth)acrylate-based furfuryl derivates (Gheneim et al. 2002; Kavitha and Singha 2007a,b). In some cases, however (Kavitha and Singha 2007a,b), thermal analysis carried out by DSC clearly showed a broadening of the endotherm associated with the DA reaction as function of the number of thermal cycles, thus suggesting that the system is actually not fully reversible.

The synthesis of a polyamide with different amounts of pending maleimide groups (Figure 9.3e) was documented by Liu et al. (2006). The polymers were cross-linked with a trisfuran compound. Cross-linking was performed at 20°C–60°C and de-cross-linking was performed at 120°C–160°C. The reactions were monitored using FTIR. It was found that there is a positive correlation between the cross-linking degree and the temperature needed for de-cross-linking. This agrees with the equilibrium between the second-order DA reaction and the first-order rDA reaction, where higher concentrations shift the equilibrium toward the DA reaction, which results in a higher de-cross-linking temperature.

The earlier considerations clearly indicate a crucial concept regarding the nature of thermally reversible systems: perfect chemical reversibility (thus at molecular level) does not automatically imply full reversibility at the macroscopic scale. An exception in this respect seems to be constituted by furan-modified polyketones (Zhang et al. 2009) (Figure 9.4), synthesized by the Paal–Knorr reaction with furfurylamine up to a maximum carbonyl conversion of 80%. The obtained polymers were cross-linked using (methylene-di-*p*-phenylene)bismaleimide, the reaction being followed by FTIR. The products were studied with solubility tests, dynamic mechanical thermal analysis (DMTA), three-point bending tests and break surfaces were studied using SEM. Mechanical tests show that the reversibility of this system is very high, with remended materials even showing higher maximum loads than the initial sample during bending tests.

Bisfuranic terminated poly(ethylene adipate)s (Figure 9.3f) with a M_n of 8700 have been cross-linked (Watanabe and Yoshie 2006) using bismaleimides and trismaleimides at 60°C. The linear product (using a bismaleimide) has an M_n that is only twice that of the starting material. The material returns to M_n of about 9000 upon heating at 145°C for 20 min. This behavior is consistent for four cycles. The molecular weight in the reaction equilibrium was dependent on the chain length of the prepolymer, but almost independent of the reaction temperature. A cross-linked network is obtained when a trismaleimide is used. This network behaves as a rubber at room temperature, and it can be stretched up to 650% before breaking. Eight DA and rDA reaction cycles have been performed. The rDA products are soluble and GPC shows that the molecular weight is similar to the one of the starting material. Polydispersity of the material does increase slightly over multiple cycles, whereas the mechanical behavior shows contrasting data. Indeed, the tensile strength and elongation at break remain practically constant upon thermal treatment, whereas the tensile modulus significantly decreases every cycle. Such discrepancy could indicate limitations in the degree of thermal reversibility for this network, thus probably

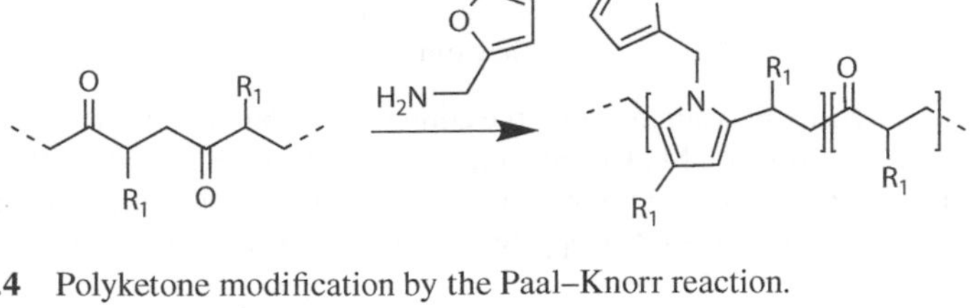

FIGURE 9.4 Polyketone modification by the Paal–Knorr reaction.

affecting the self-healing ability of the material. Indeed, in a subsequent publication (Yoshie et al. 2010), a 1 mm plate of the cross-linked material was cut, but could be repaired by placing the two pieces in contact at 60°C for 1 week. The Young's modulus, stress at yield, break strength, and max elongation of the repaired materials decreases about 20%–40% after mending. This might be because the surfaces of the material during remending are not completely connected, and is not necessarily the result of the DA–rDA reaction reversibility, as already suggested earlier.

The heat needed for the rupture of the DA adducts can also be generated locally instead of having to be applied from an external source. Indeed, it is possible (Adzima et al. 2008), by dispersing Cr_2O_3 in a trifuran/(methylene-di-*p*-phenylene) bismaleimide system, to locally generate heat upon the application of electric field. This can induce self-healing while at the same time avoiding relatively high temperature that could lead to the polymer degradation.

Although the thermal reversibility seems to be linked only to the kind of reactions (e.g., DA) taking place in the system, the use of different cross-linkers, generally leading to the formation of network structures with different rigidity, could also have an effect. A copolymer of butylmethacrylate and furfurylmethacrylate, with concentrations up to 30 mol% of the latter, was presented as a remendable coating by Wouters et al. (2009). The polymers were cross-linked with a rigid aromatic cross-linker ((methylene-di-*p*-phenylene)bismaleimide) and a more flexible aliphatic one (1,12-dimaleimidododecane). The cross-linked polymers were studied using mechanical and rheological methods. The hardness and modulus seem to increase with increasing cross-linking. Viscosity measurements were carried out between 50°C and 175°C over three cycles, showing a very high reversibility in terms of viscosity (Figure 9.5), independently on the kind of cross-linker used.

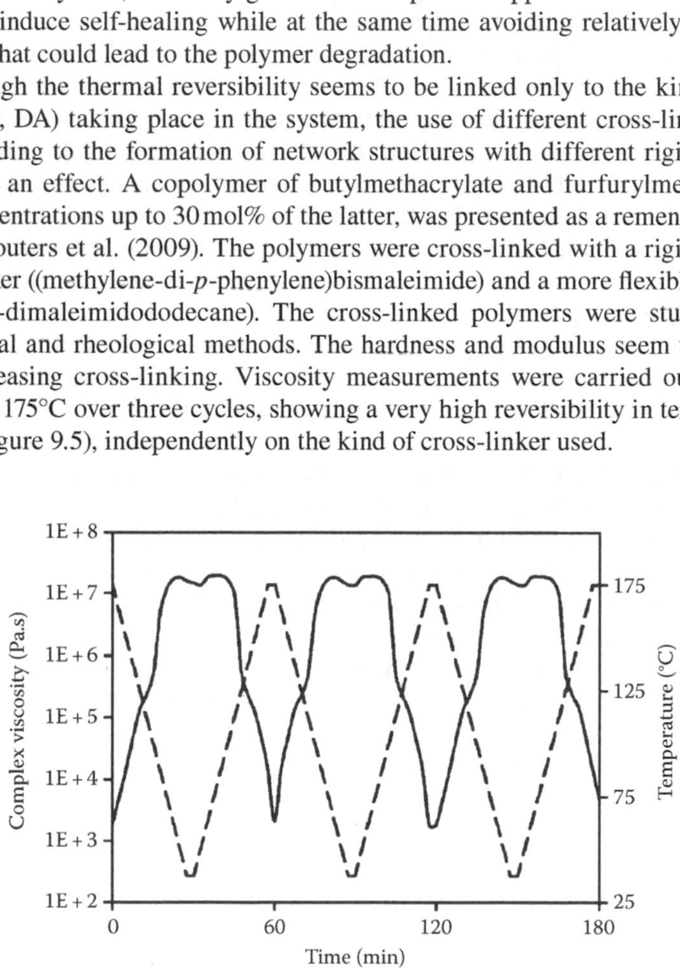

FIGURE 9.5 Viscosity measurements on the copolymer of butylmethacrylate and furfurylmethacrylate. (From Wouters, M., Craenmehr, E., Tempelaars, K., Fischer, H., Stroeks, N., and van Zanten, J.: Preparation and properties of a novel remendable coating concept. *Prog. Organ. Coat.* 2009. 64. 156–162. Copyright Wiley-VCH Verlag GmbH & Co. KGaA. With permission.)

A recent publication (Canadell et al. 2010), moreover, reported on a study between furfurylmethacrylate and three different kinds of maleimide cross-linker (Figure 9.3g through i). Two separate exothermic peaks for the DA adducts stereoisomers were observed by DSC and attributed to the endo (adduct with [g] T = 117°C and adduct with [h], T = 126°C) and exo (adduct with [g], T = 138°C and adduct with [h], T = 142°C) adducts. Finally, the rDA of the aromatic maleimide (Figure 9.3i) was observed at 158°C as a single transition. This suggests that by using aromatic bismaleimide the steric hindrance of the network increases and only the most thermodynamically stable adduct is formed.

Another study (Inoue et al. 2009) reported an ester exchange between sorbitol and poly(lactic acid), which was then modified with furan end groups to create a thermoreversible network with hexamethylenebismaleimide, trismaleimide, and dodecylbismaleimide (Figure 9.6). The obtained materials display shape memory. Indeed, a programmed shape can be set at 150°C, and a temporary one at 60°C. Upon heating to 60°C, the material goes back to the programmed shape. Upon heating to 150°C, a newly programmed shape can be set. This clearly indicates the rearrangement of cross-links at 150°C, which implies thermoreversibility at this temperature. By comparing hexamethylenbismaleimide with dodecylenbismaleimide as a cross-linker, it was found that the reduction in the inner stress of the polymer structure increased the strength of this polymer. Indeed, dodecylenbismaleimide, by having a longer spacer between maleimide moieties than hexamethylenbismaleimide, leads to a relatively lower rigidity of the network formed; thus, it renders the whole system mechanically relatively stronger.

The use of low-molecular-weight cross-linkers to be added to polymeric systems does not represent the only option for preparing thermoreversible polymeric systems. Indeed, "all polymeric" systems, that is, combination of two or more polymers

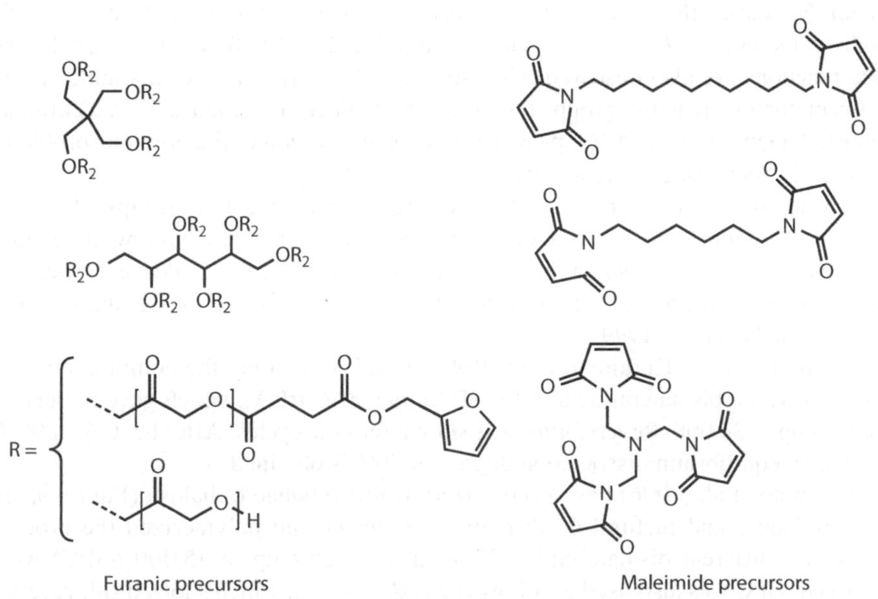

FIGURE 9.6 Lactic acid oligomers modified with furan and three different maleimides used.

displaying the presence of both reacting groups, have also been reported. A polymer with both furan and maleimide groups has been synthesized according to a procedure (Luo et al. 2003). The maleimide group was protected with furan during the reaction to prevent premature cross-linking (Figure 9.3j). The rDA reaction was used to remove the furan, which could be followed by TGA. This shows the possibility of synthesizing a single-component thermoreversible cross-linkable polymer. Unfortunately, the reversibility of the obtained system was not studied.

Cross-linked networks can also be synthesized by using (low-molecular-weight) molecules that have multiple furan and maleimide functionalities. Unlike the polymeric systems discussed until now, these ones have the potential to de-cross-link completely into low-molecular-weight compounds. In reality, the reaction does not reach full completion because of its equilibrium nature and because of the restricted mobility of the chains at relatively high cross-linking degrees. This implies that at least a certain number of molecules have to possess more than two functionalities to obtain a high-molecular-weight polymer or a network. A number of publications is also available, which describe reversible linear polymerization using furan and maleimide groups. Although these experiments do not yield a cross-linked network, they share many of the same principles as thermoreversible cross-links.

2-Furfurylmaleimide is one of the most interesting molecules to form a single-component thermoreversible polymer from a low-molecular-weight compound because it represents the smallest molecule to combine a furan group and a maleimide group (Figure 9.3k). However, the synthesis of this compound is not straightforward as it polymerizes with itself during preparation. The protection of furan or maleimide group is also difficult because the temperature needed for the ring-closing step of the maleimide also results in deprotection. However, the synthesis of this compound is not straightforward as it polymerizes with itself during preparation (Mikroyannidis 1992). Moreover, the resulting linear polymer is depolymerized only at relatively high temperatures ($T > 150°C$) (Goussé and Gandini 1998), where (degradation) side reactions are also observed (Goussé et al. 1998). Insertion of a simple spacer between the two reacting groups (i.e., furan and maleimide; Figure 9.3l), as already reported (Gandini et al. 2010), provides an elegant solution to the previous problems by slightly decreasing the reactivity.

Besides molecules displaying the presence of both reactive groups, the most simple option for the preparation of thermoreversible polymers from low-molecular-weight compounds consists in the reaction of two different molecules, each bearing at least two equal groups (e.g., a difuran with bismaleimide) (Goussé and Gandini 1999; Kamahori et al. 1999).

Kuramoto et al. (Kuramoto et al. 1994) used UV to study the combination of a bisfuran with a bismaleimide at 60°C (DA) and 90°C (rDA), which gave molecular weights up to 30,000. Reversibility was shown for nine cycles. After heating at 90°C for 1 h, an equilibrium dissociation degree of 80% is obtained.

Teramoto et al. (2006) have synthesized difurfurylidene trehalose (Figure 9.3m) from trehalose and furfural (both natural products), and polymerized the product using two different bismaleimides. Molecular weights up to 15,000 (GPC) were obtained. GPC was also used to follow the rDA reaction, which showed full reversal at 140°C, followed by degradation of the bismaleimide.

It is clear from the earlier discussion how the use of multifunctional monomers might represent an advantage over simple bifunctional ones. Indeed, even if the reaction is not perfectly reversible at the molecular level, an excess functionality might still ensure network formation. The search for multifunctional compounds has led in recent years to publications describing thermoreversible behavior of dendritic structures (McElhanon and Wheeler 2001; Szalai et al. 2007; Kose et al. 2008; Aumsuwan and Urban 2009). Indeed, very recently, larger bisfurans with two pending dendrimers of the first to third generation (Figure 9.7) have been described by Polaske et al. (2010). These compounds were polymerized with (methylene-di-*p*-phenylene) bismaleimide at 50°C. GPC results for these materials were extremely clear, even though the polymerization is slow and it ended after 240h. The rDA reaction was also studied, which shows a complete return to the starting material in 1 h at 110°C.

A very thorough study of a thermoreversible network using the furan and maleimide system is published by Chen et al. (2002). They synthesized a tetrafuran that was cross-linked with a trismaleimide at 25°C–75°C (Figure 9.8). Reversible cross-linking was achieved at 130°C–150°C. Five cycles of cross-linking and de-cross-linking were followed using solid-state ^{13}C-NMR, which showed very high reversibility. The obtained materials were hard, colorless, and transparent. The mechanical properties at room temperature are equal to those of commercial epoxy resins, as indicated by

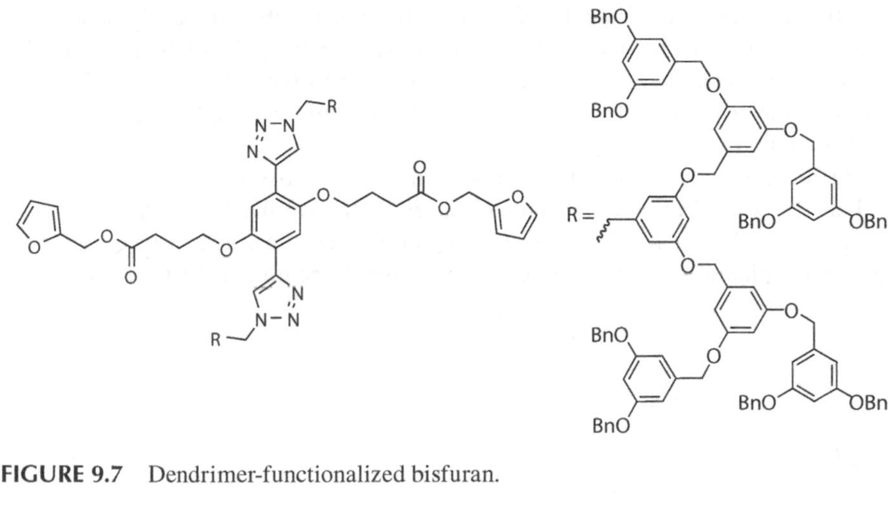

FIGURE 9.7 Dendrimer-functionalized bisfuran.

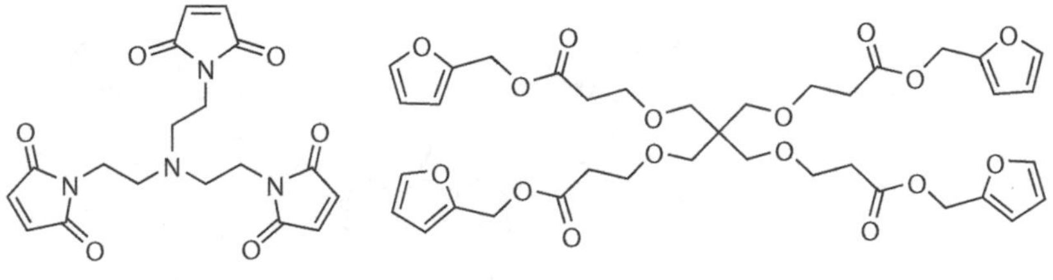

FIGURE 9.8 Tetrafuran and trismaleimide.

ASTM tests for tensile, compression, and flexural strength. A crack in the material was healed as well and the recovered breaking strength was 53%. A later publication (Chen et al. 2003) by the same group describes the application of more flexible bismaleimides instead of the trismaleimide. However, the final mechanical properties displayed only a slight dependence on the kind of maleimide used.

An interesting application of a thermoreversible network was presented by Gotsmann et al. (2006). They have synthesized cross-linked films from a trifuran and protected *p*-phenylene-bismaleimide, which can be used for data storage and lithographic applications. Thermomechanical properties were analyzed on a nanometer scale by indentation experiments with heated tips. The authors note that because the material can switch between two states of material properties, interesting new perspectives in lithography, and data storage arise. The material can, for example, store data at a capacity of 1 TB per square inch.

In a work by Liu and Hsieh (2006), trismaleimide and trisfuran-modified components (Figure 9.9) have been mixed and a higher rDA temperature (T = 185°C) has been found. The intensity changes of the FTIR absorption peak of adduct at 1560 cm^{-1} (–C=C– stretching in cyclic adduct) with respect to the band of C=O group at 1715 cm^{-1}, used as reference, was used to monitor the performance of the DA reaction. An activation energy of 56 kJ/mol for the curing reaction, as measured by DSC, has been determined and a broad endothermic peak ranging from 90°C to 160°C (centering at about 145°C) was registered for the adduct, indicating the occurrence of rDA reaction. In order to test self-healing, debonding occurred at 120°C to provide ability of chain reformation, whereas rebonding occurred at 50°C to reform the cross-linked structure. A minimum time for complete healing of 24 h has been found.

Furthermore, a tetrafuran and a bismaleimide derivative have been cross-linked by Plaisted et al. (Plaisted and Nemat-Nasser 2007), similar to the system already described by Chen et al. (2003), and healing properties of a crack surface was investigated by uniaxially crack-forming and successively healing at 85°C. This analysis

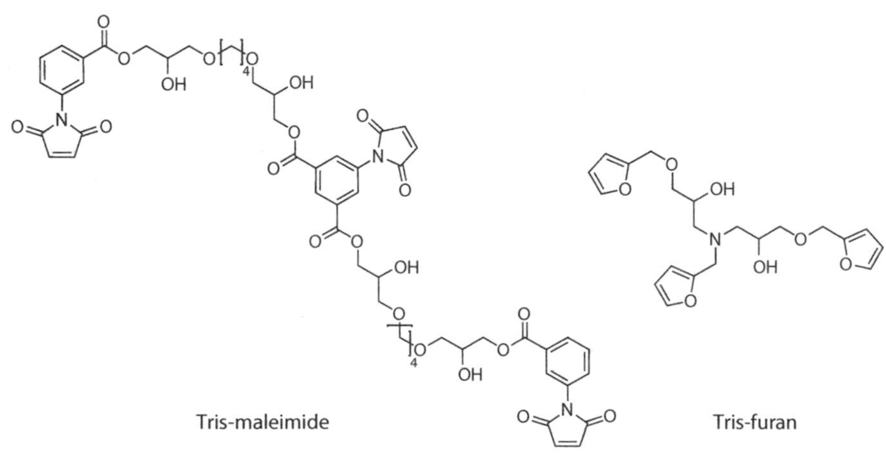

FIGURE 9.9 Trisfuran and trimaleimide compound.

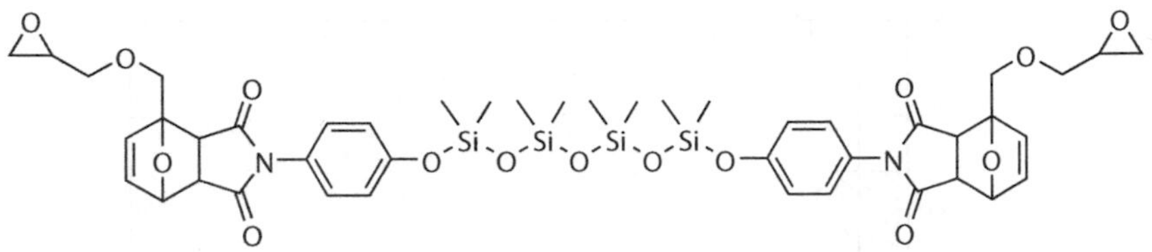

FIGURE 9.10 De-cross-linkable epoxy.

suggests that the samples slightly became more resistant to fracture with successive healing treatments and that the degree of healing does not increase with increasing the treatment time beyond 1 h. Physical bonding resulted in 28.4% recovery of the initial strength, whereas covalent bonding resulted in an additional 41.6% healing efficiency. Average healing efficiencies of 70% were observed, with two specimens recovering >100% of the initial strength.

The use of furan and maleimide-reacting groups has not been exclusively limited to systems interesting from a purely academic point of view. Indeed, epoxy (McElhanon et al. 2002) resins have been often used as a matrix modified with DA reactant (Figure 9.10). The latter forms after cross-linking a spongy network that can be solubilized in *n*-butanol at 90°C. The obtained polymers were studied using SEM. Moduli, compressive, and tensile strengths were measured to get a clear view of the material properties. Although the cross-links can be dissolved, the material cannot be efficiently re-cross-linked.

In a similar work, Palmese and Peterson (2008) reported the synthesis of an epoxy network containing furan moieties. After breaking, the specimens could be healed by applying a very concentrated solution of (methylene-di-*p*-phenylene)bismaleimide, pushing the parts together, quickly heating to 90°C and healing for 12 h at room temperature. Seven cycles of this procedure were tested, each showing a decrease in material properties, with the first cycle already losing 60% of its strength (Figure 9.11). Such loss of mechanical properties does not constitute an exception since it has been reported also by other authors for similar systems (Tian et al. 2009; Magana et al. 2010).

Lately, Peterson et al. (2010) used a different way to attach furan group to epoxy resin, by mixing a diamine, a furan-modified linear epoxy and bisphenol A diepoxy telechelic prepolymer and healing properties of the thermoreversible network have been evaluated through double cantilever beam, flexural strength, and compact tension analysis. Low T_g (55.6°C) and a high level of swelling in the healing agent solvent were assumed necessary to maximize the DA bonding across a crack surface. This polymer was found to have fracture toughness G_{IC} = 920 J/m, tangent modulus of elasticity E_B = 3.3 GPa, and flexural strength σ = 73 MPa.

9.2.1.2 Anthracene–Maleimide

Anthracene can act as a diene in the DA reaction. It reacts with a large number of dienophiles and the rDA of anthracenes was extensively studied by Chung et al. (1989). The kinetics of the rDA reaction of various 9,10-substituted anthracenes, and of various dienophiles are therefore well documented.

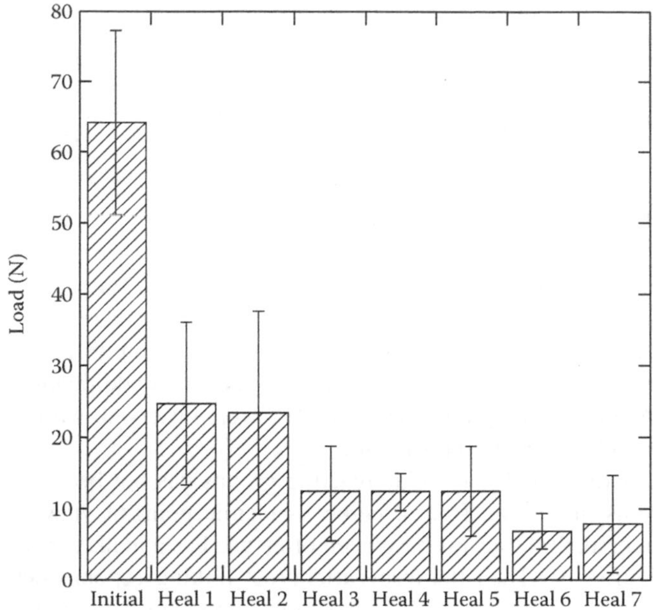

FIGURE 9.11 Load recovery of a cross-linked epoxy healed with (methylene-di-*p*-phenylene)bismaleimide over seven cycles. (From Palmese, G.R. and Peterson, A.M., Paper presented at *26th Army Science Conference*, Orlando, FL, 2008.)

In particular, the one with maleimides has been used as modification tool mainly for the introduction of chromophoric moieties on several different polymeric systems (Kim et al. 2006). Anthracenes have been used with bismaleimides to synthesize low-molecular-weight polymers (Grigoras and Colotin 2001; Grigoras et al. 2007) (Figure 9.12).

The reaction has also been used to create a block-copolymer via click chemistry, or to graft PEG and methyl methacrylate chains onto anthracene-modified polystyrene (Durmaz et al. 2005; Gacal et al. 2006). Unfortunately, these systems were not studied for reversibility. The only mention in the literature of cross-linking via this system is by Jones et al. (1999). Their publication details the synthesis of a copolymer of polyethylene terephthalate (PET) and 2,6-anthracenedicarboxylate (Figure 9.13). The polymer was prepared with 2, 4, and 18 mol% of anthracene. A cross-linked

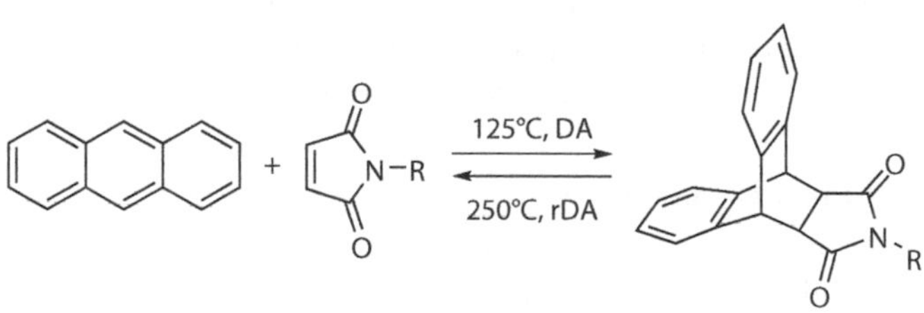

FIGURE 9.12 Diels–Alder reaction between anthracene and maleimide.

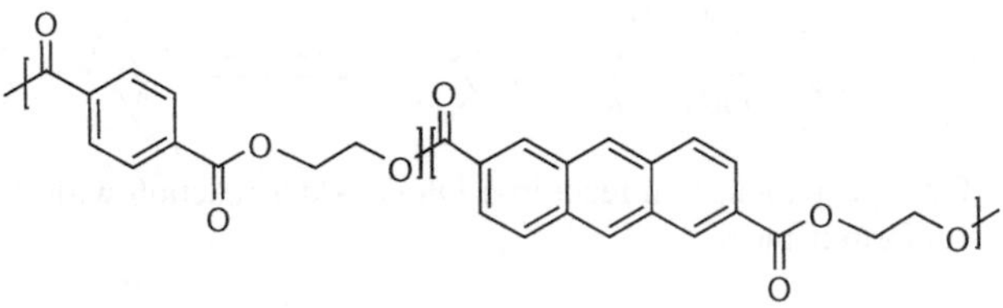

FIGURE 9.13 The copolymer of PET and anthracenedicarboxylate.

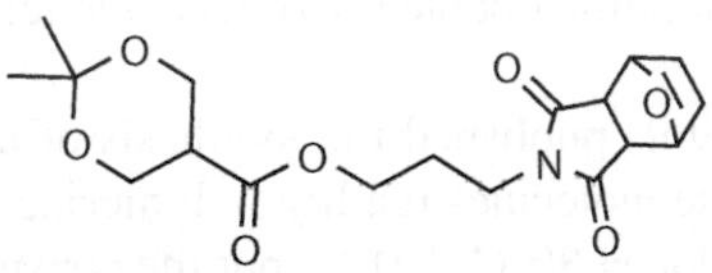

FIGURE 9.14 First-generation dendron.

network was obtained after a reaction of 15 h at 125°C with hexamethylenedimaleimide. The cross-linked network with 2 mol% of anthracene groups was soluble and could be studied by ^{1}H-NMR. Up to 27% of de-cross-linking of model compounds was achieved at 250°C in 7 h, which was shown by ^{1}H-NMR. Unfortunately, de-cross-linking of the polymer network with 2 mol% anthracene was unsuccessful, as indicated by ^{1}H-NMR. The samples with higher anthracene contents were not tested for thermoreversibility. Interestingly, DSC analysis of the polymer network shows an endotherm at 220°C, which decreased in size during three cycles of heating to 300°C and subsequent cooling to 50°C, but this peak was assigned to a melting point rather than the rDA reaction by the authors.

Lately, Tonga et al. (2010) used poly(styrene)-supported anthracene in order to build up star-arm morphology by using furan–maleimide system of first–third generations dendrons (Figure 9.14). Once at 110°C the furan–maleimide adduct begins to undergo rDA reaction. Characteristic aromatic proton peaks of anthracene between 7.3 and 8.4 ppm completely disappeared as expected and a new peak corresponding to the bridgehead proton appeared at around 4.7 ppm as a broad peak because of the overlapping with the peaks corresponding to methylene protons of CH_2 adjacent to anthracene and phenyl ring. Surprisingly, the generation of dendrons did not seem to have any effect at the rate of cycloaddition.

In summary, the anthracene system has only scarcely been used to create thermoreversible networks. The only real network obtained using anthracene did not show thermoreversibility at 250°C, but there is more room for research toward this topic. A big advantage of the system is that it is applicable in a very wide temperature range with DA reactions between 25°C and 125°C and rDA reactions between 50°C and 250°C, depending on the substituents on the anthracene and dienophile. The main disadvantage is the difficult modification of anthracene, which makes introduction of the molecule into polymers challenging.

9.2.1.3 Cyclopentadiene

Cyclopentadiene is another well-known diene and dienophile in DA reactions. Cyclopentadiene itself readily dimerizes by serving both as the diene and

FIGURE 9.15 Cyclopentadiene can react in a Diels–Alder reaction with itself, and subsequent reactions can form trimers.

dienophile. Therefore, it is often supplied as the DA adduct of two cyclopentadienes. Cyclopentadiene is then obtained via the rDA reaction by heating the adduct. Since the adduct contains a dienophilic double bond, tris- and tetraadducts are also formed (Figure 9.15).

Stille and Plummer (1961) published the synthesis of molecules with two cyclopentadiene moieties. These molecules readily polymerize with *p*-benzoquinone and hexamethylenebismaleimide at 80°C–110°C, but the conversion becomes lower with increasing temperature. This is the result of the rDA reaction, which becomes more significant at higher temperatures. The bulk homopolymerization yielded an insoluble thermosetting polymer. The authors note that this is probably because of radical polymerization, and have unsuccessfully tried to inhibit the effect with radical inhibitors. Instead, it is very likely that trimers of cyclopentadiene are responsible for cross-linking in the bulk polymer. No evidence of thermoreversibility was reported in the work, but a recent publication by Murphy et al. (2008) describes and analyzes a similar polymer and found that trimers of cyclopentadiene are indeed responsible for cross-linking of that system. Thus, although not specifically reported by Stille and Plummer, their polymer is most likely thermoreversibly cross-linked.

Kennedy and Castner (1979a) synthesized a cross-linked network using isopropylene isobutylene rubbers (IIR) and ethylene propylene rubbers (EPM) with pending cyclopentadiene groups, which formed an insoluble gel after 72 h at room temperature. Upon the addition of maleic anhydride and heating to 215°C, the material could be dissolved again. The authors noted that bulk material behaves like a cross-linked polymer, and the polymer can be remolded at high temperature (170°C), indicating that (at least partially) de-cross-linking takes place. Since then, a number of publications (Chen and Ruckenstein 2000a; Ruckenstein and Chen 2000a) have described the use of cyclopentadiene as cross-linkers for chlorine-containing polymers such as PVC, EPM, and IIR rubbers. Thermoreversibility of the obtained cross-linked materials was not studied in any of these publications.

The same group published (Chen and Ruckenstein 1999) a work on chlorine polymers where thermoreversibility was found at 150°C–170°C and confirmed by viscosimetric analysis (Figure 9.16).

The cross-linking reaction occurred at a higher rate in a polar solvent, such as dimethylformamide, than in a nonpolar one, such as toluene, and was affected by the nature of the chlorine-containing polymer. A decrease in viscosity confirms the occurrence of rDA. Use of a DA catalyst ($BzMe_3NBr$) increased the DA second-order kinetics.

In a work of Choi et al. (2000), cyclopentadiene has been used in order to protect alkene during polymerization. In fact, by mixing the reactants shown in Figure 9.17, a conjugate polymer with enyne unit has been synthesized through rDA of cyclopentadiene at 165°C. rDA kinetics have been checked through UV–vis

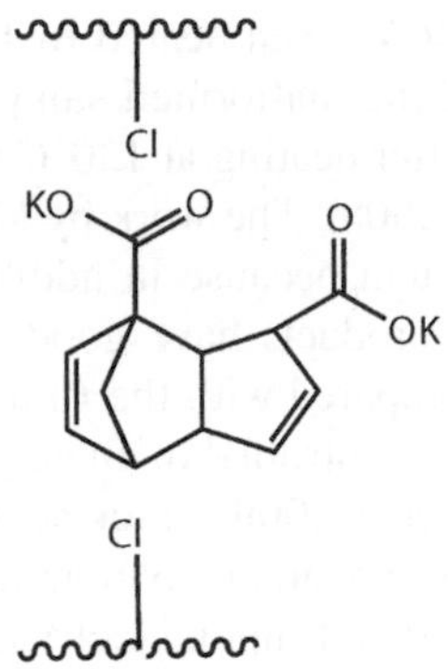

FIGURE 9.16 Dicyclopentadiene cross-linker for chlorine polymers.

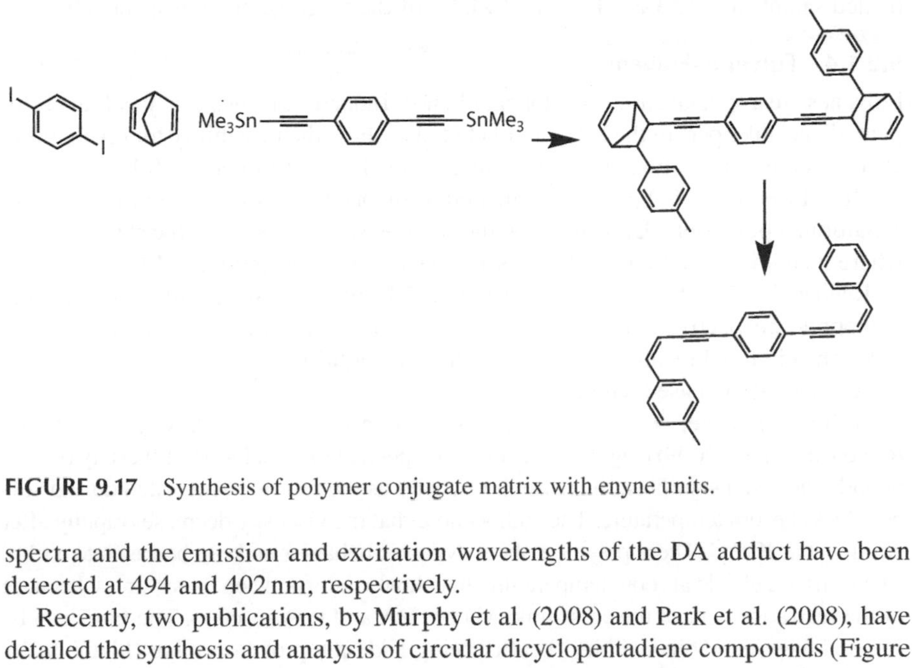

FIGURE 9.17 Synthesis of polymer conjugate matrix with enyne units.

spectra and the emission and excitation wavelengths of the DA adduct have been detected at 494 and 402 nm, respectively.

Recently, two publications, by Murphy et al. (2008) and Park et al. (2008), have detailed the synthesis and analysis of circular dicyclopentadiene compounds (Figure 9.18) that form single-component cross-linked polymer networks upon heating to 150°C and slowly cooling. A network arises because a bisadduct can react with another cyclopentadiene to a trisadduct. The obtained materials are clear, colorless, and hard. Healing tests by breaking the sample and healing the fracture gave an

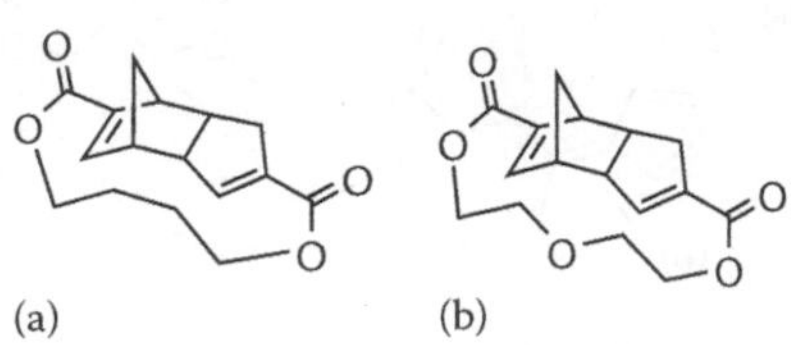

FIGURE 9.18 Dicyclopentadiene compounds (a) monomer 400 (b) monomer 401.

average healing efficiency of 46% when heated to 120°C for 20 min. The material can be remolded at 150°C and the malformed samples from the compression tests return to their original shape after heating at 120°C for 12 h (i.e., the material has a shape memory) (Murphy et al. 2008). The work by Murphy et al. shows great promise for the cyclopentadiene system because in addition to the advantages of using cyclopentadiene, the obtained products have good mechanical properties and are transparent. A disadvantage compared with the furan–maleimide system is that the amount of cross-links cannot be controlled in an easy way.

In a recent publication (Park et al. 2009), crack healing has been checked in a mendomer system (polymer with cyclopentadiene moieties). The microcrack induced by three-point bending healed itself well upon electrical resistive heating and became almost invisible on an optical microscope. The strain energy of the virgin sample up to 14 mm deflection is 99.34 mJ. The strain energies of the first, second, and third healed sample are 92.3%, 94.2%, and 94.3% of the virgin sample, respectively.

9.2.1.4 Fulvene–Fulvene

Fulvenes are a special case of cyclopentadienes. Indeed, fulvenes can, just like cyclopentadiene, take part in the DA reaction but also show the possibility for reversibility even at room temperature (Kennedy and Castner 1979b; Boul et al. 2005).

Hurwitz et al. (1966) patented a method to incorporate DA-coupled fulvenes into unsaturated polyesters. Upon heating, the fulvene was released via the rDA reaction, where it could be used for a cross-linking reaction (Kwart and King 1968).

Reeder (1967) experimented the use of bisfulvene and bismaleimide as diene and dienophile, respectively. Reversibility at higher temperatures was not studied, but, given the obtained results, it is expected that the equilibrium shifts to the de-cross-linked state at increased temperature.

A linear polymer was synthesized using tetrafulvene and biscyanoethylene by Reutenauer et al. (2009) (Figure 9.19a and b, respectively). A solution of these two compounds showed a significant increase in viscosity as the reaction proceeded during several days at room temperature. The authors note that the viscosity decreases rapidly after treatment at 100°C. These polymers show dynamic behavior and can be cast into a film. The films are elastic at room temperature, but slowly deform when under stress, because of the rearrangement of the cross-links via the DA and rDA reaction. The films can be cut, and they recombine within 10 s after being laid on top of each other (Figure 9.20). It is important to notice how the films do not stick to anything else but themselves.

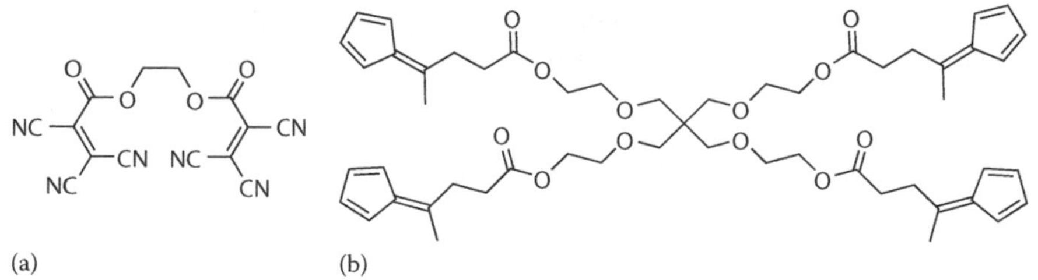

FIGURE 9.19 A room temperature dynamic network from fulvenes and cyanoethylenes (a) tie-cyanide substituted diene (b) tetra-fulvene end-capped dendron.

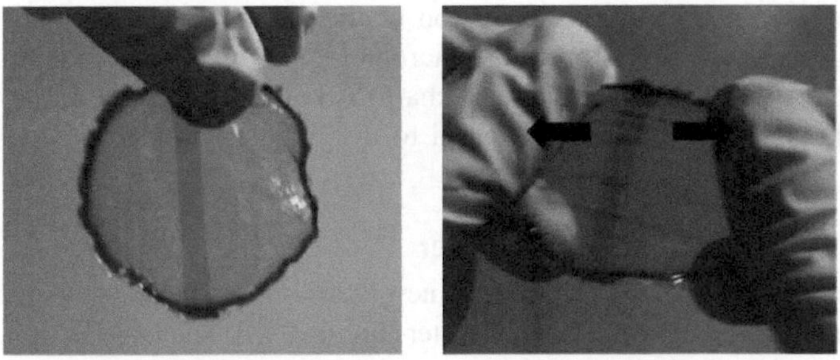

FIGURE 9.20 Stretching of a thin film of the dynamer. The darker area is where the two pieces were superposed. (From Reutenauer, P., Buhler, E., Boul, P., Candau, S., and Lehn, J.-M.: Room temperature dynamic polymers based on Diels–Alder chemistry. *Chem. Eur. J.* 2009. 15. 1893–1900. Copyright Wiley-VCH Verlag GmbH & Co. KGaA. With permission)

9.2.1.5 Fullerene–Diene

Buckminsterfullerene (C_{60}) is the smallest fullerene molecule in which no two pentagons share an edge (which can be destabilizing, as in pentalene). It is also the most common in terms of natural occurrence, as it can often be found in soot. The structure of C_{60} is a truncated icosahedron, which resembles a soccer ball of the type made of 20 hexagons and 12 pentagons, with a carbon atom at the vertices of each polygon and a bond along each polygon edge. The van der Waals diameter of a C_{60} molecule is about 1.1 nm (Qiao et al. 2010), while the nucleus to nucleus diameter of a C_{60} molecule is about 0.71 nm.

The first attempt in order to create a fullerene polymer supported by DA reaction was applied by Guhr et al. (1994), where polystyrene with cyclopentadiene side moieties was mixed with buckminsterfullerene at 250°C. Upon heating in decalin to 180°C for 8 h in the presence of maleic anhydride, the resin releases 48% of the bound C_{60}, thus indicating the thermoreversible character of the bond. One year later, the same group published a work (Nie et al. 1995) in which they used furan-supported polystyrene as polymer matrix. The capacity of this resin, as determined by prolonged reaction, is 55 mg C_{60}/g of resin. Furthermore, this resin shows efficient reversibility (80% recovery of C_{60} upon heating [180°C] with excess maleic anhydride). By replacing furan with benzene, the uptake of C_{60} reduces to zero, thus indirectly confirming the establishments of other interaction than hydrogen or π-stacking.

Linear polymer with buckminsterfullerene as backbone moieties has been developed first by Gogel et al. (1996) and later by Ilhan and Rotello (1999). In the first work, they used a mono-*o*-quinodimethane as a divalent cross-linker. Up to eight fullerenes were linked together as proved by MALDI-TOF analysis. The solubility obviously decreases with the molecular weight and a mixture of *cis–trans* conformation has been found for the diadduct.

In a later work (Kennedy and Castner 1979a), the authors used dianthracene as a cross-linker and the reaction proceeds at room temperature for 168 h. The resulting material is stable at room temperature but undergoes reversion to the monomeric species upon heating. This thermal depolymerization process was observed at temperatures above 60°C, with activation energy of 17.1 kcal/mol, thus suggesting perfect recyclability without degradation of the reactive functionalities. However, also in this case, the reversibility at macromolecular level (i.e., the amount of DA adducts that are broken per polymeric chain) is not perfect. Indeed, the cycloreversion for this polymer is mainly internal to the polymer strand, producing smaller polymers rather than monomers.

9.2.1.6 Cyclopentadiene–Dithioester

One of the last developments in terms of a new thermoreversible DA switch is the couple diene–dienophile cyclopentadiene–thioester (Figure 9.21). The DA reaction occurred at 25°C and thermoreversibility has been found between 80°C and 95°C. The reversibility temperature can in this case be checked with UV spectroscopy as in the work Paulöhrl et al. (2010), which thus produced a modular polymeric color switch (Figure 9.22).

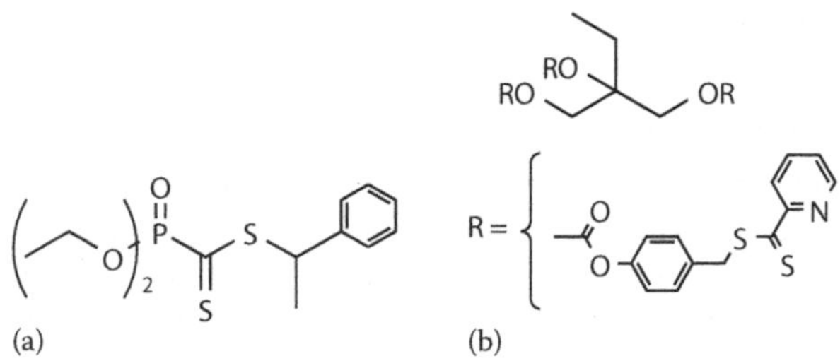

FIGURE 9.21 Thioester diene for Diels–Alder reaction (a) mono-functional thioester (b) tetra-functional thioester.

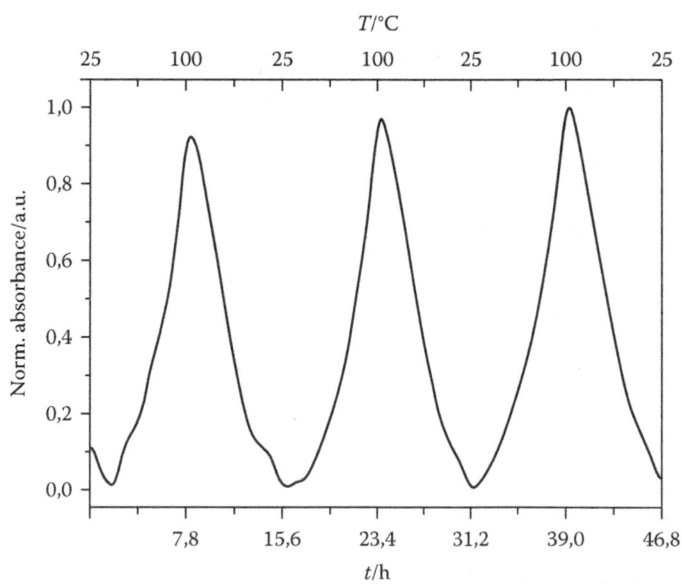

FIGURE 9.22 Thermoreversibility checked by UV–vis studies at a fixed wavelength of 574 nm in the work by Paulöhrl et al. (From Paulöhrl, T., Inglis, A.J., and Barner-Kowollik, C., Reversible Diels–Alder chemistry as a modular polymeric color switch. *Adv. Mat.* 2010. 22. 2788–2791. Copyright Wiley-VCH Verlag GmbH & Co. KGaA. With permission)

Along the same line, in a publication by Inglis et al. (2010), the optimal onset debonding temperature has been evaluated by UV at 522 nm. After a reaction time of 10 min, the molecular weight of the system markedly increases to 12,700 g/mol. It is also noted that the polydispersity of the system increases from 1.21 to 2.57.

9.2.1.7 Discussion

The different DA systems presented earlier represent a wide choice, from a purely chemical point of view, for the molecular designer who wishes to develop new thermoreversible materials. In this respect, the definition of the product thermal properties relies in first instance on the characteristics of the reaction itself and namely on the temperature of DA adducts formation and for their rupture. Already at this stage, however, a set of general design rules is difficult to find since the values of these characteristic temperatures are also a function of the polymeric matrix in which the DA and rDA reactions are taking place. This results in general in the definition of a temperature interval (rather than exact values) at which the desired reactivity manifests itself (Table 9.1).

Such spreading of the temperature values makes an a priori design rather difficult so that in most cases the synthesis of the desired materials in lab-scale quantities followed by determination of the thermal behavior is actually the most preferred route. However, once the reversibility temperature is established in a relatively narrow interval, a rough estimation of the reaction kinetics can be obtained by the use of the Arrhenius equation:

$$k = A \cdot e^{-\frac{\Delta E}{RT}} \tag{9.1}$$

where

k is the kinetic constant for the reaction
A is the so-called pre-exponential factor
R is the ideal gas constant
T is the temperature
ΔE is the energy barrier for the reaction (either DA or rDA)

The latter is related, for DA reaction and its reverse, to the difference in energy between HOMO and LUMO of diene and dienophile, which in turn can be simply estimated with simple theoretical calculations (Table 9.2).

The establishment of the reaction path at molecular level constitutes, however, only the first step toward a complete design of thermoreversible materials. Indeed, the behavior at molecular level must be properly framed in the general macromolecular structure of the material. In particular, as seen in many earlier examples, a perfect thermoreversibility at molecular level does not automatically ensure the same property for the polymeric material. In this respect, factors such as the influence of the polymeric matrix on the reaction course and the presence of (degradation) side reactions (especially for rDA step) have a definite influence on the materials' final properties and must be taken into account when studying the (desired) thermal reversibility. In order to correctly quantify these effects, one would like ideally to

TABLE 9.1
Overview of the Most Used Reversible Diels–Alder Systems

Diene	Dienophile	Adduct	T_{DA}	T_{rDA}	Source
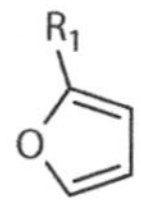	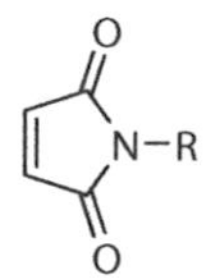	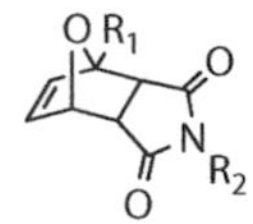	50–80	110–170	Chen et al. (2002), Wouters et al. (2009), Kuramato et al. (1994), Gandini et al. (2008)
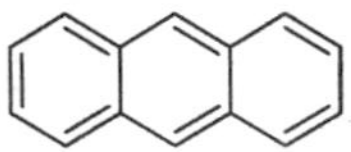	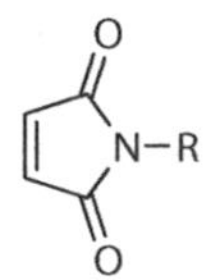	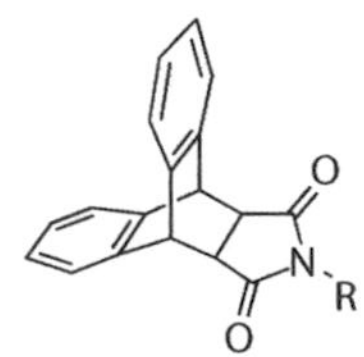	125	250	Chung et al. (1989), Gacal et al. (2006).
		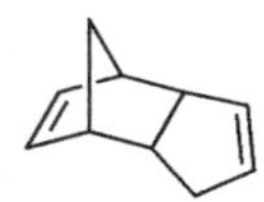	25–120	150–215	Stille and Plummer (1961), Murphy et al. (2008), Peterson and Palmese (2009)
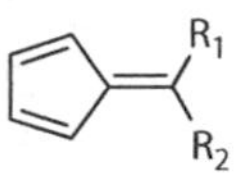	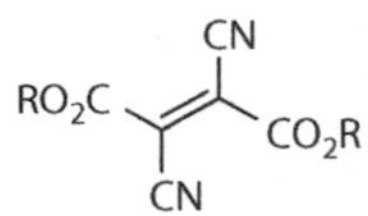	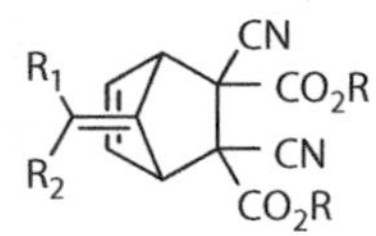	RT	50–100	Park et al. (2009)
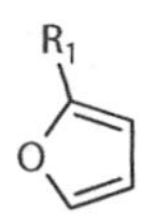		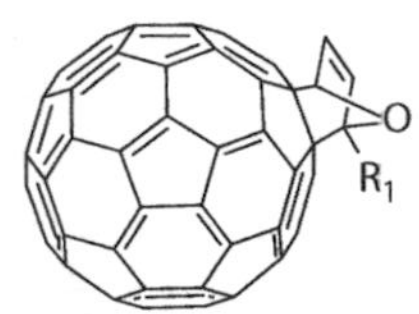	25	180	Guhr et al. (1994).

		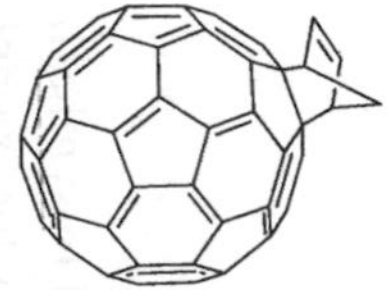	25	95	Nie et al. (1995)
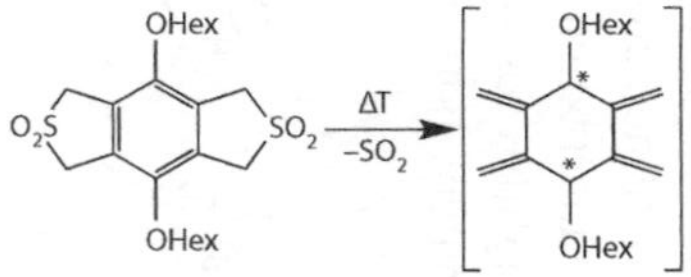		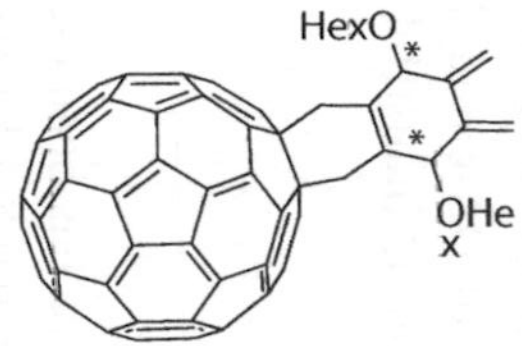	214	n.a	Gogel et al. (1996)
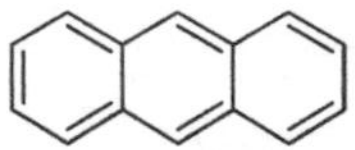			RT	60–82	Illhan and Rotello (1999)
	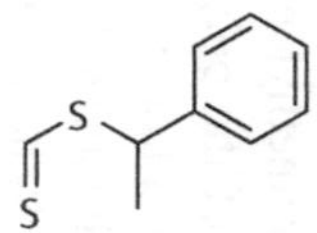	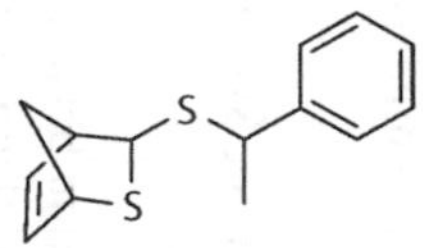	25	80–95	Paulöhrl et al. (2010), Inglis et al. (2010)

TABLE 9.2
ΔE and $\Delta E'$ (in eV) Calculations for Diels–Alder Cycloaddition Reaction in Reversible System Calculated by CNDO Simulation

	Cyclopentadiene	Maleimide	Benzylthioester
ΔE (LUMO diene–HOMO dienophile)			
Cyclopentadiene	0.53	×	0.47
Furan	×	0.61	×
Anthracene	×	0.47	×
$\Delta E'$ (LUMO dienophile–HOMO diene)			
Cyclopentadiene	0.53	×	0.39
Furan	×	0.45	×
Anthracene	×	0.37	×

prepare different polymeric structures displaying the presence of different groups (see, e.g., Table 9.1) along the same backbone and subsequently to change the chemical structure of the latter. Such study is to the best of our knowledge still missing in the open literature. This is paradigmatic for the relatively long and laborious synthetic steps that are often involved in the synthesis of DA-based thermoreversible materials, which has been hindering so far applications at industrial level.

9.2.2 Other Chemical Systems

9.2.2.1 Ionene

The Menshutkin reaction between amines and halides has been used to synthesize both thermoplastic and thermosetting ionomers. Leir and Stark (1989), for example, have studied the polymerization, depolymerization, and repolymerization of linear ionomers from polytetramethyleneoxide diamines and dihalides. They have discovered that degradation of the polymer during depolymerization is very significant, and that only by using a low depolymerization temperature, and a short time, the material could be repolymerized.

Rembaum et al. (1969) have reported on the synthesis of ionene containing cross-linkers for radical polymerization. They synthesized both di(meth)acrylate and diallylammonium cross-linkers, shown in Figure 9.23. The obtained cross-linkers were used to synthesize cross-linked polymers via radical polymerization.

Similarly, styrene–butadiene rubber (SBR) has been lightly cross-linked using the Menshutkin reaction by Buckler et al. (1978). Unfortunately, both publications did not report on the reversibility of these cross-links.

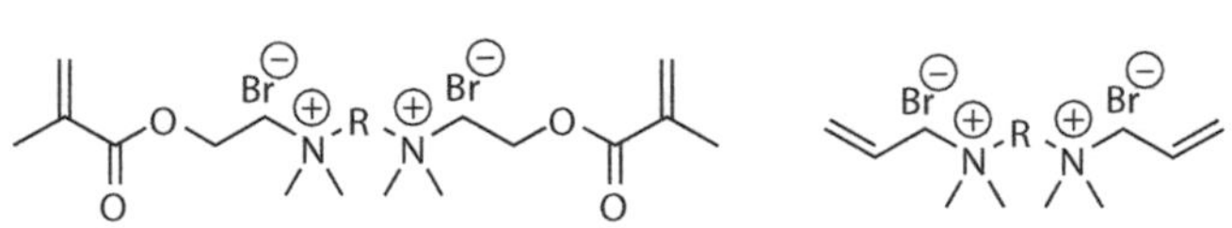

FIGURE 9.23 Ionene cross-linkers.

More recently, Ruckenstein and Chen (2000b) have reported the cross-linking reactions between Cl-containing polymers with bifunctional tertiary amines; tertiary amines containing polymers with dihalide compounds; and the reaction between Cl-containing polymers and polymers with pending tertiary amines. It was shown that the cross-linked polymers can be partially de-cross-linked, but they were not soluble without the addition of a trap to shift the equilibrium to the un-cross-linked state. Nevertheless, the polymers exhibited flowability at 215°C, and could thus be reworked at high temperature.

Ruckenstein (Ruckenstein and Chen 2001) and Chen (Chen and Ruckenstein 2000b) have reported on the synthesis of emulsions containing either chlorine-functionalized polymers via copolymerization of styrene and butylacrylate with chloromethylstyrene or tertiary amine functionalized polymers via the copolymerization of 2-(dimethylamino)ethylacrylate with butylacrylate and styrene. After mixing these emulsions, and casting and drying them at room temperature, a cross-linked film is obtained. It was shown that the cross-links in these materials are the result of quaternization of the available amines (i.e., the Menshutkin reaction). The films can be remolded by thermocompression. Similar material properties (swelling and soluble percentage) were obtained for three cycles (including the initial cycle). Long-term stability tests of the latex showed no significant changes in material properties of the latex after storage for 48 months (Chen et al. 2005).

Thermoreversible cross-linking via the Menshutkin reaction is interesting from several points of view. First, the relatively high reactivity leads to rapid cross-linking, even at moderate temperatures. Second, the relatively high de-cross-linking temperature of about 200°C creates the possibility to use the polymer in a wide range of temperatures especially as compared with thermoreversible networks based on the DA reaction. Finally, the Menshutkin reaction can easily be tuned by modification of either the tertiary amine or the halogen. Moreover, the degradation of the polymer at these high de-cross-linking temperatures limits the applicability of the system. And, even though the Menshutkin reaction is well known, literature on the use of the reaction for thermoreversible networks is less available than for the discussed DA systems.

9.2.2.2 Ester–Amide Bond Rupture

Although the breakage of an ester bond usually involves hydrolysis or exchange reactions (Engle and Wagener 1993), it may also be purely dependent on temperature as demonstrated, for example, by depolymerization of isophthalates to yield the corresponding cyclic monomeric units (Hall et al. 2000). Due to the nature of this bond, it should not be used for cross-linking of other condensation polymers (i.e., polyesters, polyamides, etc.). The occurrence of side reactions connected to the hydrolysis of the ester group (mainly hydrolysis/exchange reactions involving the polymer backbone) may dramatically change the overall macromolecular structure and consequently the material properties (Nakano et al. 1997). Although ester formation from anhydride and hydroxyl groups is a very straightforward reaction, it has been demonstrated (Nowak et al. 1976) that a catalyst should be added to the reaction mixture to achieve good cross-linking rates (Zimmermann et al. 1972). This approach has been applied to cross-link nitrile rubber (NBR) (Campbell 1974), styrene–maleic anhydride copolymers (Muskat 1963), functionalized polyolefins (Moore 1968), and ethylene copolymers containing anhydride groups (copolymerization of ethylene with bicyclo[2.2.1]hept-2-ene-5,6-dicarboxylic anhydride) (Withworth and Zutty 1967). It must be stressed that in some cases, although claimed differently, the use of primary hydroxyl functionality probably leads to an ester bond not purely reversible as a function of temperature.

A very similar reaction is the amide formation (usually from anhydride and amine groups), which is also thermoreversible provided that the amine is sterically hindered (e.g., tertiary) (Heyboer 1995; Decroix et al. 1975). This last remark is actually a very general one both for amine and ester formation: the reaction is thermally reversible in the presence of a steric hindrance, no matter if on the nucleophilic (–OH or NR_2) or electrophilic groups.

In 1996, two publications showed the reversibility of spiro-ester (Figure 9.24). In the first, Endo et al. (1996a) found out that a suspension of IIa in dichloromethane, upon treatment with 5 mol% of trifluoroacetic acid at room temperature for 1 h, changed to a homogeneous solution, from which dithiol-linked bifunctional spiro-orthoester (IIIa, yield 66%) was obtained. Under similar conditions, the cross-linked polymers IIb and IIc depolymerized to the bifunctional monomers IIIb and IIIc in 69% and 82% yields, respectively. In the second work (Miyagawa et al. 2005), the GPC profiles before cross-linking and after depolymerization confirmed the acid-catalyzed reversible reaction. Finally in 2005, a publication (Endo et al. 1996b) of the same research group reported the use of polystyrene with spiro-ester groups as a polymer matrix.

As general discussion point, it must be noted here that the limited amount of literature available on thermoreversible amide–ester bonds reflects probably two seriously complication factors. In first instance, the possible competition of the de-cross-linking reaction with the depolymerization one if the same kind of groups is present along the backbone. This severely limits the applicability of such reaction. A second point is taken by the rather laborious synthetic steps that are actually needed to design and prepare materials with the desired chemical structure. Although good evidences for perfect thermal reversibility at molecular level are present for these systems, the two hindering factors mentioned earlier play probably a crucial role in limiting the interest in them at both academic and industrial levels.

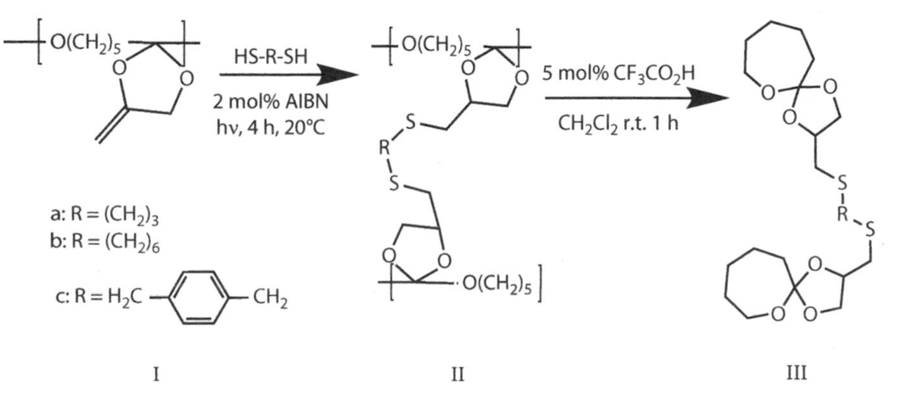

FIGURE 9.24 Synthesis of thermoreversible spiro-ester.

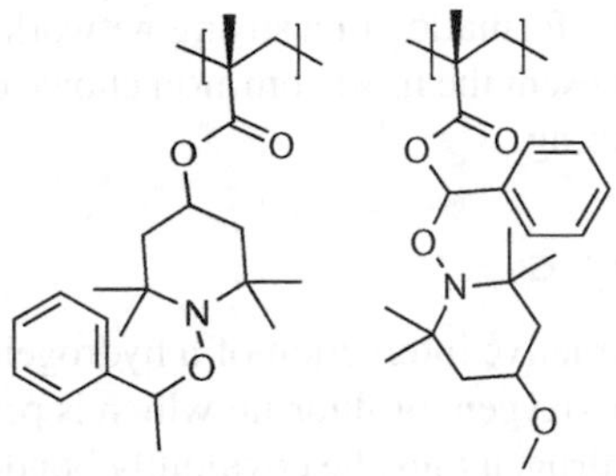

FIGURE 9.25 Structure of polystyrene moieties with radical thermobreakable functional groups.

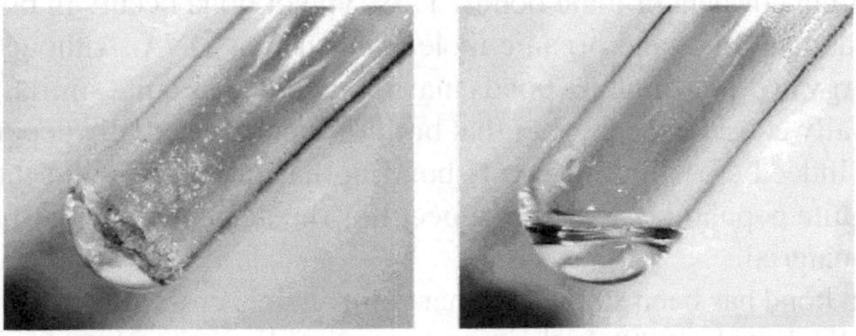

FIGURE 9.26 Cross-linked (right) and linear (left) polymer in the work discussed by Higaki et al. (From Higaki, Y. et al., *Macromolecules*, 39, 2121, 2006.)

9.2.2.3 Radical Coupling

A new innovative method based on radical coupling has been presented by Higaki et al. (2006) in which a TEMPO-sterically hindered phenol couple was anchored to a polystyrene matrix (Figure 9.25). The de-cross-linking reaction was carried out by heating the cross-linked polymer in the presence of an excess amount of alkoxyamine. Although the alkoxyamine units on the side chain dissociated homolytically at 100°C producing carbon and nitroxide radicals, no detectable intra- and intermolecular carbon–carbon coupling reactions occurred due to the rapid capping of carbon radicals by nitroxide radicals. Molecular weight increased with increasing reaction time at a $T = 100°C$ and became constant after 24 h, indicating that the reaction reached equilibrium at 24 h under this concentration condition. The cross-linked polymer was swelled in anisole solution with an excess amount of alkoxyamine (20 equiv./alkoxyamine unit) and heated at 100°C for 48 h to obtain back the linear polymer (Figure 9.26).

9.3 PHYSICAL INTERACTIONS

Physical interactions involve a different kind of association with respect to chemical bond formation. Indeed, the relative weakness of these interactions renders them attractive from the point of view of the relatively lower reversibility temperature, which should be able to prevent significant degradation reactions. Moreover, the same factor (i.e., relative weakness of the interaction) could also represent a problem since it might not ensure the formation of a strong network. Electrostatic interactions and hydrogen bonding represent the most common choice of physical interactions for thermoreversible cross-linking.

9.3.1 Hydrogen Bonding

A hydrogen bond is the attractive interaction of a hydrogen atom with an electronegative one, such as nitrogen, oxygen, or fluorine which is present on another molecule or chemical group. The hydrogen must be covalently bonded to another electronegative atom to create the bond. These bonds can occur between molecules (intermolecularly), or within different parts of a single molecule (intramolecularly) (Nic et al. 1997). The hydrogen bond (5–30 kJ/mol) is stronger than a van der Waals interaction, but weaker than covalent or ionic bonds. This type of bond occurs in both inorganic molecules such as water and organic molecules such as DNA. Although the differences in energy are quite high, H-bonds may be regarded as quite similar to chemical ones, especially considering the fact that both of them are actually dependent on the orientation. Indeed cross-linking via H-bonding has been studied in the past and it is actually quite popular because of the peculiar rheological behavior that it imparts to the final material.

Hydrogen bond has been studied as thermoreversible cross-linking system for the first time by Klok et al. (1999). In their work, a poly(siloxane) with carboxylic pendent groups was synthesized and the cross-link already occurred at room temperature.

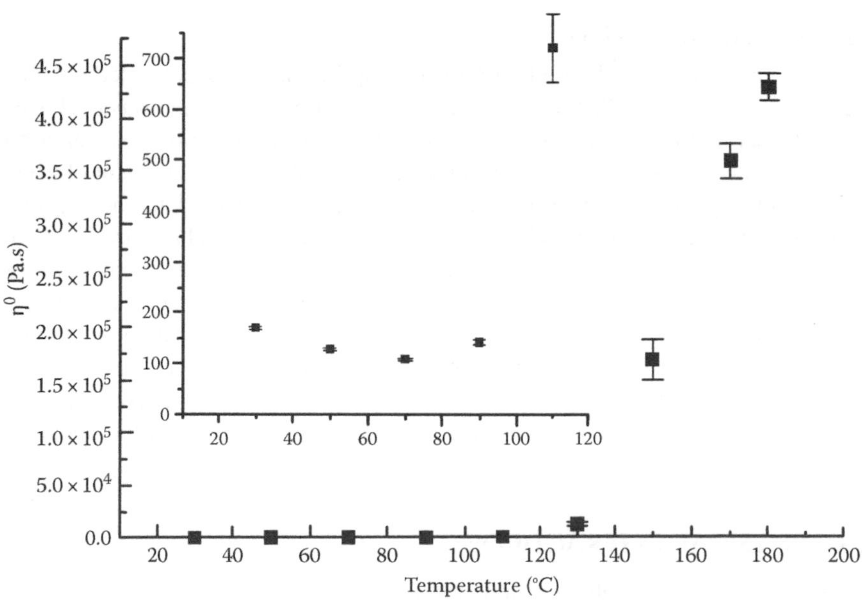

FIGURE 9.27 The zero-shear viscosity (η^0) as a function of temperature for a PDMS-COOH sample heated from 30°C to 180°C. The zero-shear viscosities are obtained by extrapolation of the Ellis fit of the η' (ϖ) curves to $\varpi = 0$. (After Klok, H.-A. et al., *J. Polym. Sci. B Polym. Phys.*, 37, 485, 1999.)

However, at temperatures higher than 70°C, the zero-shear viscosities increased strongly with temperature (Figure 9.27), thus probably suggesting the occurrence of side reactions. Furthermore, it has been demonstrated that water facilitated the rupture of the network. At low frequencies, the slope of the storage and loss modulus curves reached values of 2 and 1, respectively, which are typical for a viscous polymer melt. The frequency dependence of DMTA analysis demonstrated the physical, that is, reversible nature of the temperature-induced gelation.

A thermoreversible system with triazole rings was further developed by Chino and Ashiura (2001) by chemical modification of polyisoprene rubber with first maleic anhydride and then triazine amine (Figure 9.28). In particular, 3-amino-1,2,4-triazole (ATA) solidified the liquid rubber at room temperature. Conversion of the absorption peaks of acid anhydride at 1864 and 1788 cm^{-1} to the peaks of amide at 1635 and 1526 cm^{-1} was observed. The absorption peak of carboxylic acid was observed at a lower region (1728 cm^{-1}) than that of general carboxyl acid (1760 cm^{-1}). In addition to the glass transition, the modified polyisoprene shows a second endothermic transition around 185°C in the DSC curves. This may indicate the cleavage of the hydrogen bonding. The mechanical properties of the polymer with triazine pendent groups (3.8 mol%) (namely tensile stress; Figure 9.29) are much higher than those of uncured IR containing 30 phr (parts per hundred rubbers) of carbon black (CB). Mechanical properties of the resulting rubber resulted to be similar to those of sulfur-vulcanized rubber.

Polybutadiene has been used as a matrix to hydrogen-sensitive functional group in two publications by Peng and coworkers (Peng and Abetz 2005; Peng et al. 2005) (Figure 9.30). In the first publication, it has been found by FTIR that hydrogen-bonding strength of C=O H–N is higher than that of S=O H–N. S=O asymmetric stretching shifts from 1362 to 1349, whereas C=O shifts partly from 1751 to 1718. Due to the formation of strong hydrogen bonding network, the glass transition was greatly shifted from −103°C to −4.1°C upon a 20 mol% modification. With increasing degree of modification up to 2 mol%, crystals were formed at a $T_c = -49$°C. In the second publication, the authors conclude that the filler–rubber HB interaction was more sensitive to temperature than the filler–filler HB interaction, meaning that the relative HB stability of the former was enhanced more than the latter as the temperature dropped.

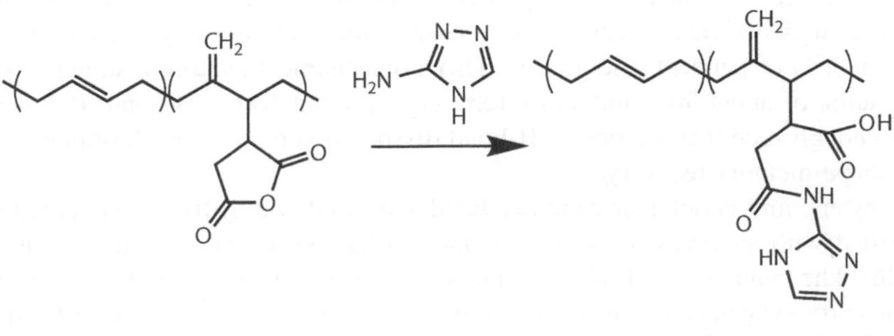

FIGURE 9.28 Reaction of 2-amino-1,2,4-triazole with maleic anhydride–modified poly(isoprene).

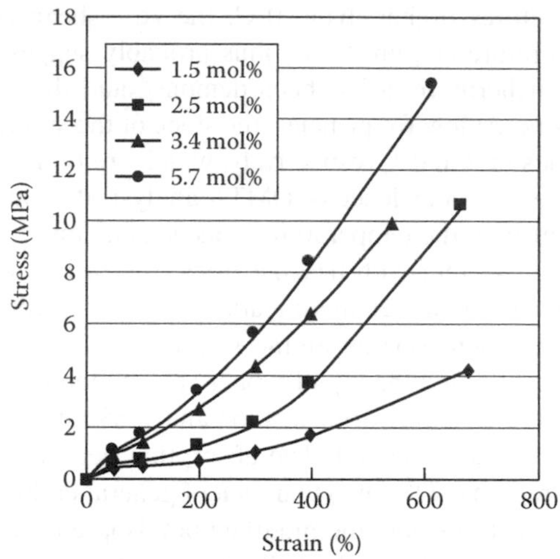

FIGURE 9.29 Stress–strain curves for TRI adducts with various modification ratio. (After Chino, K. and Ashiura, M., *Macromolecules*, 34, 9201, 2001.)

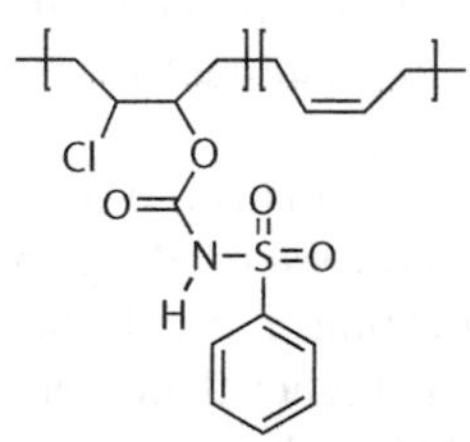

FIGURE 9.30 Structure of polybutadiene-modified rubber.

Akiba and Akiyama (2001) used a blend of polyvinylpyrrolidone (PVPr) and polyvinylphenol (PVPh) as a matrix and the hydrogen-formed network was proved by a shift to 1660 cm^{-1} for the carbonyl stretching of PVPr. The thermoreversible gel–sol transition temperature increased with increasing contents of PVPh as cross-linking agents in the PVPr/PVPh blends and the decreasing ratio of water against the PVPr/PVPh blends.

Shape memory was also found for pyrazone-functionalized polymer synthesized by Li et al. (2007). Shape-memory effects were observed in a lightly cross-linked polymer network that is easy to synthesize and contains only a small fraction (ca. 2 mol%) of pendent side groups. Thermomechanical cycling resulted in strain elongation of about 90% and strain recovery of about 100%. H-bond lifetimes are long enough such that the rate of H-bond dissociation dominates elastomeric creep and shape-memory recovery.

Ethylene and copolymer functionalized with malic anhydride and successively modified with an excess of amino compound has been investigated by Sun et al. (2006). The band around 1710 cm^{-1} can be assigned to the C=O stretching vibration of the carboxylic acid and the bands around 1640 and 1555 cm^{-1} to the C=O stretching vibration (amide I) and the NH-stretching vibration (amide II) of the amide proved the forming of a supramolecular structure. Dependence of aliphatic chain in amine compound has been analyzed and they found two opposite effects that affected the mechanical properties. The first effect is that longer apolar tails will disturb the aggregate formation, leading to poorer properties, which explains the decrease in properties for the hexylamide acid compared with the propylamide acid. Second, the longer tails may organize themselves in a crystalline-like order, which will improve the properties. A better and stronger packing can be expected for the longest alkyl tails, which may explain the trend $C_{18} > C_{14} > C_{10} > C_6$. An important concern is the occurrence of imide formation for all amide acids and amide salts after compression molding at temperatures of 120°C and higher, resulting in disappearance of the aggregates and poor mechanical properties.

Furthermore, terpolymer at different ratio of isopropylacrylamide, acrylic acid, and 1-vinylimidazole were synthesized by Ogawa et al. (2003). Aggregation at 41°C was noticed for the terpolymer with 3% and 13% of cation and anion, whereas aggregation at 52°C has been reached for hydrogel with 14% and 12% cation and anion.

Poly(acrylamide) has also been used by Xue et al. (2005) in order to create a thermoreversible hydrogen-bonding network. At a fixed content of *N,N'*-methylene bisacrylamide (BIS), values of ν_e and the elastic moduli exhibit an unusual increase with temperature, the cross-linking thus being thermally reversible. In all cases, an increase in nominal content of cross-linker leads to an increase in elastic moduli and effective cross-linking density.

Lately, Montarnal et al. (2010) explored the properties of an epoxy resin modified with imidazolidone (Figure 9.31). While the $\nu_{C=O}$ characteristic wave number of the ester linkage (1735 cm^{-1}) is independent of temperature, strong shifts are observed from 30°C to 150°C for the $\nu_{C=O}$ and ν_{N-H} characteristic bands of the amidoethylimidazolidone

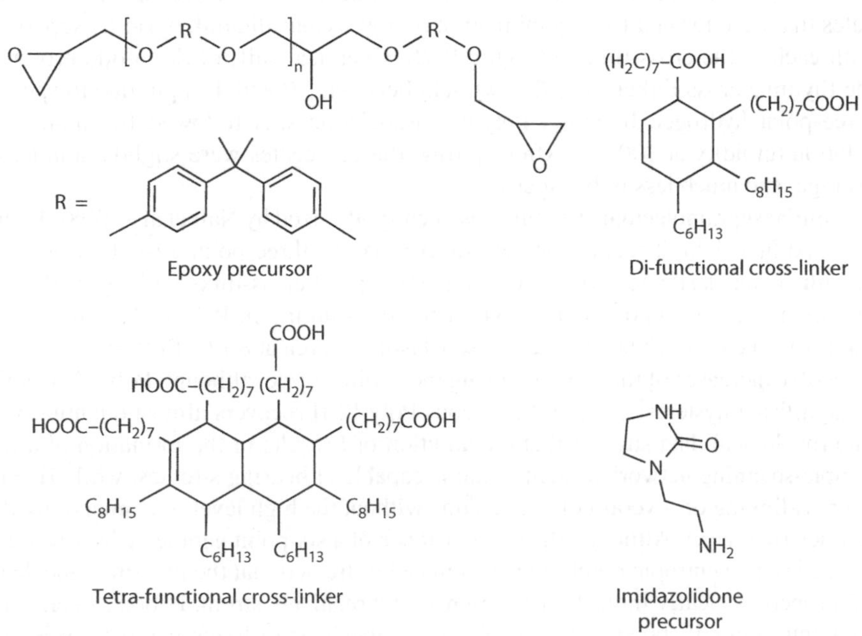

FIGURE 9.31 Modified epoxy resin precursors.

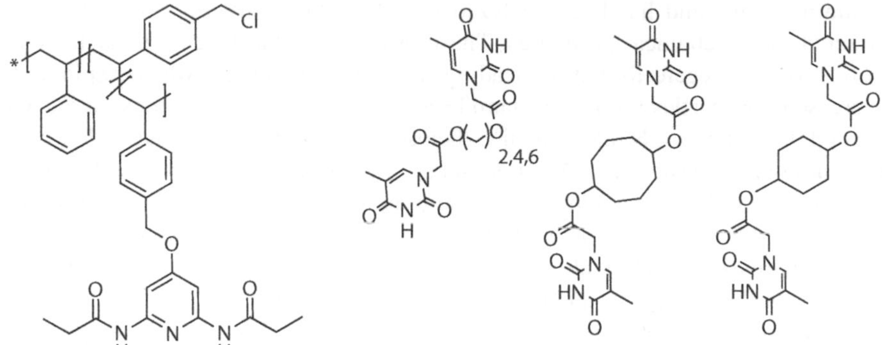

I II III

Di-amido pyridine-supported polystyrene

Thymine cross-linkers

FIGURE 9.32 Diamidopyridine-supported polystyrene with three different thymine cross-linkers (I, II, and III).

moiety ($\nu_{C=O}$ amide shifts from 1653 to 1660 cm^{-1}, $\nu_{C=O}$ imidazolidone shifts from 1682 to 1693 cm^{-1}, and ν_{N-H} shifts from 1549 to 1540 cm^{-1}). These data imply that hydrogen-bonding groups undergo a steady transition from associated to free state from 30°C to 150°C. The fact that no endotherms are shown by DSC measurements in the same temperature range suggests that the hydrogen-bonding groups do not form long-range crystalline aggregates but rather associate pairwise or in small aggregates.

Thibault et al. (2003) synthesized well-defined, micron-sized spherical aggregates that were formed from combination of polystyrene diamidopyridine supported with each of the cross-linkers (Figure 9.32). Chemical shift of the imide proton of the thymine cross-linker moved downfield between 3.0 and 4.8 ppm due to specific three-point hydrogen bonding. Polymer association was followed by monitoring solution turbidity at 700 nm. After cycling, the aggregates were slightly smaller on average, and much less polydisperse.

Similar supramolecular structure has been synthesized by Nair et al. (2008) (Figure 9.33). At first sight, it seems counterintuitive that the three-point cross-linking agent I results in an elastic gel, while the stronger six-point cross-links with agent II result in a high-viscosity fluid (Figure 9.34). The cross-linking in Poly-12-I is sufficiently stable that the network behaves as a viscoelastic gel even at 80°C. Both samples show reversible increases of the moduli during the cooling cycle, although Poly-12-3 exhibits significant hysteresis, while the viscous Poly-12-II recovers almost quantitatively. The rheological data suggest that the addition of I results in the formation of a true sample-spanning network structure that is capable of bearing stresses, while II leads to cross-linking of several polymer chains without the high level of connectivity that characterizes a gel. Although the real existence of a six-point geometry may be questioned from an entropic point of view, it must be stressed that the measured mechanical properties clearly indicate the presence of a relatively strong H-bond interaction. Such interaction is shown to be completely reversible up to 10 temperature cycles and it has also been applied on an industrial level (Chino et al. 2001).

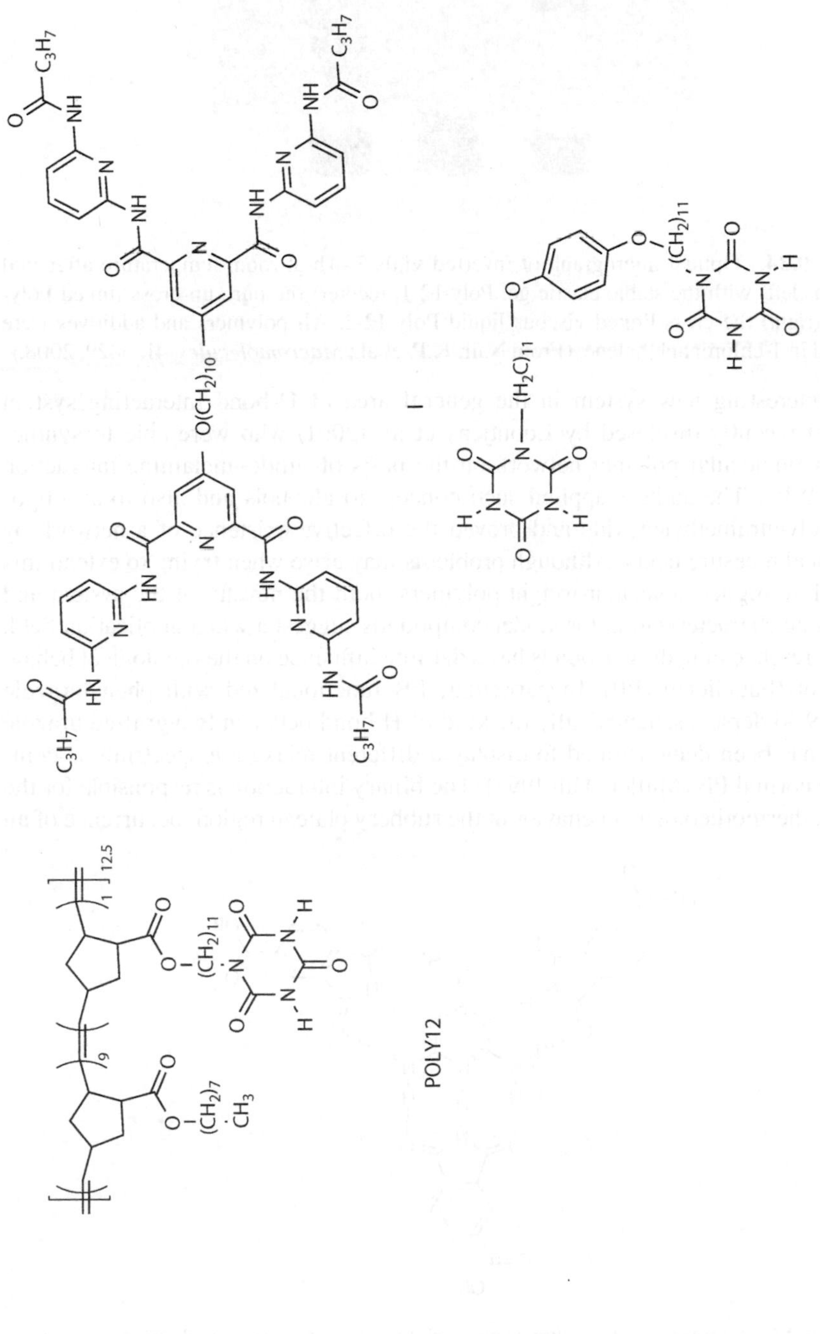

FIGURE 9.33 Thymine-supported polymer (Poly-12) and two different kinds of cross-linkers (I and II).

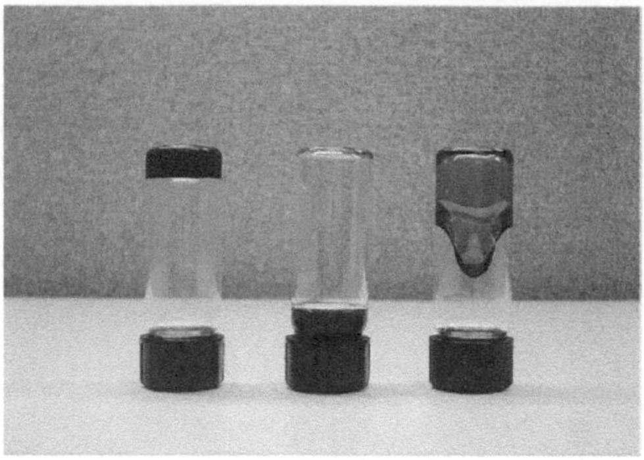

FIGURE 9.34 Optical micrograph of inverted vials 3–4 h at room temperature after vial inversion (left) with the stable elastic gel Poly-12-1, (center) the pure un-cross-linked Poly-12, and (right) the cross-linked viscous liquid Poly-12-2. All polymers and additives were dissolved in 1-chloronaphthalene. (From Nair, K.P. et al., *Macromolecules*, 41, 3429, 2008.)

An interesting new system in the general area of H-bond interacting system has been recently disclosed by Loontjens et al. (2001) who were able to synthesize supramolecular polymer network on the basis of imide–melamine interaction (Figure 9.35). The authors applied such concept to alcohols and also to an oligomeric polytetramethylenoxide and proved the effective existence of a network by rheological measurements. Although problems may arise when trying to extend this approach to higher-molecular-weight polymers, both the novelty of the system and its accurate characterization for model compounds suggest a wider application field.

The presence of hydrogen bonds has a definite influence on the rheological behavior of poly(butadiene) (PB). In particular, PB functionalized with phenylurazole (Figure 9.36 depicts schematically the kind of H-bond between two grafted urazole groups) has been demonstrated to display a different relaxation spectrum as compared to normal PB (Müller et al. 1995). The binary interaction is responsible for the complex thermorheological behavior in the rubbery plateau region (occurrence of an

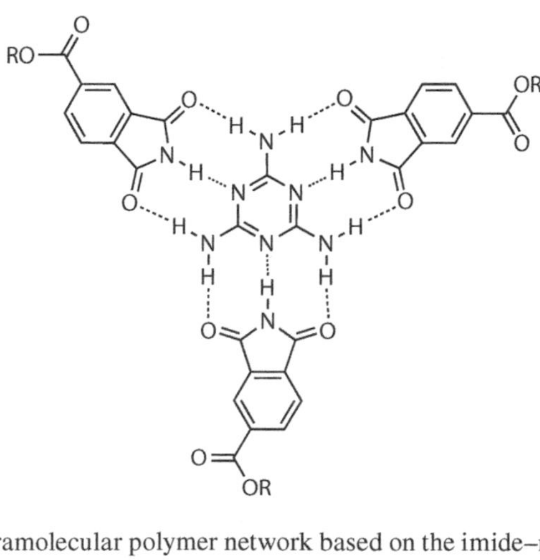

FIGURE 9.35 Supramolecular polymer network based on the imide–melamine interaction.

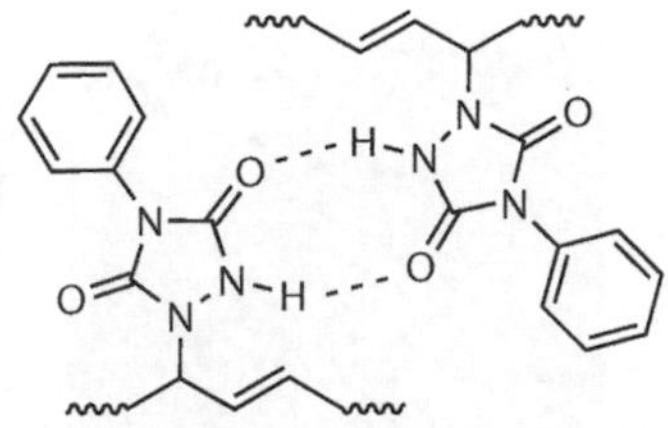

FIGURE 9.36 H-bond interaction between two PB chains functionalized with phenylurazole.

additional relaxation process associated with the association equilibrium) and for the increase in the relaxation time within the terminal relaxation zone.

Supramolecular polymers generated from heterocomplementary monomers linked through multiple hydrogen-bonding arrays formation were synthesized by Berl et al. (2002) (Figure 9.37). The rupture of the hydrogen bonds occurs visibly above 50°C, as suggested by the shifting of the amide proton signals (10.0 and 9.2 ppm) of the receptor toward higher fields NMR proton-relaxation time studies supported the formation of higher-molecular-weight aggregates when bisreceptor 1 (Figure 9.37) was mixed with stoichiometric amounts of 3. The further broadening of the spectra of a stoichiometric mixture of 1 and 3 (2 mM in tetrachloroethane) occurs if the tritopic branching receptor 2 is added. The volumes of a molecule of bisreceptor 1 and biswedge 3 were found to be approximately 1140 Å and 860 Å, respectively. The molecular volume calculated by the relaxation data would thus correspond to an average trimeric aggregate [1:3]. It is interesting to mention the observation of a rapid, reversible regeneration of the highly viscous aggregates under equilibrium conditions, after temporary destruction by the application of a high shear rate. The aggregate only redissolves in the presence of a polar/protic solvent that competes with the hydrogen-bonding sites and disrupts the polymer. The

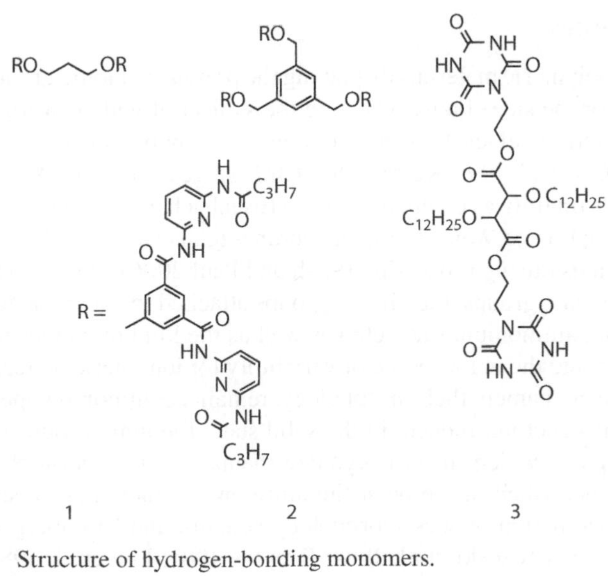

1 2 3

FIGURE 9.37 Structure of hydrogen-bonding monomers.

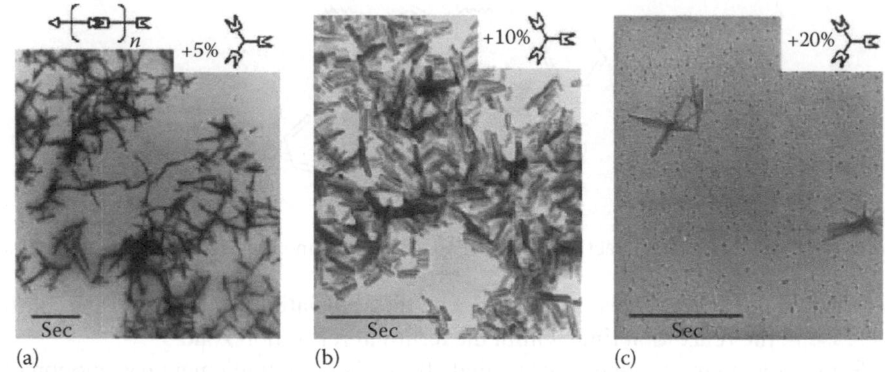

FIGURE 9.38 Electron microscopy studies revealing the strong influence of the addition of cross-linking agent 2 on the aggregate morphology of $[1:3]_n$ in $C_2H_2Cl_4$. (From Berl, V., Schmutz, M., Krische, M.J., Khoury, R.G., and Lehn, J.-M.: Supramolecular polymers generated from heterocomplementary monomers linked through multiple hydrogen-bonding arrays—Formation, characterization, and properties. *Chem. Eur. J.* 2002. 8(5). 1227–1244. Copyright Wiley-VCH Verlag GmbH & Co. KGaA. With permission)

morphological structure of the aggregate changes with the ratio between the cross-linker and the matrix (Figure 9.38).

The use of hydrogen bonds for thermoreversible cross-linking represents one of the most interesting routes between the ones involving purely physical interactions and chemical bonds. The relative weakness of the interaction must however be conceptually coupled, in the design stage, with the proper amount of groups interacting with each other in order to ensure the formation of a network of appropriate strength. Moreover, the practically perfect reversibility character of the interaction and its specificity (only certain groups along the backbone can interact) are both attractive factors for these systems.

9.3.2 Ionomers

Ionomers are polymeric materials displaying the presence of ionic groups either along the backbone or the side chains. They are the subject of wide investigations at both academic and industrial levels because of their wide application fields, which include fuel cells (Wilson et al. 1995; Kerres 2001; Chikashige et al. 2005; Asano et al. 2006), polymer blends (mainly as compatibilizers) (Eisenbach et al. 1994; Dutta et al. 1996; Feng et al. 1996; Lu and Weiss 1996), membranes (Cui et al. 1998; Kerres et al. 1998), as exfoliation aids during processing (Shah and Paul 2006). Besides variation in the position of the ionic groups, the kind of groups attached to backbone (e.g., sulfonate, carboxylic acid, ammonium salts, etc.) as well as the kind of counterion attached to the latter constitute the main source of variability for ionomeric materials. However, one characteristic, namely their morphology, remains common independently of the exact chemical structure. Indeed, in the solid state, the ionic groups tend to associate into multiplets, which in turn organize themselves into clusters. The majority of the studies published in the open literature on ionomers have been focusing on the exact determination of such morphology (Bazuin and Eisenberg 1981; Yarusso and Cooper 1983; Brozoski et al. 1984; Dreyfus 1985; Lee et al. 1988; Moore and

Martin 1989; Venkateshwaran et al. 1992; Lu et al. 1993; Grady 1999; Nguyen et al. 1999; Zhang et al. 2000; Meresi et al. 2001; Tsujita et al. 2001; Huang et al. 2002; Essafi et al. 2004; Wang et al. 2009) through a variety of spectroscopic techniques for ionomers based on polyurethanes (Lee et al. 1988; Visser and Cooper 1991a; Hourston et al. 1998; Jayakumar et al. 2005), perfluorinated polymers (Moore and Martin 1988; Rubatat et al. 2002), ethylene–methyl methacrylate copolymers (Nagao et al. 1992; Kutsumizu et al. 2000, 2002; Winey et al. 2000), and sulfonated polystyrene (Kirkmeyer et al. 2001). The ionomers morphology has a clear and established influence on the several material properties (Kim et al. 1991; Quiram and Register 1998) such as thermal (e.g., crystallization) and mechanical behavior (Hirasawa et al. 1989; Visser and Cooper 1991b, 1992; Weiss et al. 1991a,b; Lei and Zhao 1993; Tachino et al. 1993; Kim et al. 2000, 2002; Hana et al. 2003; Page et al. 2005; Zhu et al. 2006a; Osborn et al. 2007). For example, the crystallization rate for sulfonated syndiotactic polystyrene was found (Orler et al. 1996) to be inversely proportional to the ionic radius of the counterion. As the activation energy for ion-hopping increased with decreasing counterion size, longer periods of time were required to achieve the same degree of crystallinity. The mechanical behavior is consequently affected as seen in the case of sulfonated SEBS-based ionomers (Zhu et al. 2006a) (Figure 9.39), for which a clear dependence of the T_g as well as of the mechanical properties on the amount and kind of ionic groups has been reported.

In addition, the melt viscosity (and in general the rheological behavior) (Greener et al. 1993; Kim et al. 1994; Orler et al. 1996; Vanhoorne and Register 1996; Fang et al. 2003) is affected by the presence of the ionic domains and their structure. Indeed, copolymers of ethylene and methyl methacrylate (in their neutralized form with Na, Zn, and Mg) display an elongational viscosity in the melt that is a function of the kind of counterion (i.e., Zn ≫ Na and Mg) (Takahashi et al. 1994).

The general effect of ionic domains on the mechanical properties of a given base materials (thus in neutral form) consists in two separated effects (Zhu et al. 2006a)

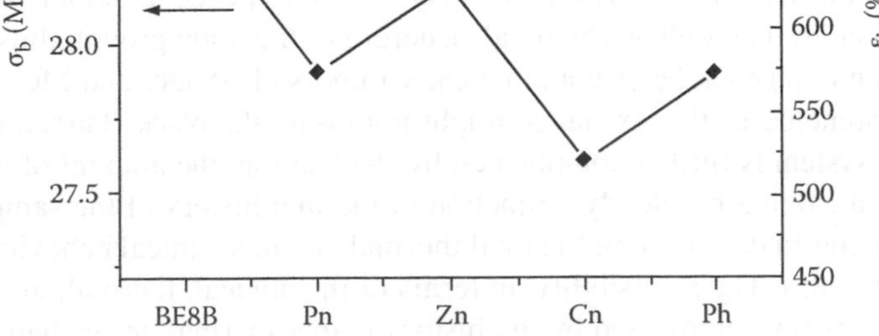

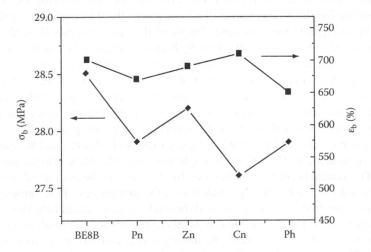

FIGURE 9.39 Dependence of stress (σ_b, left) and strain (ε_b, right) at break on the kind of counterion for SEBS-based ionomers with 11–12 mol% of sulfonic groups along the backbone.

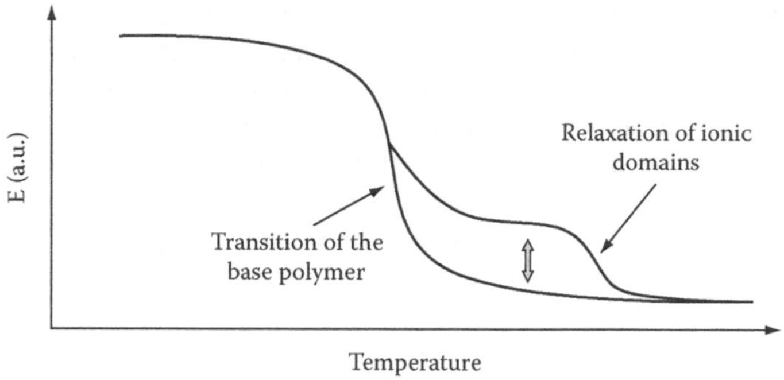

FIGURE 9.40 General effect of ionic groups on thermomechanical behavior (e.g., modulus E) as function of temperature.

(Figure 9.40): the appearance of an extra transition (usually at relatively high temperature, generally named "secondary melting") due to the relaxation of the ionic domains and the increase of the modulus (double arrow in Figure 9.40) in between the transition of the base polymer (e.g., a T_g) and the new one.

In the present work, we aim to discuss only those aspects of the ionic aggregates structure that could exert an influence on the thermal reversibility. The reader is referred to other excellent reviews in the general field of ionomeric materials for a more general description of these materials properties (Takahashi et al. 1994; Capek 2005).

When studying the thermal reversibility of ionomers, one must take into account the new relaxation induced by the presence of ionic groups and in particular the physical effects that such transition has on the morphology (Benetatos and Winey 2005) (and in turn on the properties). Indeed, ionic aggregation (either in the form of multiplets of cluster) is in reality a dynamical process as function of temperature. Indeed, for neutralization degrees lower than 100%, the ionic domain is surrounded by chain segments bearing the neutral form of the ionic group, the thickness of this extra layer being almost equal to the persistence length of the given polymer (Eisenberg et al. 1990; Vanhoorne et al. 1994). At relatively high temperature (i.e., in the melt), there is a dynamic exchange of macromolecular segments, which might have an influence on the final properties of the material (Coleman et al. 1990; Hara et al. 1991). Unfortunately, the nature of such exchanges and thus the eventual influence on the thermal and mechanical behavior is strongly system-dependent (namely on acid and neutralization level as well as chemical structure of the ionic group), thus rendering the definition of a general concept practically impossible (Sauer and McLean 2000). Indeed, in some cases this exchange might not even take place (Farrell and Grady 2000). The system is further complicated by the fact that the amount of water present in the ionic domains (clearly a function of thermal history of the sample) is also a key parameter in determining the final thermal and mechanical behavior (Yarusso and Cooper 1985). The reversibility, in terms of rheological, thermal, and mechanical behavior, is further affected by the history (either thermal or mechanical) of the material. The latter could actually hinder the thermal reversibility character of ionomers. Indeed, Akimoto et al. (2001) recently reported that after a mechanical stress,

Na and Zn salts of ethylene–MMA copolymers do not show the secondary melting anymore, thus suggesting that the system is not reversible.

Despite such limitations, ionomers have been shown to display self-healing properties as well as shape-memory effects (Kima et al. 1998; Han et al. 2007; Zhu et al. 2007). Shape-memory polymers (SMPs) remember a set shape, which can recover upon a specific action (e.g., by applying heat or light). Although there are different kinds of SMPs, the memorized shape can only be set for physically "cross-linked" SMPs. Chemically cross-linked SMPs are fixed to the shape in which they were cross-linked (Liu et al. 2007; Inoue et al. 2009). Using thermally reversible cross-links, the memorized shape can also be changed for chemically cross-linked polymers by heating above the temperature where cross-linking reverses. In a typical shape-memory test (Zhu et al. 2006b) (Figure 9.41), any given sample is subjected to a cyclic loading–unloading treatment superimposed to a thermal one.

The sample is placed in a tensile machine, loaded, and stretched at high temperature (1). Once the maximum strain (ε_m) is reached (2), the sample is cooled down to low temperature at constant ε_m and then the clamp distance is set back to the original value (3). By maintaining the strain at 0%, the sample is heated up to the high temperature again (4), when the next cycle is started (5). The shape memory ability of a given material is then usually quantified for instance by the strain recovery ratio (R_N):

$$R_N = \frac{\varepsilon_m - \varepsilon_p(N)}{\varepsilon_m - \varepsilon_p(N-1)} \tag{9.2}$$

where N is an integer denoting the thermal cycle and the strain are defined in Figure 9.41.

For polyurethane-based ionomers, application of the test described earlier (Zhu et al. 2006b) yielded average R_N values of 90%, although the reason for such nice performance could not be directly attributed to the thermal reversibility of the ionic domains, but to the interplay of the latter with the van der Waals forces acting between the hard and soft bocks of the PU matrix (Zhu et al. 2008).

The same kind of interaction (at least on a conceptual level) between physical and electrostatic forces acting between polymeric chains of ionomeric materials renders the

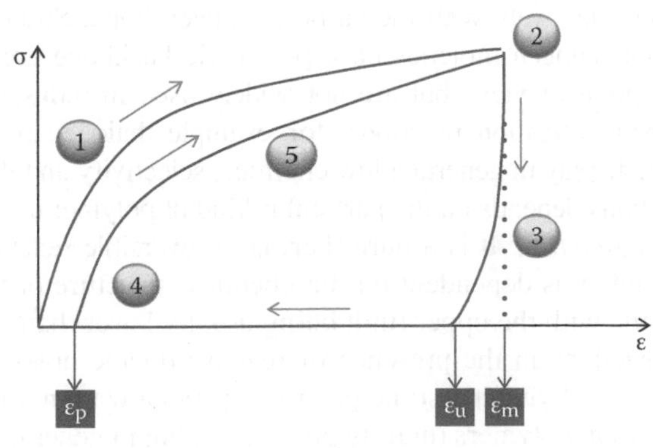

FIGURE 9.41 Schematic representation of shape-memory test.

interpretation of self-healing experiments rather difficult. For example, ionomers derived for ethylene–MMA copolymers have been shown to be able to repair themselves from projectile punctures. A detailed mechanical analysis clearly showed that for these materials the mechanism for healing at a microscopic was completely new and different for what reported on similar systems (Kalista and Ward 2007). However, when trying to relate such mechanism to the chemical structure of the polymer different works disagree on the relative importance of the different forces (van der Waals, H-bonding, and electrostatic interactions) on the final healing capability (Varleya and van der Zwaag 2010).

9.4 CONCLUSIONS

The choice of a particular kind of chemical reaction or physical interaction to achieve a thermally controlled cross-linked system is a particularly difficult task. The reaction and/or interaction, as applied to low-molecular-weight compounds, must in the first place be reversible in a temperature window that is basically wider than the working window of the final polymeric material. Second, the reaction–interaction itself should be as selective as possible, avoiding any possible interaction (unless otherwise desired) with the functional group on the polymer backbone. Finally, an available synthetic route to insert, when not already present, suitable functional groups on the polymer backbone has to be defined.

Given the aforementioned conditions, very few chemical reactions and physical interactions have been applied to polymeric systems. It must be stressed that every one of them displays advantages and disadvantages connected not only to the reaction itself, but also to the chemical structure of the polymer:

- *Ester–amide formation*: Thermal control is achieved only in special cases, the most popular strategy being the increase of the steric hindrance close to the ester–amide moiety. Temperature window and kinetics are therefore dependent on the chemical structure. In principle, it cannot be applied to polymers obtained by condensation reactions. Several synthetic strategies are available to insert acid group derivatives onto a polymer backbone.
- *Ionene formation*: It is a pure thermally reversible system with an upper temperature limit at about 200°C. It should be used in systems where the amine groups react selectively with the carbon–halogen bond. Strategies to insert amine and/or halogen moieties on a polymeric backbone are known from a scientific point of view, but are not widely used in industrial processes. Simple functionalization reactions, for example, halogenation via radical mechanism, display in general a low chemical selectivity and the occurrence of side reactions depends on the particular kind of polymer used.
- *Diels–Alder reaction*: It is a pure thermally reversible reaction. The temperature window is dependent on the chemical structure of the diene and the dienophile with the upper limit being usually lower than 200°C. It may give side reactions in the presence of reactive double bonds on the polymeric backbone. Synthetic strategies for functionalization are well known for condensation polymers (mainly polyesters), but in general quite complicated for vinyl homo- and copolymers.

- *Hydrogen bond*: It is a pure thermally reversible interaction. The temperature window strongly depends on the interaction strength and therefore on the chemical structure. The interaction is very selective. Synthetic strategies for functionalization are known mostly at a purely scientific level.

The thermal, mechanical, and rheological properties of the final material are usually dependent on the kind of chemistry as well as on the macromolecular structure. Even simple correlations between the kind of chemistry and the final properties are known in the literature only for very few systems, but suggest the possibility for achieving a very elegant control of properties as a function of the structural parameters.

REFERENCES

Adachi, K., A. K. Achimuthu, and Y. Chujo. 2004. Synthesis of organic–inorganic polymer hybrids controlled by Diels–Alder reaction. *Macromolecules* 37: 9793–9797.

Adzima, B. J., H. A. Aguirre, C. J. Kloxin, T. F. Scott, and C. N. Bowman. 2008. Rheological and chemical analysis of reverse gelation in a covalently cross-linked Diels–Alder polymer network. *Macromolecules* 41: 9112–9117.

Akiba, I. and S. Akiyama. 2001. Thermoreversible gels of aqueous solution of poly(*N*-vinylpyrrolidone)-poly(vinylphenol) blends. *Journal of Macromolecular Science: Physics* B40 (2): 157–169.

Akimoto, H., T. Kanazawa, M. Yamada, S. Matsuda, G. O. Shonaike, and A. Murakami. 2001. Impact fracture behavior of ethylene ionomer and structural change after stretching. *Journal of Applied Polymer Science* 81: 1712–1720.

Asano, N., M. Aoki, S. Suzuki, K. Miyatake, H. Uchida, and M. Watanabe. 2006. Aliphatic/aromatic polyimide ionomers as a proton conductive membrane for fuel cell applications. *Journal of the American Chemical Society* 128: 1762–1769.

Aumsuwan, N. and M. W. Urban. 2009. Reversible releasing of arms from star morphology polymers. *Polymer* 50: 33–36.

Bazuin, C. G. and A. Eisenberg. 1981. Modification of polymer properties through ion incorporation. *Industrial & Engineering Chemistry Product Research and Development* 20 (2): 271–286.

Benetatos, N. M. and K. I. Winey. 2005. Ionic aggregates in Zn- and Na-neutralized poly(ethylene-ran-methacrylic acid) blown films. *Journal of Polymer Science Part B: Polymer Physics* 43: 3549–3554.

Berl, V., M. Schmutz, M. J. Krische, R. G. Khoury, and J.-M. Lehn. 2002. Supramolecular polymers generated from heterocomplementary monomers linked through multiple hydrogen-bonding arrays—Formation, characterization, and properties. *Chemistry European Journal* 8 (5): 1227–1244.

Boul, P. J., P. Reutenauer, and J.-M. Lehn. 2005. Reversible Diels–Alder reactions for the generation of dynamic combinatorial libraries. *Organic Letters* 7: 15–18.

Brozoski, B. A., M. M. Coleman, and P. C. Painter. 1984. Local structures in ionomer multiplets. A vibrational spectroscopic analysis. *Macromolecules* 17: 230–234.

Buckler, E. J., G. J. Briggs, J. R. Dunn, E. Lasis, and Y. K. Wei. 1978. Green strength in emulsion SBR. *Rubber Chemistry & Technology* 51: 872–888.

Campbell, D. S. 1974. Thermolabile crosslinks for natural rubber. *Chemistry & industry*. 279–280.

Canadell, J., H. Fischer, G. De With, and R. A. T. M. Van Benthem. 2010. Stereoisomeric effects in thermo-remendable polymer networks based on Diels–Alder crosslink reactions. *Journal of Polymer Science Part A: Polymer Chemistry* 48: 3456–3467.

Canary, S. A. and M. P. Stevens. 1992. Thermally reversible crosslinking of polystyrene via the furan–maleimide Diels–Alder reaction. *Journal of Polymer Science Part A: Polymer Chemistry* 30: 1755–1760.

Capek, I. 2005. Fate of excited probes in micellar systems. *Advances in Colloid and Interface Science* 118: 73–112.

Chen, X., M. A. Dam, K. Ono, A. Mal, H. Shen, S. R. Nutt, K. Sheran, and F. Wudl. 2002. A thermally remendable cross-linked polymeric material. *Science* 295: 1698–1702.

Chen, X. and E. Ruckenstein. 1999. Thermally reversible linking of halide-containing polymers by potassium dicyclopentadienedicarboxylate. *Journal of Polymer Science Part A: Polymer Chemistry* 37: 4390–4401.

Chen, X. and E. Ruckenstein. 2000a. Thermally reversible covalently bonded linear polymers prepared from a dihalide monomer and a salt of di-cyclopentadiene dicarboxylic acid. *Journal of Polymer Science Part A: Polymer Chemistry* 38: 1662–1672.

Chen, X. and E. Ruckenstein. 2000b. Emulsion procedures for thermally reversible covalent crosslinking of polymers. *Journal of Polymer Science Part A: Polymer Chemistry* 38: 4373–4384.

Chen, X., E. Ruckenstein, and R. Pelton. 2005. Long-term stability of an ambient self-curable latex based on colloidal dispersions in water of two reactive polymers. *Journal of Polymer Science Part A: Polymer Chemistry* 43: 2598–2605.

Chen, X., F. Wudl, A. K. Mal, H. Shen, and S. R. Nutt. 2003. New thermally remendable highly cross-linked polymeric materials. *Macromolecules* 36: 1802–1807.

Chikashige, Y., Y. Chikyu, K. Miyatake, and M. Watanabe. 2005. Poly(arylene ether) ionomers containing sulfofluorenyl groups for fuel cell applications. *Macromolecules* 38: 7121–7126.

Chino, K. and M. Ashiura. 2001. Thermoreversible cross-linking rubber using supramolecular hydrogen-bonding networks. *Macromolecules* 34: 9201–9204.

Chino, K., M. Ikawa, and J. Natori. 2001. Thermisch reversible vernetzung von polymeren. German Patent DE10046024, filed 2001.

Choi, C.-K., I. Tomita, and T. Endo. 2000. Synthesis of novel ð-conjugated polymer having an enyne unit by palladium-catalyzed three-component coupling polymerization and subsequent retro-Diels–Alder reaction. *Macromolecules* 33: 1487–1488.

Chujo, Y., K. Sada, and T. Saegusa. 1990. Reversible gelation of polyoxazoline by means of Diels–Alder reaction. *Macromolecules* 23: 2636–2641.

Chung, Y., B. Duerr, and T. McKelvey. 1989. Structural effects controlling the rate of the retro-Diels–Alder reaction in anthracene cycloadducts. *Journal of Organic Chemistry* 54: 1018–1032.

Coleman, M. M., J. Y. Lee, and P. C. Painter. 1990. Acid salts and the structure of monomers. *Macromolecules* 23: 2339–2345.

Craven, J. 1969. Cross-linked thermally reversible polymers produced from condensation polymers with pendant furan groups cross-linked with maleimides. U.S. Patent 3435003, filed 1969.

Cui, W., J. Kerres, and G. Eigenberger. 1998. Development and characterization of ion-exchange polymer blend membranes. *Separation and Purification Technology* 14: 145–154.

Decroix, J. C., J. M. Bouvier, R. Roussel, A. Nicco, and C. M. Bruneau. 1975. Etude de la reticulation thermoreversible de copolymeres porteurs de fonctions anhydride. *Journal of Polymer Science: Polymer Symposia* 52: 299–309.

Diels, O. and K. Alder. 1929. Synthesen in der hydro-aromatischen reihe, 2. mitteiling: Uber cantharidin. *Chemische Berichte* 62: 554–562.

Dreyfus, B. 1985. Properties of partially cured networks. 2. The glass transition. *Macromolecules* 18: 284–292.

Durmaz, H., B. Colakoglu, U. Tunca, and G. Hizal. 2005. Preparation of block copolymers via Diels–Alder reaction of maleimide- and anthracene-end functionalized polymers. *Journal of Polymer Science Part A: Polymer Chemistry* 44: 1667–1675.

Dutta, D., R. A. Weiss, and J. He. 1996. Compatibilization of blends containing thermotropic liquid crystalline polymers with sulfonate ionomers. *Polymer* 37: 429–435.

Eisenbach, C. D., J. Hofmann, and W. J. Mc Knight. 1994. Dynamic mechanical and spectroscopic study of ionomer blends based on carboxylated or sulfonated flexible polystyrene and rigid poly(diacetylenes) with functional side groups. *Macromolecules* 27: 3162–3165.

Eisenberg, A., B. Hirdt, and R. B. Moore. 1990. A new multiplet-cluster model for the morphology of random ionomers. *Macromolecules* 23: 4098–4107.

Endo, T., T. Suzuki, F. Sanda, and T. Takata. 1996a. A novel approach for the chemical recycling of polymeric materials: The network polymer a bifunctional monomer reversible system. *Macromolecules* 29: 3315–3316.

Endo, T., T. Suzuki, F. Sanda, and T. Takata. 1996b. A novel network polymer a linear polymer reversible system: A new cross-linking system consisting of a reversible cross-linking-depolymerization of a polymer having a spiro orthoester moiety in the side chain. *Macromolecules* 29: 4819–4826.

Engle, L. P. and K. B. Wagener. 1993. A review of thermally controlled covalent bond formation in polymer chemistry. *Journal of Macromolecular Science Reviews Macromolecular Physics* 33 (3): 239–257.

Essafi, W., G. Gebel, and R. Mercier. 2004. Sulfonated polyimide ionomers: A structural study. *Macromolecules* 37: 1431–1440.

Fang, Z., S. Wang, S. Q. Wang, and J. P. Kennedy. 2003. Novel block ionomers. III. Mechanical and rheological properties. *Journal of Applied Polymer Science* 88: 1516–1525.

Farrell, K. V. and B. P. Grady. 2000. Effects of temperature on aggregate local structure in a zinc-neutralized carboxylate ionomer. *Macromolecules* 33: 7122–7126.

Feng, Y., A. Schmidt, and R. A. Weiss. 1996. Compatibilization of polymer blends by complexation. 1. Spectroscopic characterization of ion–amide interactions in ionomer/polyamide blends. *Macromolecules* 29: 3909–3917.

Gacal, B., H. Durmaz, M. A. Tasdelen, G. Hizal, U. Tunca, Y. Yagci, and A. L. Demirel. 2006. Anthracene–maleimide-based Diels–Alder "click chemistry" as a novel route to graft copolymers. *Macromolecules* 39: 5330–5336.

Gandini, A., D. Coelho, and A. J. D. Silvestre. 2008. Reversible click chemistry at the service of macromolecular materials. Part 1: Kinetics of the Diels–Alder reaction applied to furan–maleimide model compounds and linear polymerizations. *European Polymer Journal* 44: 4029–4036.

Gandini, A., A. J. D. Silvestre, and D. Coelho. 2010. Reversible click chemistry at the service of macromolecular materials. 2. Thermoreversible polymers based on the Diels–Alder reaction of an A-B furan/maleimide monomer. *Journal of Polymer Science Part A: Polymer Chemistry* 48: 2053–2056.

Gheneim, R., C. Perez-Berumen, and A. Gandini. 2002. Diels–Alder reactions with novel polymeric dienes and dienophiles: Synthesis of reversibly cross-linked elastomers. *Macromolecules* 35: 7246–7253.

Gogel, A., P. Belik, M. Walter, A. Kraus, E. Harth, M. Wagner, J. Spickermann, and K. Mtillen. 1996. The repetitive Diels–Alder reaction: A new approach to fullerene main-chain polymers. *Tetrahedron* 52 (14): 5007–5014.

Goiti, E., M. B. Huglin, and J. M. Rego. 2001. Some observations on the copolymerization of styrene with furfuryl methacrylate. *Polymer* 42: 10187–10193.

Goiti, E., M. B. Huglin, and J. M. Rego. 2003. Thermal breakdown by the retro Diels–Alder reaction of crosslinking in poly[styrene-*co*-(furfuryl methacrylate)]. *Macromolecular Rapid Communications* 24: 692–696.

Gotsmann, B., U. Duerig, J. Frommer, and C. J. Hawker. 2006. Exploiting chemical switching in a Diels–Alder polymer for nanoscale probe lithography and data storage. *Advanced Functional Materials* 16: 1499–1505.

Goussé, C. and A. Gandini. 1998. Synthesis of 2-furfurylmaleimide and preliminary study of its Diels–Alder polycondensation. *Polymer Bulletin* 40: 389–394.
Goussé, C. and A. Gandini. 1999. Diels–Alder polymerization of difurans with bismaleimides. *Polymer International* 48: 723–731.
Goussé, C., A. Gandini, and P. Hodge. 1998. Application of the Diels–Alder reaction to polymers bearing furan moieties. 2. Diels–Alder and retro-Diels–Alder reactions involving furan rings in some styrene copolymers. *Macromolecules* 31: 314–321.
Grady, B. P. 1999. Effect of coneutralization on internal aggregate structure in ethylene-based ionomers. *Macromolecules* 32: 2983–2988.
Greener, J., J. R. Gillmor, and R. C. Daly. 1993. Melt rheology of a class of polyester ionomers. *Macromolecules* 26: 6416–6424.
Grigoras, M. and G. Colotin. 2001. Copolymerization of a bisanthracene compound with bismaleimides by Diels–Alder cycloaddition. *Polymer International* 50: 1375–1378.
Grigoras, M., M. Sava, G. Colotin, and C. I. Simionescu. 2007. Synthesis and thermal behavior of some anthracene-based copolymers obtained by Diels–Alder cycloaddition reactions. *Journal of Applied Polymer Science* 107: 846–853.
Grossman, R. B. 2003. *The Art of Writing Reasonable Organic Reaction Mechanisms*, 2nd edn. New York: Springer-Verlag.
Guhr, K. I., M. D. Greaves, and V. M. Rotello. 1994. Reversible covalent attachment of C_{60} to a polymer support. *Journal of the American Chemical Society* 116: 5997–5998.
Hall, A. J., P. Hodge, C. S. McGrail, and J. Rickerby. 2000. Synthesis of a series of cyclic oligo(alkylidene isophtalate)s by cyclo-depolymerisation. *Polymer* 41: 1239–1249.
Han, S.-I., B. H. Gu, K. H. Nam, S. J. Im, S. C. Kim, and S. S. Im. 2007. Novel copolyester-based ionomer for a shape-memory biodegradable material. *Polymer* 48: 1830–1834.
Hana, S.-I., S. S. Ima, and D. K. Kimb. 2003. Dynamic mechanical and melt rheological properties of sulfonated poly(butylene succinate) ionomers. *Polymer* 44: 7165–7173.
Hara, M., P. Jar, and J. A. Sauer. 1991. Effect of sample history on ionic aggregate structures of sulphonated polystyrene ionomers. *Polymer* 32 (8): 1380–1383.
Hennink, W. E. and C. F. Van Nostrum. 2002. Novel crosslinking methods to design hydrogels. *Advanced Drug Delivery Reviews* 54: 13–36.
Heyboer, N. 1995. Rotary tool with balancing rings. U.S. Patent 5810527, filed 1995.
Higaki, Y., H. Otsuka, and A. Takahara. 2006. A thermodynamic polymer cross-linking system based on radically exchangeable covalent bonds. *Macromolecules* 39: 2121–2125.
Hirasawa, E., Y. Yamamoto, K. Tadano, and S. Yano. 1989. Formation of ionic crystallites and its effect on the modulus of ethylene ionomers. *Macromolecules* 22: 2776–2780.
Hourston, D. J., G. Williams, R. Satguru, J. D. Padget, and D. Pears. 1998. A structure–property study of IPDI-based polyurethane anionomers. *Journal of Applied Polymer Science* 67: 1437–1448.
Huang, Z., Y. Yu, and Y. Huang. 2002. Ion aggregation in the polysiloxane ionomers bearing pendant quaternary ammonium groups. *Journal of Applied Polymer Science* 83: 3099–3104.
Hurwitz, M. J., E. Park, and D. M. Fenton. 1966. Fulvene-containing unsaturated polyester resins. U.S. Patent 3262990, filed 1966.
Ilhan, F. and V. M. Rotello. 1999. Thermoreversible polymerization. Formation of fullerene–diene oligomers and copolymers. *Journal of Organic Chemistry* 64: 1455–1458.
Imai, Y., H. Itoh, K. Naka, and Y. Chujo. 2000. Thermally reversible in organic–inorganic polymer hybrids utilizing the Diels–Alder reaction. *Macromolecules* 33: 4343–4346.
Inglis, A. J., L. Nebhani, O. Altintas, F. G. Schmidt, and C. Barner-Kowollik. 2010. Rapid bonding/debonding on demand: Reversibly cross-linked functional polymers via Diels–Alder chemistry. *Macromolecules* 43: 5515–5520.
Inoue, K., M. Yamashiro, and M. Iji. 2009. Recyclable shape-memory polymer: Poly(lactic acid) crosslinked by a thermoreversible Diels–Alder reaction. *Journal of Applied Polymer Science* 112: 876–885.

Jayakumar, R., S. Nanjundan, and M. Prabaharan. 2005. Developments in metal-containing polyurethanes, co-polyurethanes and polyurethane ionomers. *Journal of Macromolecular Science Part C: Polymer Reviews* 45: 231–261.

Jones, J. R., C. L. Liotta, D. M. Collard, and D. A. Schiraldi. 1999. Cross-linking and modification of poly(ethylene terephthalate-*co*-2,6-anthracenedicarboxylate) by Diels–Alder reactions with maleimides. *Macromolecules* 32: 5786–5792.

Kalista, S. J. and T. C. Ward. 2007. Thermal characteristics of the self-healing response in poly(ethylene-*co*-methacrylic acid) copolymers. *Journal of the Royal Society Interface* 4: 405–411.

Kamahori, K., S. Tada, K. Ito, and S. Itsuno. 1999. Optically active polymer synthesis by Diels–Alder polymerization with chirally modified Lewis acid catalyst. *Macromolecules* 32: 541–547.

Kavitha, A. A. and N. K. Singha. 2007a. A tailor-made polymethacrylate bearing a reactive diene in reversible Diels–Alder reaction. *Journal of Polymer Science Part A: Polymer Chemistry* 45: 4441–4449.

Kavitha, A. A. and N. K. Singha. 2007b. Atom-transfer radical copolymerization of furfuryl methacrylate (FMA) and methyl methacrylate (MMA): A thermally-amendable copolymer. *Macromolecular Chemistry and Physics* 208: 2569–2577.

Kennedy, J. P. and K. F. Castner. 1979a. Thermally reversible polymer systems by cyclopentadienylation. 2. The synthesis of cyclopentadiene-containing polymers. *Journal of Polymer Science: Polymer Chemistry Edition* 17: 2055–2070.

Kennedy, J. P. and K. F. Castner. 1979b. Thermally reversible polymer systems by cyclopentadienylation. 2. The synthesis of cyclopentadiene-containing polymers. *Journal of Polymer Science: Polymer Chemistry Edition* 17: 2055–2070.

Kerres, J. A. 2001. Development of ionomers membranes for fuel cells. *Journal of Membrane Science* 185: 3–27.

Kerres, J., W. Cui, R. Disson, and W. Neubrand. 1998. Development and characterization of crosslinked ionomer membranes based upon sulfinated and sulfonated PSU crosslinked PSU blend membranes by disproportionation of sulfinic acid groups. *Journal of Membrane Science* 139: 211–225.

Kim, J.-S., M.-C. Hong, and Y. H. Nah. 2002. Effects of two ionic groups in an ionic repeat unit on the properties of styrene ionomers. *Macromolecules* 35: 155–160.

Kim, C. K., B. K. Kim, and H. M. Jeong. 1991. Aqueous dispersion of polyurethane ionomers from hexamethylene diisocyanate and trimellitic anhydride. *Colloid Polymer Science* 269: 895–900.

Kim, T.-D., J. Luo, J.-W. Ka, S. Hau, Y. Tian, Z. Shi, N. M. Tucker, S. H. Jang, J. W. Kang, and A. K. Y. Jen. 2006. Ultralarge and thermally stable electro-optic activities from Diels–Alder crosslinkable polymers containing binary chromophore systems. *Advanced Materials* 18: 3038–3042.

Kim, J.-S., Y. H. Nah, S.-S. Jarng, W. Kim, Y. Lee, and Y.-W. Kim. 2000. Clustering in poly(methyl acrylate) ionomers. *Polymer* 41: 3099–3102.

Kim, J.-S., K. Yoshikawa, and A. Eisenberg. 1994. Molecular weight dependence of the viscoelastic properties of polystyrene-based ionomers. *Macromolecules* 27: 6347–6357.

Kima, B. K., S. Y. Leea, J. S. Leea, S. H. Baeka, Y. J. ChoP, J. O. Lee, and M. Xu. 1998. Polyurethane ionomers having shape memory effects. *Polymer* 39 (13): 2803–2808.

Kirkmeyer, B. P., R. A. Weiss, and K. I. Winey. 2001. Spherical and vesicular ionic aggregates in Zn-neutralized sulfonated polystyrene ionomers. *Journal of Polymer Science Part B: Polymer Physics* 39: 477–483.

Klok, H.-A., E. A. Rebrov, A. M. Muzafarov, W. Michelberger, and M. Moller. 1999. Reversible gelation of poly(dimethylsiloxane) with ionic and hydrogen-bonding substituents. *Journal of Polymer Science Part B: Polymer Physics* 37: 485–495.

Kose, M. M., G. Yesilbag, and A. Sanyal. 2008. Segment block dendrimers via Diels–Alder cycloaddition. *Organic Letters* 10 (12): 2353–2356.

Kuramoto, N., K. Hayashi, and K. Nagai. 1994. Thermoreversible reaction of Diels–Alder polymer composed of difurufuryladipate with bismaleimidodiphenylmethane. *Journal of Polymer Science Part A: Polymer Chemistry* 32: 2501–2504.

Kutsumizu, S., M. Goto, S. Yano, and S. Schlick. 2002. Structure and dynamics of ionic aggregates in ethylene ionomers and their effect on polymer dynamics: A study by small-angle X-ray scattering and electron spin resonance spectroscopy. *Macromolecules* 35: 6298–6305.

Kutsumizu, S., K. Tadano, Y. Matsuda, M. Goto, H. Tachino, H. Hara, E. Hirasawa, H. Tagawa, Y. Muroga, and S. Yano. 2000. Investigation of microphase separation and thermal properties of noncrystalline ethylene ionomers. 2. IR, DSC, and dielectric characterization. *Macromolecules* 33: 9044–9053.

Kwart, H. and K. King. 1968. The reverse Diels–Alder or retrodiene reaction. *Chemical Reviews* 68: 415–447.

Ladd, E. C. 1967. Dithiobis (N-phenylmaleimides). U.S. Patent 3297713, filed 1967.

Lee, D.-C., R. A. Register, C.-Z. Yang, and S. L. Cooper. 1988. MDI-based polyurethane ionomers. 1. New small-angle x-ray scattering model. *Macromolecules* 21: 998–1004.

Lei, H. and Y. Zhao. 1993. An easy-way of preparing side-chain liquid crystalline monomers. *Polymer Bulletin* 31: 645–649.

Leir, C. M. and J. E. Stark. 1989. Ionene elastomers from polytetramethylene oxide diamines and reactive dihalides. I. Effect of dihalide structure on polymerization and thermal reversibility. *Journal of Applied Polymer Science* 38: 1535–1547.

Li, J., J. A. Viveros, M. H. Wrue, and M. Anthamatten. 2007. Shape-memory effects in polymer networks containing reversibly associating side-groups. *Advanced Materials* 19: 2851–2855.

Liu, Y.-L. and Y.-W. Chen. 2007. Thermally reversible cross-linked polyamides with high toughness and self-repairing ability from maleimide- and furan-functionalized aromatic polyamides. *Macromolecular Chemistry and Physics* 208: 224–232.

Liu, Y.-L. and C.-Y. Hsieh. 2006. Crosslinked epoxy materials exhibiting thermal remendability and removability from multifunctional maleimide and furan compounds. *Journal of Polymer Science Part A: Polymer Chemistry* 44: 905–913.

Liu, Y.-L., C.-Y. Hsieh, and Y.-W. Chen. 2006. Thermally reversible cross-linked polyamides and thermo-responsive gels by means of Diels–Alder reaction. *Polymer* 47: 2581–2586.

Liu, J., E. N. Kadnikova, Y. Liu, M. D. McGehee, and J. M. J. Fréchet. 2004. Polythiophene containing thermally removable solubilizing groups enhances the interface and the performance of polymer–titania hybrid solar cells. *Journal of the American Chemical Society* 126 (31): 9486–9487.

Liu, C., H. Qin, and P. T. Mather. 2007. Review of progress in shape-memory polymers. *Journal of Materials Chemistry* 17: 1543–1558.

Loontjens, T., J. Put, B. Coussens, R. Lange, J. Palmen, T. Sleijpen, and B. Plum. 2001. Novel supramolecular polymer networks based on melamine- and imide-containing oligomers. *Macromolecular Symposia* 174: 357–371.

Lu, X., W. P. Steckle, and R. A. Weiss. 1993. Ionic aggregation in a block copolymer ionomer. *Macromolecules* 26: 5876–5884.

Lu, X. and R. A. Weiss. 1996. Development of miscible blends of bisphenol A polycarbonate and lightly sulfonated polystyrene ionomers from intrapolymer repulsive interactions. *Macromolecules* 29: 1216–1221.

Luo, J., M. Haller, H. Li, T. Kim, and A. Jen. 2003. Highly efficient and thermally stable electro-optic polymer from a smartly controlled crosslinking process. *Advanced Materials* 15: 1635–1638.

Magana, A., S. Zerroukhi, C. Jegat, and N. Mignard. 2010. Thermally reversible cross-linked polyethylene using Diels–Alder reaction in molten state. *Reactive & Functional Polymers* 70: 442–448.

McElhanon, J. R., E. M. Russick, D. R. Wheeler, D. A. Loy, and J. H. Aubert. 2002. Removable foams based on an epoxy resin incorporating reversible Diels–Alder adducts. *Journal of Applied Polymer Science* 85: 1496–1502.

McElhanon, J. R. and D. R. Wheeler. 2001. Thermally responsive dendrons and dendrimers based on reversible furan–maleimide Diels–Alder adducts. *Organic Letters* 3 (17): 2681–2683.

Meresi, G., Y. Wang, A. Bandis, P. T. Inglefield, A. A. Jones, and W.-Y. Wen. 2001. Morphology of dry and swollen perfluorosulfonate ionomer by fluorine-19 MAS, NMR and xenon-129 NMR. *Polymer* 42: 6153–6160.

Mikroyannidis, J. A. 1992. Synthesis and Diels–Alder polymerization of furfurylidene and furfuryl-substituted maleamic acids. *Journal of Polymer Science Part A: Polymer Chemistry* 30: 125–132.

Miyagawa, T., M. Shimizu, F. Sanda, and T. Endo. 2005. Six-membered cyclic carbonate having styrene moiety as a chemically recyclable monomer. Construction of novel cross-linking–de-cross-linking system of network polymers. *Macromolecules* 38: 7944–7949.

Montarnal, D., F. O. Tournilhac, M. Hildalgo, and L. Leibler. 2010. Epoxy-based networks combining chemical and supramolecular hydrogen-bonding crosslinks. *Journal of Polymer Science Part A: Polymer Chemistry* 48: 1133–1141.

Moore, E. R. 1968. Process for removal of cross-linking agent from a thermoset-thermoplastic. U.S. Patent 3408337, filed 1968.

Moore, R. B. and C. R. Martin. 1988. Chemical and morphological properties of solution-cast. Perfluorosulfonate ionomers. *Macromolecules* 21: 1334–1339.

Moore, R. B. and C. R. Martin. 1989. Morphology and chemical properties of the Dow perfluorosulfonate ionomers. *Macromolecules* 22: 3594–3599.

Müller, M., U. Seidel, and R. Stadler. 1995. Influence of hydrogen bonding on the viscoelastic properties of thermoreversible networks: Analysis of the local complex dynamics. *Polymer* 36 (16): 3143–3150.

Murphy, E. B., E. Bolanos, C. Schaffner-Hamann, F. Wudl, S. R. Nutt, and M. L. Auad. 2008. Synthesis and characterization of a single-component thermally remendable polymer network: Staudinger and Stille revisited. *Macromolecules* 41: 5203–5209.

Muskat, I. E. 1963. Fabrication of shaped articles cross-linked by chemical addition reactions. U.S. Patent 3085986, 1963.

Nagao, I., M. Hattori, E. Hirasawa, and K. Tadanot. 1992. Dielectric relaxations of ethylene ionomers. *Macromolecules* 25: 368–376.

Nair, K. P., V. Breedveld, and M. Weck. 2008. Complementary hydrogen-bonded thermoreversible polymer networks with tunable properties. *Macromolecules* 41: 3429–3438.

Nakano, M., M. Mouri, A. Usuki, and A. Okada. 1997. Recyclable cross-linked polymer, method for producing a molded article, and method for recycling the same. U.S. Patent 5654368, filed 1997.

Nguyen, D., J.-S. Kim, M. D. Guiver, and A. Eisenberg. 1999. Clustering in carboxylated polysulfone ionomers: A characterization by dynamic mechanical and small-angle X-ray scattering methods. *Journal of Polymer Science Part B: Polymer Physics* 37: 3226–3232.

Nic, M., J. Jirat, and B. Kosata. 1997. *IUPAC. Compendium of Chemical Terminology* (*the "Gold Book"*), 2nd edn. Oxford, U.K.: Blackwell Scientific Publications.

Nie, B., K. Hasan, M. D. Greaves, and V. M. Rotello. 1995. Reversible covalent attachment of C_{60} to a furan-functionalized resin. *Tetrahedron Letters* 36 (21): 3617–3618.

Nowak, R. M., K. J. Guilette, and E. R. Moore. 1976. Reversible cross-linking of maleic anhydride copolymers. U.S. Patent 3933747, filed 1976.

Ogawa, K., A. Nakayama, and E. Kokufuta. 2003. Preparation and characterization of thermosensitive polyampholyte nanogels. *Langmuir* 19: 3178–3184.

Orler, E. B., B. H. Calhoun, and R. B. Moore. 1996. Crystallization kinetics as a probe of the dynamic network in lightly sulfonated syndiotactic polystyrene ionomers. *Macromolecules* 29: 5965–5971.

Osborn, S. J., M. K. Hassan, G. M. Divoux, D. W. Rhoades, K. A. Mauritz, and R. B. Moore. 2007. Glass transition temperature of perfluorosulfonic acid ionomers. *Macromolecules* 40: 3886–3890.

Page, K. A., K. M. Cable, and R. B. Moore. 2005. Molecular origins of the thermal transitions and dynamic mechanical relaxations in perfluorosulfonate ionomers. *Macromolecules* 38: 6472–6484.

Palmese, G. R. and A. M. Peterson. 2008. *Remendable Polymeric Materials Using Reversible Covalent Bonds* Paper presented at *26th Army Science Conference*, Orlando, FL.

Park, J. S., H. S. Kim, and H. T. Hahn. 2009. Healing behavior of a matrix crack on a carbon fiber/mendomer composite. *Composites Science and Technology* 69: 1082–1087.

Park, J. S., K. Takahashi, Z. Guo, Y. Wang, E. Bolanos, C. Hamann-Schaffner, E. Murphy, and H. T. Hahn. 2008. Towards development of a self-healing composite using a mendable polymer and resistive heating. *Journal of Composite Materials* 42: 2869–2881.

Paulöhrl, T., A. J. Inglis, and C. Barner-Kowollik. 2010. Reversible Diels–Alder chemistry as a modular polymeric color switch. *Advanced Materials* 22: 2788–2791.

Peng, C. C. and V. Abetz. 2005. A simple pathway toward quantitative modification of polybutadiene: A new approach to thermoreversible cross-linking rubber comprising supramolecular hydrogen-bonding networks. *Macromolecules* 38: 5575–5580.

Peng, C.-C., A. Gopfert, M. Drechsler, and V. Abetz. 2005. "Smart" silica–rubber nanocomposites in virtue of hydrogen bonding interaction. *Polymer for Advanced Technologies* 16: 770–782.

Peterson, A. M., R. E. Jensen, and G. R. Palmese. 2010. Room-temperature healing of a thermosetting polymer using the Diels–Alder reaction. *Applied Materials & Interfaces* 2 (4): 1141–1149.

Peterson, A. M. and G. R. Palmese, eds. 2009. *Click Chemistry for Biotechnology and Materials Science*. Chichester, Weat Sussex: Wiley.

Plaisted, T. A. and S. Nemat-Nasser. 2007. Quantitative evaluation of fracture, healing and re-healing of a reversibly cross-linked polymer. *Acta Materialia* 55: 5684–5696.

Polaske, N. W., D. V. McGrath, and J. R. McElhanon. 2010. Thermally reversible dendronized step-polymers based on sequential Huisgen 1,3-dipolar cycloaddition and Diels–Alder "click" reactions. *Macromolecules* 43: 1270–1276.

Qiao, R., A. P. Roberts, A. S. Mount, S. J. Klaine, and P. C. Ke. 2010. Translocation of C60 and its derivatives across a lipid bilayer. *Nano Letters* 7 (3): 614–619.

Quiram, D. J. and R. A. Register. 1998. Crystallization and ionic associations in semicrystalline ionomers. *Macromolecules* 31: 1432–1435.

Reeder, J. A. 1967. Polyimides from dimaleimides and bisfulvenes. U.S. Patent 3334071, filed 1967.

Rembaum, A., S. Singer, and H. Keyzer. 1969. Ionene polymers. III. Dicationic crosslinking agents. *Journal of Polymer Science Part B: Polymer Letters* 7: 395–402.

Reutenauer, P., E. Buhler, P. Boul, S. Candau, and J.-M. Lehn. 2009. Room temperature dynamic polymers based on Diels–Alder chemistry. *Chemistry European Journal* 15: 1893–1900.

Rubatat, L., A. L. Rollet, G. Gebel, and O. Diat. 2002. Evidence of elongated polymeric aggregates in nafion. *Macromolecules* 35: 4050–4055.

Ruckenstein, E. and X. Chen. 2000a. Crosslinking of chlorine-containing polymers by dicyclopentadiene dicarboxylic salts. *Journal of Polymer Science Part A: Polymer Chemistry* 38: 818–825.

Ruckenstein, E. and X. Chen. 2000b. Covalent cross-linking of polymers through ionene formation and their thermal de-cross-linking. *Macromolecules* 33: 8992–9001.

Ruckenstein, E. and X. Chen. 2001. An ambient self-curable latex based on colloidal dispersions in water of two functionalized polymers and the thermally reversible crosslinked films generated. *Journal of Polymer Science Part A: Polymer Chemistry* 39: 389–397.

Sauer, B. B. and R. S. McLean. 2000. AFM and X-ray studies of crystal and ionic domain morphology in poly(ethylene-*co*-methacrylic acid) ionomers. *Macromolecules* 33: 7939–7949.
Saxegaard, H. 2003. Crack self-healing properties of asphalt concrete: Laboratory simulation. *The International Journal on Hydropower & Dams* 10 (3): 106–109.
Shah, R. K. and D. R. Paul. 2006. Comparison of nanocomposites prepared from sodium, zinc, and lithium ionomers of ethylene/methacrylic acid copolymers. *Macromolecules* 39: 3327–3336.
Stevens, M. P. and A. D. Jenkins. 1979. Crosslinking of polystyrene via pendant maleimide groups. *Journal of Polymer Science: Polymer Chemistry Edition* 17: 3675–3685.
Stille, J. and L. Plummer. 1961. Polymerization by the Diels–Alder reaction. *The Journal of Organic Chemistry* 26: 4026–4029.
Sun, C. X., M. A. J. van der Mee, J. G. P. Goossens, and M. van Duin. 2006. Thermoreversible cross-linking of maleated ethylene/propylene copolymers using hydrogen-bonding and ionic interactions. *Macromolecules* 39: 3441–3449.
Szalai, M., D. McGrath, D. Wheeler, T. Zifer, and J. R. McElhanon. 2007. Dendrimers based on thermally reversible furan–maleimide Diels–Alder adducts. *Macromolecules* 40: 818–823.
Tachino, H., H. Hara, E. Hirasawa, S. Kutsumizu, K. Tadano, and S. Yano. 1993. Dynamic mechanical relaxations of ethylene ionomers. *Macromolecules* 26: 752–757.
Takahashi, T., J. Watanabe, K. Minagawa, and K. Koyama. 1994. Effect of ionic interaction on elongational viscosity of ethylene based ionomer melts. *Polymer* 35 (26): 5722–5728.
Teramoto, N., Y. Arai, and M. Shibata. 2006. Thermo-reversible Diels–Alder polymerization of difurfurylidene trehalose and bismaleimides. *Carbohydrate Polymers* 64: 78–84.
Thibault, R. J., P. J. Hotchkiss, M. Gray, and V. M. Rotello. 2003. Thermally reversible formation of microspheres through non-covalent polymer cross-linking. *Journal of the American Chemical Society* 125: 11249–11252.
Tian, Q., Y. C. Yuan, M. Z. Rong, and M. Q. Zhang. 2009. A thermally remendable epoxy resin. *Journal of Material Chemistry* 19: 1289–1296.
Tonga, M., N. Cengiz, M. M. Kose, T. Dede, and A. Sanyal. 2010. Dendronized polymers via Diels–Alder "click" reaction. *Journal of Polymer Science Part A: Polymer Chemistry* 48: 410–416.
Tsujita, Y., M. Yasuda, M. Takei, T. Kinoshita, A. Takizawa, and H. Yoshimizu. 2001. Structure of ionic aggregates of ionomers. 1. Variation in the structure of ionic aggregates with different acid content and degree of neutralization of ethylene and styrene ionomers. *Macromolecules* 34: 2220–2224.
Vanhoorne, P., R. Jerome, P. Teyssie, and F. Laupretre. 1994. Direct NMR evidence for a local restriction of chain segment mobility in a model ionomer. *Macromolecules* 27: 2548–2552.
Vanhoorne, P. and R. A. Register. 1996. Low-shear melt rheology of partially-neutralized ethylene–methacrylic acid ionomers. *Macromolecules* 29: 598–604.
Varleya, R. J. and S. van der Zwaag. 2010. Autonomous damage initiated healing in a thermo-responsive ionomer. *Polymer International* 59: 1031–1038.
Venkateshwaran, L. N., G. A. Yorkt, C. D. DePorter, J. E. McGrath, and G. L. Wilkes. 1992. Morphological characterization of well defined methacrylic based di- and triblock ionomers. *Polymer* 33 (11): 2277–2286.
Visser, S. A. and S. L. Cooper. 1991a. Analysis of small-angle x-ray scattering data for model polyurethane ionomers: Evaluation of hard-sphere models. *Macromolecules* 24: 2584–2593.
Visser, S. A. and S. L. Cooper. 1991b. Comparison of the physical properties of carboxylated and sulfonated model polyurethane ionomers. *Macromolecules* 24: 2576–2583.
Visser, S. A. and S. L. Cooper. 1992. Effect of neutralizing cation type on the morphology and properties of model polyurethane ionomers. *Polymer* 33 (5): 920–929.
Wang, W., T.-T. Chan, A. J. Perkowski, S. Schlick, and K. I. Winey. 2009. Local structure and composition of the ionic aggregates in Cu(II)-neutralized poly(styrene-*co*-methacrylic acid) ionomers depend on acid content and neutralization level. *Polymer* 50: 1281–1287.

Watanabe, M. and N. Yoshie. 2006. Synthesis and properties of readily recyclable polymers from bisfuranic terminated poly(ethylene adipate) and multi-maleimide linkers. *Polymer* 47: 4946–4952.

Weiss, R. A., A. Sen, L. A. Pottick, and C. L. Willis. 1991b. Block copolymer ionomers: 2. Viscoelastic and mechanical properties of sulphonated poly(styrene–ethylene/butylene–styrene). *Polymer* 32 (15): 2785–2792.

Weiss, R. A., A. Sen, C. L. Willis, and L. A. Pottick. 1991a. Block copolymer ionomers: 1. Synthesis and physical properties of sulphonated poly(styrene–ethylene/butylene–styrene). *Polymer* 32 (10): 1867–1874.

Wilson, M. S., J. A. Valerio, and S. Gottesfeld. 1995. Low platinum loading electrodes for polymer electrolyte fuel cells fabricated using thermoplastic ionomers. *Electrochimica Acta* 40 (3): 355–363.

Winey, K. I., J. H. Laurer, and B. P. Kirkmeyer. 2000. Ionic aggregates in partially Zn-neutralized poly(ethylene-ran-methacrylic acid) ionomers: Shape, size, and size distribution. *Macromolecules* 33: 507–513.

Withworth, C. J. and N. L. Zutty. 1967. Method for processing fusible crosslinked ethylene/dicarboxylic anhydride copolymers. U.S. Patent 3299184, filed 1967.

Wouters, M., E. Craenmehr, K. Tempelaars, H. Fischer, N. Stroeks, and J. van Zanten. 2009. Preparation and properties of a novel remendable coating concept. *Progress in Organic Coatings* 64: 156–162.

Xue, W., M. B. Huglin, and T. G. J. Jones. 2005. Swelling and network parameters of crosslinked thermoreversible hydrogels of poly(*N*-ethylacrylamide). *European Polymer Journal* 41: 239–248.

Yarusso, D. J. and S. L. Cooper. 1983. Microstructure of ionomers: Interpretation of small-angle x-ray scattering data. *Macromolecules* 16: 1871–1880.

Yarusso, D. J. and S. L. Cooper. 1985. Analysis of SAXS data from ionomer systems. *Polymer* 26: 371–378.

Yoshie, N., M. Watanabe, H. Araki, and K. Ishida. 2010. Thermo-responsive mending of polymers crosslinked by thermally reversible covalent bond: Polymers from bisfuranic terminated poly(ethylene adipate) and tris-maleimide. *Polymer Degradation and Stability* 95: 826–829.

Zhang, Y., A. A. Broekhuis, and F. Picchioni. 2009. Thermally self-healing polymeric materials: The next step to recycling thermoset polymers? *Macromolecules* 42: 1906–1912.

Zhang, G., L. Liu, H. Wang, and M. Jiang. 2000. Preparation and association behavior of diblock copolymer ionomers based on poly(styrene-*b*-ethylene-*co*-propylene). *European Polymer Journal* 36: 61–68.

Zhu, Y., J. Hu, K.-F. Choi, Q. Meng, S. Chen, and K.-W. Yeung. 2008. Shape memory effect and reversible phase crystallization process in SMPU ionomer. *Polymer for Advanced Technologies* 19: 328–333.

Zhu, Y., J. Hu, K.-W. Yeung, K.-F. Choi, Y. Liu, and H. Liem. 2007. Effect of cationic group content on shape memory effect in segmented polyurethane cationomer. *Journal of Applied Polymer Science* 103: 545–556.

Zhu, Y., J.-L. Hu, K.-W. Yeung, H.-J. Fan, and Y.-Q. Liu. 2006b. Shape memory effect of PU ionomers with ionic groups on hard-segments. *Chinese Journal of Polymer Science* 24 (2): 173–186.

Zhu, Y., J. L. Hu, K. W. Yeung, Y. Q. Liu, and H. M. Liem. 2006a. Influence of ionic groups on the crystallization and melting behavior of segmented polyurethane ionomers. *Journal of Applied Polymer Science* 100: 4603–4613.

Zimmermann, R. L., K. S. Dennis, and E. R. Moore. 1972. Fabrication of shaped articles cross-linked by chemical addition reaction. U.S. Patent 3678016, filed 1972.

10 Self-Repairing by Damage-Triggered Smart Containers

Dmitry G. Shchukin and Dmitry O. Grigoriev

CONTENTS

10.1 INTRODUCTION

The material losses and performance failure costs connected with corrosion are estimated to be 3%–4% of GDP (Koch et al., 2002; McCafferty, 2010). Preventing and slowing corrosion is, therefore, one of the most challenging problems facing modern industrial society that includes various scientific, technical, and engineering issues.

Broad spectrum of approaches for controlling corrosion is employed today depending on the specific industry branch and corresponding material under protection. The most frequently used method is the application of different organic, ceramic, and metallic protective coatings: the use of corrosion-resistant materials such as alloys (Institute of Materials (Great Britain), 2002; Ahluwalia and Uhlenkamp, 2008), polymers (Pacitti, 1964; Pritchard, 1995; Schweitzer, 2000), and composites (Bogner, 2005; McConnell, 2005); utilization of diverse corrosion inhibitors (Raja and Sethuraman, 2008; Zhao et al., 2010); and cathodic (von Baeckmann et al., 1997) or anodic (Roberge, 2000) protection. Each of these approaches in the corrosion protection can be further subdivided into the manifold of special solutions determined by particular requirements for protection efficiency, conditions at which protective performance has to be achieved, and many other factors.

Until recently, the protective coatings for metal substrates possessed conversion layer fabricated using compounds of hexavalent chromium (Osborne et al., 2001; Kendig and Buchheit, 2003). Due to strong oxidative effect of chromate anion on the metal under protection, the tight solid layer of $Cr_2O_3 \cdot nH_2O$ is formed on the substrate surface (Xia and McCreery, 1998; Sinko, 2001). On the other hand, strong oxidizing ability of chromates is responsible also for their extremely high environmental toxicity and can cause various cruel diseases such as cancer or mutagenic damages (Twite and Bierwagen, 1998). For these reasons, chromates have been banned from being used in many corrosion protection applications since early twenty-first century (Directive 2000/53/EC).

Currently, almost all the main industries have stopped using chromates for corrosion protection and search intensively for more environmental-friendly substitutions. Various solutions are proposed. Compounds of several transition metals in high oxidation state are used for chemical conversion coatings—titanium (Tsai et al., 2010), vanadium (Zou et al., 2011), zirconium and niobium (Ardelean et al., 2008), manganese (Hughes et al., 2006; Zhao et al., 2006), cobalt (Hughes et al., 2004), molybdenum (Magalhaes et al., 2004), tungsten and silicon (Li et al., 2007), cerium (Kobayashi and Fujiwara, 2006; Lin and Li, 2006; Hosseini et al., 2007; O'Keefe et al., 2007; Ardelean et al., 2008), and other rare earth metals (Yang et al., 2009; Kong et al., 2010). Electrochemical chromate-free oxidation treatment (anodizing) (Knudsen et al., 2004; Niu et al., 2006; Alanazi et al., 2010), diverse organic–inorganic pretreatments (Hansal et al., 2006; Liu et al., 2006), application of organic inhibitors directly on the metal substrate surface (Dufek and Buttry, 2008; Hernandez-Alvarado et al., 2009) present other numerous attempts to find appropriate alternative to chromates.

10.2 APPROACHES FOR THE DEVELOPMENT OF FUNCTIONAL ACTIVE COATINGS

Development of a new generation of self-repairing coatings and bulk materials, which have both passive mechanical characteristics originated from matrix material and active response sensitive to the changes in local environment or to the integrity of the passive matrix, opens an avenue for the fabrication of future high-tech functional surfaces. Novel feedback active surfaces can be composed of a passive matrix inherited from a "classical" approach for coatings and active structures for fast response of the coating properties to the outer environmental impacts. Active corrosion protection aims at a decrease of the corrosion rate when the main barrier (main coating matrix) is damaged and corrosive species come in contact with the substrate (Brooman, 2002).

The active agents (e.g., corrosion inhibitors) can be introduced in different components of the coating: pretreatment, primer, and topcoat. The agents are effective only if their solubility in the defect environment is in the right range. Very low solubility leads to lack of active agent at the substrate interface and consequently to weak feedback activity. If the solubility is too high the substrate will be protected, but for only a relatively short time since the active agent will be rapidly leached out from the coating. Another drawback, which can appear due to high solubility, is the osmotic pressure that leads to blistering and delamination of active surface. The osmotic pressure can stimulate water to be transported through the coating, which acts as a semipermeable membrane, causing the destruction of the passive matrix.

Recent developments in surface science and technology offer new opportunities for modern engineering concepts for fabrication of active surfaces of the "passive"

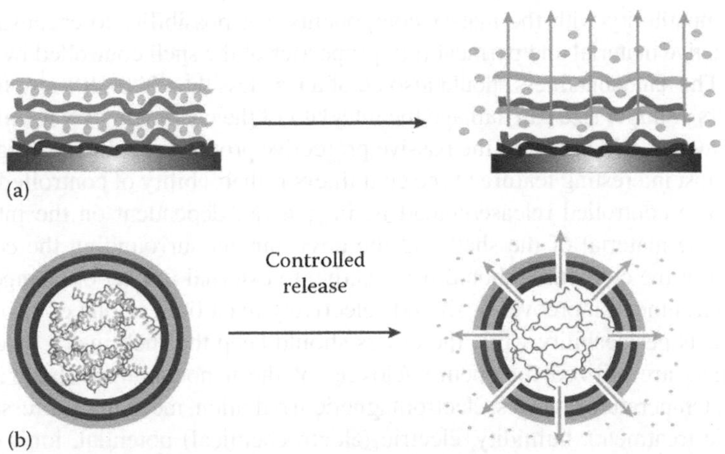

FIGURE 10.1 Schematic illustration of the entrapment/release of active materials. (a) Active material is embedded in the "passive" matrix of the coating and (b) active material is encapsulated into nanocontainers with a shell possessing controlled permeability properties. (From Shchukin, D.G. and Möhwald, H.: Self-repairing coatings containing active nanoreservoirs. *Small*. 2007. 3. 926–943. Copyright Wiley-VCH Verlag GmbH & Co. KGaA. With permission.)

host–"active" guest structure through either the fabrication of active composite layered systems in which the passive coating matrix is alternated with layers bearing an active coating component (e.g., inhibitor, lubricant) or the integration of the nanoscale containers (carriers) loaded with the active components into existing "classical" coatings (Figure 10.1a and b) (Shchukin and Möhwald, 2007). The first approach (Figure 10.1a) involves the doping passive matrix with active components to be released upon environmental triggering (some examples of this approach are described in the following) (Kasten et al., 2001; Osborne et al., 2001; Garcia-Heras et al., 2004; Sheffer et al., 2004). However, interaction of the active materials with the passive matrix leads to significant shortcomings in the stability and self-repairing activity of the coating. Moreover, free inhibitor inside the active matrix is often subjected to the spontaneous leakage from the surface during aging. The rate of release of the active load is dependent on the structure, chemical, and acidic/alkaline properties of the passive matrix (Vreugdenhil and Woods, 2005).

The main idea of the second approach is to load active compounds into nanocontainers with a shell possessing controlled permeability (Figure 10.1b) and then to introduce them into the coating matrix. As a result, nanocontainers are uniformly distributed in the passive matrix keeping active material in "trapped" state, thus avoiding the undesirable interaction between active component and passive matrix and spontaneous leakage. When the local environment undergoes changes or if the active surface is affected by the outer impact, the nanocontainers respond to this signal and release encapsulated active material.

The containers can have inorganic, organic, or composite origin and be applied for encapsulation of drugs, vitamins, quantum dots, bactericides, oils, biopolymers, corrosion inhibitors, low-molecular-weight compounds, and many other active molecules. The most important task for the second approach is to develop nanocontainers with good compatibility with the matrix components, the possibility to encapsulate and upkeep active material and permeability properties of the shell controlled by external stimuli. The nanocontainers should also be of a size less than 300–400 nm; the nanocontainers of larger size can damage the integrity of the coating matrix forming large hollow cavities, which reduce the passive protective properties of the coating.

The most interesting feature of the containers is their ability of controlled release/upload. The controlled release/upload is, in general, dependent on the interaction between the material of the shell and the environment surrounding the container. The shell of the containers should be sensitive to external impact or changes in the local environment; moreover, it should selectively react only to one or two triggers changing its permeability while the others should keep the shell intact. The following triggers are utilized for opening/closing of the nanocontainer shell: local pH changes, temperature changes, electromagnetic irradiation, mechanical pressure (also ultrasonic treatment), humidity, electric (electrochemical) potential, ionic strength (e.g., with concentration of NaCl above 0.5 M), and dielectric permeability of the solvent. The simplest trigger for opening/closing of the container shell is the pH shift in the local environment. Polyelectrolyte capsules, hydrogels, and emulsions with weak acidic or basic functional groups in the shell are sensitive to it demonstrating reversible and (or) irreversible changes of the shell permeability in a wide pH range (e.g., at low $pH < 4$ or high $pH > 9$) (Shchukin and Sukhorukov, 2004). The other important triggers are external electromagnetic irradiation and mechanical impact. For opening by electromagnetic irradiation, the container shell should have sensitive components such as metal (silver) nanoparticles for infrared light (Javier et al., 2008), dyes for visible light (Tao et al., 2004), and semiconductors (TiO_2 nanoparticles) for UV light (Shchukin and Sviridov, 2006). Mechanical impact requires certain level of rigidity or brittleness of the shell because the elastic shell can undergo deformation under pressure but not rupture (Fery and Weinkamer, 2007); containers with diameter <100 nm are hardly destroyed with reasonable mechanical force because they tend to escape from the force direction. On the contrary, ultrasonic treatment can be applied to irreversibly open containers of any size (Shchukin et al., 2006a). Other triggers for capsule opening were demonstrated presumably for hydrogels (temperature, electric potential, high ionic strength) and polyelectrolyte capsules (temperature, high ionic strength, dielectric permeability of the solvent) (Antipov and Sukhorukov, 2004; Koehler et al., 2005). These triggers are usually involved in feedback active systems of specific functionality (e.g., coatings with electrochemically reversible permeability) (Shchukin et al., 2006b).

10.3 NANOCONTAINERS BASED ON ION-EXCHANGE EFFECT

Hydrophobic layers (Mitchon and White, 2006) of surfactant molecules on the surface have recently been proposed as corrosion inhibitors but suffer from the drawbacks that the layers have limited stability and molecule-sized defects, which allow water to reach the underlying surface. These problems should be mitigated if the surfactant could be incorporated in an inorganic host matrix, a thin film of which has been previously strongly bonded to metal, for example, aluminum, surface. Layered double hydroxides (LDHs) are one such potential inorganic host. They can be expressed by the general formula $[M^{2+}_{1-x}M^{3+}_{x}(OH)_2]A^{n-}_{x/n} \cdot mH_2O$, where the cations M^{2+}(Mg^{2+}, Zn^{2+}, Fe^{2+}, Co^{2+}, Cu^{2+}, and others) and M^{3+} (Al^{3+}, Cr^{3+}, Fe^{3+}, Ga^{3+}, and others) occupy the octahedral holes in a brucite-like layer and the anion An^- is located in the hydrated interlayer galleries (Williams and O'Hare, 2006). The ability to vary the composition over a wide range allows materials with a wide variety of properties to be prepared. The LDHs have been applied as catalysts and catalyst supports (Albertazzi et al., 2004), polymer stabilizers (Sorrentino et al., 2005), and traps of anionic pollutants (Palmer et al., 2009). The application of hydrotalcite-like compounds in corrosion science has covered different aspects. In some studies, the LDHs have been produced *in situ*, on the top of metallic substrates as protective films (Buchheit et al., 2002). Hydrotalcite-based conversion films have demonstrated good corrosion protection, and some research groups have been trying to improve the interaction between these conversion films and organic coatings (Leggat et al., 2002). A different perspective is to use these anionic clays as containers for corrosion inhibitors and incorporate them into the organic coatings. In this case, the aims are twofold: not only to release the species that impart active protection but also to trap the corrosive agents (Cl^-, SO_4^{2-}). Regarding this topic, several works (Williams and McMurray, 2004; Mahajanarn and Buchheit, 2008) can be found in the literature. Within the LDH family, a class of materials with emerging importance is that constituted by the LDHs loaded with organic anions (Theng, 1974). Two short studies on these materials are available in the literature. Williams and McMurray prepared LDHs with different organic species (benzotriazole, ethyl xanthate, and oxalate) by rehydration of commercial hydroltalcite $[Mg_6Al_2(OH)_{16} \cdot CO_3 \cdot 4H_2O]$. The resulting layered systems were inserted into a poly(vinylbutyral) coating, prepared by a bar cast on the top of AA2024-T3. In another report, Kendig and Hon (2005) prepared LDHs intercalated with 2,5-dimercapto-1,3,4-thiadiazolate in a similar way and studied the inhibiting properties of this anion with respect to the oxygen reduction reaction on copper. Another recently developed possibility (Zheludkevich et al., 2010) used Zn–Al and Mg–Al LDHs loaded with quinaldate and 2-mercaptobenzothiazolate in which anions were synthesized via anion-exchange reaction (Figure 10.2). Spectrophotometric measurements demonstrated that the release of organic anions from these LDHs into the bulk solution is triggered by the presence of chloride anions, evidencing the anion-exchange nature of this process. A significant reduction of the corrosion rate was observed when the LDH nanopigments are present in the corrosive media. The mechanism by which the inhibiting anions can be released from the LDHs underlines the versatility of these environmental-friendly structures and their potential application as nanocontainers in self-healing coatings.

10.4 NANOCONTAINERS WITH POLYMER SHELL

Self-organizing block copolymers and lipid nanocontainers allow the possibility of entrapping hydrophobic drugs in the core, while the outer shell confers water solubility. The self-assembly of amphiphilic block copolymers in water is not only limited to the formation of nanocontainers. The ratio between the blocks, the concentration of the polymer, and the ionic strength of the medium influences the aggregation of block copolymers into vesicles, lamellae, rods, and other related

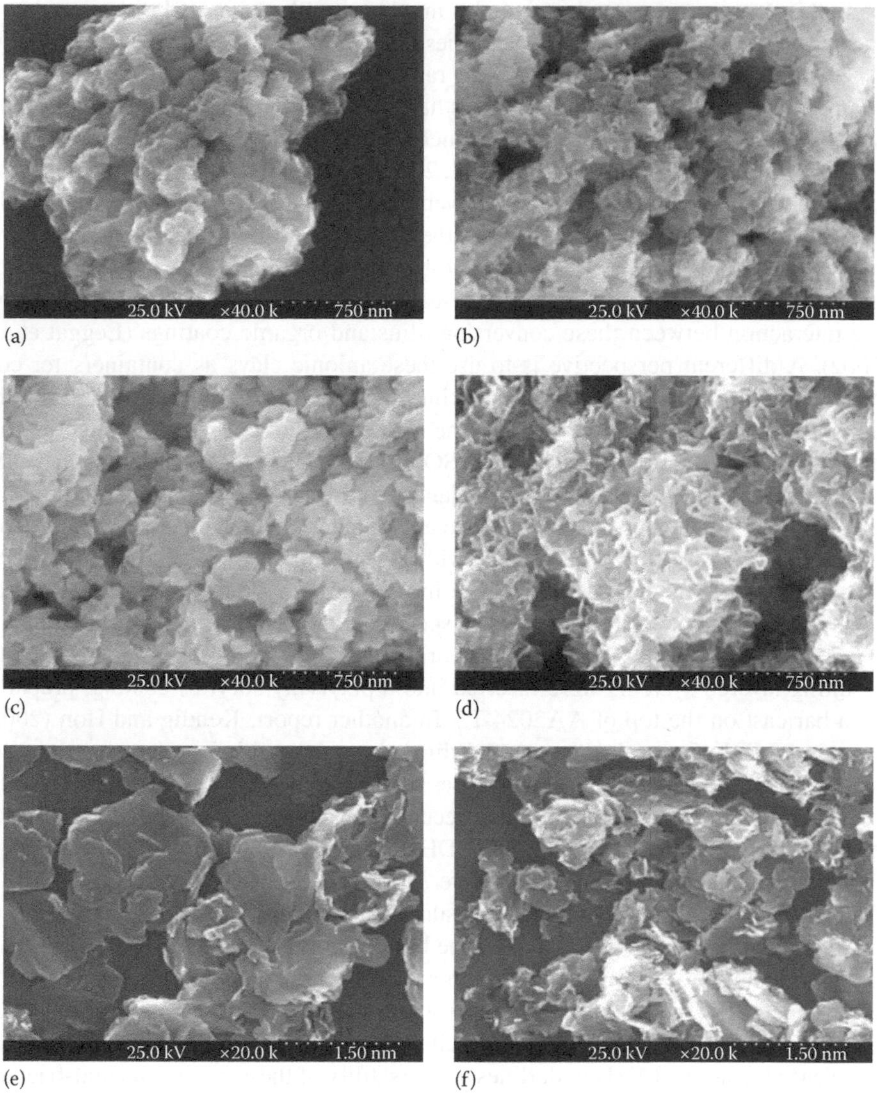

FIGURE 10.2 SEM micrographs of the LDH particles: (a) Mg_2AlVO_3 prepared by direct synthesis (as deposited), (b) Mg_2AlVO_3 prepared by direct synthesis (after heating at 65°C), (c) Zn_2AlVO_3 prepared by direct synthesis (as deposited), (d) Zn_2AlVO_3 prepared by direct synthesis (after heating at 65°C), (e) Zn_2AlNO_3 (after heating at 65°C), and (f) Zn_2AlVO_3 prepared from Zn_2AlNO_3 by anion exchange. (Reprinted from *Corr. Sci.*, 52, Zheludkevich, M.L., Poznyak, S.K., Rodrigues, L.M., Raps, D., Hack, T., Dick, L.F., Nunes, T., Ferreira, M.G.S., Active protection coatings with layered double hydroxide nanocontainers of corrosion inhibitor, 602–611, Copyright 2010, with permission from Elsevier.)

structures (Choucair et al., 2004). It is possible to induce a transition between aggregate structures by slight perturbations of any given system (Soo and Eisenberg, 2004). The sizes of nanocontainers can be controlled by molecular weight of the polymer and the ratio between the block sizes (Förster et al., 1996). The diffusion of (polymeric) amphiphiles in these vesicles is very slow compared with liposomes and for high-molecular-weight chain entanglements even make it possible to trap near-equilibrium and metastable morphologies. Additionally, in contrast to liposomes, the thickness of the block copolymer shells is higher and can exceed 200 nm. As a consequence, this increased thickness, in combination with the conformational freedom of the polymer chains, results in a much lower permeability for water of block copolymer micro- and nanocontainers as compared with liposomes.

Although the basic principles for the construction of block copolymer nanocontainers are now in the process of being formulated, applications of these systems in the fields of delivery of encapsulated materials look promising. Combining the block copolymer nanocontainers with biomacromolecules opens a new area of coupling self-organization properties of block copolymers with the functionality of enzymes. Initial steps in this direction were taken by preparing vesicle-forming block copolymers having a polypeptide block (Chécot et al., 2002). Using both covalent and noncovalent coupling procedures, proteins (e.g., lipase, streptavidin, horseradish peroxidase) were connected to the block copolymer nanocontainer.

Several amphiphilic block copolymers were investigated up to date as components of the nanocontainers: poly(ethylene oxide), poly(*N*-isopropylacrylamide), polypyrrolidones, biodegradable polymers such as poly(caprolactone), poly(D,L-lactide), or poly(D,L-lactic acid-*co*-glycolic acid), and so on. "Intelligent" nanocontainer vehicles can be designed by utilizing stimuli-responsive polymers sensitive to the alteration of the temperature or pH value (Jeong and Gutowska, 2002). Poly(*N*-isopropyl acrylamide) undergoes a phase transition at 32°C and can be used for fabrication of thermosensitive nanocontainers. The pH-sensitive micelles are usually made from double hydrophilic block copolymers that can dissolve molecularly in water in a certain pH range and aggregate spontaneously upon an appropriate change in the pH value (Bangham et al., 1965).

However, the application of self-organizing structures is limited because of their instability against aggregation and collapse. Attempts have been made to stabilize micellar structures by cross-linking either the core or the shell. Typical methods to cross-link involve the polymerization of macromonomers with amphiphilic block structures in their side chain (Iijima et al., 1999) or the attachment of polymerizable groups along the hydrophobic block (Won et al., 1999). Unfortunately, such cross-linking decreases the capacity of the inner volume. An alternative approach is to cross-link the shell after aggregation of the block copolymers in aqueous solution (Joralemon et al., 2004). Another example is poly(organosiloxane) containers that can be prepared with defined core-shell structures introducing functionalities and stabilizing the nanocontainer shell (Jungmann et al., 2003) after the formation of the particles.

New optically addressable microcapsules were prepared combining photochemistry approach and self-assembly (Yuan et al., 2005). Chlorobenzyl-functionalized poly(organosiloxane) nanoparticles were introduced in water-in-oil-in-water emulsion and their photochemical cross-linking was performed by a 200 W Hg–Xe lamp for several hours. The inner aqueous void contained cyclodextrin. The thickness, mechanical stability, and light resistance of the container walls can be controlled simply by regulating the number of photoreactive chlorobenzyl-functionalized poly(organosiloxane) nanoparticles in the shell. Importantly, these nanocontainers can also be destroyed by photocleavage, thereby opening a way to the optically controlled decomposition and release of entrapped cyclodextrin.

A protocol to construct microcontainers of novel type was introduced by Donath et al. (1998). This procedure consists in templating of layer-by-layer (LbL)-assembled polyelectrolyte films on the surface of micron and submicron-sized colloidal particles. A suspension of the micro- or nanoparticles is mixed with charged (e.g., positive) polyelectrolyte solution. The polyelectrolyte adsorbs and the surface of the template particle is recharged. After washing out the excess of the polyelectrolyte, the particle/polyelectrolyte composites are mixed next with oppositely charged polyelectrolyte (e.g., negative) and the surface recharges again. By repeating the cycle, a multilayer polyelectrolyte capsules can be obtained and, depending on the internal content of the capsule core, capsules can have pH gradient across the polyelectrolyte capsule shell (Figure 10.3). The universal character of the method does not have any restriction on the type of the charged species employed for shell construction. The layer-by-layer (LbL) deposition method has been used for >50 various charged macromolecules including synthetic polyelectrolytes, conductive polymers and biopolymers (proteins and nucleic acids), carbon nanotubes, viruses, lipid vesicles, and nanoparticles (e.g., Fe_3O_4, CdSe). The precision of one adsorbed layer thickness is about 1 nm. Core decomposition leads to formation of hollow structures with size and shape determined by the initial colloidal core and shell composed of polyelectrolyte multilayers tunable in the nanometer range. It should be noted that the polyelectrolyte capsules completely repeat the shape of the templating colloids as was shown by means of an example of echinocyte cells, which have star-like shape (Neu et al., 2001).

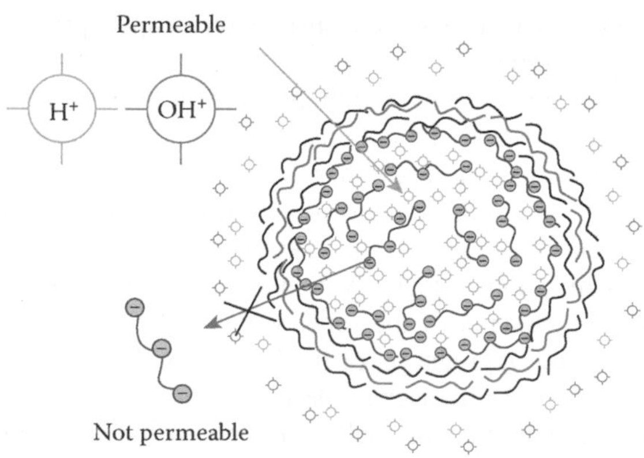

FIGURE 10.3 Schematic illustration of the pH gradient established across the wall of polyelectrolyte capsule loaded with polyanion (due to electroneutrality [Poly⁻]≈[H⁺]). (From Shchukin, D.G. and Sukhorukov, G.B.: Nanoparticle synthesis in engineered organic nanoscale reactors. *Adv. Mater.* 2004. 16. 671–682. Copyright Wiley-VCH Verlag GmbH & Co. KGaA. Reprinted with permission.)

The shell of the polyelectrolyte capsules is semipermeable and sensitive to a variety of physical and chemical conditions of the surrounding media, which might dramatically influence the structure of polyelectrolyte complexes and the permeability of the capsules. High-molecular-weight compounds are excluded by the polyelectrolyte shell, whereas small molecules, such as dyes and ions, can readily penetrate the capsule wall (Sukhorukov et al., 2000). Usually, the polyelectrolyte capsule shell is permeable for macromolecules and nanoparticles at low pH (<3) or high ionic strength whereas it is in "closed" state at high pH (>8). A possible explanation for the permeability properties of polyelectrolyte shell can be provided by considering the polyelectrolyte interactions in the shell wall. At the pH of capsule formation (pH 7), the charge densities on both polyelectrolytes determine their stoichiometric ratio during adsorption. Since the polymers are irreversibly adsorbed in the shell wall, a pH decrease does not induce polymer desorption. However, charging of one of the polyelectrolytes may occur, which would induce positive (negative) excess charge into the shell wall. This may alter the shell wall morphology by enhancing the repulsion, which could lead to defects in the polyelectrolyte shell.

Organic solvents can also induce permeation through the LbL-assembled shell (Sukhorukov et al., 2001). Using a 1:1 ethanol/water mixture, Lvov et al. (2001) encapsulated urease (5-nm diameter globules) in poly(allylamine)/poly(styrene sulfonate) capsules. Ethanol might partially remove the hydration water between the polyelectrolytes, leading to a segregation of the polyion network and the formation of pores.

Polyelectrolyte capsules composed of poly(allylamine)/poly(styrene sulfonate) multilayers preserve their integrity after heating at 120°C for 20 min in aqueous solution but show a considerable decrease in size (50%–70% depending on the diameter of the initial capsules) (Koehler et al., 2005). The diameter decrease is accompanied by a strong increase of layer thickness and decrease of permeability, which lead to the enhanced entrapment capabilities. The capsules become impermeable even for low-molecular-weight compounds. The driving force for this polyelectrolyte rearrangement process is the entropy gain through the more coiled state of the polyions and the decreased interface between polyelectrolytes and water. Probably some water, which fills the pores of the multilayers, is expelled during temperature treatment.

10.5 INTRODUCTION OF THE CONTAINERS INTO THE COATINGS

There are two main application areas of container-based active coatings: (a) bioactive coatings and (b) various self-healing protective coatings (first of all, anticorrosion coatings), and most publications up to date are devoted to them. Systems, capable of smart self-repair, consist of several parts: a stimulus to release the repairing chemical such as the cracking of a microcontainer; a microcontainer; a repair chemical monomer or inhibitor carried inside; and a method of hardening the microcontainer load in the coating matrix. Typical macroscopic damage involves destruction of a material due to impact. On the microscale, damage such as matrix microcracking alters mechanical properties such as strength, stiffness, and dimensional stability depending on the material type. Thermal, electrical, and acoustical properties are changed as matrix cracks. Microdefects act as sites for environmental degradation, as well as for nucleation of the macrodefects. The samples in which prior release of a cross-linking adhesive into a cracked matrix is allowed to set up over a time have greater strength and ability to repair than the specimens, which had no cross-linking adhesive internally released (Dry, 1996). The internal delayed release of adhesive improves impact strength and ability to deflect while carrying a load.

Microencapsulated dicyclopentadiene healing agent and highly porous Ru catalyst are incorporated into an epoxy matrix to produce a polymer composite capable of self-healing (Figure 10.4) (Brown et al., 2004). Both the virgin and healed fracture toughness depend strongly on the size and concentration of microcapsules added to the epoxy. Addition of dicyclopentadiene-filled urea–formaldehyde microcapsules into epoxy samples yields up to 127% increase in coating toughness. The increased toughening associated with fluid-filled microcapsules is attributed to the healing of subsurface microcracking. Overall the embedded microcapsules provide two independent effects: the increase in virgin fracture toughness from general toughening and the ability to self-heal the virgin fracture event. Healed fracture toughness increases steadily with microcapsule concentration until reaching a plateau at about 20 vol%. The maximum healing efficiency for 180 μm microcapsules occurs at low concentrations (5 vol%). For 50 μm microcapsules, high healing efficiency only occurs at higher microcapsule concentrations (20 vol%) since more capsules are required to deliver the same volume of DCPD healing agent to the fracture plane.

Diene monomers (dicyclopentadiene, 5-ethylidene-2-norbornene, etc.) and their blends were investigated as candidate self-healing agents with Grubbs' catalyst (Liu et al., 2006). It was found that the reaction becomes faster with the increase of 5-ethylidene-2-norbornene content at lower catalyst loading. Rigidity after 120 min cure was the highest in a dicyclopentadiene:5-ethylidene-2-norbornene blend when it was cured on the epoxy resin coating. Considering requirements for effective self-healing (i.e., fast reaction during cure, high rigidity after cure, reduction of catalyst amount, and lower temperature capabilities), dicyclopentadiene:5-ethylidene-2-norbornene blends are very potential candidates for self-healing agents. The inherent shortcomings of the method based on diene monomers are the potential side reactions with the polymer matrix and air.

FIGURE 10.4 Urea–formaldehyde microcapsules containing dicyclopentadiene. (Reprinted from Brown, E.N. et al., *J. Mater. Sci.*, 39, 1703, 2004. Copyright 2004 Kluwer Academic Publishers.)

Other examples in the following demonstrate once more how productive can be emulsion route in the fabrication of micro- and nanocontainers for anticorrosion coatings. Organosiloxanes form covalent bonds with the surface of metal substrate bearing hydroxyl groups and impart coupling (in case of usual siloxanes) or water-repelling (for long chain–terminated siloxanes) functionality to it (Mittal, 1992). Moreover, organosiloxanes with multiple SiOR moieties can undergo lateral polycondensation reaction at the substrate surface forming 2D network with excellent protective ability against corrosion (Mittal, 1992). These properties of organosiloxanes were taken into account by Latnikova et al. (2011) where authors have proposed to encapsulate a mixture of emulsified organosiloxanes and then incorporate them into coating matrix. Micro- and nanocontainers with core-shell morphology and polyurethane/polyurea shells were successfully synthesized by emulsion interfacial polyaddition and then embedded in the corrosion protective coatings on the epoxy basis. Comparative study of protective efficiency showed much better performance for the coatings with organosiloxane-loaded containers.

Healable polymeric systems may, for example, contain encapsulated monomers and polymerization catalysts, or latent functionalities, which are able to participate in thermally reversible (Cho et al., 2006), covalent bond–forming reactions (Adzima et al., 2008). It has also been shown that noncovalent interactions, specifically hydrogen bonds (Fouquey et al., 1990), may be used to effect healing within a supramolecular polymer blend. In the latter system, it is proposed that fracture propagates via dissociation of the weak supramolecular interactions rather than by scission of covalent bonds, so that reassembly of the supramolecular network restores the original physical properties of the material.

Urea–formaldehyde microcapsules filled with drying linseed oil were used for the healing of cracks in an epoxy coating (Suryanarayana et al., 2008). Microcapsules were synthesized by *in situ* polymerization in an oil-in-water (o/w) emulsion. Initially, fully water-compatible urea and formaldehyde react in continuous aqueous medium to form poly(urea–formaldehyde). As molecular weight of this polymer increases the fraction of polar groups gradually decreases till the polymer molecules become hydrophobic and get deposited on the surface of o/w emulsion droplets. Microcapsules obtained were then incorporated in an epoxy resin coating. Since outer shell surface of microcapsules was very rough, a good binding with the coating matrix was provided. The encapsulated linseed oil is released by the coating crack and fills the crack in a coating matrix. Finally, oxidation of linseed oil by atmospheric oxygen led to the formation of continuous film inside the crack. Superior corrosion resistance of such self-healing coating was demonstrated by comparative exposure of two specimens in salt spray cabinet. Sample coated by film containing capsules remained free from corrosion at artificial scratch up to 72 h of exposure. In contrast, control specimen suffered from corrosion already after 48 h.

In our group, nano- and microcapsules with polyurea/polyurethane shell and "oil" core were recently developed using interfacial polymerization (Latnikova et al., 2011). Depending on the application purpose, encapsulated oil can contain either water-repelling agent forming dewetted spot around the damaged site or sealant covering this site with the protective polymeric film. In former case, hydrophobic compounds with ability to be bound covalently to the substrate under protection are used whereas the sealing agent or its solution in a volatile diluent polymerizable upon action of environmental triggers is employed in the latter one.

10.6 COATINGS WITH pH-RESPONSIVE INORGANIC NANOCONTAINERS

The next promising step is now to proceed toward a construct the response that, in turn, affects the stimulus. The stimuli have predominantly been of mechanical (as described earlier) and chemical nature, for example, concentration, pH (for polyelectrolyte shell), electrochemical potential. To improve corrosion protection properties of sol–gel-derived hybrid coatings on the aluminum AA2024 alloy, two organic corrosion inhibitors (mercaptobenzothiazole and mercaptobenzimidazole) have been encapsulated within the coating matrix in either the presence or absence of cyclodextrin (Aramaki, 2003). Superior corrosion protection properties have been found for formulations that contain β-cyclodextrin and can be explained by the act of slow release of the inhibitor from the cyclodextrin/inhibitor inclusion complexes and by the self-healing of corrosion defects. The inclusion complexes are more easily trapped within the cross-linked coating material making the inhibitor more difficult to leach out. The encapsulation of organic corrosion inhibitors into the coating host material as inclusion complexes with cyclodextrin can be considered as effective delivery systems of organic inhibitors in active corrosion protection applications. The slow release of organic corrosion inhibitor from the molecular cavity of cyclodextrin ensures the long-term delivery of corrosion inhibitor and thus the healing of a damaged coating.

Deposition of polyelectrolyte multilayers via LbL assembly onto emulsion droplets can also be applied for the droplets of Pickering emulsions as initial templates yielding finally microcontainers with significantly improved stability (Li and Stöver, 2010). Similar containers were used by Haase et al. (2010) for loading of corrosion inhibitor 8-hydroxyquinoline (8HQ) in their interior. It is interesting that in this case the active agent 8HQ played simultaneously the role of hydrophobizing agent for silica nanoparticles forming the container shells. Due to amphoteric character of 8HQ, its charge and solubility increase in the ranges of low and high pH values (<4 and >9) leading, therefore, to enhanced electrostatic repulsion between molecules on the particles surface and, finally, to disturbing of particulate shells around Pickering emulsion droplets. This, in turn, causes the breakup of containers and release of the encapsulated inhibitor.

The strategy of a preparation of nanocontainers based on mesoporous oxide core presented in Skorb et al. (2009). The fabrication of a polyelectrolyte shell around the container can possibly be done by LbL assembly of oppositely charged species and allows one to prevent the spontaneous release of loaded corrosion inhibitor. The precision of one adsorbed layer thickness is about 2 nm. Polyelectrolyte nanocontainers completely repeat the shape of the templating colloids. The polyelectrolyte shell lends controlled release properties to the nanocontainers. The opening of the shell can be induced only by changing the surrounding pH value to the acidic or alkali region (Sukhorukov et al., 2000), while in neutral pH the polyelectrolyte shell remains intact preventing undesirable leakage of the entrapped inhibitor.

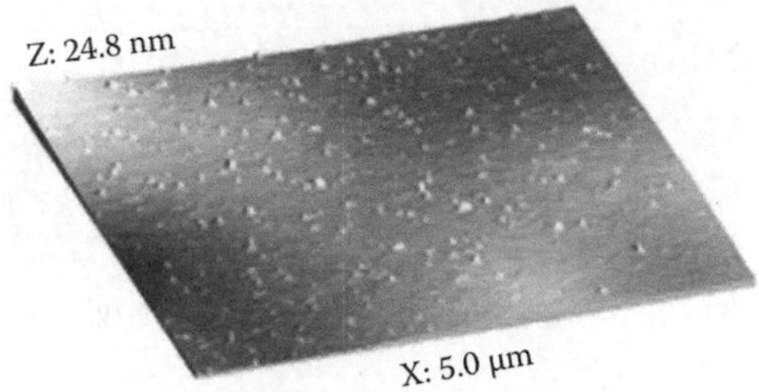

FIGURE 10.5 SiO_2/ZrO_2 coatings with benzotriazole-loaded nanocontainers. (From Shchukin, D.G., Zheludkevich, M.L., Yasakau, K.A., Lamaka, S.V., Ferreira, M.G.S., Möhwald, H. Layer-by-layer assembled nanocontainers for self-healing corrosion protection. *Adv. Mater.* 2006. 18. 1672–1678. Copyright Wiley-VCH Verlag GmbH & Co. KGaA. Reprinted with permission.)

Strong self-healing and long-term active corrosion protection of aluminium was shown by means of the example of benzotriazole-loaded SiO_2 nanocontainers impregnated into a ZrO_x/SiO_x hybrid film (Shchukin et al., 2006c; Zheludkevich et al., 2007). To produce SiO_2 nanocontainers for loading into a silica–zirconia hybrid film, the LbL deposition procedure was employed involving both large polyelectrolyte molecules and small benzotriazole ones. The suspension of benzotriazole-loaded nanocontainers is mixed following the sol–gel protocol and is deposited onto aluminium alloy by a dip-coating procedure. The uniformly distributed nanoparticles of a diameter about 100 nm are impregnated into the sol–gel film formed on aluminium substrate (Figure 10.5) with average concentration of the nanocontainers ~70 per μm^2. AFM does not show any signs of nanocontainer agglomeration. Optical photos of the two aluminum samples show the drastic difference between the nanocontainer-impregnated and initial sol–gel film. Many pit-like defects formed on the surface were found after aging even in diluted 0.005 M NaCl solution for a pure sol–gel film. The film with nanocontainers does not exhibit any visible signs of the corrosion attack even after 14 days in 100 times more concentrated 0.5 M NaCl solution. This pronounced difference shows the advantages of the "nanocontainer" approach over the direct introduction of inhibitor to the sol–gel coating.

The self-healing efficiency of the ZrO_x/SiO_x films impregnated with inhibitor-loaded nanocontainers was demonstrated by the scanning vibrating electrode technique (Figure 10.6). Scanning vibrating electrode technique (SVET) permits the mapping of the current density in an electrolyte close to the substrate surface. The electrode vibrates and indicates the current flow in the electrolyte due to the corrosion processes occurring on the substrate surface. The measured current density vectors permit mapping of both the magnitude and direction of current flow immediately above the substrate surface. This capability permits correlation of anodic and cathodic current processes with visual features on the substrate surface. The SVET is capable of providing detailed spatially resolved information not readily obtainable by other techniques. Artificial defects were formed and well-defined cathodic activity appears in the place of the induced defect on aluminium coated with undoped hybrid film. Impressively different behavior was revealed after defect formation on the substrate coated with ZrO_x/SiO_x film doped by benzotriazole-loaded nanocontainers.

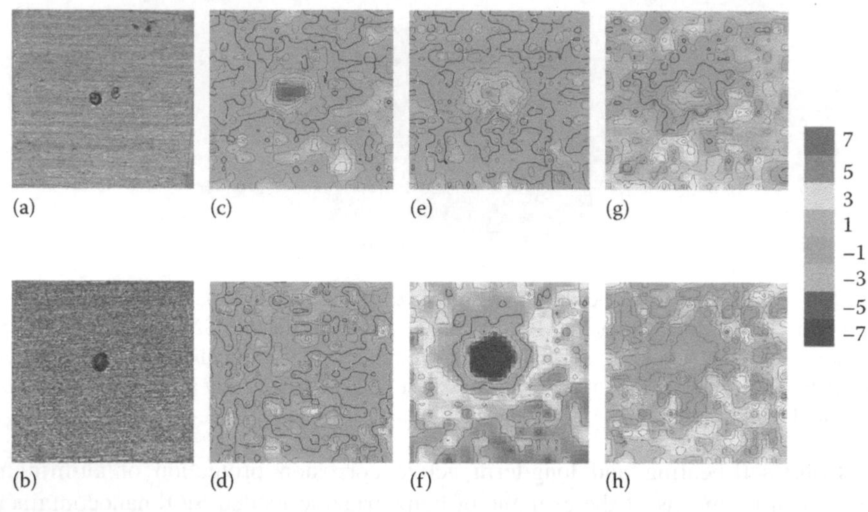

FIGURE 10.6 (See color insert.) SVET maps of the ionic currents measured above the surface of the defected (a, b) AA2024 coated with undoped sol–gel pretreatment (c, e, g) and with that impregnated by nanoreservoirs (d, f, h). The maps were obtained in 5h (c, d), 24h (e, f), and 26h (g, h) after defects formation. Scale units: μA/cm². Scanned area: 2mm × 2mm. (From Shchukin, D.G., Zheludkevich, M.L., Yasakau, K.A., Lamaka, S.V., Ferreira, M.G.S., Möhwald, H. Layer-by-layer assembled nanocontainers for self-healing corrosion protection. *Adv. Mater.* 2006. 18. 1672–1678. Copyright Wiley-VCH Verlag GmbH & Co. KGaA. Reprinted with permission.)

No corrosion activity appears in this case after 4h following the defect formation. Only after about 24h, a well-defined cathodic activity appears in the zone of the induced defect. However, the defect becomes passivated again 2h later. The most probable mechanism is based on the local change of the pH in the damaged area due to the corrosion processes. When the corrosion processes are started, the pH value is changed in the neighboring area, which opens the polyelectrolyte shell of the nanocontainers in a local area followed by the release of benzotriazole. Then, the released inhibitor suppresses the corrosion activity and the pH value recovers closing the polyelectrolyte shell of nanocontainers and terminating further release of the inhibitor.

Halloysite is an economically viable raw material that can be mined from the corresponding deposit as a raw mineral (Lvov et al., 2002). As for most natural materials, the size of halloysite particle varies within 1–15 μm of length and 10–100 nm of inner diameter depending on the deposits. Embedding of the corrosion inhibitor (e.g., benzotriazole) inside the inner volume of the halloysite G nanotubes was performed according to the adapted procedure described by Price et al. (2001). To attain controlled release properties in the halloysite nanotubes, the surface of the nanotubes could be modified by LbL deposition of polyelectrolyte bilayers (Lvov et al., 2008). The bare metals alloy exhibited several regions of anodic activity according to the SVET maps, reflecting sites of localized corrosion. One peak of corrosion activity is visible on the SVET map of the alloy panel coated with individual sol–gel film; consequently, this film results in higher protection ability of the alloy. However, sol–gel coating impregnated with benzotriazole-loaded halloysites exhibits superior anticorrosion efficiency, which is visible from the practical absence of anodic activity on SVET maps. Thus, it is visible from SVET measurements that during the corrosion process the corrosion centers appear in all samples. Simultaneously, the highest corrosion occurs in the case of bare Al alloy. The second place in the range of the protection ability has the sol–gel coating. The highest protection was provided by the halloysite-loaded sol–gel film.

10.7 CONCLUSIONS

After nearly 10 years of development, fabrication of the functional micro- and nanocontainers is still a field of growing importance, especially for the envisaged practical application in functional coatings, cosmetics, medicine, and biotechnology. Functional micro- and nanocontainers represent an unique class of materials, in which a spatially confined inner nanovolume is combined with controllable permeability and protective properties of the shell.

Controlled permeability of the nanocontainer shell can be employed in a new generation of "smart"-sensitive self-healing coatings rapidly responding to changes of the ambient environment (e.g., increasing temperature, humidity) via immediate or prolonged release of the species entrapped inside the nanocontainer volume. It is shown that there are conceptual solutions but still many technical and economical issues to be solved to arrive at practical solutions. The stimuli have predominantly been of mechanical and electrochemical nature but in future one may envision more of these, for example, friction, biocide. One should, however, also be aware that feedback loops are ubiquitous in nature and technology, and they are mostly not used to remove a stimulus but to keep it at a defined level, for example, concentration, pH, potential.

The demonstrated universal approach for the fabrication of active coatings on the other hand is a great challenge to develop multifunctional organic and composite nanocontainers able to encapsulate active material, retain it in the inner volume for a long period, and immediately release it on demand. A lot of research work still remains to be done, especially for the better understanding of the detailed mechanism of the shell permeability and structure of the inner void. Notwithstanding to the fact that perspectives of nanocontainer applications are already demonstrated, the necessary up-scaling technologies for fabrication of micro- and nanocontainers in large quantities are not yet comprehensively developed. Moreover, the active coating can have several active functionalities (e.g., antibacterial, anticorrosion, and antistatic) when several types of nanocontainers loaded with corresponding active agent will be incorporated simultaneously into a coating matrix.

REFERENCES

Adzima J, Aguirre H, Kloxin CJ, Scott TF, Bowman CN (2008), Rheological and chemical analysis of reverse gelation in a covalently cross-linked Diels–Alder polymer network, *Macromolecules*, 41, 9112–9117.

Ahluwalia H, Uhlenkamp BJ (2008), The importance of quality in corrosion-resistant alloys in biopharmaceutical manufacturing, *Pharmaceut Tech*, 3. Available from: http://pharmtech.findpharma.com/pharmtech/Peer-Reviewed+Research/The-Importance-of-Quality-in-Corrosion-Resistant-A/ArticleStandard/Article/detail/500410 (accessed February 5, 2011).

Alanazi NM, Leyland A, Yerokhin AL, Matthews A (2010), Substitution of hexavalent chromate conversion treatment with a plasma electrolytic oxidation process to improve the corrosion properties of ion vapour deposited AlMg coatings, *Surf Coat Tech*, 205, 1750–1756.

Albertazzi S, Basile F, Vaccari A (2004), Catalytic properties of hydrotalcite-type anionic clays, in Wypych F, Satyanarayana KG, Eds., *Clay Surfaces: Fundamentals and Applications*, Elsevier, Amsterdam, the Netherlands, pp. 496–546.

Antipov AA, Sukhorukov GB (2004), Polyelectrolyte multilayer capsules as vehicles with tunable permeability, *Adv Colloid Interface Sci*, 111, 49–61.

Aramaki K (2003), Improvement in the self-healing ability of a protective film consisting of hydrated cerium(III) oxide and sodium phosphate layers on zinc, *Corr Sci*, 45, 451–464.

Ardelean H, Frateur I, Marcus P (2008), Corrosion protection of magnesium alloys by cerium, zirconium and niobium-based conversion coatings, *Corr Sci*, 50, 1907–1918.

Bangham AD, Standish MM, Watkins JC (1965), Diffusion of univalent ions across the lamellae of swollen phospholipids, *J Mol Biol*, 13, 238–252.

Bogner B (2005), Composites for chemical resistance and infrastructure applications, *Reinf Plast*, 49, 30–34.

Brooman EW (2002), Modifying organic coatings to provide corrosion resistance: Part II—Inorganic additives and inhibitors, *Met Finish*, 100, 42–53.

Brown EN, White SR, Sottos NR (2004), Microcapsule induced toughening in a self-healing polymer composite, *J Mater Sci*, 39, 1703–1710.

Buchheit RG, Mamidipally SB, Schmutz P, Guan H (2002), Active corrosion protection in Ce-modified hydrotalcite conversion coatings, *Corrosion*, 58, 3–14.

Chécot F, Lecommandoux S, Gnanou Y, Klok H-A (2002), Water-soluble stimuli-responsive vesicles from peptide-based diblock copolymers, *Angew Chem Int Ed*, 41, 1339–1343.

Cho SH, Andersson HM, White SR, Sottos NR, Braun PV (2006), Polydimethylsiloxane-based self-healing materials, *Adv Mater*, 18, 997–1000.

Choucair A, Lavigueur C, Eisenberg A (2004), Polystyrene-*b*-poly(acrylic acid) vesicle size control using solution properties and hydrophilic block length, *Langmuir*, 20, 3894–3900.

Directive 2000/53/EC of the European Parliament and the council of 18.09.2000 on end-life vehicles, *Off J Eur Commun*, L269/34–L269/42. Available from: http://www.bmu.de/files/pdfs/allgemein/application/pdf/vehiclesdir.pdf (accessed January 1, 2011).

Donath E, Sukhorukov GB, Caruso F, Davis S, Möhwald H (1998), Novel hollow polymer shells by colloid-templated assembly of polyelectrolytes, *Angew Chem Int Ed*, 37, 2202–2205.

Dry C (1996), Procedures developed for self-repair of polymer matrix composite materials, *Comp Struct*, 35, 263–269.

Dufek EJ, Buttry DA (2008), Inhibition of O_2 reduction on AA2024-T3 using a Zr(IV)–octadecyl phosphonate coating system, *Electrochem Solid State Lett*, 11, C9–C12.

Fery A, Weinkamer R (2007), Mechanical properties of micro- and nanocapsules: Single-capsule measurements, *Polymer*, 48, 7221–7235.

Förster S, Zisenis M, Wenz E, Antonietti M (1996), Micellization of strongly segregated block copolymers, *J Chem Phys*, 104, 9956–9970.

Fouquey C, Lehn J-M, Levelut A-M (1990), Molecular recognition directed self-assembly of supramolecular liquid crystalline polymers from complementary chiral components, *Adv Mater*, 2, 254–257.

Garcia-Heras M, Jimenez-Morales A, Casal B, Galvan JC, Radzki S, Villegas MA (2004), Preparation and electrochemical study of cerium–silica sol–gel thin films, *J Alloys Compd*, 380, 219–224.

Haase MF, Grigoriev D, Moehwald H, Tiersch B, Shchukin DG (2010), Encapsulation of amphoteric substances in a pH-sensitive pickering emulsion, *J Phys Chem C*, 114, 17304–17310.

Hansal WEG, Hansal S, Polzler M, Kornherr A, Zifferer G, Nauer GE (2006), Investigation of polysiloxane coatings as corrosion inhibitors of zinc surfaces, *Surf Coat Tech*, 200, 3056–3063.

Hernandez-Alvarado LA, Hernandez LS, Miranda JM, Dominguez O (2009), The protection of galvanised steel using a chromate-free organic inhibitor, *Anti-Corros Method M*, 56, 114–120.

Hosseini M, Ashassi-Sorkhabi H, Ghiasvand H (2007), Corrosion protection of electro-galvanized steel by green conversion coatings, *J Rare Earth*, 25, 537–543.

Hughes AE, Gorman JD, Harvey TG, Galassi A, McAdam G (2006), Development of permanganate-based coatings on aluminum alloy 2024-T3, *Corrosion*, 62, 773–780.

Hughes AE, Gorman J, Harvey TG, McCulloch D, Toh SK (2004), SEM and RBS characterization of a cobalt based conversion coating process on AA2024-T3 and ALA7075-T6, *Surf Interface Anal*, 36, 1585–1591.

Iijima M, Nagasaki Y, Okada T, Kato M, Kataoka K (1999), Core-polymerized reactive micelles from heterotelechelic amphiphilic block copolymers, *Macromolecules*, 32, 1140–1146.

Institute of Materials (Great Britain) (2002), *Corrosion Resistant Alloys for Oil and Gas Production: Guidance on General Requirements and Test Methods for H2S Service*, 2nd edn., Maney Publishing, Leeds, U.K.

Javier M, de Pino P, Bedard MF et al. (2008), Photoactivated release of cargo from the cavity of polyelectrolyte capsules to the cytosol of cells, *Langmuir*, 24, 12517–12520.

Jeong B, Gutowska A (2002), Lessons from nature: Stimuli responsive polymers and their biomedical applications, *Trends Biotechnol*, 20, 305–311.

Joralemon MJ, Murthy K, Shanmugananda R, Remsen EE, Becker ML, Wooley KL (2004), Synthesis, characterization, and bioavailability of mannosylated shell cross-linked nanoparticles, *Biomacromolecules*, 5, 903–913.

Jungmann N, Schmidt M, Ebenhoch J, Weis J, Maskos M (2003), Dye loading of amphiphilic poly(organosiloxane) nanoparticles, *Angew Chem Int Ed*, 42, 1714–1717.

Kasten LS, Grant JT, Grebasch N, Voevodin N, Arnold FE, Donley MS (2001), An XPS study of cerium dopants in sol–gel coatings for aluminum 2024-T3, *Surf Coat Tech*, 140, 11–15.

Kendig MW, Buchheit RG (2003), Corrosion inhibition of aluminum and aluminum alloys by soluble chromates, chromate coatings, and chromate-free coatings, *Corrosion*, 59, 379–400.

Kendig MH, Hon M (2005), A hydrotalcite-like pigment containing an organic anion corrosion inhibitor, *Electrochem Solid-State Lett*, 8, B10–B11.

Knudsen OO, Tanem BS, Bjorgum A, Mardalen J, Hallenstvet M (2004), Anodising as pretreatment before organic coating of extruded and cast aluminium alloys, *Corros Sci*, 46, 2081–2095.

Kobayashi Y, Fujiwara Y (2006), Corrosion protection of cerium conversion coating modified with a self-assembled layer of phosphoric acid mono-*n*-alkyl ester, *Electrochem Solid State Lett*, 9, B15–B18.

Koch GH, Brongers MPH, Thompson NG, Virmani YP, Payer JH (2002), Corrosion costs and preventive strategies in the United States, *Suppl Mater Perform*, 42, 2–11.

Koehler K, Shchukin DG, Moehwald H, Sukhorukov GB (2005), Thermal behavior of polyelectrolyte multilayer microcapsules. 1. The effect of odd and even layer number, *J Phys Chem B*, 109, 18250–18259.

Kong G, Liu RB, Lu JT, Che CS, Zhong Z (2010), Study on growth mechanism of lanthanum salt conversion coating on galvanized steel, *Acta Metall Sin*, 46, 487–493.

Latnikova A, Grigoriev DO, Hartmann J, Möhwald H, Shchukin DG (2011), Polyfunctional active coatings with damage-triggered water-repelling effect, *Soft Matter*, 7, 369–372.

Leggat RB, Taylor SA, Taylor SR (2002), Adhesion of epoxy to hydrotalcite conversion coatings: II. Surface modification with ionic surfactants, *Colloids Surf A*, 210, 83–94.

Li Z, Dai C, Liu Y, Zhu J (2007), Study of silicate and tungstate composite conversion coatings on magnesium alloy, *Electroplating Pollut Control*, 1, 16–19.

Li J, Stöver HDH (2010), Pickering emulsion templated layer-by-layer assembly for making microcapsules, *Langmuir*, 26, 15554–15560.

Lin CS, Li WJ (2006), Corrosion resistance of cerium-conversion coated AZ31 magnesium alloys in cerium nitrate solutions, *Mater Trans*, 47, 1020–1025.

Liu JR, Guo YN, Huang WD (2006), Study on the corrosion resistance of phytic acid conversion coating for magnesium alloys, *Surf Coat Technol*, 201, 1536–1541.

Liu X, Lee JK, Yoon SH, Kessler MR (2006), Characterization of diene monomers as healing agents for autonomic damage repair, *J Appl Polym Sci*, 101, 1266–1272.

Lvov Y, Antipov AA, Mamedov A, Möhwald H, Sukhorukov GB (2001), Urease encapsulation in nanoorganized microshells, *Nanoletters*, 1, 125–128.

Lvov Y, Price R, Gaber B, Ichinose I (2002), Thin film nanofabrication via layer-by-layer adsorption of tubule halloysite, spherical silica, proteins and polycations, *Colloids Surf A*, 198, 375–382.

Lvov Y, Shchukin D, Moehwald H, Price P (2008), Halloysite clay nanotubes for controlled release of protective agents, *ACS Nano*, 2, 814–820.

Magalhaes AAO, Margarit ICP, Mattos OR (2004), Molybdate conversion coatings on zinc surfaces, *J Electroanal Chem*, 572, 433–440.

Mahajanarn PV, Buchheit RG (2008), Characterization of inhibitor release from Zn–Al–$[V_{10}O_{28}]^{6-}$ hydrotalcite pigments and corrosion protection from hydrotalcite-pigmented epoxy coatings, *Corrosion*, 64, 230–240.

McCafferty E (2010), *Introduction to Corrosion Science*, Springer, New York.

McConnell VP (2005), Resurgence in corrosion-resistant composites, *Reinf Plast*, 49, 20–25.

Mitchon LN, White JM (2006), Growth and analysis of octadecylsiloxane monolayers on Al_2O_3 (0001), *Langmuir*, 22, 6549–6554.

Mittal KL (1992), *Silanes and Other Coupling Agents*, VSP, Utrecht, the Netherlands.

Neu B, Voigt A, Mitlöhner R, Leporatti S, Donath E, Gao CY, Kiesewetter H, Möhwald H, Meiselman HJ, Bäumler H (2001), Biological cells as templates for hollow microcapsules, *J Microencapsul*, 18, 385–395.

Niu LY, Jiang ZH, Li GY, Gu CD, Lian JS (2006), A study and application of zinc phosphate coating on AZ91D magnesium alloy, *Surf Coat Tech*, 200, 3021–3026.

O'Keefe MJ, Geng S, Joshi S (2007), Cerium-based conversion coatings as alternatives to hex chrome, *Met Finish*, 105, 25–28.

Osborne JH, Blohowiak KY, Taylor SR, Hunter C, Bierwagen GP, Carlson B, Bernard D, Donley MS (2001), Testing and evaluation of non-chromated coating systems for aerospace applications, *Prog Organ Coat*, 41, 217–225.

Pacitti J (1964), Plastics for corrosion-resistance applications, *Anti-Corr Methods M*, 11, 18–24.

Palmer SJ, Frost RL, Nguyen T (2009), Hydrotalcites and their role in coordination of anions in Bayer liquors: Anion binding in layered double hydroxides, *Coord Chem Rev*, 253, 250–267.

Price R, Gaber B, Lvov Y (2001), In-vitro release characteristics of tetracycline HCl, khellin and nicotinamide adenine dineculeotide from halloysite; a cylindrical mineral, *J Microencapsul*, 18, 713–722.

Pritchard G (1995), *Anti-Corrosion Polymers: PEEK, PEKK and Other Polyaryls*, ChemTec Publishing, Toronto, Ontario, Canada.

Raja PB, Sethuraman MG (2008), Natural products as corrosion inhibitor for metals in corrosive media—A review, *Mater Lett*, 62, 113–116.

Roberge PR (2000), *Handbook of Corrosion Engineering*, McGraw-Hill, New York.

Schweitzer PA (2000), *Mechanical and Corrosion-Resistant Properties of Plastics and Elastomers*, Marcel Dekker, New York.

Shchukin DG, Gorin DA, Möhwald H (2006), Ultrasonically induced opening of polyelectrolyte microcontainers, *Langmuir*, 22, 7400–7404.
Shchukin DG, Köhler K, Möhwald H (2006), Microcontainers with electrochemically reversible permeability, *J Am Chem Soc*, 128, 4560–4561.
Shchukin DG, Möhwald H (2007), Self-repairing coatings containing active nanoreservoirs, *Small*, 3, 926–943.
Shchukin DG, Sukhorukov GB (2004) Nanoparticle synthesis in engineered organic nanoscale reactors, *Adv Mater*, 16, 671–682.
Shchukin DG, Sviridov DV (2006), Photocatalytic processes in spatially confined micro- and nanoreactors, *J Photochem Photobiol C*, 7, 23–39.
Shchukin DG, Zheludkevich ML, Yasakau KA, Lamaka SV, Ferreira MGS, Möhwald H (2006), Layer-by-layer assembled nanocontainers for self-healing corrosion protection, *Adv Mater*, 18, 1672–1678.
Sheffer M, Groysman A, Starosvetsky D, Savchenko N, Mandler D (2004), Anion embedded sol–gel films on Al for corrosion protection, *Corros Sci*, 46, 2975–2985.
Sinko J (2001), Challenges of chromate inhibitor pigments replacement in organic coatings, *Prog Organ Coat*, 42, 267–282.
Skorb EV, Fix D, Andreeva DV, Moehwald H, Shchukin DG (2009), Surface-modified mesoporous SiO_2 containers for corrosion protection, *Adv Funct Mater*, 19, 2373–2379.
Soo PL, Eisenberg A (2004), Preparation of block copolymer vesicles in solution, *J Polym Sci B*, 42, 924–938.
Sorrentino A, Gorrasi G, Tortora M, Vittoria V, Constantino U, Marmottini F, Padella F (2005), Incorporation of Mg–Al hydrotalcite into a biodegradable poly(3-caprolactone) by high energy ball milling, *Polymer*, 46, 1601–1608.
Sukhorukov GB, Antipov AA, Voigt A, Donath E, Möhwald H (2001), pH-Controlled macromolecule encapsulation in and release from polyelectrolyte multilayer nanocapsules, *Macromol Rapid Commun*, 22, 44–46.
Sukhorukov GB, Donath E, Moya S, Susha AS, Voigt A, Hartmann J, Möhwald H (2000), Microencapsulation by means of step-wise adsorption of polyelectrolytes, *J Microencapsul*, 17, 177–185.
Suryanarayana C, Chowdoji Rao K, Kumar D (2008), Preparation and characterization of microcapsules containing linseed oil and its use in self-healing coatings, *Prog Org Coat*, 63, 72–78.
Tao X, Li JB, Moehwald H (2004), Self-assembly, optical behavior, and permeability of a novel capsule based on an azo dye and polyelectrolytes, *Chem Eur J*, 10, 3397–3403.
Theng BKG (1974), *The Chemistry of Clay—Organic Reactions*, Wiley, New York.
Tsai YT, Hou KH, Bai CY, Lee JL, Ger MD (2010), The influence on immersion time of titanium conversion coatings on electrogalvanized steel, *Thin Solid Films*, 518, 7541–7544.
Twite RL, Bierwagen GP (1998), Review of alternatives to chromate for corrosion protection of aluminum aerospace alloys, *Prog Organ Coat*, 33, 91–100.
von Baeckmann W, Schwenk W, Prinz W, Eds. (1997), *Handbook of Cathodic Corrosion Protection: Theory and Practice of Electrochemical Protection Processes*, Gulf Publishing, Houston, TX.
Vreugdenhi AJ, Woods ME (2005), Triggered release of molecular additives from epoxy-amine sol–gel coatings, *Prog Organ Coat*, 53, 119–125.
Williams G, McMurray HN (2004), Inhibition of filiform corrosion on polymer coated AA2024-T3 by hydrotalcite-like pigments incorporating organic anions, *Electrochem Solid State Lett*, 7, B13–B15.
Williams GR, O'Hare D (2006), Towards understanding, control and application of layered double hydroxide chemistry, *J Mater Chem*, 16, 3065–3074.
Won Y-Y, Davis HT, Bates FS (1999), Giant wormlike rubber micelles, *Science*, 283, 960–963.

Xia L, McCreery RL (1998), Chemistry of a chromate conversion coating on aluminum alloy AA2024-T3 probed by vibrational spectroscopy, *J Electrochem Soc*, 145, 3083–3089.

Yang X, Wang G, Dong G, Gong F, Zhang M (2009), Rare earth conversion coating on Mg-8.5Li alloys, *J Alloy Compd*, 487, 64–68.

Yuan X, Fischer K, Schärtl W (2005), Photocleavable microcapsules built from photoreactive nanospheres, *Langmuir*, 21, 9374–9380.

Zhao M, Wu SS, Luo JR, Fukuda Y, Nakae H (2006), A chromium-free conversion coating of magnesium alloy by a phosphate–permanganate solution, *Surf Coat Technol*, 200, 5407–5412.

Zhao XD, Yang J, Fan X (2010), Review on research and progress of corrosion inhibitors, *Appl Mech Mater*, 44–47, 4063–4066.

Zheludkevich ML, Poznyak SK, Rodrigues LM, Raps D, Hack T, Dick LF, Nunes T, Ferreira MGS (2010), Active protection coatings with layered double hydroxide nanocontainers of corrosion inhibitor, *Corr Sci*, 52, 602–611.

Zheludkevich ML, Shchukin DG, Yasakau KA, Möhwald H, Ferreira MGS (2007), Anticorrosion coatings with self-healing effect based on nanocontainers impregnated with corrosion inhibitor, *Chem Mater*, 19, 402–411.

Zou ZL, Li N, Li DY, Liu HP, Mu SL (2011), A vanadium-based conversion coating as chromate replacement for electrogalvanized steel substrates, *J Alloy Compd*, 509, 503–507.

11 Self-Healing Process in Glassy Materials

François O. Méar, Daniel Coillot, Renaud Podor, and Lionel Montagne

CONTENTS

11.1 INTRODUCTION

Self-healing materials are polymers, metals, ceramics, and their composites that when damaged through thermal, mechanical, ballistic, or other means have the ability to heal and restore their material to its original set of properties. Ghosh (2009a) defined the self-healing as the ability of a material to heal damages automatically and autonomously. Few materials intrinsically posses this ability, and self-healing capacity in manmade materials generally cannot be obtained without an external trigger. As schematically described in Figure 11.1, self-healing can be autonomic (or extrinsic) or nonautonomic (or intrinsic).

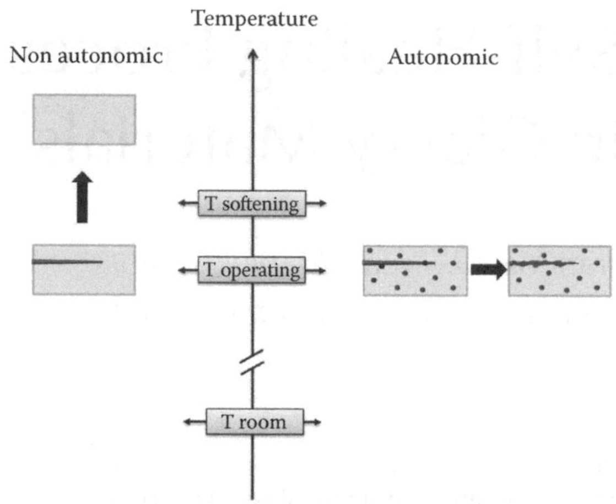

FIGURE 11.1 Schematic presentation of nonautonomic (intrinsic) and autonomic (extrinsic) self-healing processes. Nonautonomic self-healing involves overheating to enable crack healing by glass softening; autonomic self-repairing is obtained through the oxidation of active healing particles dispersed within the glassy matrix, and thus without the need to increase the operating temperature. The active particle is selected for its capacity to oxidize at a lower temperature than the softening point of the glass.

Nonautonomic self-healing requires an external constraint (mechanical, chemical, or thermal) such as temperature increase. On the other hand, autonomic self-healing involves healing agents encapsulated in the material, and their healing effect is obtained at the operating temperature through a chemical reaction.

In this contribution, we will first make an overview of the methods used for the characterization of self-healing process in glassy materials. Then, we will show that applications of self-healing concept concern mainly nonautonomic self-healing of glasses and glass ceramics, and we will report a recently developed method for obtaining autonomic self-healing in these materials.

11.2 CHARACTERIZATION OF SELF-HEALING IN GLASSY MATERIALS

Different methods have been used to characterize the self-healing behavior of glassy materials at high temperature. They were generally developed for a dedicated application (e.g., for the sealing of high-temperature solid oxide fuel cells [SOFCs] or solid oxide electrolyzer cells [SOECs]).

11.2.1 Hermiticity Test under Thermal Cycling

This test provides the possibility to assess the seal hermiticity when submitted to thermal cycling between service temperature (700°C–900°C) and room temperature. The aim is to make evidence that seal properties can be recovered after several thermal cycles in which cracks are expected to occur. Different setups have been

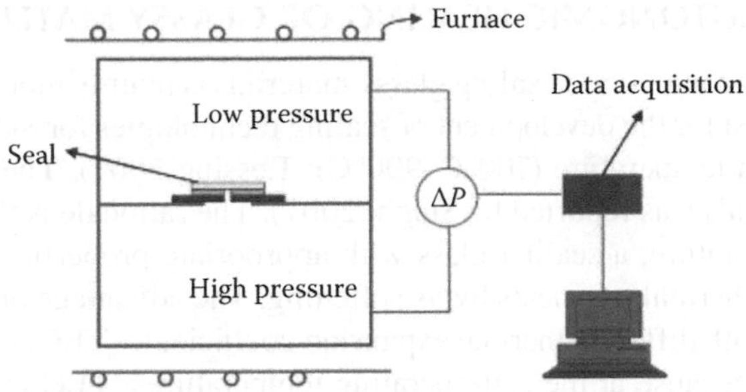

FIGURE 11.2 Simplified diagram of the seal testing setup. (Reprinted from Singh, R.N., *Int. J. Appl. Ceram. Technol.*, 4, 134, 2007.)

developed (Deng et al. 2007; Singh 2007); a simplified diagram of the seal testing setup is reported in Figure 11.2 (Singh 2007).

The seal was attached to steel housing to which a steel tube is connected. This assembly is placed inside a tube furnace with one closed side. The other side of the tube is cleared by a metal flange with openings for vacuum pumping, thermocouple, and gas entry into the furnace. This setup enables to control the temperature and the pressure difference (ΔP) experienced by the seal. Self-healing can be demonstrated by the recovery of gas leakage that occurs during the thermal cycles from room temperature to operating temperature.

11.2.2 *Ex Situ* Self-Healing Characterization

Cracks at the glass surface are obtained by Vickers hardness tester indentation on a polished surface. This method enables obtaining reproducible cracks from one sample to another.

After indentation, the glass is heat-treated in a furnace, cooled down, and observed by optical microscopy or scanning electron microscopy (SEM). Transmission optical microscopy allows top views and side views of the cracks (Hrma et al. 1988), whereas reflection optical microscopy allows measuring crack lengths (Girard et al. 2011). SEM imaging enables high-resolution observation of crack morphologies (Wilson and Case 1997).

11.2.3 *In Situ* Self-Healing Characterization

High-temperature stages can be associated with optical microscopes (Parihar 2006) or environmental SEM (ESEM) (Wilson and Case 1997). They enable *in situ* observation of the crack-healing process. Furthermore, continuous image recording allows continuous characterization of self-healing (Singh and Parihar 2009; Coillot 2010). When the experiments are performed in an ESEM, the composition of the atmosphere in the observation chamber can be easily modified and its influence on self-healing mechanisms can be characterized (Coillot et al. 2010a). These observations can also be coupled with energy-dispersive spectroscopy (EDS) analyses in order to determine the local phase compositions (Coillot et al. 2010b, 2011).

11.3 NONAUTONOMIC HEALING OF GLASSY MATERIALS

The design of nonautonomic healing glassy materials is mainly motivated by the current large interest for the development of sealing technologies for SOFCs and SOECs working at high temperature (700°C–900°C) (Lessing 2007). The concept of self-healing glass seals was reported by Singh (2007). The rationale is that at the SOFCs operating temperature, a sealing glass with appropriate properties can heal cracks created during thermal transients by its softening. The advantage of this approach is that materials with different thermal expansion coefficients (TEC) can potentially be used for seals, because at the cell-operating temperatures, TEC mismatch-induced thermomechanical stresses can be relaxed out. However, there are still a number of challenges in making a seal with nonautonomic self-healing properties. Among them, Liu et al. (2011a) recently pointed out the importance of the geometric stability and structural integrity of the glass seal system for its successful application in SOFCs. Many studies also reported the modification of healing properties due to the glassy material aging (Singh 2007; Liu et al. 2008, 2010; Zhang et al. 2011).

Other material development strategies and/or applications are currently under study. Two of them must be noted because they illustrate the new uses of self-healing glassy materials: 10 μm thick glass layers are proposed to be used as dielectrics in high temperature (140°C) capacitors. The self-healing properties of glasses are here required for long-term operation (the objective is >10,000 h of module operation) (Tuncer 2009; Lanagan 2010). Temporary glass coatings are proposed to be used as an oxidation protective layer during slab reheating process of stainless steel (Liu et al. 2011b). The self-healing property of the glass is used to transform the initial glass slurry into a bubble-free glass coating during stainless steel reheating. Then, the glass coating is removed before hot rolling.

11.3.1 Nonautonomic Self-Healing Glasses

Even if many different glass-based compositions for SOFCs sealing were formulated, only few were specifically designed for their self-healing properties (Mahapatra and Lu 2010). For the study of glass self-healing properties, Parihar (2006) selected glass compositions regarding their TECs (9–12 ppm/°C) and their stability against crystallization. Coillot (2010) designed specific compositions aiming at obtaining the best self-healing efficiency taking TEC and crystallization parameters into account, as well as the glass viscosity in the application temperature range. Hrma and coworkers (Hrma et al. 1988; Wilson and Case 1997; Girard et al. 2011) used soda lime–silicate glasses, whereas Wilson and Case (1997) have selected a borosilicate glass.

Coillot (2010) and Singh (2007) have characterized the self-healing properties of their glasses using *in situ* optical microscopy (Figure 11.3) and *in situ* HT-ESEM (Figure 11.4) (Coillot et al. 2010a).

Parihar (2006) and Singh (2007) described the crack healing of a glass at 570°C, which can be divided into three stages. In the first stage, crack tip blunting is observed, which result in cylinderization of the crack. In the second stage, this cylindrical crack becomes filled by viscous flow of the glass and results in spheroidization of the crack. In the third stage, the spherical crack cavity also decreases with time and results in a crack-free surface (complete healing). This description is in good agreement with previous

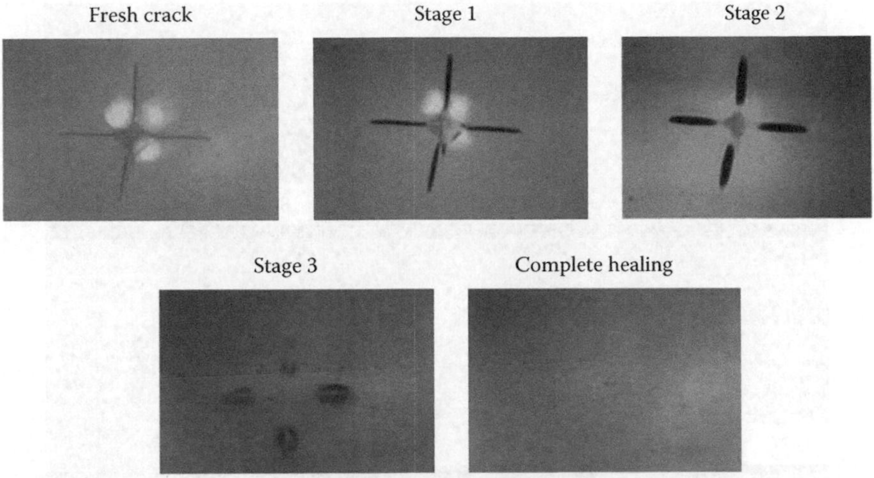

FIGURE 11.3 Optical micropictures of a glass showing several stages of self-healing of cracks produced by the microindentation technique. (Reprinted from Singh, R.N., *Int. J. Appl. Ceram. Technol.*, 4, 134, 2007.)

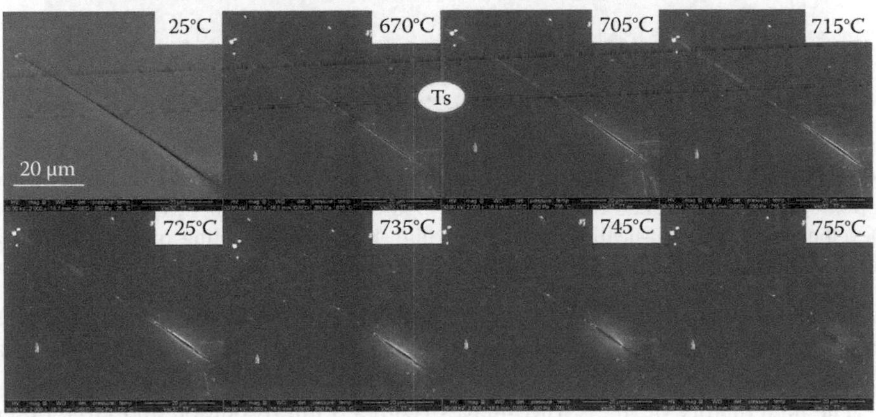

FIGURE 11.4 HT-ESEM micrographs of radial cracks generated by Vickers indentation as a function of treatment temperature (indicated in the figure) under air atmosphere. The softening point of the glass (indicated in the figure) is close to 680°C. Healing starts at 670°C (corresponding to a viscosity η of $10^{11.0}$ Poise) and is completed at 755°C (corresponding to viscosity η of $10^{8.9}$ Poise). (From Coillot, D., Development of the self-healing concept for high temperature sealing of electrochemical cells, PhD dissertation, Lille University, Lille, France, 2010.)

experimental study performed by Hrma et al. (1988) as well as with the crack evolution models proposed by Nichols and coworkers (Nichols and Mullins 1965; Nichols 1976). These authors stated that a surface of revolution evolves due to capillary induced surface diffusion, and their results indicate that a semi-infinite cylindrical shape will either evolve into a large sphere (spheroidization) or into a string of smaller spherical pores (ovulation), followed by the shrinkage of isolated pores. Experimental proof for the presence of residual persistent bubbles in the bulk glass was obtained by Hrma et al. (1988).

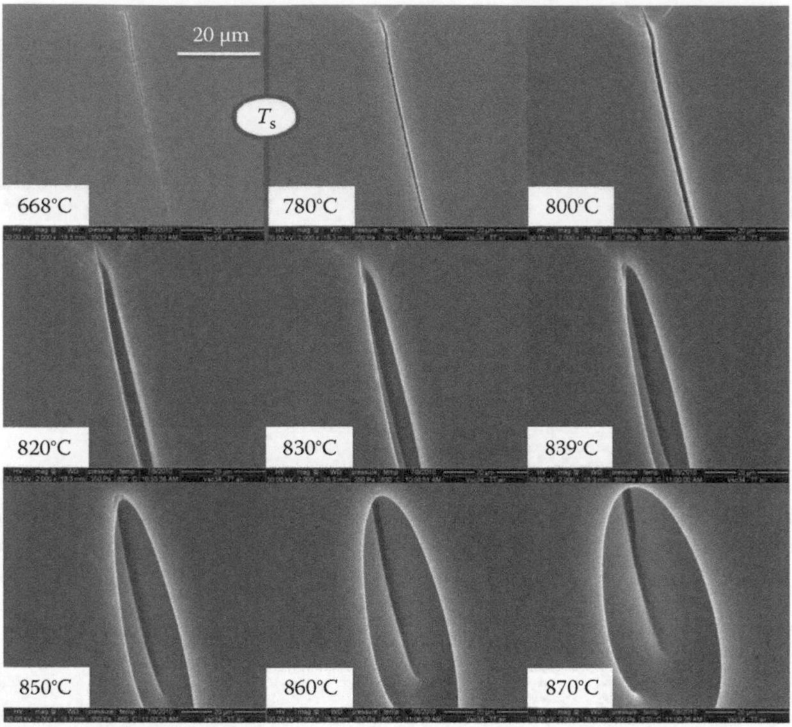

FIGURE 11.5 HT-ESEM micrographs of radial cracks generated by Vickers indentation as a function of treatment temperature (indicated in the figure) under air atmosphere. The softening point of the glass (indicated in the figure) is close to 780°C. (From Coillot, D., Development of the self-healing concept for high temperature sealing of electrochemical cells, PhD dissertation, Lille University, Lille, France, 2010.)

A more complex healing process occurs with a high T_s (softening temperature) glass corresponding to a pinching and healing of the bottom of the crack while the superficial part of the crack enlarges itself is reported by Coillot (2010) and shown in Figure 11.5.

The high quality of the ESEM images (with a 3 nm resolution and large observation depth), reported by Wilson and Case (1997) and Coillot (2010), allowed the observation of multiple areas of crack pinch-off along the length of the crack (Figure 11.6). The presence of crack debris was observed to hinder the crack-healing process (Wilson and Case 1997).

Singh and Parihar (2009) have developed a mathematical model for the description of the kinetics of each stage of the self-healing process. This model allows predicting the time necessary for crack healing, for a given glass composition and operating temperature, using a time coefficient t_c defined as

$$t_c = K \cdot \eta \tag{11.1}$$

where

η is the glass viscosity at the healing temperature
K is a constant that depends on glass composition

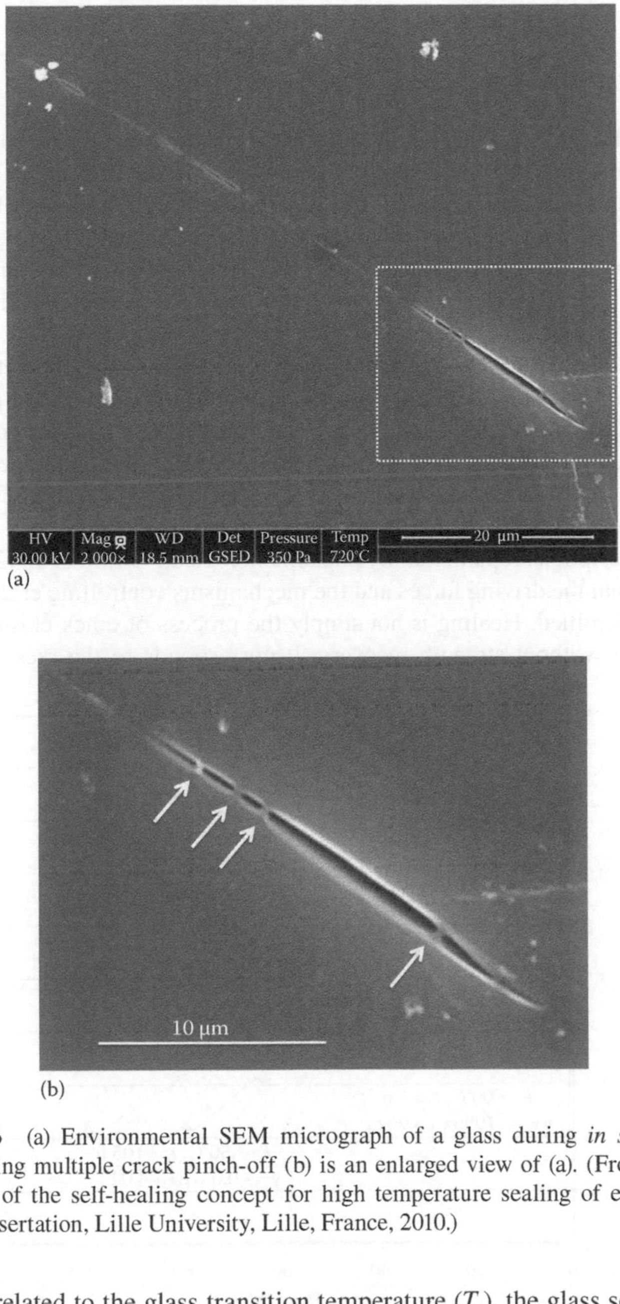

FIGURE 11.6 (a) Environmental SEM micrograph of a glass during *in situ* heat treatment illustrating multiple crack pinch-off (b) is an enlarged view of (a). (From Coillot, D., Development of the self-healing concept for high temperature sealing of electrochemical cells, PhD dissertation, Lille University, Lille, France, 2010.)

The t_c is related to the glass transition temperature (T_g), the glass softening temperature (T_s), and the experimental temperature (T):

$$\ln t_c = \ln K + \frac{6.44}{\left[(1/T_g)-(1/T_s)\right]}\left[\frac{1}{T}-\frac{1}{T_g}\right] + 26.01 \tag{11.2}$$

The time t necessary for healing a crack of length L is expressed as

$$\ln L = t_c \cdot t \tag{11.3}$$

From these data, it is concluded that self-healing of a crack can be obtained within very short time, but that the healing kinetics strongly depend on healing temperature. Singh and Parihar (2009) reported that the healing rate is very low when the operating temperature is lower than the softening point while the healing can be highly accelerated when operating temperature is larger than the softening temperature. An increase of the processing temperature by 75°C yields to an increase of the healing rate by a factor 100 (Figure 11.7).

The crack-healing kinetics is used to determine the relationship between softening temperature and healing temperatures (Figure 11.8). Both sets of data reported by Parihar (2006) and Coillot (2010) are concordant and yield to the conclusion that the healing temperature of glasses is very close to the softening temperature (Singh and Parihar 2009). It must also be noted that self-healing can be obtained over a large range of temperatures, depending on glass composition.

These descriptions remain mainly phenomenological. Girard et al. (2011) recently pointed out that the driving forces and the mechanisms controlling crack healing are not clearly identified. Healing is not simply the process of crack closure, i.e., crack disappearance without strength recovery. It corresponds to the reestablishment of

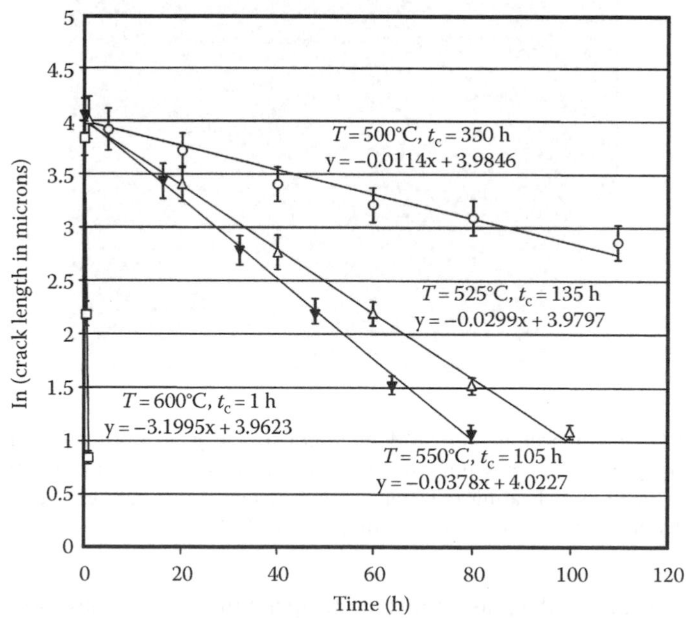

FIGURE 11.7 Evolution of crack length versus time at the temperature indicated. t_c is the characteristic healing time. (From Singh, R.N. and Parihar, S.S.: Self healing behavior of glasses for high temperature seals in solid oxide fuel cells, in *Advances in Solid Oxide Fuel Cells III: Ceramic and Engineering Science Proceeding*, eds. N.P. Bansal, J. Salem, and D. Zhu, pp. 325–332, 2009. Copyright Wiley-VCH Verlag GmbH & Co. KGaA. Reproduced with permission.)

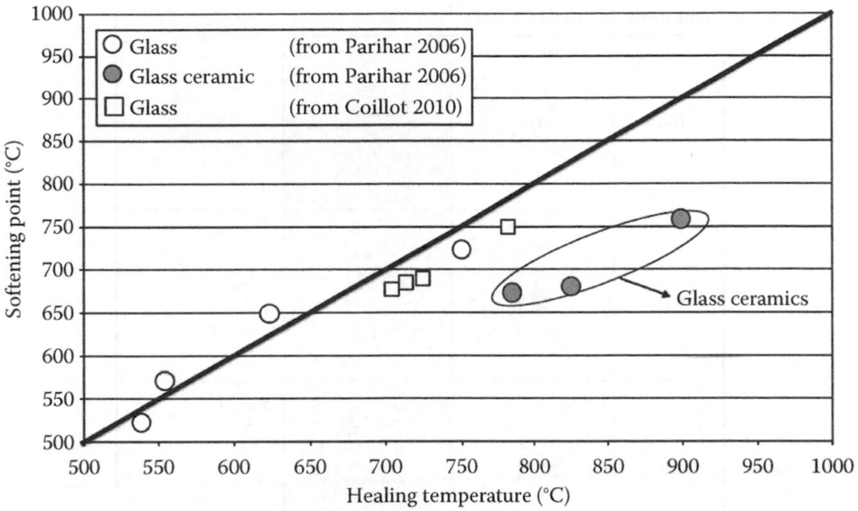

FIGURE 11.8 Softening point versus healing temperature determined for glasses and glass ceramics. (Modified from Parihar, S.S., High temperature seals for solid oxide fuel cells, PhD dissertation, Cincinnati University, Cincinnati, OH, 2006.)

the mechanical continuum that has been broken during fracture process, and consequently leads to strength recovery. Healing process can be driven by different physical and chemical mechanisms which depend on the temperature and atmosphere.

These authors explored the influence of the atmosphere composition (mainly water content) on crack healing. They clearly evidenced the importance of the humidity level but also of the initial water content of the glass at the vicinity of the crack. Two distinct phenomena were observed. On the one hand, direct crack closure only occurred when the hydration level of the glass is very low. The initial viscosity is quite large around the cracks and the viscous flow driven by capillary forces is impeded; the authors assume that direct crack closure is then due to release of residual tensile stresses induced by indentation. On the other hand, the crack morphological changes (enlargement, shortening, and rounding of radial cracks) were attributed to the smoothing of the crack tip through viscous flow driven by capillary forces. They suggest that the viscosity of glass surrounding cracks can be significantly reduced or increased by hydration/dehydration of the sample, which is strongly dependent on the water vapor pressure in the furnace. It is noteworthy that these morphological changes also depend on capillary forces, that is, on the curvature radius of the crack.

The effect of moisture on crack-healing kinetics is thus clearly evidenced in this study (Girard et al. 2011), as reported in Figure 11.9a and b. Wilson and Case (1997) performed their *in situ* study using ESEM under water vapor pressure. The sample morphologies that are reported are equivalent to those performed under air. Similarly, Coillot (2010) observed no significant difference between the experiments performed under air and water pressure. But these experiments were performed under ambient air (and not dry air) and during dynamic sample heating, contrary to those performed by Girard et al. (2011); it can thus be supposed that the water content in air during these experiments was sufficient to modify the healing mechanism.

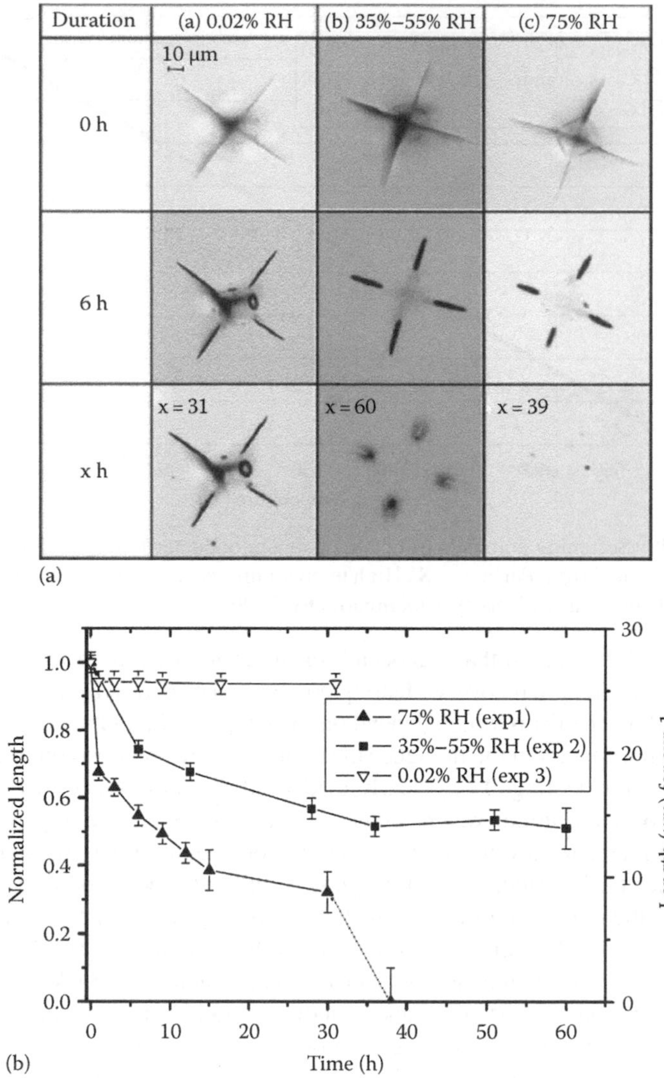

FIGURE 11.9 (a) Optical micrographs of Vickers radial cracks as a function of treatment time (indicated in the figure) when heat-treated at 620°C under (a) argon, (b) air, or (c) humid atmospheres. Prior to indentation, the sample was pretreated under air at 620°C. (b) Normalized radial crack length versus time at 620°C. (Reprinted from Girard, R. et al., *J. Am. Ceram. Soc.*, 94, 2404–2405, 2011.)

Furthermore, Girard et al. (2011) demonstrated that the influence of water content in the glass on the healing mechanism is due to the modification of the glass viscosity (the higher the water content, the lower the glass viscosity).

As stated by Jin and Lu (2010), a seal must possess long-term stability, as well as a good chemical stability versus cell components (Lessing 2007), in a wide range of oxygen partial pressure (fuel gas or air) and wet environment, with hundreds of thermal cycles >40,000 operating hours. High devitrification resistance in the aforementioned conditions is necessary in order to avoid thermophysical property mismatch at the sealing interfaces. When using a self-healing glass as a seal, the service temperature is close to the softening temperature of the glass, the diffusion of elements in the glass can be accelerated and yield to glass crystallization. Consequently, self-healing glasses are prone to detrimental crystallization under the SOFCs operating condition (Lu and Mahapatra 2008).

11.3.2 Nonautonomic Self-Healing Glass Ceramics

The use of glass ceramics can offer an interesting solution to the crystallization problem of self-healing glasses (Figure 11.10). The behavior at room temperature of glass ceramics can be described as being brittle. During thermal cycling, the tensile residual stresses caused by the mismatch of TEC of various cell components can potentially generate many small cracks and voids in the brittle seal material after cool down. If left untreated/unhealed, these small cracks in the seal could be detrimental to the overall stack reliability for the next operating cycle. Glass ceramics exhibit a healing rate that is strongly dependent on the amount of residual glass phase. Glass ceramics with large crystalline phase content shows a very slow healing response. For the glass-ceramic composition reported by Parihar (2006), considerably high temperature (950°C) is required for complete healing of cracks in a reasonable time of 4 h, and at

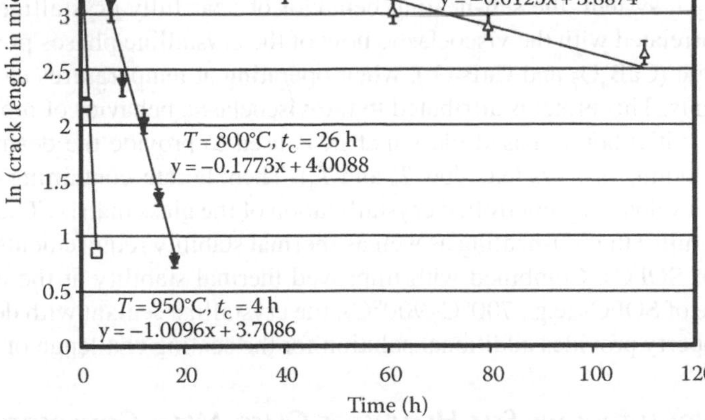

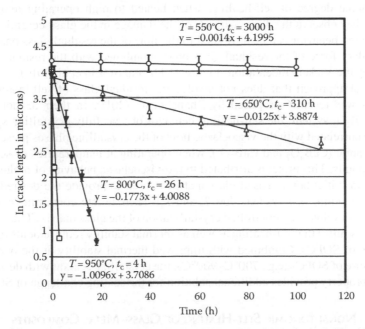

FIGURE 11.10 Evolution of crack length versus time at the temperature indicated. t_c is the characteristic healing time. (From Singh, R.N. and Parihar, S.S.: Self healing behavior of glasses for high temperature seals in solid oxide fuel cells, in *Advances in Solid Oxide Fuel Cells III: Ceramic and Engineering Science Proceeding*, ed. N.P. Bansal, J. Salem, and D. Zhu, pp. 325–332, 2009. Copyright Wiley-VCH Verlag GmbH & Co. KGaA. Reproduced with permission.)

(a) (b)

FIGURE 11.11 SEM pictures of glass ceramic (a) after indentation at room temperature and (b) after 0.5 h at 750°C. (Reprinted from Liu, W. et al., *J. Power Sources*, 185, 1193, 2008.)

typical SOFCs operating temperature of 800°C, >25 h are required for completion of healing. Similar tests were performed by Zhao et al. (2011) who concluded that only limited self-healing is obtained using glass ceramics as seals and that the recrystallization effect is more important than the self-healing due to the viscous flow of the glass. The conclusion of Parihar (2006) is that it seems that glass ceramics are not good choice for self-healing glassy seals, because their crack-healing response is too slow.

However, Liu et al. (2008) claimed that glass-ceramic sealant material does exhibit some degree of self-healing: when heated to high operating temperature because they observe the disappearance of the damage in the glass-ceramic sealant. This healing behavior can be attributed to the flow of the residual glass phase, or to the capillary force of the residual glass phase combined with the residual stresses caused by the Vickers indentation (Figure 11.11a and b) (Liu et al. 2010).

A novel approach that does not involve the residual glass of self-healing glass ceramics was recently proposed by Zhang et al. (2011). In their 33 mol% CaO-66 mol% B_2O_3 system, the self-healing behavior of near fully crystalline specimen could be correlated with the viscoelastic flow of the crystalline phases present in the glass ceramic (CaB_4O_7 and CaB_2O_4), when operating at temperatures close to their melting points. This effect is attributed to the viscoelastic behavior of molten borate phases. An initial borate-based glass matrix is used to provide the desired properties for the joining process (e.g., low T_g and T_d). Then, borate-containing crystalline phases are developed by controlled crystallization of the glass matrix. These crystalline phases fulfill the self-healing as well as thermal stability requirements in routine operation of SOFCs. Combined with improved thermal stability at the operational temperature of SOFCs (e.g., 700°C–900°C), the crystalline sealant with desired self-healing property provides additional solution for the sealing challenge of SOFCs.

11.3.3 Nonautonomic Self-Healing of Glass–Metal Composites

Silver–glass composites were designed for high-temperature sealing of fuel cells and sensors by Deng et al. (2007). The silver phase constitutes the matrix of the composite and provides ductility for sealing applications. The composite requires an appropriate amount of glass (softening temperature of which is lower than 800°C) that fills the small voids, adheres to the silver, and bonds to the NiO–YSZ ceramic parts of the cell. The optimized compositions, containing approximately 68–86 wt% metallic silver, exhibit good sealing integrity with a leak rate of <0.02 SCCM/cm (standard cubic centimeter per minute per centimeter of seal length) at 800°C, excellent thermal shock resistance, and self-healing ability through cycling between 800°C and room temperature.

11.3.4 Nonautonomic Self-Healing Properties of Metallic Glasses

An innovative application of self-healing is found in metallic glasses to reduce the surface defects (Moreira et al. 2008). When these glasses are heated to the supercooled liquid region above T_g, the healing kinetics is high enough such that defects with dimensions of the order of a micron have sufficient time to disappear prior to the onset of crystallization.

11.4 AUTONOMIC HEALING OF GLASSY MATERIALS

White et al. (2001), who introduced major innovations in self-healing polymeric materials, predicted that their approach would be extended to brittle ceramics in the future. Coillot et al. (2010b) indeed reported recently the first application of the autonomic extrinsic self-healing concept to glasses and glass ceramics. Appropriate selection of the healing agent enabled the self-repair of cracks that form during high-temperature operation at a temperature below the softening point of the material. This process thus enables self-healing without risk of deformation or thermal degradation of the structure.

This extrinsic self-healing concept, derived from that developed for polymeric materials, is illustrated in Figure 11.12. Healing particles of vanadium boride (VB) are dispersed within the glass or glass-ceramic matrix; they are called active particles. The choice of VB was justified on the basis of its reactivity with atmosphere and the thermal characteristics of the matrix. When a crack occurs on the surface of the sample (Figure 11.12a) and propagates in the bulk (Figure 11.12b), VB particles react in contact with atmospheric oxygen to produce a new glass that fills the crack and finally closes it (Figure 11.12c). This process can be operated on any glass or glass-ceramic composition; we describe in the following text its application to a Na_2O-K_2O-ZrO_2-SiO_2 system glass composite.

11.4.1 Selection of Efficiency Conditions for the Active Particles

Active particles must be stable at the working temperature in the absence of air, but must oxidize rapidly in the presence of air. An active particle can be as follows:

- A metallic element, which forms an oxide by direct oxidation: $a\mathrm{M} + b\mathrm{O}_2 \rightarrow \mathrm{M}_a\mathrm{O}_{2b}$.
- A ceramic (non-oxide), which generates oxides by oxidation. One can cite the borides that form B_2O_3 and a second oxide. Similarly, one can imagine using phosphide generating P_2O_5 and a second oxide; however, they are easily hydrolyzed in the presence of humid atmosphere to give a hydroxide and phosphine PH_3, a very dangerous gas. For example, VB, B, B_4C, V, and VC were shown to be efficient healing particles (Coillot et al. 2011).

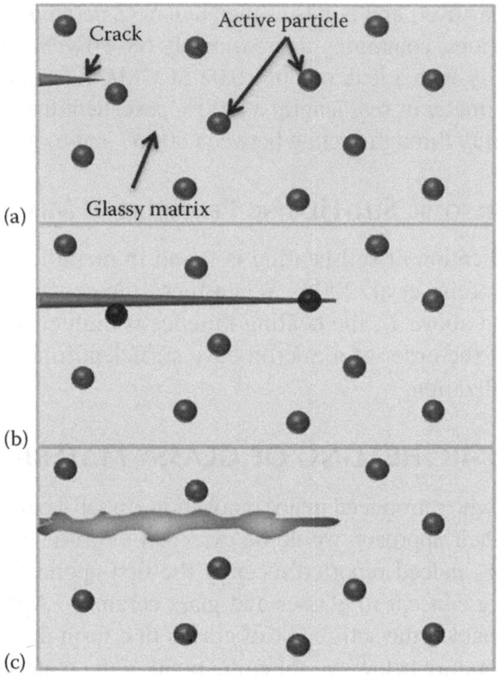

FIGURE 11.12 Schematic description of the autonomic (extrinsic) self-healing of glassy materials: active particles (healing agent) such as VB are embedded in a glassy matrix. (a) Crack forms in the matrix wherever damage occurs. (b) Oxidation of the active particle located in the crack occurs by reaction with oxygen. (c) V_2O_5 and B_2O_3 produced by the oxidation of VB have a low viscosity at the operating temperature, which enables them to spread and leads to crack healing. (Reprinted from Coillot, D. et al., *Adv. Funct. Mater.*, 20, 4374, 2010.)

The choice of precursors also depends on the melting temperature of the oxides they produce by their reaction, because at least one of them must be fluid at the working temperature. Only oxides having a melting temperature below this temperature would be able to flow into the crack during the autonomic self-healing.

As an example, thermal properties of VB active particles are reported in Figure 11.13. The differential thermal analysis (DTA) curve relative to VB indicates that it oxidizes in the temperature range from 400°C (T_1) to 700°C (T_2). This reaction is also revealed on the thermogravimetric (TG) curve by a weight increase due to the reaction with oxygen. The glass transition T_g temperature of the glass matrix is close to 750°C; Figure 11.13 thus shows that the oxidation of the VB particles occurs as required at a temperature lower than the T_g of the glass, meaning that the healing effect can be obtained without deformation of the glass piece. This confirms that the process can be classified as an autonomic (extrinsic) healing effect.

Since the healing must operate within a reasonable delay, the oxidation kinetic of VB was followed by the weight increase due to the reaction with oxygen. VB is oxidized only after 30 min of exposition to oxygen at 700°C, meaning a healing effect within a short period of time.

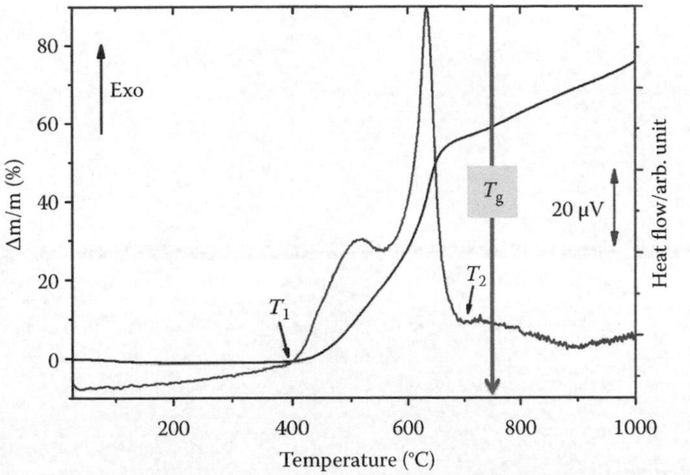

FIGURE 11.13 Thermogravimetric analysis (TG) and differential thermal analysis (DTA) curves of VB particles. They show that VB oxidation begins at 400°C (T_1) and is almost completed at 700°C (T_2), although TG indicates that some residual oxidation still occurs above 700°C, probably owing to limitation of oxygen diffusion by V_2O_5 and B_2O_3 formation. The glass T_g is close to 750°C.

11.4.2 ESEM *In Situ* Observation of the Autonomous Self-Healing Effect

HT-ESEM pictures of the sample heated at $T = 700$°C under oxidizing atmosphere, recorded at different run durations ($t = 0$, 11, 21, 30, 42, and 57 min), are reported in Figure 11.14.

From these micrographs, the oxidation of the VB particles and the formation of oxidized species are clearly evidenced. Post-experiment EDS analyses performed on the particles indicate that vanadium and boron oxides were formed during the heat treatment. Moreover, an incorporation of some elements initially present in the glass (Ca and Ba) was observed in the healed crack. Furthermore, due to the relatively low P_{O_2} and low level of gas renewal in the ESEM chamber, the kinetic of active particle oxidation is probably lowered in comparison with oxidation in an open medium. It was previously shown that VB is oxidized into B_2O_3 and V_2O_5, which are molten at 700°C (Coillot et al. 2010b). Their low viscosity enables spreading of the reaction products, which fill in the crack. Several vanadium oxides (V_2O_3, $T_{flow} = 1970$°C; VO_2, $T_{flow} = 1967$°C; and V_2O_5, $T_{flow} = 690$°C) are probably formed successively. Among them, only V_2O_5 is sufficiently fluid to pour into the cracks with B_2O_3.

The series of photomicrographs (Figure 11.14) indicates that healing process occurs first in the zones where the active VB particles are oxidized. After the healing process has begun, the cracks can be completely filled relatively rapidly. One way to characterize the healing behavior of the composite is to determine the so-called "healing ratio" defined as the ratio between the length of the crack that is filled with oxidation products and the initial length of the crack. The filling of the cracks by the VB oxidation products is illustrated in Figure 11.15 where the "healing ratio" is reported as

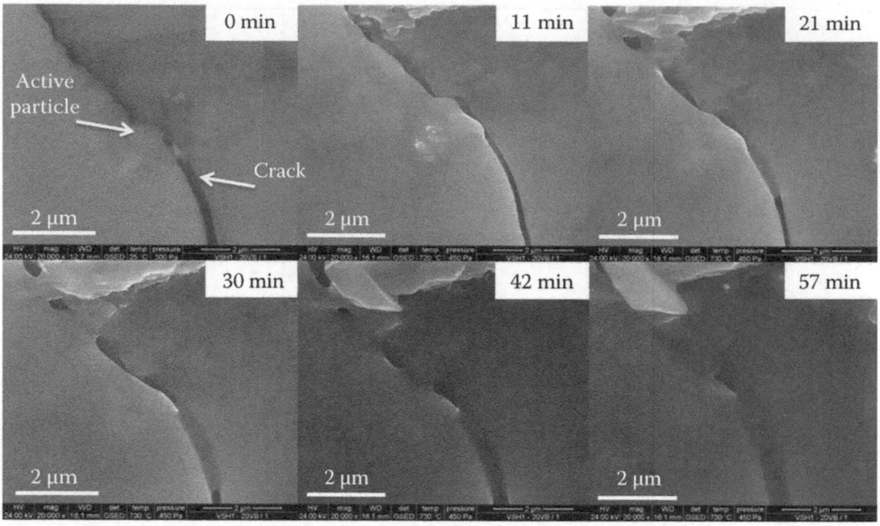

FIGURE 11.14 HT-ESEM images taken after different oxidation time. A crack was generated by Vickers indentation. An isothermal treatment at 700°C in air led to the oxidation of VB particles and the formation of V_2O_5 and B_2O_3, which flowed and filled the crack. (Reprinted from Coillot, D. et al., *Adv. Eng. Mater.*, 13, 433, 2011.)

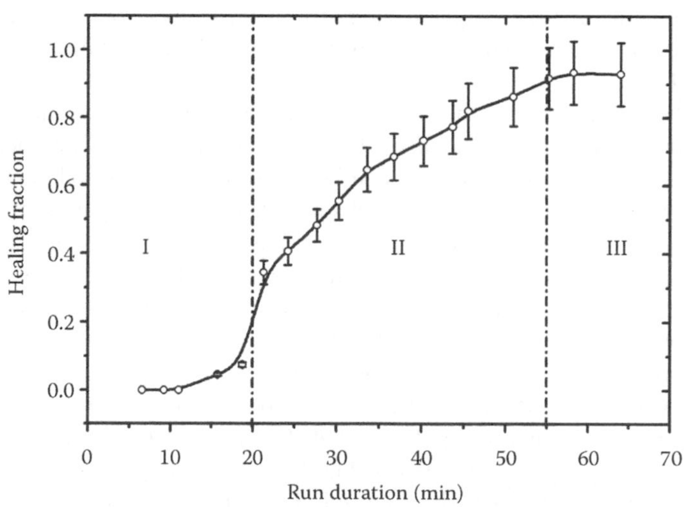

FIGURE 11.15 "Healing fraction" plotted as a function of heat treatment duration. The healing ratio is defined as the length of healed crack divided by the full length of the crack. Three domains can be distinguished: from 0 to 10 min corresponding to an incubation period (I), from 10 to 58 min where the crack healing occurs up to a ratio of 1 (II), the crack is fully repaired (III). (Reprinted from Coillot, D. et al., *J. Electron Microsc.*, 59, 366, 2010.)

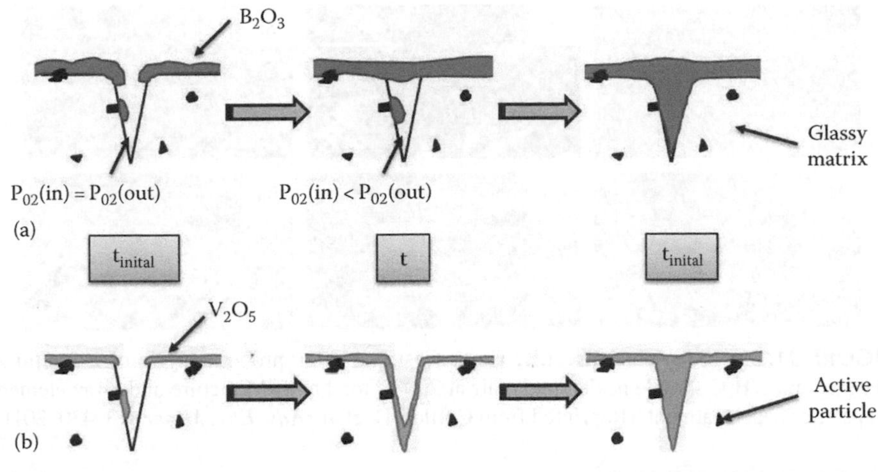

FIGURE 11.16 Schematization of the crack-healing mechanism from B_2O_3 (a) and V_2O_5 (b). (Reprinted from Coillot, D. et al., *Adv. Eng. Mater.*, 13, 433, 2011.)

a function of heat treatment duration. Three steps were distinguished by the authors (Coillot et al. 2010a). The first one, from 0 to 10 min, corresponds to an incubation period (I). During this period, the oxidation products slowly form, pour onto the crack, and begin to fill it. After a given run duration (II), the filling of the crack is considered to begin when the oxidation products level have reached the upper part of the crack. The filling process begins simultaneously at several points of the crack and develops up to the collapsing of neighboring zones. The healing ratio slowly and regularly increases, according to a sigmoid-like curve, up to the value of 1. When this value is reached, the crack is completely filled (III). When looking at the zone that corresponds to the initial Vickers indent, the volume of the indent mark is totally healed.

On the basis of the HT-ESEM results, two distinct crack healing mechanisms have been proposed (Figure 11.16) by Coillot et al. (Coillot et al., 2011). These mechanisms mainly depend on the viscosity of the oxydes formed from the active particles oxidation that controls the filling of the cracks.

11.4.3 Microprobe *Ex Situ* Observation of the Autonomous Self-Healing Effect

Complementary *ex situ* analyses were carried out on glass ceramic (barium–lime–silicate-based glass) mixed with B_4C healing particles, at the interface of a crack generated in the sample (Coillot 2010). X-ray maps of B, Ca, Ba, and Si were recorded after 1 h heat treatment at $T = 700°C$ (Figure 11.17). The element maps indicate that the molten products that have filled the crack are rich in B. One can observe that only the active particles in contact with oxygen in the atmosphere along the crack are oxidized and lead to B_2O_3, whereas B_4C particles that remain in the bulk of the samples remain unaltered. Moreover, the reaction of B_4C with glass produces locally a new glass that contributes to the crack healing and shows excellent compatibility with the glass matrix.

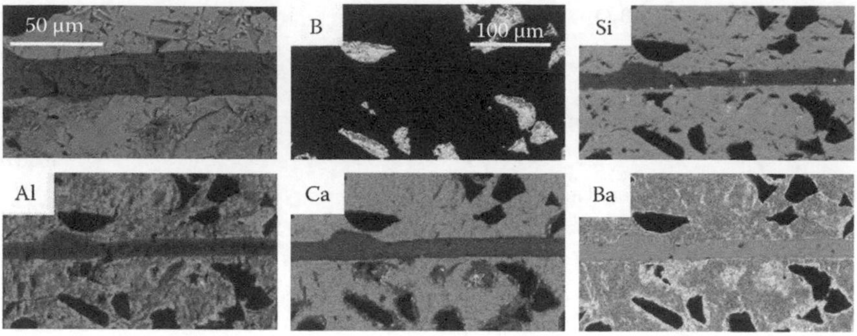

FIGURE 11.17 **(See color insert.)** *Ex situ* Castaing microprobe analysis of a fractured glass-ceramic/B_4C sample healed under air at 700°C for 1 h: SEM picture and x-ray element maps after heat treatment. (Reprinted from Coillot, D. et al., *Adv. Eng. Mater.*, 13, 430, 2011.)

11.4.4 Autonomous Self-Healing of Surface Cracks in Structural Ceramics

The autonomous self-healing of cracks in ceramics was reported in 1970 by Lange (1970). The process was based on the oxidation of polycrystalline SiC (Lange 1970) or Si_3N_4 (Easler et al. 1982) particles by heat treatment in air. The temperature required to achieve this crack-healing mechanism is lower than that required for the healing by a simple sintering (intrinsic healing). Many investigations on crack healing of ceramics have been reported, mainly for aeronautic applications at high temperature (Thompson et al. 1995; Niihara and Nakahira 1998; Chu et al. 2005). This type of crack healing is driven by the oxidation of >10 vol% SiC. It is triggered when the material is damaged and occurs under service conditions. However, the strength recovery is not high enough.

After 1999, a new generation of materials was proposed by Ando and coworkers (Chu et al. 1995; Ando et al. 1998, 2004). They developed ceramic-based composites containing >15 vol% SiC particles that exhibit crack-healing property allowing a complete recovery of the material strength. Viricelle et al. (2001) have developed a multilayered matrix deposited by chemical vapor infiltration (Figure 11.18a and b). The matrix is composed of three different constituents: a phase containing silicon, boron, and carbon noted SiBC and referred to as matrix 1; a phase containing boron and carbon noted B_4C, called matrix 2; and a third phase containing silicon and carbon noted SiC, called matrix 3.

Such a material exhibits an efficient protection against oxidation from 650°C to 1200°C, under dry or wet atmosphere. In the 650°C–900°C temperature range, protection is mainly related to the oxidation of boron carbide into boron oxide. At higher temperatures, SiC and SiBC oxidize and yield to the formation of silica or of a borosilicate glassy phase. In each temperature domain, the oxides are sufficiently fluid to fill in cracks and to generate the healing process.

The healing efficiency was reported to depend on three conditions: atmosphere, temperature, and stress (Ghosh 2009b):

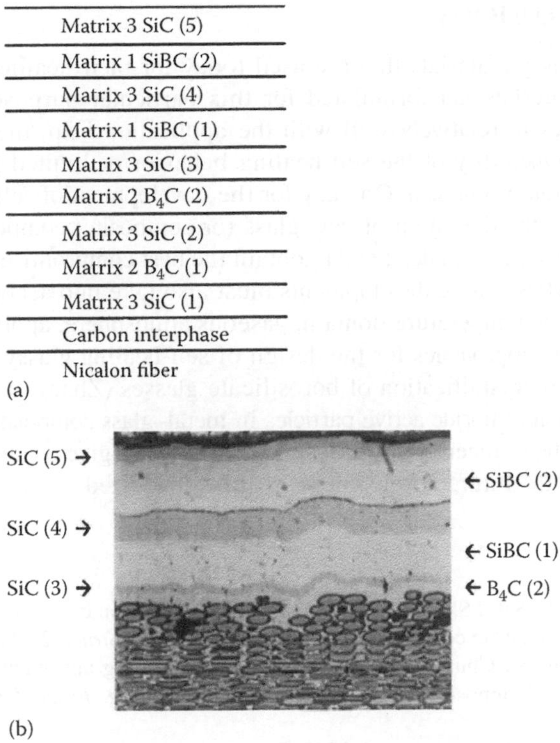

FIGURE 11.18 (a) Schematic illustration of the self-healing composite architecture. (Reprinted from Viricelle, J.P. et al., *Comp. Sci. Technol.*, 61, 614, 2001.) (b) Optical micrograph of the self-healing composite. (Reprinted from Wilson, B.A. and Case, E.D., *J. Mater. Sci.*, 32, 3163, 1997. With permission.)

1. The presence of oxygen in the atmosphere is mandatory since it produces the self-healing phenomenon, as crack healing is driven by the oxidation of SiC.
2. A relation between the temperature and the strength recovery follows an Arrhenius equation. Thus it is important to determine the temperature range for self-healing of cracks.
3. The stress applied to the components is also an important factor to determine the condition of self-healing since it is reported to influence crack growth.

11.4.5 Repeatability of the Self-Healing Process

Many efforts in the research of new self-healing processes are still needed. The microencapsulate-based self-healing approach using healing particles has indeed the major disadvantage of uncertainty in achieving complete and/or multiple healings, since it involves a limited amount of healing agent and it is not known when the healing agent will be fully consumed.

11.5 CONCLUSIONS

Most of the glassy materials that are used today for their healing properties are commercial materials not formulated for this particular purpose. Even if their initial properties fit relatively well with the application field, they often present a limited reproducibility of the self-healing behavior or limited lifetime (due to the crystallization of glasses). One key for the development of self-healing glassy materials lies in the definition of new glass (or composite) compositions that are specifically developed and designed to obtain the best compromise between all the required properties. These developments must also take into account the application regarding the temperature domain, gaseous atmosphere, application lifetime, and so on. Novel approaches for the design of self-healing glassy materials, such as the controlled crystallization of borosilicate glasses (Zhang et al. 2011) or the selective oxidation of boride active particles in metal–glass composites (Coillot et al. 2010b,c), have been recently proposed. They probably give a good indication in what directions the future developments will be conducted.

REFERENCES

Ando, K., Ikeda, T., Sato, S. et al. 1998. A preliminary study on crack healing behaviour of Si_3N_4/SiC composite ceramics. *Fatig. Fract. Eng. Mater. Struct.* 21:119–122.

Ando, K., Kim, B. S., Chu, M. C. et al. 2004. Crack-healing and mechanical behaviour of Al_2O_3/SiC composites at elevated temperature. *Fatig. Fract. Eng. Mater. Struct.* 27:533–541.

Chu, M. C., Cho, S. J., Yoon, K. J. et al. 2005. Crack repairing in alumina by penetrating glass. *J. Am. Ceram. Soc.* 88:491–493.

Chu, M. C., Sato, S., Kobayashi, Y. et al. 1995. Damage healing and strengthening behaviour in intelligent mullite/SiC ceramics. *Fatig. Fract. Eng. Mater. Struct.* 18:1019–1029.

Coillot, D. 2010. Development of the self-healing concept for high temperature sealing of electrochemical cells. PhD dissertation. Lille University, Lille, France.

Coillot, D., Méar, F. O., and Montagne, L. 2010c. Self-healing vitreous composition, method for preparing same, and uses thereof. WO2010/136721.

Coillot, D., Méar, F. O., Podor, R. et al. 2010b. Autonomic self-repairing glassy materials. *Adv. Funct. Mater.* 20:4371–4374.

Coillot, D., Méar, F. O., Podor, R. et al. 2011. Influence of the active particles on the self-healing efficiency in glassy matrix. *Adv. Eng. Mater.* 13:426–435.

Coillot, D., Podor, R., Méar, F. O. et al. 2010a. Characterization of self-healing glassy composites by high-temperature environmental scanning electron microscopy (HT-ESEM). *J. Electron Microsc.* 59:359–366.

Deng, X., Duquette, J., and Petric, A. 2007. Silver–glass composite for high temperature sealing. *Int. J. Appl. Ceram. Technol.* 4:145–151.

Easler, T. E., Bradt, R. C., and Tressler, R. E. 1982. Effects of oxidation and oxidation under load of strength distributions of Si_3N_4. *J. Am. Ceram. Soc.* 65:317–320.

Ghosh, S. K. 2009a. Self-healing materials: Fundamentals, design strategies, and applications. In: *Self-Healing Materials*, ed. S. K. Ghosh, pp. 1–28. Wiley-VCH, Weinheim, Germany.

Ghosh, S. K. 2009b. Self-healing of surface cracks in structural ceramics. In: *Self-Healing Materials*, ed. S. K. Ghosh, pp. 183–217. Wiley-VCH, Weinheim, Germany.

Girard, R., Faivre, A., and Despetis, F. 2011. Influence of water on crack self-healing in soda-lime silicate glass. *J. Am. Ceram. Soc.* 94:2402–2407.

Hrma, P., Han, W. T., and Cooper, A. R. 1988. Thermal healing of cracks in glass. *J. Non-Cryst. Solids* 102:88–94.

Jin, T. and Lu, K. 2010. Thermal stability of a new solid oxide fuel/electrolyzer cell seal glass. *J. Power Sources* 195:195–203.

Lanagan, M. 2010. Glass dielectrics for DC bus capacitors. DOE hydrogen program and vehicle technologies program annual merit review and peer evaluation meeting, June 7–11, Washington, DC.

Lange, F. F. 1970. Healing of surface crack in SiC by oxidation. *J. Am. Ceram. Soc.* 53:290.

Lessing, P. A. 2007. A review of sealing technologies applicable to solid oxide electrolysis cells. *J. Mater. Sci.* 42:3465–3476.

Liu, W., Sun, X., and Khaleel, M. A. 2008. Predicting Young's modulus of glass/ceramic sealant for solid oxide fuel cell considering the combined effects of aging, micro-voids and self-healing. *J. Power Sources* 185:1193–1200.

Liu, W. N., Sun, X., and Khaleel, M. A. 2011a. Study of geometric stability and structural integrity of self-healing glass seal system used in solid oxide fuel cells. *J. Power Sources* 196:1750–1761.

Liu, W. N., Sun, X., Koeppel, B., and Khaleel, M. 2010. Experimental study of the aging and self-healing of the glass/ceramic sealant used in SOFCs. *Int. J. Appl. Ceram. Technol.* 7:22–29.

Liu, P., Wei, L., Ye, S. et al. 2011b. Protecting stainless steel by glass coating during slab reheating. *Surf. Coat. Technol.* 205:3582–3587.

Lu, K. and Mahapatra, M. K. 2008. Network structure and thermal stability study of high temperature seal glass. *J. Appl. Phys.* 104:074910–074918.

Mahapatra, M. K. and Lu, K. 2010. Glass-based seals for solid oxide fuel and electrolyzer cells—A review. *Mater. Sci. Eng. R* 67:65–85.

Moreira, Jorge Jr., A., Inoue, A., and Yavari, A. R. 2008. Surface self-healing of metallic glasses in the super-cooled liquid region $\Delta T = T_x - T_g$. *Rev. Adv. Mater. Sci.* 18:193–196.

Nichols, F. A. 1976. Spheroidization of rod-shaped particles of finite length. *J. Mater. Sci.* 11: 1077–1182.

Nichols, F. A. and Mullins, W. W. 1965. Morphological changes of a surface of revolution due to capillarity-induced surface diffusion. *J. Appl. Phys.* 36:1926–1935.

Niihara, K. and Nakahira, A. 1998. Strengthening of oxide ceramics by SiC and Si_3N_4 dispersions. In *Proceeding of the Third International Symposium on Ceramic Materials and Components for Engines*. American Ceramics Society, Westerville, OH, pp. 916–926.

Parihar, S. S. 2006. High temperature seals for solid oxide fuel cells. PhD dissertation. Cincinnati University, Cincinnati, OH.

Singh, R. N. 2007. Sealing technology for solid oxide fuel cells (SOFC). *Int. J. Appl. Ceram. Technol.* 4:134–144.

Singh, R. N. and Parihar, S. S. 2009. Self healing behavior of glasses for high temperature seals in solid oxide fuel cells. In: *Advances in Solid Oxide Fuel Cells III: Ceramic and Engineering Science Proceeding*, eds. N. P. Bansal, J. Salem, and D. Zhu, pp. 325–332. John Wiley & Sons.

Thompson, A. M., Chan, H. M., and Harmer, M. P. 1995. Crack healing and stress relaxation in Al_2O_3–SiC "nanocomposites." *J. Am. Ceram. Soc.* 78:567–571.

Tuncer, E. 2009. Well defined structures for capacitor applications. US12/351121.

Viricelle, J. P., Goursat, P., and Bahloul-Hourlier, D. 2001. Oxidation behaviour of a multi-layered ceramic-matrix composite $(SiC)_f/C/(SiBC)_m$. *Compos. Sci. Technol.* 61:607–614.

White, S. R., Sottos, N. R., Geubelle, P. H. et al. 2001. Autonomic healing of polymer composites. *Nature* 409:794–797.

Wilson, B. A. and Case, E. D. 1997. *In-situ* microscopy of crack healing in borosilicate glass. *J. Mater. Sci.* 32:3163–3175.

Zhang, T., Tanga, D., and Yang, H. 2011. Can crystalline phases be self-healing sealants for solid oxide fuel cells? *J. Power Sources* 196:1321–1323.

Zhao, Y., Malzbender, J., and Gross, S. M. 2011. The effect of room temperature and high temperature exposure on the elastic modulus, hardness and fracture toughness of glass ceramic sealants for solid oxide fuel cells. *J. Eur. Ceram. Soc.* 31:541–548.

12 Basic Principles of Self-Healing in Polymers, Ceramics, Concrete, and Metals*

Martin D. Hager and Ulrich S. Schubert

CONTENTS

12.1 INTRODUCTION

Biological materials are prime examples for evolutionarily optimized functional systems. One of their most outstanding properties is the ability of regeneration of function upon the infliction of damage by external negative influences, in particular, mechanical loads. In nature, self-healing can take place either at the molecular level (e.g., repair of DNA) or at the macroscopic level: closure and healing of injuries of blood vessels and merging of broken bones. These processes are familiar, even self-evident (e.g., healing of a small cut in the finger). In contrast, man-made engineering materials generally do not possess this healing ability. Instead, they were and are still developed on the basis of the "damage prevention" paradigm rather than a "damage management" concept (van der Zwaag 2007). The materials developed should stand more and more stress before they start to fail, but they cannot "manage" damage, that is, they cannot compensate smaller damages in order to elongate their lifetime.

* This chapter is based on the review of Hager et al. (2010). (From Hager, M.D., Greil, P., Leyens, C., van der Zwaag, S., and Schubert, U.S., *Adv. Mater.* 2010, *22*, 5424–5430. Copyright Wiley-VCH Verlag GmbH & Co. KGaA. Reproduced with permission.)

It is, however, undisputed that self-healing materials would offer enormous opportunities, in particular, for applications where long-term reliability in poorly accessible areas is important, for instance, in off-shore wind parks or tunnels. Furthermore, applications that require a high-level reliability (e.g., aviation, space flight) would profit immensely from self-healing materials; a fatal failure of a material could be circumvented. In addition, self-healing would be in particular suited for applications that are prone to damage, such as surface coatings. It has to be emphasized that, in general, it is the functionality rather than the exact external or internal microstructure that is to be repaired.

Different strategies and approaches to devise self-healing materials in important material classes in engineering have been investigated, in particular metals, ceramics, concrete, and polymers. These materials possess different intrinsic properties; nevertheless, for all of them self-healing is based on the same common general principle and uses the same underlying concepts (Hager et al. 2010). Prerequisite for a self-healing of a (mechanical) damage is always the generation of a mobile phase, which can close this crack (Figure 12.1).

In case a damage is inflicted on the material (a and b), a crack can occur. Self-healing can take place at the microscopic to macroscopic level. The common principle is the subsequent generation of a "mobile phase" (c) triggered either by the occurrence of damage (in the ideal case) or by external stimuli. Subsequently, the damage can be removed due to the directed mass transport toward the damage site and the subsequent local mending reaction (d). The latter assures the (re)connection of the crack planes by physical interactions and/or chemical bonds. After the healing of the damage the previously mobile material is immobilized again, resulting in the best case in fully restored mechanical properties (e). This general principle is not limited to a single material class. However, the required temperature can vary, depending on the materials: ambient temperature for concrete, low temperatures (<120°C) for polymers (and their composites), high temperatures for metals (<600°C) and ceramics (>800°C), due to their respective intrinsic properties. Moreover, the size of damage able to be healed can vary substantially according to the size and number of species being transported and the mobility that can be achieved.

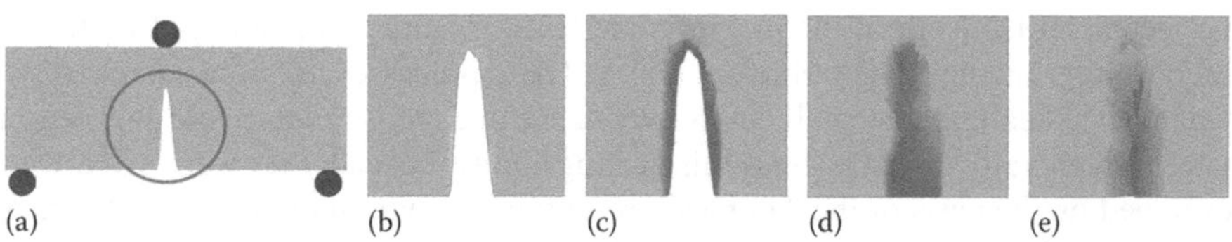

FIGURE 12.1 Common basic principle of self-healing materials. (a) The mechanical load induces a crack; (b) detailed view of the crack; (c) a "mobile phase" is induced; (d) closure of the crack by the "mobile phase"; and (e) immobilization after healing. (From Hager, M.D., Greil, P., Leyens, C., van der Zwaag, S., and Schubert, U.S.: Self-healing materials. *Adv. Mater.* 22, 5424, 2010. Copyright Wiley-VCH Verlag GmbH & Co. KGaA. Reproduced with permission.)

12.2 BASIC DEFINITIONS

Self-healing materials have been intensely investigated over the last 10 years, accompanied by a significant increase in the number of scientific publications (from <20 in 2001 to nearly 160 in 2010). As a result of the multitude of different approaches used and materials studied, self-healing materials can be divided into two different classes, depending on the required trigger and the nature of the self-healing process: Nonautonomic and autonomic. Nonautonomic self-healing materials require a modest external trigger, such as heat or light. The (additional) activation energy for the healing process can be supplied by the prevailing operating conditions as well as by targeted external stimuli (e.g., laser beam, inductive or resistive heating). In the case of polymers, these materials are also denoted as mendable polymers (Bergman and Wudl 2008). In contrast, autonomic self-healing materials do not require any additional external trigger; the damage itself is the stimulus for the healing. This concept corresponds to an adaptive structure, because the detection of the damage (by a sensor) as well as the repair (by an actuator) proceeds autonomically within the material structure.

Still another property of the respective self-healing process could be used to distinguish material subclasses, leading to the terms "extrinsic" and "intrinsic" self-healing (van der Zwaag 2007; Yuan et al. 2008; Ghosh 2009). Extrinsic self-healing materials themselves do not owe a hidden intrinsic capability for self-healing; rather, the healing process is based on external healing components, such as micro- or nanocapsules, intentionally embedded into the matrix materials to make them self-healing. The content of these capsules becomes the mobile phase upon damage. On the other hand, intrinsic self-healing requires no separate healing agents, which is to be preferred but, depending on the material class and healing mechanism, not always feasible. Formation of (secondary or primary) chemical bonds as well as physical interactions between the interfaces of the crack (adhesion, wetting) is a successful example to achieve the self-healing via this route, provided that the crack width is below a certain limit.

An ideal self-healing material (van der Zwaag 2007) could heal the damage completely, that is, the material has equal or better properties after healing than before the crack formed. The healing process would be autonomous, that is, without addition of a healing agent and while the component is in service. Lastly, the material could heal multiple times. However, such a material is still to be discovered.

12.3 KEY DEVELOPMENTS CONCERNING SELF-HEALING MATERIALS

The intense research during the last decade has led to the development of several different concepts for self-healing materials, covering different material classes (e.g., polymers, polymer composites, ceramics, concrete materials, and metals). The key developments regarding general concepts and principles will be discussed shortly in the following sections.

12.3.1 Self-Healing Polymers

Currently, polymers (and composites) are by far the most-studied material class in the context of self-healing behavior (for recent reviews see, e.g., Bergman and Wudl 2008; Wu et al. 2008; Yuan et al. 2008; Blaiszik et al. 2010; Hager et al. 2010; Fischer 2010; Murphy and Wudl 2010; Syrett et al. 2010; Brochu et al. 2011). This may be due to the ease by which functionalization and modification of polymeric systems can be achieved, the rather low temperatures required to induce mobility and the large volume of mobile molecules in comparison with the volume of mobile atoms. Initially, research was focused on extrinsic self-healing by embedded liquid-healing agents: Microcapsules filled with a monomer or another healing agent release their content after mechanical damage; afterward, the monomer polymerizes in the crack plane with the help of a catalyst and the healing agent closes the crack, respectively, which results in a rebinding of the crack faces (Figure 12.2 left) (White et al. 2001). The concept, though undoubtedly working very well (White et al. 2001; Brown et al. 2003, 2004; Jones et al. 2006; Anderson et al. 2007), has the intrinsic drawback that locally the regeneration of function can happen only once. The capsules can release only once the embedded healing agent. A repeated crack/damage at the same location cannot be healed. However, due to the straightforward principle, a large variety of different healing agents have been proposed and studied, for example, pure solvents (e.g., chlorobenzene) (Caruso et al. 2007) as well as combinations of reactive monomers (Blaiszik et al. 2009), linseed oil (Suryanarayana et al. 2008), isocyanates (Yang et al. 2008), polydimethylsiloxane and polydiethoxysiloxane (tin-catalyzed polycondensation) (Cho et al. 2006), polymeric azides and alkynes for "click chemistry" (Gragert et al. 2011), or epoxy resins (Yin et al. 2007; Xiao et al. 2009). Moreover, binary systems have been developed, which contain additional Pickering stabilizers (Mookhoek et al. 2008). Semicommercial microcapsule systems for polymer coatings (Cho et al. 2009; Mehta and Bogere 2009) and bulk polymers are already commercially available (company: Autonomic Materials™).

The healing agent can also be embedded in hollow (glass) fibers enabling the long-range transport of the agent (Dry 1992, 1996; Dry and McMillan 1996; Motuku 1999; Trask and Bond 2006; Trask et al. 2007; Bond et al. 2008) (Figure 12.2 middle). This one-dimensional system was extended to two-dimensional ordering of the fibers filled with healing agent (Williams et al. 2008a,b). A subsequent biomimetic

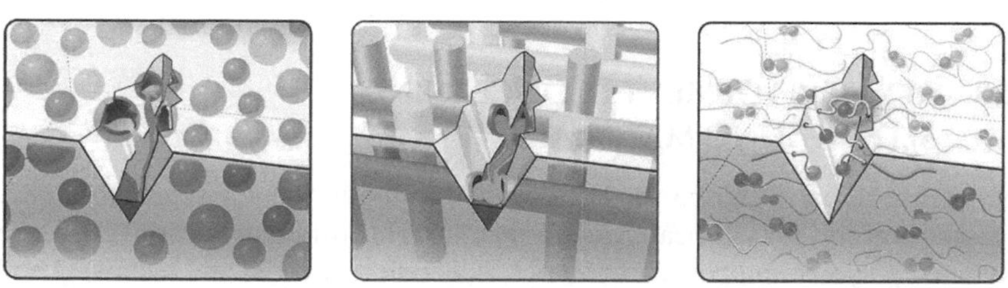

FIGURE 12.2 **(See color insert.)** Concepts for self-healing polymers based on microcapsules (left), vascular system (middle), and reversible interactions (right). (Reproduced from Blaiszik, B.J. et al., *Annu. Rev. Mater. Res.*, 40, 211, 2010. With permission from Annual Reviews.)

development is represented by three-dimensional microvascular networks (Toohey et al. 2007; Hansen et al. 2009). The systems described can also be used for the simultaneous transport of two healing agents (Toohey et al. 2009). A recent numerical study indicated that neither the spherical capsule nor the hollow fiber geometry are ideal for obtaining a good healing efficiency, and much better healing efficiencies can be obtained using elongated capsules with aspects ratios up to a value of 10 (Mookhoek et al. 2009). Finally, a recent example of filled microcapsules extended the biomimetic approach by the mimicking of white blood cells. A healing agent can be released from microcapsules that function as microcarrier ("repair-and-go" system) (Kolmakov et al. 2010). The target specificity was increased and a high level of control could be achieved. A further extension of the concept might be feasible by linking the generic healing capabilities to specific recognition capabilities given by nature (e.g., enzyme/substrate, antibody/antigen) (White and Geubelle 2010).

The examples of extrinsic self-healing described up to now concern only materials that show autonomic self-healing: the damage itself initiates the healing process, mainly by the release of a healing agent after breakage of a container, that is, microcapsule or hollow fiber (Yuan et al. 2008).

In contrast, intrinsic self-healing in most cases still requires a targeted external trigger (mostly heat). Also, other stimuli (Murphy and Wudl 2010) can be applied for the healing of polymers (besides the well-known mechanical stimulus, also electrical, electromagnetic, magnetic [Corten and Urban 2009], ballistic, and photostimuli [Pastine et al. 2009]). In the simplest case, intrinsic self-healing is achieved by physical interactions, that is, (modest) heating of the materials results in diffusion of the polymer chains followed by formation of new entanglements, resulting in the closure of the crack (Jud and Kaush 1979; Jud et al. 1981; Kim and Wool 1983; Lin et al. 1990). In case of a ballistic damage, sufficient heat is generated to allow self-healing processes, via the formation of secondary chemical bonds (DuPont, Surlyn®, and Nucrel®; already used at shooting stands) (Kalista and Ward 2007; Varley 2007, 2008; Wu et al. 2008). For damage involving the cleavage of the (covalent) polymer chains, self-healing can be achieved by the formation of new chemical bonds (Figure 12.2 right). Polycarbonates, for example, possess reactive end groups (Takeda et al. 2004) that can be used for the reconnection of polymer chains. In addition, functional side groups of polymers can be utilized for reversible thermal cross-linking by Diels–Alder reactions (Chen et al. 2002, 2003; Liu and Hsieh 2006; Murphy et al. 2008; Wouters et al. 2009) and Michael additions (Liu and Chen 2007), respectively. However, frequently higher temperatures (>100°C) are required for the self-healing process. The reversible binding units (e.g., cycloadduct furan maleimide or dicyclopentadiene) will open upon heating and new bonds are formed upon cooling, resulting in healing of a crack (Figure 12.3). Furthermore, photochemical cycloadditions have been investigated in the context of self-healing materials. A large variety of olefins can undergo a photochemical [2+2] cycloaddition. Some of these systems were utilized for polymers; however, only scarce examples are reported utilizing this approach for self-healing materials (Murphy and Wudl 2010). Cinnamoyl groups were incorporated into polymers; in this case, only a low healing efficiency was observed (Chung et al. 2004). A recent example is based on hyperbranched polymers, which bear anthracene units, capable of a photochemical cycloaddition (Froimowicz et al. 2011).

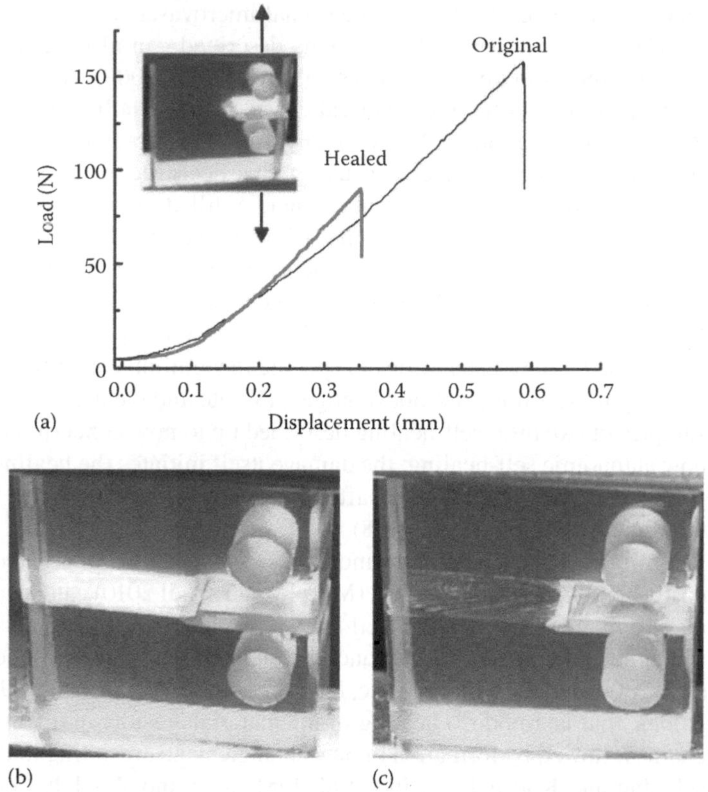

FIGURE 12.3 Self-healing (mendable) polymer based on reversible Diels–Alder cross-links. (a) Mending efficiency. Pictures of broken specimen (b) and after thermal treatment (c). (Reprinted from Chen, X. et al., *Science*, 295, 1702, 2002. With permission from American Association for the Advancement of Science.)

By this approach, a healable hydrogel was obtained. Constitutional dynamic polymers (dynamers) (Lehn 2010), which consist of monomeric components that are linked through reversible connections, revealed interesting potential for self-healing materials (e.g., in polymer networks) (Deng et al. 2010), hitherto not yet properly explored.

As compared to covalent bonds, weaker interactions, such as hydrogen bonding (Sijbesma et al. 1997; Berl et al. 2002; Scherman et al. 2006; Ohkawa et al. 2007; Wietor et al. 2009), seem to offer promising chances for the development of potential intrinsic self-healing materials, provided the spatial density of the bonds is high enough. A prominent current example of an autonomic intrinsic self-healing polymer, which may lead to a large-scale application (started production by ARKEMA) (Wietor and Sijbesma 2008; Tullo 2009), was developed recently by Leibler et al.: an oligomeric, thermoplastic elastomer consisting of fatty acids and diethylene diamine functionalized with urea (Cordier et al. 2008; Montarnal et al. 2008). During a damage event, mainly hydrogen bonds will break which, since being reversible, are able to rebuild after pressing the surfaces of the cut together. The material properties can be influenced by changing the ratio of the monomers (Montarnal et al. 2009). The same mechanism has now been used by AkzoNobel coatings to create commercial self-healing automotive repair coatings, which are capable of smoothening surface cracks under the influence of a modest temperature rise. Besides the earlier described hydrogen bonds, metal complexes or metal–ligand interactions (e.g., terpyridine metal complexes) (Schubert and Eschbaumer 2002; Andres and Schubert 2004; Fustin et al. 2007; Ott 2008; Wild et al. 2011) represent promising candidates for self-healing by noncovalent interactions (Murphy and Wudl 2010; Whitell et al. 2011). However, only a very limited number of examples are described up to now. For example, a polymer network, which contained metal complexes (bifunctional Pd(II) or Pt(II) compounds), was synthesized. The incorporation of the metal ions into the polymer networks led to an increase in the network's strength (Kersey et al. 2007). Metallopolymers based on *N*-heterocyclic carbenes have also been utilized as self-healing materials that, in addition, show electrical conductivity (Williams et al. 2007a,b).

Furthermore, ionic interactions (Varley 2008) and π–π interactions (Burattini et al. 2009a,b) are auspicious noncovalent interactions for potential self-healing behavior. The resulting materials behave like a molecular hook-and-loop fastener: chemical bonds (covalent and noncovalent) will break during a damage event, but the noncovalent ones can form again, similar to the macroscopic analogue.

Another new field in self-healing materials is "mechanochemistry": mechanical stress can activate, for example, a catalyst inducing a chemical reaction (Caruso et al. 2009; Piermattei et al. 2009). Furthermore, mechanophores are capable of visualization of the mechanical stress in a polymeric material; mechanical forces induce a conversion between a colorless and a colored structure (Davis et al. 2009).

12.3.2 Self-Healing Ceramics and Concrete

In principle, self-healing hydraulic concrete is known since Roman times (Jonkers 2007). However, currently applied concrete does not feature this ability. Recently, the autogenous self-healing of hydraulic concrete materials has been the object of intensive scientific and application-oriented research. In damaged concrete, water permeates through the cracks and induces the local formation of hydrates, which lead, depending on the water pressure, pH value, crack morphology, and crack width, to a detectable crack healing (Ramm and Biscoping 1998). As with all materials of interest here, crack width plays an important role in the self-healing process. For this reason, engineered cementitious composites (ECC) have been developed, which control the crack width to values below 60 μm even for very high strains well in excess of the strain leading to the first observable crack (Li and Yang 2007). In addition, a bioinspired approach to self-healing concrete is being investigated: Bacteria are immobilized in the concrete and will be activated if water permeates into fresh cracks, where they start to precipitate minerals (Jonkers 2007; Jonkers et al. 2010) (Figure 12.4). Calcium lactate is applied as nutrition for the bacteria. Encapsulation of healing agents in concrete, which will close the crack upon damage, has also been investigated (Tittelboom et al. 2011; Yang et al. 2011). This highlights the applicability of general approaches for self-healing materials for different material classes.

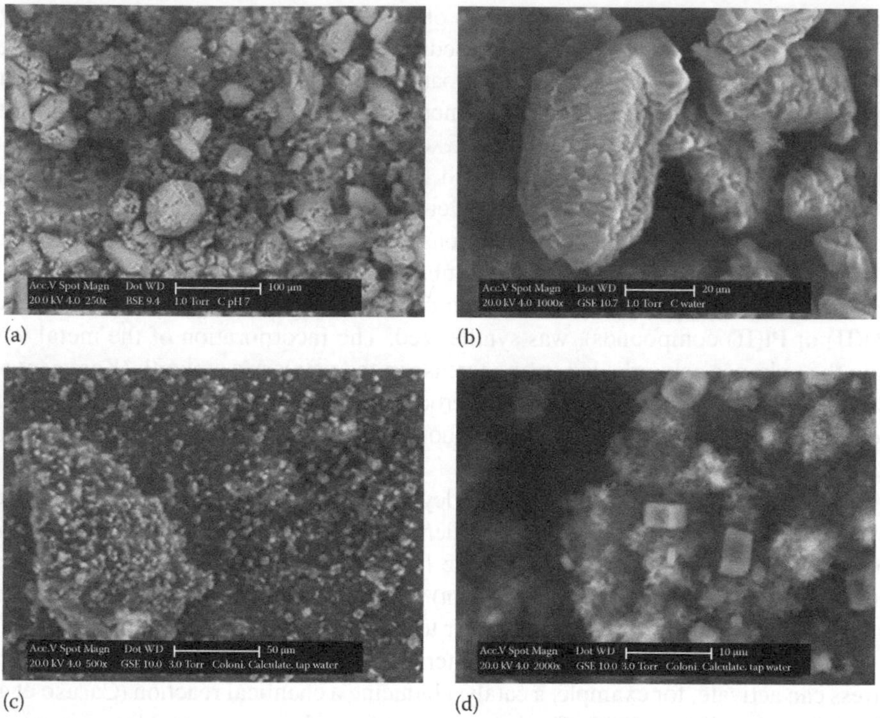

FIGURE 12.4 Specimens of cement stone with incorporated bacteria (*Bacillus cohnii* spores) and calcium lactate. Cracked after 7 (panel a: 250× and panel b: 1000× magnification) or 28 days curing (panel c: 500× and panel d: 2000× magnification). Mineral precipitates visible on crack surfaces of young specimen (a and b) are presumably due to bacterial conversion of calcium lactate to calcium carbonate. (Reprinted from *Ecol. Eng.*, 36, Jonkers, H.M., Thijssen, A., Muyzer, G., Copuroglu, O., and Schlangen, E., Application of bacteria as self-healing agent for the development of sustainable concrete, 230–235, Copyright 2010, with permission from Elsevier.)

In contrast, little is known on the regeneration processes in cracked engineered ceramic materials. Thus, the maximum crack length and, more importantly, the width of the mendable cracks, the optimal conditions (preferably low temperatures) as well as the local heat transport properties that change with crack growth and shrinkage still remain as open questions. With most current ceramics, very high local temperatures are required for repair, due to the high activation energies of the diffusive mass transport in the covalent or ionic structures of ceramics (e.g., SiC/$Al_6Si_2O_{13}$ composite: 1300°C) (Ando et al. 2002a,b). On the other hand, it could be shown that Al_2O_3- and $Al_6Si_2O_{13}$ ceramics reinforced by SiC whiskers are able to completely heal cracks up to a length >100 µm (Sugiyama et al. 2008). The extensive regeneration of the original strength could be demonstrated in SiC- and Al_2O_3 ceramics (sintering temperature >1400–1600°C) (Mitomo and Nishimura 1996). Considerably lower reaction temperatures can be expected if the grain boundary is wetted by an intergranular glass phase. Depending on the glass transition temperature T_g (Na–Ca silicate glass 500°C, and SiO_2 1100°C), stress relaxation processes, viscoelastic flow processes, and diffusion can be initiated at considerably lower activation temperatures. The healing reaction starts by grain boundary and surface diffusion at contact points in the region of crack faces (Mitomo and Nishimura 1996). Further energy supply results in the formation of spherical and cylindrical pores that can collapse at high temperatures provided they have small radii. The driving force is the reduction in surface energy (Rayleigh instability). Model calculations taking into account the surface diffusion–controlled collapse in Al_2O_3 have shown that a pore radius smaller than 10–25 nm offers the possibility for closing the pore (Travitzky et al. 2008). In addition, the oxidation of non-oxidic phases in oxidizing atmosphere (air) was shown to be another effective healing mechanism (Jun et al. 2004; Song et al. 2008). Furthermore, in multicomponent and multiphase ceramic materials the formation of an eutectic melt as well as the local particle rearrangement induced by a phase transition ($ZrO_{2,mkl} \rightarrow ZrO_{2,tetr}$) are considered as promising healing mechanisms (Jun et al. 2004).

While most of the work in the field of ceramics focuses on crack healing processes at temperatures above the working temperature, few efforts have been made on healing below these working conditions. As an example, a glass phase containing Si_3N_4/SiC composite already features a significant healing at 1000°C (also under cyclic stress), which leads to a distinct increase in the static and the dynamic fatigue strength (Ando et al. 2002a,b). Further investigations on this material have demonstrated a significant influence of the stress frequency (Takahashi et al. 2006). Other examples are self-healing oxidation and corrosion protection coatings on SiC ceramics (burner nozzle), where the combustion heat initiates the healing reaction. A reactive filler ($MoSi_2$) embedded in a SiOC matrix is oxidized by permeating oxygen, whereby the formed SiO_2 closes the crack completely (Figure 12.5) (Müller et al. 1998).

12.3.3 Self-Healing Metals

Self-healing (by physical and chemical effects) pursues the objective to recover the function or surface of the metallic material. However, this goal is, unfortunately, still far from being reached, as self-healing in metals is intrinsically more difficult than in other material classes (van der Zwaag 2007). Given the small size of metal atoms

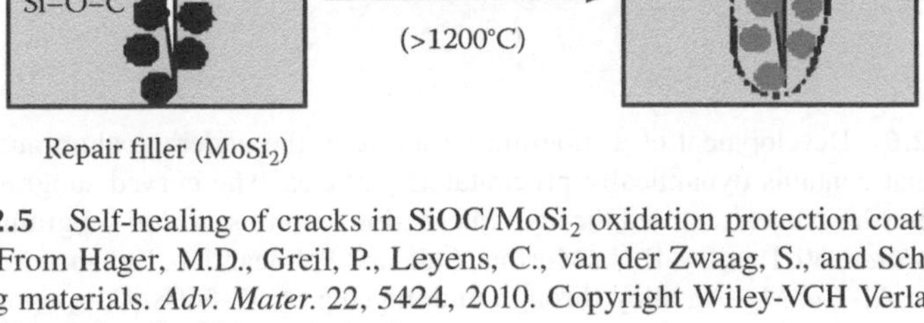

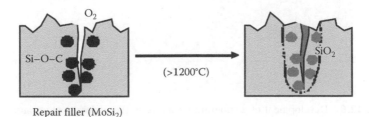

FIGURE 12.5 Self-healing of cracks in SiOC/$MoSi_2$ oxidation protection coatings on SiC ceramics. (From Hager, M.D., Greil, P., Leyens, C., van der Zwaag, S., and Schubert, U.S.: Self-healing materials. *Adv. Mater.* 22, 5424, 2010. Copyright Wiley-VCH Verlag GmbH & Co. KGaA. Reproduced with permission.)

and the absence of directionality in the chemical bonds, currently known mechanisms of healing in metals only lead to healing of defects with a rather small volume. Hence, examples of successful work on this class of materials are yet scarce. An effect known and utilized in technology since long is the reaction of metallic surfaces with surrounding material, which can lead to surface layers protecting the metal against harmful effects by the atmosphere. In principle, creep and diffusion processes can contribute to closing microcracks and pores at elevated temperatures. Recently, the self-healing properties of Al–Cu–Mg alloys were investigated (Lumley et al. 2002, 2003; Hautakangas et al. 2008). The deformed underage aluminum alloy AA2024 was aged at room temperature and the influence of the precipitation of fine Guinier–Preston zones on the mechanical properties was determined (Figure 12.6). It could be shown by positron annihilation spectroscopy that the deformation-induced defects due to diffusion and clustering of Cu atoms could be healed by secondary precipitations. In the context of metallic composite materials, effects of the yield stress anomaly and the development of oxygen diffusion barriers on the mechanical properties of high-performance materials are being investigated. In other studies, shape-memory alloys (SMA) were integrated into resins with filled microcapsules (Kirkby et al. 2009).

Three other examples of self-healing in metal composites or alloys are worth to be mentioned: (1) Ti–Bi alloys in the form of wires were arranged in parallel to the stress direction, so that a potential crack would proceed vertically to the direction of these wires. The wires perform two tasks: They contribute to crack bridging, and

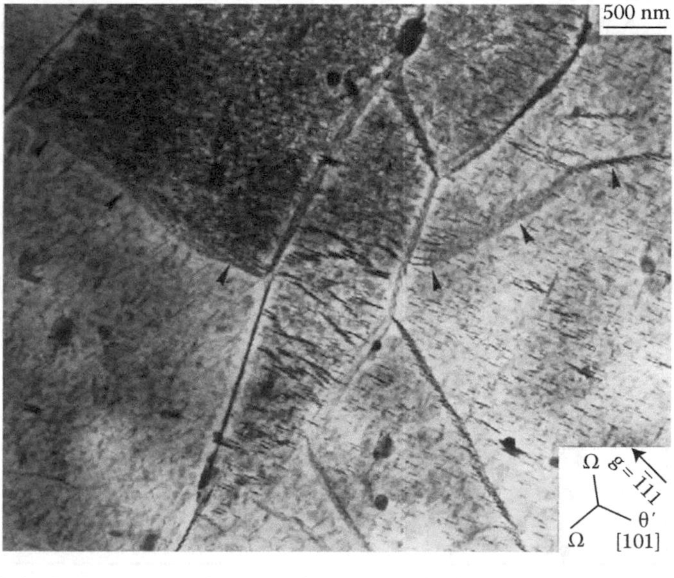

FIGURE 12.6 Development of a subgrain structure in the underage aluminum alloy during creep that contains dynamically precipitated particles. The curved subgrain boundary (arrowed) may have been retarded through interaction with this central subgrain. (Reprinted from *Acta Mater.*, 50, Lumley, R.N., Morton, A.J., and Polmear, I.J., Enhanced creep performance in an Al–Cu–Mg–Ag alloy through underageing, 3597–3608, Copyright 2002, with permission from Elsevier.)

they function as tie rods. As the material is able to undergo a martensitic transformation, the wire can withstand high strains and, thus, enable the bridging of the crack. If the damaged composite material is heated above the martensite finish temperature, the wire will adopt its shorter austenitic form and, consequently, pull the crack planes back. (2) Experimentally, the so-called creep steels have demonstrated the potential for self-healing creep cavities usually caused by nucleation, growth, and coalescence, which can lead to creep fracture (Manuel 2009; Shinya 2009; He et al. 2010). Therefore, autonomously self-healing cavities could prevent material failure, for example, in high-temperature applications of steel. The precipitation of copper, boron, and nitrogen, which are intentional alloying elements in these steels, within the creep cavities improved the long-term stability of steel significantly, provided a heat treatment was applied which would safeguard the mobility of these alloying elements. The heat treatment imposed deviates significantly from that used to give commercial creep steels not having a self-healing capability in their optimal properties. (3) Incorporation of capsules filled with liquid into an electroplated coating was recently successfully demonstrated. The coating was 15 μm thick, while the polymer capsules are a few hundred nanometers in diameters. While the self-healing properties have not yet been proven, it is considered a major success to include liquid-filled capsules into an electroplated coating since the electroplating process typically tends to destroy the capsules. In the event of surface damage, the capsules are designed to release a liquid capable of closing cracks or inhibiting corrosion.

Another topic of special interest is the incorporation of intermetallic phases featuring yield stress anomalies into metallic or intermetallic matrices. The research includes theoretical considerations and experimental verification by fabrication of composites, mechanical testing, and structure characterization. However, the quantitative evaluation of the efficiency of self-healing in these materials will have to await a full understanding of the interaction of phase composition, size distribution and topology, matrix (e.g., ductile metal, intermetallic phase, or high-performance polymer), fabrication method, microstructure, and mechanical properties.

12.4 FUNDAMENTAL PRINCIPLES AND MECHANISM

The earlier overview on the self-healing behavior of three distinctly different material classes described some of the prevailing concepts to obtain self-healing materials. These concepts, developed for a certain material, could in principle be also applied to other material classes, considering their intrinsic properties (Table 12.1). The general overview reveals that most concepts apparently can be applied more easily to polymers than to ceramics or in particular metals, due to the special molecular architecture of the polymers and the temperature range in which self-healing can proceed in polymers. In contrast, metals require much higher temperatures and only small atoms (in contrast to large polymer chains) are available. Nevertheless, also metals and ceramics offer good chances, whereby concepts developed either exclusively for them or originally for polymers could be applied.

The most general concepts for self-healing in the context of a wide applicability to different material classes are the concepts of *expanding phases* and *temperature increase*, which provide a higher mobility. Latter mechanism can also be combined

TABLE 12.1
Selected Potential Mechanisms to Give Rise to Self-Healing Behavior for the Discussed Material Classes

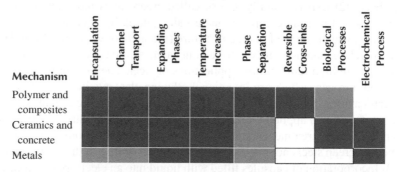

Mechanism	**Encapsulation**	**Channel Transport**	**Expanding Phases**	**Temperature Increase**	**Phase Separation**	**Reversible Cross-links**	**Biological Processes**	**Electrochemical Process**
Polymer and composites	dark gray	dark gray	dark gray	dark gray	dark gray	dark gray	light gray	
Ceramics and concrete	dark gray	dark gray	dark gray	dark gray	light gray	white	dark gray	dark gray
Metals	light gray	light gray	dark gray	dark gray	light gray	white	white	dark gray

Source: Hager, M.D., Greil, P., Leyens, C., van der Zwaag, S., and Schubert, U.S.: Self-healing materials. *Adv. Mater.* 22, 5424, 2010. Copyright Wiley-VCH Verlag GmbH & Co. KGaA. Reproduced with permission.

The color code indicates the expected or demonstrated level of success (dark gray—positive; light gray—potential; white—unlikely).

with other concepts, in particular with the concept of *expanding phases*. The *encapsulation* of healing agents in microcapsules or the *transport in channels*/vascular networks is also widely applicable concepts. In this manner extrinsic self-healing can be achieved, that is, materials that do not feature the ability for self-healing could obtain these features. The materials are "only" the inert matrix material. Due to the universality of these approaches (already used for polymers, concrete, and metals), they are the most-studied concepts in the field of self-healing materials, in particular considering polymers. *Reversible cross-links* are up to now limited to polymers. Due to the special mechanism, which requires the reversible opening and closing of bonds/interactions combined with a sufficient mobility of the binding units, it will presumably very difficult to transfer this concept successfully to other material classes. *Biological processes* for self-healing are well-known for concrete (self-healing by bacteria). Their application is favored by the low healing temperature, that is, room temperature. In contrast, ceramics and metals seem to be unsuitable for healing by *biological processes*, due to the normally high temperatures required for the healing. The application of this mechanism in polymers is imaginable. Lastly, *electrochemical processes* are well-suited for ceramics, and in particular, metals. However, polymers are not suited for the healing by *electrochemical processes* in general. Conjugated polymers could be considered as an exception, considering the restoration of conductivity, doping level, or optical properties by redox processes.

In Figure 12.7, the general mechanisms are depicted in the context of autonomic and nonautonomic self-healing, that is, the damage itself is the trigger, as well as extrinsic and intrinsic self-healing. It is conspicuous that autonomic self-healing

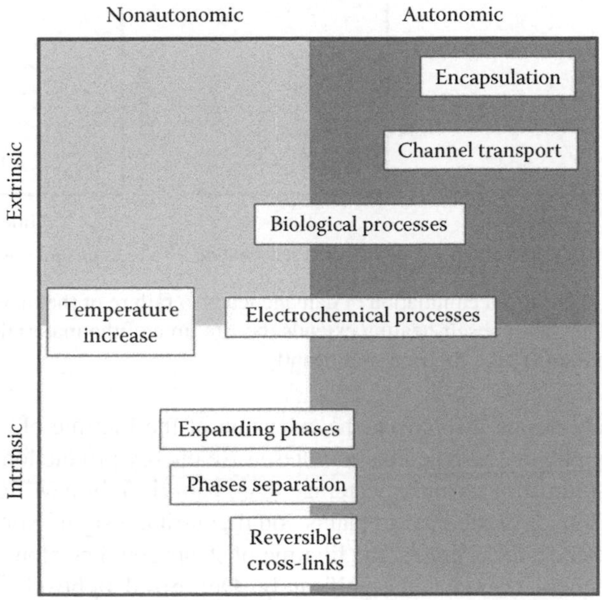

FIGURE 12.7 Schematic representation of the classification of the principle healing mechanisms in the context of extrinsic and intrinsic as well as autonomic and nonautonomic self-healing materials.

has been mainly achieved by extrinsic approaches. In contrast, mechanisms based on intrinsic self-healing are still nonautonomic; however, considering the research efforts in this area this approaches will supposably overcome this issue in the future.

12.5 POTENTIAL APPLICATIONS OF SELF-HEALING MATERIALS

During their use, all natural and synthetic materials accumulate damage. If the accumulated damage zone exceeds a critical value, this will result in component failure and, consequently, in loss of the function of the device. In contrast, self-healing materials have the ability to reverse the damage development once, several, or even multiple times and hence are able to expand the lifetime and the reliability of the material and, thus, of the device significantly ("damage management") (Figure 12.8). Such new levels of performance are of particular relevance to materials that are used only without or with limited access by men, for example, in medical applications as well as in civil, aerospace, automotive, and power engineering. If self-healing of damage caused by manufacturing or during application can significantly increase the effective lifetime and reliability of future novel materials, costs could be drastically cut down, also by reducing monitoring and controlling measures. In addition, a substantial saving of resources and energy could be anticipated due to the reduction of the hitherto necessary safety margins for the constructive geometry of mechanically, thermally, and corrosively exposed components in almost all technical areas. Self-healing materials would greatly improve materials' reliability and, thus, revolutionize component

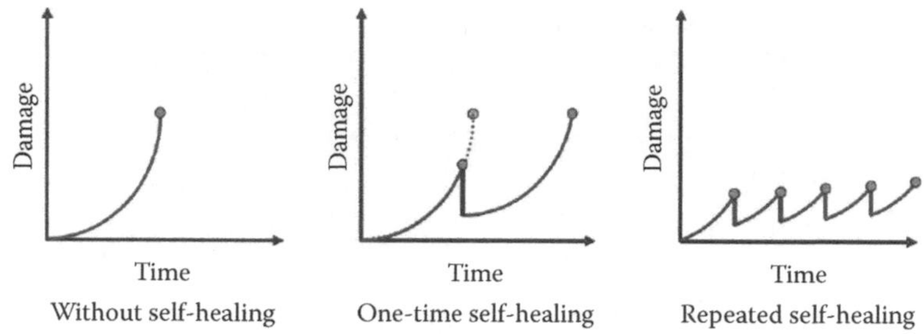

FIGURE 12.8 Left: the accumulation of damage leads to failure of the material after a certain time; middle: one time self-healing extends the life time of the material; right: multiple self-healing can increase the lifetime significantly.

construction and design. Moreover, the extension of the lifetime of critical components, for example, in constructions for alternative energy production (wind energy, photovoltaic, solar heat) for energy storage (e.g., batteries), in new lighting applications (e.g., LEDs), or of medical implants could contribute significantly to optimize economic efficiency; for example, the lifetime of stents could be elongated and, thus, therapy costs could be reduced significantly. Optimized lightweight construction accompanied with a higher structural reliability in aircraft and automotive engineering could reduce energy costs as well as environmental burden. The high potential for applications of these materials is also manifested in the started production of self-healing elastomers by Arkema and by self-healing coatings (e.g., Autonomic Materials) (Tullo 2009). Moreover, self-healing concrete has been applied for bridge decks and buildings in Japan. Finally, a Dutch highway was paved with self-healing asphalt recently (http://selfhealingasphalt.blogspot.com/2011/01/tu-delta-article-on-self-healing.html).

12.6 CONCLUSION

Research on self-healing materials has been a topic for almost a decade now and represents a wide interdisciplinary area, showing a large variety of different approaches. The progress achieved also becomes visible in dedicated conferences (Delft 2007, Chicago 2009, Bath 2011). Some of the concepts developed are driven by the direct application of the materials. However, to gain a deeper understanding of the basic mechanisms and principles of self-healing materials is still a challenging aim. If attained, it will undoubtedly lead to the development of a next generation of (tailor-made) self-healing materials.

Up to now, the design of new self-healing materials has mostly been inspired by nature, and nature will definitely stay an important source of inspiration for the development of new concepts. However, in copying the model the intrinsic character of engineered materials has to be taken into account (Youngblood and Sottos 2008; Amendola and Meneghetti 2009; van der Zwaag et al. 2009).

Self-healing up to now refers almost exclusively to structural materials: mechanical damages are being healed or mended, and afterward the mechanical properties are (partially) restored. However, other challenges exist, for example, to develop self-healing materials that are capable of restoring other properties (conductivity, color, fluorescence) after mechanical damage as well as other negative influences (extreme light, heat, and others). The well-established microcapsule concept could be successfully applied for the restoration of conductivity by the release of a charge-transfer salt as "healing agent" (Odom et al. 2010). Photodegradation can lead—for example—to the failure of solar cells. A recent example is the first step to overcome this issue. A complex consisting of two recombinant proteins, phospholipids and a carbon nanotube, was applied to mimic natural photosynthetic systems and their self-repair (Ham et al. 2010). Challenges also exist with respect to nanostructured systems (e.g., photonic nanomaterials/metamaterials). These systems are particularly prone to damage due to their large surface (compared with the volume) (Amendola 2009).

REFERENCES

Amendola, V. and Meneghetti, M. 2009. Self-healing at the nanoscale. *Nanoscale* 1:74–88.

Andersson, H. M., Keller, M. W., Moore, J. S., Sottos, N. R., and White, S. R. 2007. Self-healing polymers and composites. In *Self Healing Materials—An Alternative Approach to 20 Centuries Materials Science*, ed. S. van der Zwaag, pp. 19–44. Springer, Dordrecht, the Netherlands.

Ando, K., Chu, M.-C., Tsuji, K., Hirasawa, T., Kobayashi, Y., and Sato, S. 2002a. Crack healing behaviour and high-temperature strength of mullite/SiC composite ceramics. *J. Eur. Ceram. Soc.* 22:1313–1319.

Ando, K., et al. 2002b. Crack-healing behavior of Si_3N_4/SiC ceramics under stress and fatigue strength at the temperature of healing (1000°C). *J. Eur. Ceram. Soc.* 22:1339–1346.

Andres, P. R. and Schubert, U. S. 2004. New functional polymers and materials based on 2,2′:6′,2″-terpyridine metal complexes. *Adv. Mater.* 16:1043–1068.

Bergman, S. D. and Wudl, F. 2008. Remendable polymers. *J. Mater. Chem.* 18:41–62.

Berl, V., Schmutz, M., Krische, M. J., Khoury, R. G., and Lehn, J.-M. 2002. Supramolecular polymers generated from heterocomplementary monomers linked through multiple hydrogen-bonding arrays—Formation, characterization, and properties. *Chem. Eur. J.* 8:1227–1244.

Blaiszik, B. J., Caruso, M. M., McIlroy, D. A., Moore, J. S., White, S. R., and Sottos, N. R. 2009. Microcapsules filled with reactive solutions for self-healing materials. *Polymer* 50:990–997.

Blaiszik, B. J., Kramer, S. L. B., Olugebefola, S. C., Moore, J. S., Sottos, N. R., and White, S. R. 2010. Self-healing polymers and composites. *Annu. Rev. Mater. Res.* 40:179–211.

Bond, I. P., Trask, R. S., Williams, G. J., and Williams, H. R. 2008. Self-healing fiber-reinforced polymer composites, *MRS Bull.* 33:770–774.

Brochu, A. B. W., Craig, S. L., and Reichert, W. M. 2011. Self-healing biomaterials. *J. Biomed. Mater. Res.* 96:492–506.

Brown, E. N., Kessler, M. R., Sottos, N. R., and White, S. R. 2003. In situ poly(urea-formaldehyde) microencapsulation of dicyclopentadiene. *J. Microencapsul.* 20:719–730.

Brown, E. N., Sottos, N. R., and White, S. R. 2004. Microcapsule induced toughening in a self-healing polymer composite. *J. Mater. Sci.* 39:1703–1710.

Burattini, S., Colquhoun, H. M., Fox, J. D., et al. 2009b. A self-repairing, supramolecular polymer system: Healability as a consequence of donor–acceptor π–π stacking interactions. *Chem. Commun.* 6717–6719.

Burattini, S., Colquhoun, H. M., Greenland, B. W., and Hayes, W. 2009a. A novel self-healing supramolecular polymer system. *Faraday Discuss.* 143:251–264.

Caruso, M. M., Davis, D. A., Shen, Q., et al. 2009. Mechanically-induced chemical changes in polymeric materials. *Chem. Rev.* 109:5755–5798.

Caruso, M. M., Delafuente, D. A., Ho, V., Sottos, N. R., Moore, J. S., and White, S. R. 2007. Solvent-promoted self-healing epoxy materials. *Macromolecules* 40:8830–8832.

Chen, X., Dam, M. A., Ono, K., et al. 2002. A thermally re-mendable cross-linked polymeric material. *Science* 295:1698–1702.

Chen, X., Wudl, F., Mal, A. K., Shen, H., and Nutt, S. R. 2003. New thermally remendable highly cross-linked polymeric materials. *Macromolecules* 36:1802–1807.

Cho, S. H., Andersson, H. M., White, S. R., Sottos, N. R., and Braun, P. V. 2006. Environmentally stable polydimethysiloxane-based self-healing of polymers. *Adv. Mater.* 18:997–1000.

Cho, S. H., White, S. R., and Braun, P. V. 2009. Self-healing polymer coatings. *Adv. Mater.* 21:645–649.

Chung, C. M., Roh, Y. S., Cho, S. Y., and Kim, J. G. 2004. Crack healing in polymeric materials via photochemical [2+2] cycloaddition. *Chem. Mater.* 16:3982–3984.

Cordier, P., Tournilhac, F., Soulié-Ziakovic, C., and Leibler, L. 2008. Self-healing and thermoreversible rubber from supramolecular assembly. *Nature* 451:977–980.

Corten, C. C. and Urban, M. W. 2009. Repairing polymers using oscillating magnetic field. *Adv. Mater.* 21:5011–5015.

Davis, D. A., Hamilton, A., Yang, J., et al. 2009. Force-induced activation of covalent bonds in mechanoresponsive polymeric materials. *Nature* 459:68–72.

Deng, G., Tang, C., Li, F., Jiang, H., and Chen, Y. 2010. Covalent cross-linked polymer gels with reversible sol–gel transition and self-healing properties. *Macromolecules* 43:1191–1194.

Dry, C. 1992. Passive tunable fibers and matrices. *Intern. J. Modern Phys. B* 6:2763–2771.

Dry, C. 1996. Procedures developed for self-repair of polymer matrix composite materials. *Compos. Struct.* 35:263–269.

Dry, C. and McMillan, W. 1996. Three-part methylmethacrylate adhesive system as an internal delivery system for smart responsive concrete. *Smart Mater. Struct.* 5:297–301.

Fischer, H. 2010. Self-repairing material systems?A dream or a reality? *Nat. Sci.* 2:873–901.

Froimowicz, P., Frey, H., and Landfester, K. 2011. Towards the generation of self-healing materials by means of a reversible photo-induced approach. *Macromol. Rapid Commun.* 32:468–473.

Fustin, C.-A., Gohy, J.-F., Guillet, P., and Schubert, U. S. 2007. Metallo-supramolecular block copolymers. *Adv. Mater.* 19:1665–1673.

Ghosh, S. K. 2009. Self-healing materials: Fundamentals, design strategies, and applications. In *Self-healing Materials: Fundamentals, Design Strategies, and Applications*, ed. S. K. Ghosh. Wiley-VCH, Weinheim, Germany.

Gragert, M., Schunack, M., and Binder, W. H. 2011. Azide/alkyne-"click"-reactions of encapsulated reagents: Toward self-healing materials. *Macromol. Rapid Commun.* 32:419–425.

Hager, M. D., Greil, P., Leyens, C., van der Zwaag, S., and Schubert, U. S. 2010. Self-healing materials. *Adv. Mater.* 22:5424–5430.

Ham, M.-H., Chio, J. H., Boghossian, A. A., et al. 2010. Photoelectrochemical complexes for solar energy conversion that chemically and autonomously regenerate. *Nat. Chem.* 2:929–936.

Hansen, C. J., Wu, W., Toohey, K. S., Sottos, N. R., White, S. R., and Lewis, J. A. 2009. Self-healing materials with interpenetrating microvascular networks. *Adv. Mater.* 21:4143–4147.

Hautakangas, S., Schut, H., van Dijk, N. H., Rivera Díaz del Castillo, P. E. J., and van der Zwaag, S. 2008. Self-healing of deformation damage in underaged Al–Cu–Mg alloys. *Scr. Mater.* 58:719–722.

He, S., van Dijk, N. H., Schut, H., and van der Zwaag, S. 2010. Thermally activated precipitation at deformation-induced defects in Fe–Cu and Fe–Cu–B–N alloys studied by positron annihilation spectroscopy. *Phys. Rev. B* 81:094103, 10pp.

Jones, A. S., Rule, J. D., Moore, J. S., White, S. R., and Sottos, N. R. 2006. Catalyst morphology and dissolution kinetics for self-healing polymers. *Chem. Mater.* 18:1312–1317.

Jonkers, H. M. 2007. Self-healing concrete: A biological approach. In *Self Healing Materials—An Alternative Approach to 20 Centuries Materials Science*, ed. S. van der Zwaag, pp. 195–204. Springer, Dordrecht, the Netherlands.

Jonkers, H. M., Thijssen, A., Muyzer, G., Copuroglu, O., and Schlangen, E. 2010. Application of bacteria as self-healing agent for the development of sustainable concrete. *Ecol. Eng.* 36:230–235.

Jud, K. and Kaush, H. H. 1979. Load transfer through chain molecules after interpenetration at interfaces. *Polym. Bull.* 1:697–707.

Jud, K., Kausch, H. H., and Williams, J. G. 1981. Fracture mechanics studies of crack healing and welding of polymers *J. Mater. Sci.* 16:204–210.

Jun, L., Zheng, Z. X., Ding, H. F., and Jin, Z. H. 2004. Preliminary study of the crack healing and strength recovery of Al_2O_3-matrix composites. *Fatigue Fract. Eng. Mat.* 27:89–97.

Kalista, S. J. and Ward, T. C. 2007. Thermal characteristics of the self-healing response in poly(ethylene-*co*-methacrylic acid) copolymers. *J. R. Soc. Interface* 4:405–411.

Kersey, F. R., Loveless, D. M., and Craig, S. L. 2007. A hybrid polymer gel with controlled rates of cross-link rupture and self-repair. *J. R. Soc. Interface* 4:373–380.

Kim, Y. H. and Wool, R. P. 1983. A theory of healing at a polymer–polymer interface. *Macromolecules* 16:1115–1120.

Kirkby, E. L., Michaud, V. J., Sottos, N. R., White, S. R., and Manson, J.-A. E. 2009. Performance of self-healing epoxy with microencapsulated healing agent and shape memory alloy wires. *Polymer* 50:5533–5538.

Kolmakov, G. V., Revanur, R., Tangirala, R., et al. 2010. Using nanoparticle-filled microcapsules for site-specific healing of damaged substrates: Creating a "repair-and-go" system. *ACS Nano.* 4:1115–1123.

Lehn, J.-M. 2010. Dynamers: Dynamic molecular and supramolecular polymers. *Aust. J. Chem.* 63:611–623.

Li, V. C. and Yang, E.-H. 2007. Self-healing in concrete materials. In *Self Healing Materials—An Alternative Approach to 20 Centuries Materials Science*, ed. S. van der Zwaag, pp. 161–194. Springer, Dordrecht, the Netherlands.

Lin, B. C., Lee, S., and Liu, K. S. 1990. Methanol-induced crack healing in poly(methyl methacrylate). *Polym. Eng. Sci.* 30:1399–1406.

Liu, Y.-L. and Chen, Y.-W. 2007. Thermally reversible cross-linked polyamides with high toughness and self-repairing ability from maleimide- and furan-functionalized aromatic polyamides. *Macromol. Chem. Phys.* 208:224–232.

Liu, Y.-L. and Hsieh, C.-Y. 2006. Crosslinked epoxy materials exhibiting thermal remendablility and removability from multifunctional maleimide and furan compounds. *J. Polym. Sci., Part A: Polym. Chem.* 44:905–913.

Lumley, R. N., Morton, A. J., and Polmear, I. J. 2002. Enhanced creep performance in an Al–Cu–Mg–Ag alloy through underageing. *Acta Mater.* 50:3597–3608.

Lumley, R. N., Polmear, I. J., and Morton, A. J. 2003. Interrupted aging and secondary precipitation in aluminium alloys. *Mater. Sci. Technol.* 19:1483–1490.

Manuel, M. V. 2009. Principles of self-healing in metals and alloys: An introduction. In *Self-Healing Materials: Fundamentals, Design Strategies, and Applications*, ed. S. K. Ghosh, pp. 251–266. Wiley-VCH, Weinheim, Germany.

Mehta, N. K. and Bogere, M. N. 2009. Environmental studies of smart/self-healing coating system for steel. *Progr. Org. Coat.* 64:419–428.

Mitomo, M. and Nishimura, T. M. 1996. Crack healing in silicon nitride and alumina ceramics. *J. Mater. Sci. Lett.* 15:1976–1978.

Montarnal, D., Cordier, P., Soulé-Ziakovic, C., Tournilhac, F., and Leibler, L. 2008. Synthesis of self-healing supramolecular rubbers from fatty acid derivatives, diethylene triamine, and urea. *J. Polym. Sci., Part A: Polym. Chem.* 46:7925–7936.

Montarnal, D., Tournilhac, F., Hidalgo, M., Couturier, J.-L., and Leibler, L. 2009. Versatile one-pot synthesis of supramolecular plastics and self-healing rubbers. *J. Am. Chem. Soc.* 131:7966–7967.

Mookhoek, S. D., Blaiszik, B. J., Fischer, H. R., Sottos, N. R., White, S. R., and van der Zwaag, S. 2008. Peripherally decorated binary microcapsules containing two liquids. *J. Mater. Chem.* 18:5390–5394.

Mookhoek, S. D., Fischer, H. R., and van der Zwaag, S. 2009. A numerical study into the effects of elongated capsules on the healing efficiency of liquid-based systems. *Comp. Mater. Sci.* 47:506–511.

Motuku, M., Vaidya, U. K., and Janowski, C. M. 1999. Parametric studies on self-repairing approaches for resin infused composites subjected to low velocity impact. *Smart Mater. Struct.* 8:623–638.

Müller, C., Greil, P., Bundschuh, K., and Schütze, M. 1998. Polymer/filler-derived high temperature oxidation resistant silicon oxycarbide/$MoSi_2$ composites. *Ceramic Trans.* 85:393–404.

Murphy, E. B., Bolanos, E., Schaffner-Hamann, C., Wudl, F., Nutt, S. R., and Auad, M. L. 2008. Synthesis and characterization of a single component thermally remendable polymer network: Staudinger and stille revisited. *Macromolecules* 41:5203–5209.

Murphy, E. B. and Wudl, F. 2010. The world of smart healable materials. *Prog. Polym. Sci.* 35:223–251.

Odom, S. A., Caruso, M. M., Finke, A. D., et al. 2010. Restoration of conductivity with TTF-TCNQ charge-transfer salts. *Adv. Funct. Mater.* 20:1721–1727.

Ohkawa, H., Ligthart, G. B. W. L., Sijbesma, R. P., and Meijer, E. W. 2007. Supramolecular graft copolymers based on 2,7-diamido-1,8-naphthyridines. *Macromolecules* 40:1453–1459.

Ott, C., Hoogenboom, R., and Schubert, U.S. 2008. Post-modification of poly(pentafluorostyrene): a versatile "dick" method to create well-defined multifunctional graft copolymers. *Chem. Commun.* 3516–3518.

Pastine, S. J., Okawa, D., Zettl, A., and Fréchet, J. M. J. 2009. Chemicals on demand with phototriggerable microcapsules. *J. Am. Chem. Soc.* 131:13586–13587.

Piermattei, A., Karthikeyan, S., and Sijbesma, R. P. 2009. Activating catalysts with mechanical force. *Nat. Chem.* 1:133–137.

Ramm, W. and Biscoping, M. 1998. Autogenous healing and reinforcement corrosion of water-penetrated separation cracks in reinforced concrete. *Nucl. Eng. Des.* 179:191–200.

Scherman, O. A., Ligthart, G. B. W. L., Sijbesma, R. P., and Meijer, E. W. 2006. A selectivity-driven supramolecular polymerization of an AB monomer. *Angew. Chem. Int. Ed.* 45:2072–2076.

Schubert, U. S. and Eschbaumer, C. 2002. Macromolecules containing bipyridine and terpyridine metal complexes: Towards metallosupramolecular polymers. *Angew. Chem. Int. Ed.* 41:2892–2926.

Shinya, N. 2009. Self-healing of metallic materials: Self-healing of creep cavity and fatigue cavity/crack. In *Self-Healing Materials: Fundamentals, Design Strategies, and Applications*, ed. S. K. Ghosh, pp. 219–250. Wiley-VCH, Weinheim, Germany.

Sijbesma, R. P., Beijer, F. H., Brunsveld, L., et al. 1997. Reversible polymers formed from self-complementary monomers using quadruple hydrogen bonding. *Science* 278:1601–1604.

Song, G. M., Pei, Y. T., Sloof, W. G., Li, S. B., De Hosson, J. Th. M., and van der Zwaag, S. 2008. Oxidation-induced crack healing in Ti_3AlC_2 ceramics. *Scr. Mater.* 58:13–16.

Sugiyama, R., Yamane, K., Nakao, W., Takahashi, K., and Ando, K. 2008. Effect of difference in crack-healing ability on fatigue behavior of alumina/silicon carbide composites. *J. Intell. Mater. Syst. Struct.* 19:411–415.

Suryanarayana, C., Chowdoji Rao, K., and Kumar, D. 2008. Preparation and characterization of microcapsules containing linseed oil and its use in self-healing coatings. *Progr. Org. Coat.* 63:72–78.

Syrett, J. A., Becer, C. R., and Haddleton, D. M. 2010. Self-healing and self-mendable polymers. *Polym. Chem.* 1:978–987.

Takahashi, K., Mizobe, Y., Ando, K., and Saito, S. 2006. Effects of frequency on the crack-healing behavior of Si_3N_4/SiC composite under cyclic stress. *JSME Int. J. Ser. A* 49:307–313.

Takeda, K., Unno, H., and Zhang, M. 2004. Polymer reaction in polycarbonate with Na_2CO_3. *J. Appl. Polym. Sci.* 93:920–926.

Tittelboom, K. V., De Belie, N., Van Loo, D., and Jacobs, P. 2011. Self-healing efficiency of cementitious materials containing tubular capsules filled with healing agent. *Cem. Concr. Compos.* 33:497–505.

Toohey, K. S., Hansen, C. J., Lewis, J. A., White, S. R., and Sottos, N. R. 2009. Delivery of two-part self-healing chemistry via microvascular networks. *Adv. Funct. Mater.* 19:1399–1405.

Toohey, K. S., Sottos, N. R., Lewis, J. A., Moore, J. S., and White, S. R. 2007. Self-healing materials with microvascular networks. *Nat. Mater.* 6:581–585.

Trask, R. S. and Bond, I. P. 2006. Biomimetic self-healing of advanced composite structures using hollow glass fibres. *Smart Mater. Struct.* 15:704–710.

Trask, R. S., Williams, G. J., and Bond, I. P. 2007. Bioinspired self-healing of advanced composite structures using hollow glass fibres. *J. R. Soc. Interface* 4:363–371.

Travitzky, N., Windsheimer, H., Fey, T., and Greil, P. 2008. Preceramic paper-derived ceramics. *J. Am. Ceram. Soc.* 91:3477–3492.

Tullo, A. 2009. Arkema to produce self-healing rubber. *Chem. Eng. News* 87:14–15.

van der Zwaag, S. 2007. An introduction to materials design principles: Damage prevention versus Damage management. In *Self Healing Materials—An Alternative Approach to 20 Centuries Materials Science*, ed. S. van der Zwaag, pp. 1–18. Springer, Dordrecht, the Netherlands.

van der Zwaag, S., van Dijk, N. H., Jonkers, H. M., Mookhoek, S. D., and Sloof, W. G. 2009. Self-healing behaviour in man-made engineering materials: Bioinspired but taking into account their intrinsic character. *Phil. Trans. R. Soc. A* 367:1689–1704.

Varley, R. 2007. Ionomers as self-healing polymers. In *Self Healing Materials—An Alternative Approach to 20 Centuries Materials Science*, ed. S. van der Zwaag, pp. 95–114. Springer, Dordrecht, the Netherlands.

Varley, R. J. and van Der Zwaag, S. 2008. Towards an understanding of thermally activated self-healing of an ionomer system during ballistic penetration. *Acta Mater.* 56:5737–5750.

White, S. R. and Geubelle, P. H. 2010. Self-healing materials: Get ready for repair-and-go. *Nat. Nanotech.* 5:247–248.

White, S. R., Sottos, N. R., Geubelle, P. H., et al. 2001. Autonomic healing of polymer composites. *Nature* 409:794–797.

Whitell, G. R., Hager, M. D., Schubert, U. S., and Manners, I. 2011. Functional soft materials from metallopolymers and metallosupramolecular polymers. *Nat. Mater.* 10:176–188.

Wietor, J.-L., Dimopoulos, A., Govaert, L. E., van Benthem, R. A. T. M., de With, G., and Sijbesma, R. P. 2009. Preemptive healing through supramolecular cross-links. *Macromolecules* 42:6640–6646.

Wietor, J.-L. and Sijbesma, R. P. 2008. A self-healing elastomer. *Angew. Chem. Int. Ed.* 47:8161–8163.

Wild, A., Winter, A., Schlütter, F., and Schubert, U. S. 2011. Advances in the field of π-conjugated 2,2′:6′,2″-terpyridines. *Chem. Soc. Rev.* 40:1459–1511.

Williams, K. A., Boydston, A. J., and Bielawski, C. W. 2007b. Towards electrically conductive, self-healing materials. *J. R. Soc. Interface* 4:359–362.

Williams, H. R., Trask, R. S., and Bond, I. P. 2007a. Self-healing composite sandwich structures. *Smart Mater. Struct.* 16:1198–1207.

Williams, H. R., Trask, R. S., and Bond, I. P. 2008a. Self-healing sandwich panels: Restoration of compressive strength after impact. *Comp. Sci. Technol.* 68:3171–3177.

Williams, H. R., Trask, R. S., Knights, A. C., Williams, E. R., and Bond, I. P. 2008b. Biomimetic reliability strategies for self-healing vascular networks in engineering materials. *J. R. Soc. Interface* 5:735–747.

Wouters, M., Craenmehr, E., Tempelaars, K., Fischer, H., Stroeks, N., and van Zanten, J. 2009. Preparation and properties of a novel remendable coating concept. *Prog. Org. Coat.* 64:156–162.

Wu, D. Y., Meure, S., and Solomon, D. 2008. Self-healing polymeric materials: A review of recent developments. *Prog. Polym. Sci.* 33:479–522.

Xiao, D. S., Yuan, Y. C., Rong, M. Z., and Zhang, M. Q. 2009. Self-healing epoxy based on cationic chain polymerization. *Polymer* 50:2967–2975.

Yang, Z., Hollar, J., He, X., and Shi, X. 2011. A self-healing cementitious composite using oil core/silica gel shell microcapsules. *Cem. Concr. Comp.* 33:506–512.

Yang, J., Keller, M. W., Moore, J. S., White, S. R., and Sottos, N. R. 2008. Microencapsulation of isocyanates for self-healing polymers, *Macromolecules* 41:9650–9655.

Yin, T., Rong, M. Z., Zhang, M. Q., and Yang, G. C. 2007. Self-healing epoxy composites—Preparation and effect of the healant consisting of microencapsulated epoxy and latent curing agent. *Comp. Sci. Technol.* 67:201–212.

Youngblood, J. P. and Sottos, N. R. 2008. Bioinspired materials for self-cleaning and self-healing. *MRS Bull.* 33:732–741.

Yuan, Y. C., Yin, T., Rong, M. Z., and Zhang, M. Q. 2008. Self healing in polymers and polymer composites. Concepts, realization and outlook: A review. *eXPRESS Polym. Lett.* 2:238–250.

Part III

Frontiers of Self-Healing at the Nanoscale

13 Design of a Repair-and-Go System for Site-Specific Healing at the Nanoscale

German V. Kolmakov, Todd Emrick, Thomas P. Russell, Alfred J. Crosby, and Anna C. Balazs

CONTENTS

13.1 INTRODUCTION

The ability to heal wounds is one of the truly remarkable properties of biological systems. A grand challenge in materials science is to design "smart" synthetic systems that can mimic this behavior by not only "sensing" the presence of a "wound" or defect, but also actively reestablishing the continuity and integrity of the damaged area. Such materials would significantly extend the lifetime and utility of a vast array of manufactured items. Nanotechnology is particularly relevant to both the utility and fabrication of self-healing materials. For example, as devices reach nanoscale dimensions, it becomes critical to establish means of promoting repair at these length scales. Operating and directing minute tools to carry out this operation is still far from trivial. An optimal solution would be to design a system that could recognize the appearance of a nanoscopic crack or fissure and then could direct agents of repair specifically to that site. Even in the manufacture of various macroscopic components, nanoscale damage is a critical issue. For instance, nanoscopic notches and scratches can appear on the surface of materials during the manufacturing process. Due to the small size of these defects, they are difficult to detect and consequently, difficult to repair. Such defects, however, can have a substantial effect on the mechanical properties of the system. For example, significant stress concentrations can occur at the tip of notches in the surface; such regions of high stress can ultimately lead to the propagation of cracks through the system and the degradation of mechanical behavior.

Thus, one of the driving forces for creating self-healing materials (Chen et al. 2002; Balazs 2007; Hickenboth et al. 2007; Trask et al. 2007; Cordier et al. 2008; Wool 2008; Wu et al. 2008; Amendola and Meneghetti 2009; Caruso et al. 2009; Kolmakov et al, 2009; White and Geubelle, 2010) is in fact the need to affect repair on the nanoscale. On the positive side, advances in nanotechnology could also provide routes for realizing the creation of these materials. In particular, scientists can now produce a stunning array of nanoscopic particles and have become highly adapt at tailoring the surface chemistry of the particles. In what follows, we review our recent computational studies on designing self-healing materials that exploit the unique properties of nanoscopic particles.

In carrying out these studies (Verberg et al. 2007; Kolmakov et al. 2010), we took our inspiration from nature: namely, the functionality of biological leucocytes, which localize at a wound and thereby facilitate the repair process. In our synthetic system, the "leukocyte" is a polymeric microcapsule, the healing agents are encapsulated solid nanoparticles, and the "wound" is a microscopic crack on a surface. In our computer simulations, the nanoparticle-filled microcapsules are driven by an imposed fluid flow to move along the cracked substrate. Our goal was to determine how the release of the encapsulated nanoparticles could be harnessed to repair damage on the underlying surface. The simulations revealed that these capsules can deliver the encapsulated materials to specific sites on the substrate, effectively generating an alternate route to repairing surface defects. Once the healing nanoparticles were deposited on the desired sites, the fluid-driven capsules could move further along the surface and for this reason, the strategy was termed "repair and go." The latter strategy could be particularly advantageous since it would have negligible impact on the precision of the nondefective regions and involves minimal amount of the repair materials.

It is noteworthy that micron-sized capsules filled with dissolved particles can encompass very high payloads, allowing them to rapidly carry and deliver large amounts of nanoparticles to a desired location. Furthermore, the continued, flow-driven motion of these microcarriers potentially allows multiple damaged regions to be healed by the capsules.

As the introduction of a synthetic microvasculature (Therriault et al. 2003; Wu et al. 2010) into structural materials becomes more developed, the use of such microcapsules as cellular mimics could expand the efficiency of the artificial circulatory systems. In addition to supplying healing reagents in the channels, it could be advantageous to encapsulate "damage markers" within the microcapsules. The microcapsules could continue to circulate in a "healthy," undamaged system, but become trapped or localized at a damaged site and thereby, deliver a chemical "marker" (i.e., a visible or fluorescent dye) through its porous shell. Such markers will enable one to nondestructively locate and track the damaged regions over time.

Computational modeling can play an important role in designing effective "repair-and-go" capsules or "damage markers" because the performance of the system depends on multiple factors, including the properties of the nanoparticles, the carrier capsules, the external fluid, and the underlying substrate. Via the simulations, we can scan the parameter space in a relatively efficient manner and thus isolate fruitful paths for the experimentalists to pursue in the fabrication of the nanoparticle-filled microcarriers. There are, however, a number of computational challenges that must be addressed in order to simulate these complex, multicomponent systems. In particular, we must capture the dynamic interactions between the compliant microcapsules and the surrounding, flowing fluid. In addition, we must simulate phenomena on both the micron and nanometer length scales.

In Section 13.2, we discuss how we addressed these computational challenges by developing a hybrid simulation method, which allowed us to capture the critical fluid–structure interactions and incorporate mobile nanoparticles into the system. Following Section 13.2, we describe our findings on the behavior of nanoparticle-filled microcapsules that are propelled by an imposed flow to move over a damaged substrate. Much as our predictions can facilitate the fabrication of repair-and-go carriers, some of our choices for the system parameters were informed by recent experiments; we discuss these choices and relevant experiments at the beginning of Section 13.3.

13.2 METHODOLOGY

The fundamental structural unit in our simulations is a three-dimensional, fluid-filled microcapsule that is localized on a substrate and immersed in a host fluid. The fluid encapsulated in the interior of the microcapsule contains a dispersion of nanoparticles. An imposed flow drives the capsule to move along the substrate, which contains a well-defined crack. To model the behavior of this multicomponent system, we take advantage of our recently developed hybrid "LBM/LSM" technique (Alexeev et al. 2005, 2006, 2007; Buxton et al. 2005; Alexeev and Balazs 2007; Usta et al. 2007). In this approach, the dynamic behavior of the encapsulated and external host fluids is captured through the lattice Boltzmann model (LBM) for hydrodynamics. The capsule's shell and the underlying surface are modeled through the lattice spring model (LSM) for the micromechanics of elastic solids. In this integrated LBM/LSM method, the fluid and solid components interact through appropriate boundary conditions at the solid–fluid interfaces. We augmented this LBM/LSM approach by introducing a Brownian dynamics simulation (Verberg et al. 2006, 2007; Usta et al. 2008) to model the diffusion of the nanoparticles within the system. In the following, we provide a brief description of each of these different techniques.

The capsule's elastic, solid shell is represented by a LSM, which consists of a triangular network of harmonic springs that connect regularly spaced mass points, or nodes (Alexeev and Balazs 2007; Alexeev et al. 2007; Usta et al. 2007). The spring force $\mathbf{F}_s$ on node $\mathbf{r}_i$ is equal to

$$\mathbf{F}_s(\mathbf{r}_i) = -\sum_j k_j \left[\frac{r_{ij} - r_{ij}^{eq}}{r_{ij}} \right] \mathbf{r}_{ij} \tag{13.1}$$

where the summation runs over all nearest- and next-nearest neighbor nodes. The quantity $\mathbf{r}_{ij} = \mathbf{r}_i - \mathbf{r}_j$ is the radius vector between ith and jth nodes, r_{ij}^{eq} is the equilibrium length of the spring and k_j is the spring constant. To capture the dynamics of the solid shell, we numerically integrate Newton's equations of motion, $M(d^2\mathbf{r}_i/dt^2) = \mathbf{F}(\mathbf{r}_i)$, where M is the mass of a node. The total force $\mathbf{F}$ acting on a node consists of the following: the sum of the spring forces between the masses (representing the elastic response of the solid shell), the force exerted by the fluid on the shell at the fluid–solid boundary, and the adhesion forces at the solid substrate (see the following).

The capsule's spherical shell is formed from two concentric layers of LSM nodes; each layer contains $N = 122$ nodes. These two layers are separated by a distance of $\Delta x_{LSM} = 1.5\ \Delta x$, where Δx_{LSM} is the lattice spacing between nearest nodes in the LSM and Δx is the spacing in the LBM (see the following). The outer radius of the shell was taken to be $R = 5\ \Delta x$. Given that k_1 and $k_{\sqrt{2}}$ are the respective spring constants in the orthogonal and the diagonal directions, for $k_1/2 = k_{\sqrt{2}} \equiv k$ and for small deformations, the LSM system obeys linear elasticity theory and results in a Young's modulus of $E = 5k/2\Delta x_{LSM}$ (Ladd et al. 1997; Buxton et al. 2001).

The LBM can be viewed as an efficient solver for the Navier–Stokes equation (Succi 2001). Specifically, this lattice-based model consists of two processes: the propagation of fluid "particles" to neighboring lattice sites, and the subsequent collisions between particles when they reach a site. Here, the fluid particles are representative of mesoscopic portions of the fluid, and are described by a particle distribution function $f_i(\mathbf{r},t)$, which characterizes the mass density of fluid particles at a lattice node $\mathbf{r}$ and time t propagating in the direction i with a constant velocity $\mathbf{c}_i$. (The velocities $\mathbf{c}_i$ in ith direction are chosen so that fluid particles propagate from one lattice site to the next in exactly one time step Δt.) The time evolution of these distribution functions is governed by a discretized Boltzmann equation (Succi 2001). In three-dimensional systems, the simulations involve a set of 19 particle velocity distribution functions at each node. The hydrodynamic quantities of interest are moments of the distribution function, that is, the mass density $\rho = \sum_i f_i$, the momentum density $\mathbf{j} = \rho\,\mathbf{u} = \sum_i \mathbf{c}_i f_i$, with $\mathbf{u}$ being the local fluid velocity, and the momentum flux $\Pi = \sum_i \mathbf{c}_i \mathbf{c}_i f_i$.

In our LBM/LSM simulations, the fluid and solid phases interact through appropriate boundary conditions (Alexeev et al. 2005, 2006; Buxton et al. 2005). In particular, lattice spring nodes that are situated at the solid–fluid interface impose their velocities on the surrounding fluids; the velocities are transmitted through a linked bounce-back rule (Lallemand and Luo 2000) to those LBM distribution functions that intersect the moving solid boundary. In turn, LS nodes at the solid–fluid interface experience forces due to the fluid pressure and viscous stresses at that boundary. We calculate the latter force based on the momentum exchange between the LBM particle and solid boundary, and then distribute this quantity as a load to the neighboring LS nodes.

The interaction between the capsule and the surface (both the undamaged region and crack wall) is modeled with a nonspecific Morse potential (Alexeev et al. 2005, 2006; Verberg et al. 2007):

$$\varphi(r) = \varepsilon\left(1 - \exp\left[-\frac{(r - r_{eq})}{l_0}\right]\right)^2, \tag{13.2}$$

where ε and l_0 characterize the respective strength and range of the interaction potential. Additionally, r is the distance between the LSM node and the substrate surface, and r_{eq} is the distance at which this force is equal to zero. In all our simulations, we set $l_0 = 1$ and $r_{eq} = 1$, while ε was varied to determine the effect of the adhesion force on the microcapsule's motion.

To capture the diffusion of nanoparticles within the fluid, we use a Brownian dynamics model for the particles (Verberg et al. 2005, 2006, 2007). The nanoparticle trajectories obey a stochastic differential equation:

$$d\mathbf{r}(t) = \mathbf{u}(\mathbf{r}, t)dt + \sqrt{2D_0}\,d\mathbf{W}(t), \tag{13.3}$$

where the first term describes the advection due to the local fluid velocity $\mathbf{u}(r,t)$ and the second term describes the particle's Brownian motion, with D_0 being the particle's diffusion coefficient and $d\mathbf{W}(t)$ being the differential of a Weiner process with unit variance. We neglect backflow effects (i.e., the impact of the particles motion on the flow field); the latter assumption is valid for submicron-sized particles at relatively low concentrations. We also neglect the interactions between the particles. We use a first-order Euler scheme method to solve Equation 13.3 (Verberg 2006). Note that an ensemble average of the particle trajectories computed from Equation 13.3 is equivalent to solving the convection–diffusion equation for the concentration of nanoparticles (Ottinger 1996; Szymchak and Ladd 2003; Verberg 2006).

We simulate the chemisorbtion of nanoparticles onto the crack walls by assigning a probability w_{dep} for a nanoparticle to deposit from the microcapsule's interior onto the surface in a given time step Δt. This probability has the form

$$w_{dep} = k\Delta t e^{-r/r_{dep}}. \tag{13.4}$$

Here r is the distance from the microcapsule's shell to the crack wall, r_{dep} is a characteristic distance over which the probability decays, and k is a characteristic deposition rate, which was set to $k = 1/\Delta t$ in our simulations. We also set the characteristic deposition distance r_{dep} in Equation 13.4 equal to the distance l_0 in the Morse potential (in Equation 13.2). A "rejection" rule (Lawn 1993; Ottinger 1996) is applied to those nanoparticles that are not deposited onto the surface at a given time step (i.e., with a probability $(1 - w_{dep})$); the variable $d\mathbf{W}$ is set equal to zero for the "rejected" particles, and their positions are then updated in accordance with Equation 13.3.

The profile of the substrate–fluid interface is updated dynamically at each time step; the local elevation of the profile due to the nanoparticle deposition is computed as $h = (4\pi/3)r_{NP}^3 n$ where n in the density of deposited nanoparticles at the wall at a given point. The radius of the nanoparticles, r_{NP}, is taken to be 10^3 times smaller than the radius of the microcapsule.

13.3 RESULTS AND DISCUSSION: SITE-SPECIFIC HEALING OF DAMAGED SUBSTRATES

In our system, it is the ability of the particle-loaded microcapsules to deliver the nanoparticles to the damaged sites, which provides the unique healing behavior. In particular, with the capsules localized in the cracks, the nanoparticles could diffuse from their carriers and ultimately fill the fissures in the surface. In this manner, incipient cracks can be mended and the material can be protected against further damage.

Using the model described earlier, we specifically determine conditions that provide maximal deposition of nanoparticles into the surface crack (Kolmakov et al. 2010). We focus on a single nanoparticle-filled capsule that interacts with a bare crack to gain insight into the initial steps involved in filling the defect and to establish optimal conditions early in the deposition process. Additional fluid-driven capsules moving along the surface are needed to continue the selective delivery of the fillers into the damaged regions. We also examine the interactions between a microcarrier and a partially filled fissure and then finally investigate the behavior of a microcapsule on surface containing a nanoparticle-filled, "healed" domain.

While the parameter space we considered is relatively general, we nonetheless were inspired by specific experimentally realizable systems (Breitenkamp and Emrick 2003; Lin et al. 2003; Balazs et al. 2006; Gupta et al. 2006; Boker et al. 2007; Revanur 2007). In particular, we assumed that the surface of the microcapsules is amphiphilic in nature and thus could have comparable interactions with hydrophilic and hydrophobic domains. Such microcapsules can be fabricated from comb copolymers that encompass a hydrophobic backbone and hydrophilic side chains (Breitenkamp and Emrick 2003). We further assume that the microcapsules encase an oil phase, which contains dispersed hydrophobic nanoparticles. Again, this aspect of the model can be experimentally realized. In particular, the encapsulated particles could be quantum dots that were functionalized with hydrophobic ligands (Lin et al. 2003; Boker et al. 2007). Within an oil/water mixture, the amphiphilic comb copolymers can out-compete the functionalized quantum dots for an oil-in-water droplet interface and thus provide an encapsulating shell for the coated nanoparticles, which are dispersed in the interior oil phase (Lin et al. 2003; Boker et al. 2007). Upon cross-linking, the copolymers form a robust microcarrier for the nanoparticles, with a very thin (~10 nm) wall (Lin et al. 2003; Boker et al. 2007) through which the nanoparticles can permeate under appropriate conditions.

In the simulation, we assume that the hydrophilic surface contains a crack and the interior of this crack is assumed to be hydrophobic. This substrate can be fabricated from poly(dimethyl siloxane) (PDMS)-based polymers with brittle surfaces. PDMS is hydrophobic, but ultraviolet/ozone (UVO) treatment effectively converts the surface of a PDMS film from hydrophobic to hydrophilic by production of a thin silica-like surface layer (having a very low water contact angle) (Berdichevsky et al. 2004). The surface silica layer has a higher modulus than the underlying PDMS (Mills et al. 2008). Subjecting such PDMS-silica composite films to strain forces (either mechanical or osmotic) that exceed the failure strain of the surface coating leads to cracking at the surface. Osmotic strain (generated by swelling the underlying PDMS in hexanes) results in formation of random cracks on the PDMS surface; mechanical stain (uniaxial extension) generates aligned cracks (Crosby et al. 2009).

With the aforementioned materials in mind, we simulate the scenario in Figure 13.1a where hydrophobic nanoparticles are encapsulated in a fluid-filled microcapsule, which is localized on a substrate in an aqueous solution. The intact portion of the surface is hydrophilic, but this substrate also encompasses cracks, which expose the underlying hydrophobic domains (e.g., the underlying PDMS). The amphiphilic nature of the capsule allows this colloid to be solubilized in water and attracted to both hydrophilic and hydrophobic portions of the surface. On the other hand, the encapsulated hydrophobic nanoparticles are only attracted to the hydrophobic cracks.

In effect, the enthalpic interactions between the nanoparticles and the hydrophobic crack provide a site-specific response of these particles. As we show in the

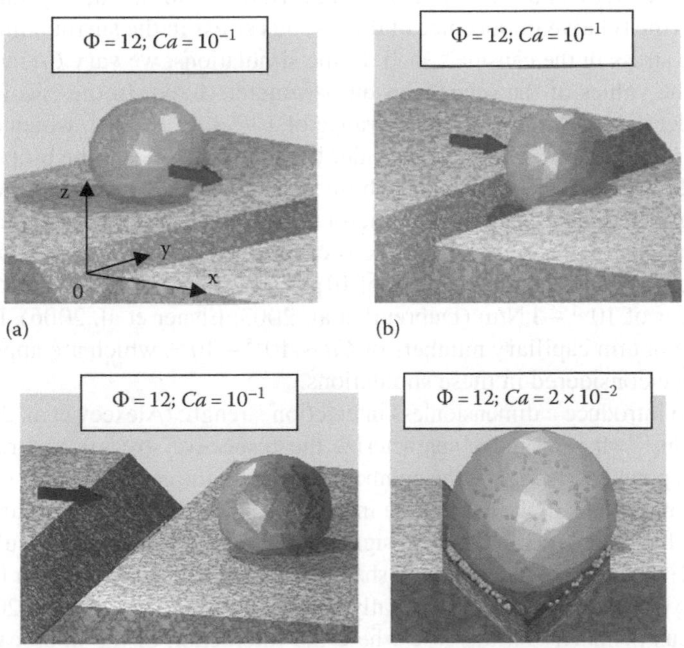

FIGURE 13.1 Graphical output from the simulation showing the motion of a capsule on a damaged surface from its initial position (a) to the interior of the crack (b) and its reemergence onto the undamaged portion of the surface (c). For this case, the capillary number $Ca = 10^{-1}$ and dimensionless adhesion strength $\Phi = 12$. The gray-shaded areas mark the substrate and the light gray points correspond to the nanoparticles. Black arrows indicate the direction of the imposed shear flow. (d) A capsule arrested within the crack at $Ca = 2 \times 10^{-2}$ and $\Phi = 12$.

following, the nanoparticles can permeate through the microcapsule's shell to bind to and fill the cracked regions. Furthermore, we find that the application of a nonsteady, pulsatile flow field could be useful in maximizing the deposition of particles into the crack and sustaining the continued motion of microcapsules, therefore crucially increasing the effectiveness of the autonomous healing of a damaged surface.

13.3.1 Effect of Imposing a Steady Shear Flow

The simulation box is $L_x \times L_y \times L_y = 100 \times 40 \times 40$ lattice Boltzmann units in size. Periodic boundary conditions are applied in the x- and y-directions. A well-defined crack is located in the center of the substrate along the y-axes, with the crack tip being located at $x = 0$ (see Figure 13.1). The capsule diameter is equal to $10\Delta x$, where Δx is equal to the lattice spacing in the LBM. The crack depth, d_c, is equal to $5\Delta x$ (i.e., comparable with the capsule radius), and the crack width is equal to $2d_c$.

The flat, upper wall of the simulation box is moved with a constant velocity U_{wall} to create a uniform shear rate of $\dot{\gamma} = U_{\text{wall}}/L_z$. We characterize the flow by specifying the dimensionless capillary number $Ca = \dot{\gamma}\mu R/Eh$, where μ is the fluid viscosity, R is the capsule radius, E is the modulus of the shell, and h is the shell thickness (Alexeev et al. 2005; Alexeev and Balazs 2007). The capillary number represents the relative importance of the viscous stress in the surrounding fluid and the elastic stress in the capsule's shell. In the simulations, we vary Ca by altering $\dot{\gamma}$, keeping the values of the other relevant parameters fixed. In the ensuing studies, we consider capillary numbers in the range of $10^{-4} \leq Ca \leq 10^{-1}$, which are experimentally reasonable values if we consider the microcapsules to be propelled by an imposed flow in an aqueous solution whose viscosity is $\mu \approx 10^{-3}$ kg/s m and density is $\rho \approx 10^3$ kg/m^3. Typical velocities in microfluidic devices are on the order of 1 cm/s and typical channel heights are hundreds of microns (Stone et al. 2004). For polymeric microcapsules with a diameter of 10 μm, the stiffness of the shell, Eh, can be on the order of $10^{-3} - 1$ N/m (Dubreuil et al. 2003; Elsner et al. 2006). From these values, we obtain capillary numbers of $Ca \sim 10^{-5} - 10^{-2}$, which are approximately in the range considered in these simulations.

We also introduce a dimensionless interaction strength (Alexeev et al. 2005, 2006) $\Phi = \varepsilon N/Eh\kappa^2$, where ε and κ characterize the respective strength and range of the interaction potential; here, N is the number of nodes on the capsule's outer surface. The parameter represents Φ the ratio of the interaction strength to the membrane stiffness. For $\Phi >> 1$, this interaction leads to significant deformation of the capsule, whereas for $\Phi << 1$, the effect on the capsule's shape is small. Herein, we consider $0 < \Phi < 25$, which also corresponds to experimentally realistic values (Alexeev et al. 2005).

We focus primarily on the case where the interaction of the microcapsule with the undamaged horizontal substrate and with the crack walls is characterized by the same interaction constant ε. As noted earlier, the surface of the microcapsule is assumed to be amphiphilic and thus it can exhibit a comparable interaction with both the intact hydrophilic surface and the hydrophobic crack. We also investigate one case where the capsule is less attracted to the hydrophobic than the hydrophilic region; we simulate this case by setting the interaction strength within the crack to half the value for the rest of the substrate. In all the simulations, we further assume that the deposited hydrophobic nanoparticles do not modify the adhesive properties of the crack walls. (The latter choice is not a necessary constraint of the method: in the simulations, the nanoparticles can dynamically modify the adhesive properties of the surface (Verberg et al. 2007).)

As shown in Figure 13.1a, the capsule is initially placed on the undamaged, horizontal region of the surface (with its center of mass positioned at $x = -15$) and in the middle of the channel in the y-direction (at $y = 20$). The capsule initially contains $N_{NP}(0) = 10^5$ nanoparticles. The particle-filled capsule is propelled toward the crack by the imposed shear flow; the black arrow in Figure 13.1a shows the direction of the imposed shear. The images in Figure 13.1b and c are for a capillary number of $Ca = 10^{-1}$ and adhesion strength of $\Phi = 12$ and correspond to points in time when the capsule is localized in the crack (Figure 13.1b) and when it has been propelled away from this region by the imposed flow (Figure 13.1c). For the image in Figure 13.1d, the adhesion strength is held fixed, but the shear rate is decreased to $Ca = 2 \times 10^{-2}$ and the figure shows that the capsule became stuck in the crack as it moved along the surface.

Figure 13.2 shows the time dependence of the capsule's center-of-mass x-coordinate for $\Phi = 2.4$ and for three shear rates, corresponding to $Ca = 2 \times 10^{-2}$, 6×10^{-2}, and 1.2×10^{-1}. The figure shows that at the lowest shear rate, the capsule becomes stuck within the crack. Additionally, in Figure 13.3, we plot a phase map for the system as a function of Ca and Φ that indicates the response of the capsules to the imposed flow and the attractive surface. To generate this phase map, we varied the dimensionless interaction strength from weak to strong adhesion, $3 \times 10^{-2} \leq \Phi \leq 24$, and the shear rates, Ca, from 2.7×10^{-4} to 1.2×10^{-1}. For each point, we averaged over three independent runs. In Figure 13.3a, the interaction strength of the amphiphilic microcapsule with the undamaged (ε_u) and damaged (ε_d) regions of

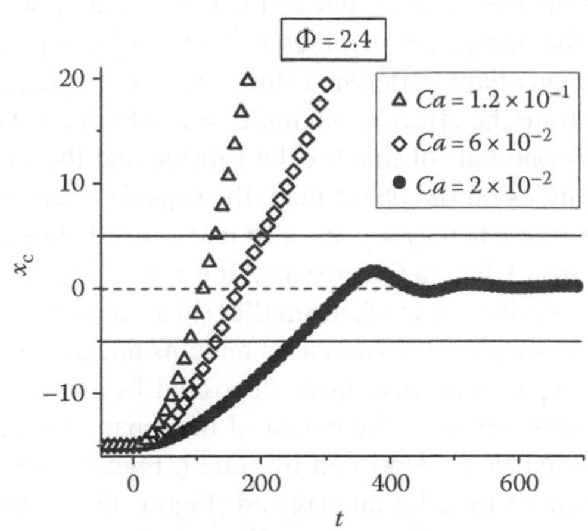

FIGURE 13.2 The center-of-mass coordinate x_c of a capsule as a function of time. The respective shear rates are labeled on the plot. Open symbols are for moving capsules and the filled symbols are for the arrested capsule. The x-coordinates of the crack boundaries are indicated by the solid horizontal lines and the x-coordinate of the crack tip is shown by a dashed horizontal line. Damped oscillations of an arrested capsule (black circles) in a crack (see Figure 13.1d) are clearly seen at $t > 350$.

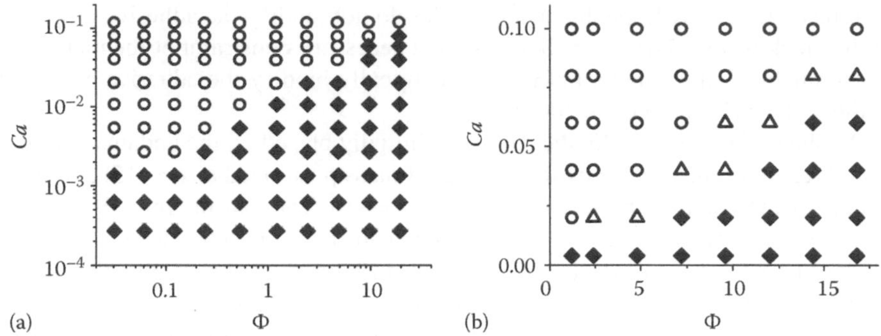

FIGURE 13.3 (a) Numerically calculated phase map as a function of capillary number *Ca* and interaction strength Φ. The unfilled circles indicate moving capsules and the solid diamonds indicate arrested capsules. The interaction strengths characterizing the intact surface and within the crack are the same. (b) Phase map computed for the case where the interaction strength within the crack is equal to half the interaction strength Φ for the rest of the substrate. Open triangles shows the points on the map at which the microcapsule changes its behavior from an arrested state (at plot [a]) to a moving state when the interaction within the crack is reduced.

the substrate are equal. These results reveal two general scenarios for the capsule's motion. For relatively high adhesion or low shear rates (indicated by the solid diamonds in the phase map), the capsule becomes arrested within the crack (as shown in Figure 13.1d). For weaker adhesion and higher shear rates (marked by the empty circles), the capsule moves into the crack, but then exits this region to continue moving along the surface (as seen in Figure 13.1c). In rare events, after the capsule leaves the crack, it bounds off the substrate due to lift forces that arise at high shear rates.

In Figure 13.3b, the interaction parameters between the capsule and the damaged and undamaged regions have different values; here, $\varepsilon_d = 1/2\varepsilon_u$. (The latter values were chosen to illustrate the effect of having $\varepsilon_d \neq \varepsilon_u$.) In other words, Φ for the capsule and the crack is one-half of that for the capsule and the undamaged substrate. For the unfilled triangles on the phase map, the capsule is arrested in the crack for $\varepsilon_d = \varepsilon_u$ but becomes free when $\varepsilon_d = 1/2\varepsilon_u$. (To make this difference more visible, we plot the map on Figure 13.3b in a linear scale.) It is clear from the plot that the microcapsule leaves the less adhesive crack at smaller shear rates. Nonetheless, the overall character of the phase map for the two cases remains unchanged.

To characterize the nanoparticle layer deposited by the microcapsule onto the crack walls, and, hence, quantify the extent of the repair, we calculated $N_{dep}(t)$, the total number of nanoparticles deposited by time *t*. Figure 13.4 shows the behavior of $N_{dep}(t)$ as a function of time for an arrested (Figure 13.4a) and a moving (Figure 13.4b) capsule; here, the quantity is normalized to the total number of nanoparticles initially encased in the microcapsule, $N_{NP}(0)$. For an arrested capsule, the ratio $N_{dep}(t)/N_{NP}(0)$ approaches unity. The total number of deposited nanoparticles provided by a moving capsule, however, is smaller. For example, we find that only 8% of the particles are deposited within the crack (i.e., $N_{dep}(t)/N_{NP}(0) \sim 0.08$) for the case of the mobile capsule shown in Figure 13.4b.

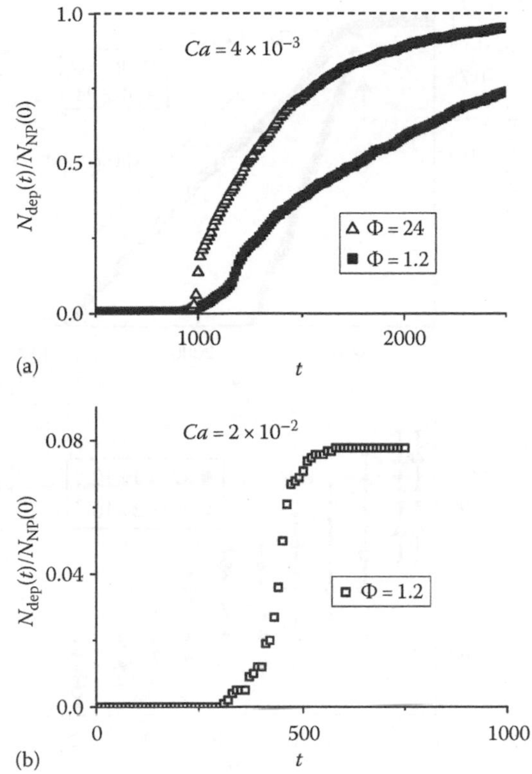

FIGURE 13.4 Time dependence of $N_{dep}(t)$, the number of nanoparticles deposited on the crack surface. The number is normalized to the initial number of nanoparticles in the capsule $N_{NP}(0)$. (a) Arrested capsules and (b) moving capsules. Respective values of Ca and Φ are marked on the plot.

To further characterize the performance of the microcarriers, we calculate $N_{NP}(t) = N_{NP}(0) - N_{dep}(t)$, the number of nanoparticles that remain within the capsule as a function of time. In Figure 13.5a, we plot, $N_{NP}(t)$ on a semilogarithmic scale for two different points (Φ, Ca) within the "arrested" region of the phase map in Figure 13.3. The parameter t_{in} indicates the time when the capsule becomes arrested in the crack, and, hence, the nanoparticles begin to be deposited in this region. Figure 13.5a indicates that at $t > t_{in}$, the temporal dependence of $N_{NP}(t)$ can be approximated by an exponentially decaying function of the form:

$$N_{NP}(t) = N_{NP}(0)\exp\left[\frac{-(t - t_{in})}{\tau_{dep}}\right] \tag{13.5}$$

with a characteristic deposition time τ_{dep}. According to Equation 13.5, approximately $(1 - e^{-1}) \times 100\% \approx 63\%$ of the nanoparticles initially placed in the capsule are deposited into the crack during the time period τ_{dep} after the capsule was arrested. An amount

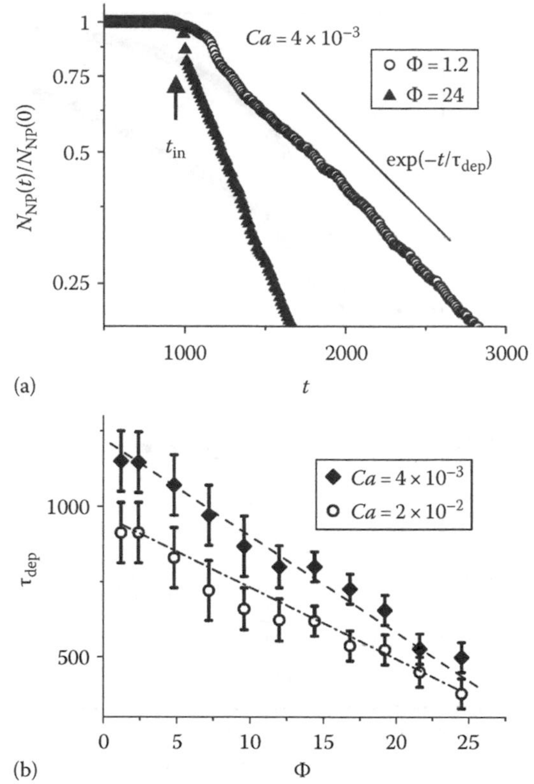

FIGURE 13.5 (a) Number of nanoparticles within an arrested capsule normalized to its initial value, $N_{NP}(t)/N_{NP}(0)$, plotted as a function of time on a semilogarithmic scale. The capillary number is $Ca = 4 \times 10^{-3}$. The parameter t_{in} (marked by an arrow) indicates the time when the capsule becomes arrested in the crack. Solid line shows an exponential dependence $N_{NP}(t) \sim \exp(-t/\tau_{dep})$ with a characteristic deposition time τ_{dep}. (b) Dependence of the deposition time τ_{dep} on the dimensionless interaction strength Φ plotted for $Ca = 4 \times 10^{-3}$ and 2×10^{-2}. Straight lines are plotted as a guide for the eye.

$(1 - e^{-2}) \times 100\% \approx 86\%$ of nanoparticles are deposited after the time period $2\tau_{dep}$ and greater than 95% of nanoparticles are deposited after the time period $>3\tau_{dep}$.

Figure 13.5b shows the dependence of the deposition time τ_{dep} on the interaction strength Φ at two shear rates for capsules that are arrested in the crack. The results are averaged over three independent runs. The deposition time τ_{dep} is seen to decrease with an increase in the interaction strength. The reason for this behavior can be explained as follows. At moderate adhesion ($\Phi \sim 1$), a spherical capsule arrested within the crack is essentially undeformed and, therefore, has only a small contact area with the crack walls. At higher adhesion ($\Phi >> 1$), the capsule is deformed by the attractive interaction with the substrate and thus the contact area and the average distance r between the shell and the walls is decreased. Since the probability for nanoparticles to be deposited onto the crack wall rises rapidly with decreasing r

(see Equation 13.4), the deposition time τ_{dep} is a decreasing function of the interaction strength. In other words, a shorter time is required for achieving higher coverage for strong adhesion strengths.

Figure 13.5b also indicates that the deposition time is shorter for relatively larger shear rates. For $Ca \geq 2 \times 10^{-2}$, the fluid acts to deform the capsule (Alexeev et al. 2005, 2006, 2007; Alexeev and Balazs 2007; Usta et al. 2007) and thus again produce an increase in the contact area between the capsule and surface, and a decrease in the distance r.

We now turn our attention to the case of a nonarrested or "free" capsule, which leaves the crack and continues to move along the surface under the imposed flow (the parameters are marked by the circles in Figure 13.3). For these cases, the number of deposited nanoparticles, N_{dep}, is relatively small (see Figure 13.4) as compared to the number obtained with an arrested capsule. The reasons for the observed decrease in the effectiveness of these moving capsules are twofold. First, the average distance r between the capsule shell and the crack walls is large in comparison with the characteristic distance κ for nanoparticle deposition. In particular, $r/l_0 \sim d_c/l_0 \sim 5$ for a capsule moving over the crack. Second, the time during which the capsule moves over the crack ($\leq$50 simulation time steps according to Figure 13.2) and the nanoparticles can be deposited into this region is short compared with the deposition time $\tau_{dep} \geq 300$. As a result, it is more difficult to obtain full coverage in the case of a "free" capsule.

To summarize, we find that full surface coverage by the nanoparticles can be more readily attained in the case of an arrested capsule, that is, in a "repair" rather than a "repair-and-go" system. In the next section, however, we demonstrate that a pulsatile fluid flow can lead to an effective "repair-and-go" scenario where the microcarriers not only deliver a high volume fraction of particles into the crack, but also leave the crack and thus, could potentially contribute to repairing additional cracks within the system with the remaining encapsulated particles.

13.3.2 Utility of Applying a Pulsatile Flow

In the following simulations, we apply a pulsatile shear where the value of the shear rate $\dot{\gamma}(t)$ (and, hence, the capillary number $Ca(t)$) depends on time in the manner shown in Figure 13.6a. In effect, we use the information from the previous section to tailor the motility of the capsule. Specifically, for a fixed Φ, we choose a low capillary number Ca_1 from the region in the phase map where the capsule is arrested within the crack. After a time interval Δt, we then increase the shear rate to Ca_2 so that for the specific Φ, the capsule is in the "free" region of phase space. (We note that the capsule can move along the undamaged portion of the surface for both Ca_1 and Ca_2.) In the time interval $t_1 < t < t_2$ during which the low shear corresponding to $Ca = Ca_1$ is applied, the capsule can become arrested in a nearby crack and if this time interval is sufficiently long, the microcarrier can deposit a high density of nanoparticles along the walls of the crack (see the following). At $t = t_2$, we increase the shear rate to $Ca = Ca_2$; in other words, we move vertically up in the (Φ, Ca) phase map. The simulations show that at some

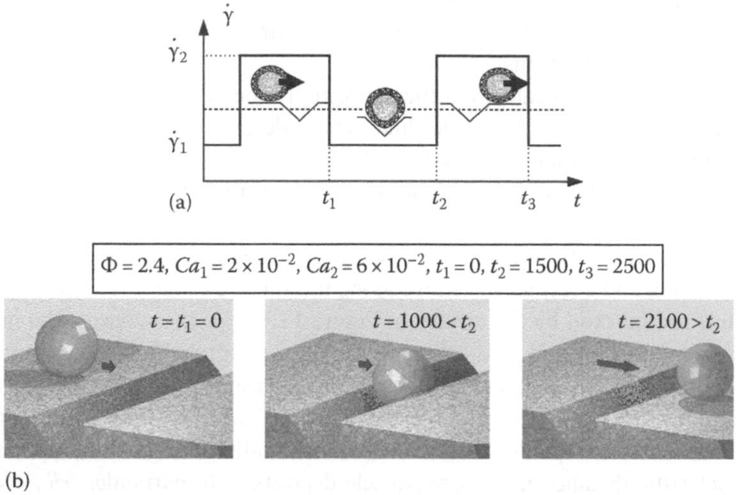

FIGURE 13.6 (a) Schematic indicating the time dependence of the shear rate $\dot{\gamma}$ for the imposed pulsatile shear flow. The t_1, t_2, and t_3 are the points in time when the shear rate is switched. A horizontal dashed line shows the ordinate of the boundary between the regions in Figure 13.3 where the capsule is free (unfilled circles) and arrested (solid diamonds) for a given value of interaction strength Φ. The shear rate $\dot{\gamma}_1$ belongs to the capillary number Ca_1 at which the capsule is in an arrested state and the shear rate $\dot{\gamma}_2$ belongs to the capillary number Ca_2 at which the capsule is in a moving state. (b) Motion of the capsule computed for $t_1 = 0$, $t_2 = 1500$, $t_3 = 2500$, $\Phi = 2.4$, $Ca_1 = 2 \times 10^{-2}$, and $Ca_2 = 6 \times 10^{-2}$. Left frame: initial position of the capsule; middle: the capsule arrested within the crack during the low shear period, $t_1 < t < t_2$; right: capsule leaves the crack at $t > t_2$. The length of the black arrow is proportional to the applied shear rate.

time $t = t_{out} > t_2$, the capsule leaves the crack due to the strong drag force from the surrounding fluid and moves further along the surface.

Figure 13.6b shows this behavior for $Ca_1 = 2 \times 10^{-2}$, $Ca_2 = 6 \times 10^{-2}$ and an adhesion strength of $\Phi = 2.4$. The inset in Figure 13.7 shows the variations with time of the center-of-mass x-coordinate of the capsule calculated for two parameter sets, $Ca_1 = 2 \times 10^{-2}$ and $Ca_2 = 6 \times 10^{-2}$ for an adhesion strength of $\Phi = 2.4$ (the case of moderate adhesion), and $Ca_1 = 6 \times 10^{-2}$ and $Ca_2 = 1.2 \times 10^{-1}$ for an adhesion strength of $\Phi = 12$ (strong adhesion). The capsule is arrested within the crack at a time $t = t_{in} \sim 100$ and leaves the crack at $t = t_{out} \sim 1500$–2000 in both cases. The main plot in Figure 13.7 shows the time dependence of the coverage computed for the two parameter sets. As can be seen, a nearly full deposition $N_{dep}/N_{NP}(0)$ (73% for moderate adhesion and 90% for strong adhesion) is obtained for a pulsatile flow. The latter values are significantly above the typical values of $N_{dep}/N_{NP}(0)$ (~10%–20%) obtained for a permanently moving or "free" capsule in a steady flow.

As indicated in Figure 13.7, the advantage of using the pulsatile shear is that we can combine high adhesion strength Φ, which provides nearly full nanoparticle coverage of the crack, with the possibility for a capsule to remain moving along the surface, therefore realizing the basic ideas of a "repair-and-go" system. Additionally, by applying a pulsatile flow, we can control the effective exposure time $t_{exp} = t_{out} - t_{in} \approx t_2 - t_1$

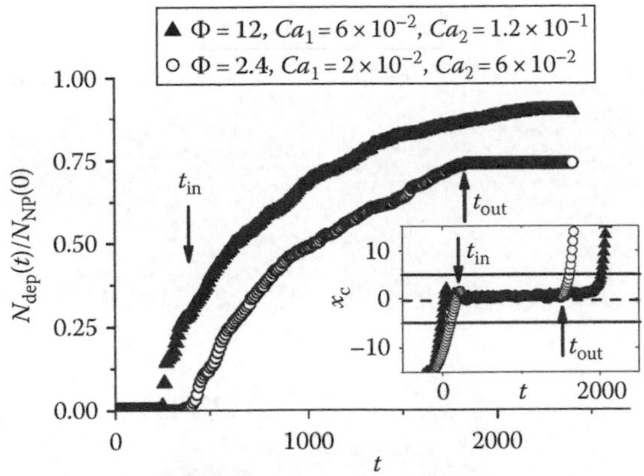

FIGURE 13.7 Time dependence of the nanoparticle coverage calculated for the pulsatile shear shown for two different values of the interaction strength. Inset shows the time dependence of the center-of-mass *x*-coordinate of the capsule calculated for the same parameter set as in the main plot. The symbols in the inset match the symbols in the main plot. Arrows mark the moments of time when the capsule was arrested in (t_{in}) and exits the crack (t_{out}) for $\Phi = 2.4$ (open circles). In the simulations, we take $t_1 = 0$, $t_2 = 1500$, and $t_3 = 2500$.

(see Figure 13.6a); that is, the exposure time can be set manually and independently, and can be varied over wide limits. (In the repair-and-go system with a pulsatile shear, the exposure time t_{exp} is defined by the duration of the low shear portion $t_2 - t_1$.) Furthermore, we can exploit the attractive features offered by the case of relatively high adhesion strength; namely, the higher Φ affords a more deformable capsule, which in turn contributes to an increased healing rate (see Figure 13.5b).

13.3.3 Microcapsule Motion on a Nanoparticle-Filled Domain

In the later stages of healing, the crack is essentially filled with nanoparticles. Figure 13.8a through c demonstrates the motion of a microcapsule on a surface where the crack is completely filled with nanoparticles. Here, we set $\Phi = 1.2$ and $Ca = 4 \times 10^{-3}$; for these parameters, the microcapsule is arrested within the empty crack (Figure 13.3). It is seen, however, that in the case of a filled crack, the microcapsule passes over the crack without being arrested. Figure 13.8d shows the capsule's center-of-mass coordinate as a function of time for the adhesion strength $\Phi = 1.2$ and for capillary numbers $Ca = 4 \times 10^{-3}$ and 2×10^{-2} (i.e., for points that lie within the different regions of the phase map in Figure 13.8) calculated for a completely filled crack. It is clearly seen that in both cases the microcapsule moves freely along the surface under the action of an imposed shear flow.

These results indicate that when the hydrophobic crack is filled with the hydrophobic particles, the amphiphilic capsules can move along the surface without hindrance. Thus, these mobile microcarriers can provide healing agents to other damaged sites further along the surface. Moreover, since the capsules do not stick to the repaired crack, they will not potentially clog the system and inhibit other capsules from reaching their targets.

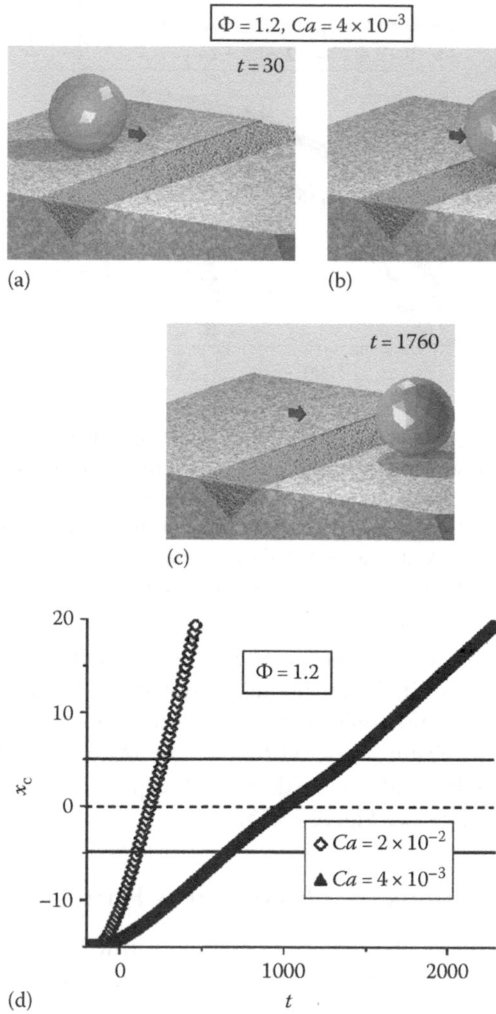

FIGURE 13.8 (a–c) Motion of the microcapsule along the surface with a crack completely filled with nanoparticles. (d) Time dependence of the center-of-mass coordinate x_c of a capsule computed for $\Phi = 1.2$, and $Ca = 2 \times 10^{-2}$ and 4×10^{-3}.

13.4 CONCLUSIONS

In summary, we reviewed our studies on devising a "repair-and-go" system where a healing agent is delivered to damaged regions on the substrate by fluid-driven, nanoparticle-filled microcapsules. This approach provides the ability to selectively repair a surface defect without affecting the rest of the surface; this is particularly attractive in cases involving precision fabricated systems.

In the design of the "repair-and-go" system, we were inspired by the functionality of leukocytes that respond to damage within the body by localizing at the affected areas and initiating the repair process. Leukocytes encase a host of biological machinery that enables these cells to sense the location of the wound and subsequently promote the cascade of dynamic events that lead to wound healing. Our aim, in essence, was to design an artificial leukocyte that could affect the repair of a synthetic wound, namely a crack in a surface.

The ability to fabricate amphiphilic capsules that encapsulate hydrophobic nanoparticles and surfaces that encase hydrophobic cracks allowed us to formulate a delivery system that could be transported by an imposed flow in an aqueous solution and specifically target the nanoparticles to the hydrophobic domains. We found that there is a range of capsule–surface interaction parameters and shear rates that maximize the deposition of the nanoparticles into the cracks. For this range of parameters, however, the capsule becomes arrested in the crack. At this stage in the experimental systems, the organic component could be removed by heating the sample and thus, burning away the polymer. In effect, this could be a way of making a "repair and go away" system.

Our ultimate goal, however, was to design a system where the nanoparticle-filled capsule would not only aid in the repair of one site, but also travel on to heal other cracks. To this end, we introduced a pulsatile shear flow that involves alternating low and high shear rates. In the low shear rate phase, the pulsatile flow enables the capsule to be localized within a crack for a time Δt. With an increase in the shear rate, however, this capsule is propelled to leave the fissure and thus, be available to carry out further healing of the damaged substrate. Hence, by utilizing the pulsatile flow field, we could achieve the desired "repair and go" functionality in our simulations. (An attractive feature of this system is that the user can tailor the flow profile in order to control the value of t_{exp}.)

We note that in the aforementioned studies, we focused on the action of a single microcapsule, allowing us to isolate the factors that control the dynamic behavior of this microcarrier. In the actual physical system, which involves a solution of microcapsules moving in a microchannel, there is a high volume fraction of microcarriers available to repair the surface and thus, the cracks could potentially be completely filled by deposited nanoparticles. As we showed in the final section of this chapter, for the chosen interaction parameters, the amphiphilic capsules would then move over these repaired areas and go onto other locations where they are needed.

ACKNOWLEDGMENTS

The authors gratefully acknowledge financial support from the DOE (for partial support of G.V.K.). G.V.K. also acknowledges partial support from NSF through TeraGrid resources provided by NCSA and TACC.

REFERENCES

Alexeev, A. and A. C. Balazs. 2007. Designing smart systems to selectively entrap and burst microcapsules. *Soft Matter* 3:1500–1505.

Alexeev, A., R. Verberg, and A. C. Balazs. 2005. Modeling the motion of microcapsules on compliant polymeric surfaces. *Macromolecules* 38:10244–10260.

Alexeev, A., R. Verberg, and A. C. Balazs. 2006. Designing compliant substrates to regulate the motion of vesicles. *Phys. Rev. Lett.* 96:148103.

Alexeev, A., R. Verberg, and A. C. Balazs. 2007. Patterned surfaces segregate compliant microcapsules. *Langmuir* 23:983–987.

Amendola, V. and M. Meneghetti. 2009. Self-healing at the nanoscale. *Nanoscale* 1:74–88.

Balazs, A. C. 2007. Modeling self-healing materials. *Mater. Today* 10:18–23.

Balazs, A. C., T. Emrick, and T. P. Russell. 2006. Nanoparticle-polymer composites: Where two small worlds meet. *Science* 314:1107–1110.

Berdichevsky, Y., J. Khandurinab, A. Guttmanb, and Y.-H. Loa. 2004. UV/ozone modification of poly(dimethylsiloxane) microfluidic channels. *Sensors Actuators B: Chem.* 97:402–408.

Boker, A., J. He, T. Emrick, and T. P. Russell. 2007. Self-assembly of nanoparticles at interfaces. *Soft Matter* 3:1231–1248.

Breitenkamp, K. and T. Emrick. 2003. Novel polymer capsules from amphiphilic graft copolymers and ring-opening cross metathesis. *J. Am. Chem. Soc.* 125:12070–12071.

Buxton, G. A., C. M. Care, and D. J. Cleaver. 2001. A lattice spring model of heterogeneous materials with plasticity. *Model. Simul. Mater. Sci. Eng.* 9:485–497.

Buxton, G. A., R. Verberg, D. Jasnow, and A. C. Balazs. 2005. Newtonian fluid meets an elastic solid: Coupling lattice Boltzmann and lattice-spring models. *Phys. Rev. E* 71:056707.

Caruso, M. M., D. A. Davis, Q. Shen et al. 2009. Mechanically-induced chemical changes in polymeric materials. *Chem. Rev.* 109:5755–5798.

Chen, X., M. A. Dam, K. Ono et al. 2002. Thermally re-mendable cross-linked polymeric material. *Science* 295:1698–1702.

Cordier, P., F. Tournilhac, C. Soulie-Ziakovic, and L. Leibler. 2008. Self-healing and thermoreversible rubber from supramolecular assembly. *Nature* 451:977–980.

Crosby, A. et al. 2009. Unpublished data.

Dubreuil, F., N. Elsner, and A. Fery. 2003. Elastic properties of polyelectrolyte capsules studied by atomic-force microscopy and RICM. *Eur. Phys. J. E* 12:215–221.

Elsner, N., V. Kozlovskaya, S. A. Sukhishvili, and A. Fery. 2006. pH-Triggered softening of crosslinked hydrogen-bonded capsules. *Soft Matter* 2:966–972.

Gupta, S., Q. L. Zhang, T. Emrick, A. C. Balazs, and T. P. Russell. 2006. Entropy-driven segregation of nanoparticles to cracks: A route to self-healing systems. *Nat. Mater.* 5:229–233.

Hickenboth, C. R., J. S. Moore, S. R. White, N. R. Sottos, J. Baudry, and S. R. Wilson. 2007. Biasing reaction pathways with mechanical force. *Nature* 446:423–427.

Kolmakov, G. V., K. Matyjaszewski, and A. C. Balazs. 2009. Harnessing labile bonds between nanogel particles to create self-healing materials, *ACS Nano* 3:885–892.

Kolmakov, G. V., R. Revanur, R. Tangirala, T. Emrick, T. P. Russell, A. J. Crosby, and A. C. Balazs. 2010. Using nanoparticle-filled microcapsules for site-specific healing of damaged substrates: Creating a "repair-and-go" system. *ACS Nano* 4:1115–1123.

Ladd, A. J. C., J. H. Kinney, and T. M. Breunig. 1997. Deformation and failure in cellular materials. *Phys. Rev. E* 55:3271–3275.

Lallemand, P. and L.-S. Luo. 2000. Theory of the lattice Boltzmann method: Dispersion, dissipation, isotropy, Galilean invariance, and stability. *Phys. Rev. E* 61:6546–6562.

Lawn, B. R. 1993. *Fracture of Brittle Solids*, 2nd edn. New York: Cambridge University Press.

Lin, Y., H. Skaff, T. Emrick, A. D. Dinsmore, and T. P. Russell. 2003. Nanoparticles at liquid–liquid interfaces: Assembly, displacement and transport. *Science* 299:226–229.

Mills, K. L., X. Y. Zhu, S. C. Takayama, and M. D. Thouless. 2008. The mechanical properties of a surface-modified layer on polydimethylsiloxane. *J. Mater. Res.* 23:37–48.

Ottinger, H. C. 1996. *Stochastic Processes in Polymeric Fluids*. Berlin, Germany: Springer.

Revanur, R., B. McCloskey, K. Breitenkamp, B. D. Freeman, and T. Emrick. 2007. Reactive amphiphilic graft copolymer coatings applied to poly(vinylidene fluoride) ultrafiltration membranes. *Macromolecules* 40:3624–3630.

Stone, H. A., A. D. Stroock, and A. Ajdari. 2004. Engineering flows in small devices: Microfluidics toward a lab-on-a-chip. *Annu. Rev. Fluid Mech.* 36:381–411.

Succi, S. 2001. *The Lattice Boltzmann Equation for Fluid Dynamics and Beyond.* Oxford, U.K.: Clarendon Press.

Szymchak, P. and A. J. C. Ladd. 2003. Boundary conditions for stochastic solutions of the convective-diffusion equations. *Phys. Rev. E* 68:036704.

Therriault, D, S. R. White, and J. A. Lewis. 2003. Chaotic mixing in three-dimensional microvascular networks fabricated by direct-write assembly. *Nat. Mater.* 2:265–271.

Trask, R. S., H. R. Williams, and I. P. Bond. 2007. Self-healing polymer composites: Mimicking nature to enhance performance. *Bioinsp. Biomim.* 2:P1–P9.

Usta, O. B., A. Alexeev, and A. C. Balazs. 2007. Fork in the road: Patterned surfaces direct microcapsules to make a decision. *Langmuir* 23:10887–10890.

Usta, O. B., A. Alexeev, G. Zhu, and A. C. Balazs. 2008. Modeling microcapsules that communicate through nanoparticles to undergo self-propelled motion. *ACS Nano* 2:471–476.

Verberg, R., A. Alexeev, and A. C. Balazs. 2006. Modeling the release of nanoparticles from mobile microcapsules. *J. Chem. Phys.* 125:224712.

Verberg, R., A. T. Dale, P. Kumar, A. Alexeev, and A. C. Balazs. 2007. Healing substrates with mobile particle-filled microcapsules: Designing a 'repair and go' system. *J. R. Soc. Interface* 4:349–357.

Verberg, R., J. M. Yeomans, and A. C. Balazs. 2005. Modeling the flow of fluid/particle mixtures in microchannels: Encapsulating nanoparticles within monodisperse droplets. *J. Chem. Phys.* 123:224706.

White, S. and P. H. Geubelle. 2010. News and Views. Self-healing materials: Get ready for repair-and-go. *Nat. Nanotechnol.* 5:247–248.

Wool, R. A. 2008. Self-healing materials: A review. *Soft Matter* 4:400–418.

Wu, W., C. J. Hansen, A. M. Aragón, P. H. Geubelle, S. R. White, and J. A. Lewis. 2010. Direct-write assembly of biomimetic microvascular networks for efficient fluid transport. *Soft Matter* 6:739–742.

Wu, D. Y., S. Meure, and D. Solomon. 2008. Self-healing polymeric materials: A review of recent developments. *Prog. Polym. Sci.* 33:479–522.

14 Autopoietic Self-Reproduction as a Distinctive Feature of Structural and Dynamic Organization in Microcompartment Systems

From Self-Assembled Micelles to Synthetic Cells

Pasquale Stano

CONTENTS

14.1 INTRODUCTION

Prompted by the pioneering studies on the rational investigation on supra- and intermolecular self-organization that led to the origin of life on Earth, in recent years the topic of constructing cell models became a very fascinating research avenue (Morowitz, 1992; Forster and Church, 2006; Luisi et al., 2006; Mansy and Szostak, 2009; Luisi and Stano, 2011). The focus of these studies is placed on structural and dynamical organization in microcompartmentalized systems, generally based on self-assembled compartments such as lipid vesicles, micelles, and reverse micelles.

Along a scale of complexity, the living cell occupies a key position. At lower hierarchical levels, we find only molecules and supramolecular aggregates, per se nonliving, whereas at higher level we find more complex structures, such as multicellular organisms, or tissues and organs in higher organisms; therefore, understanding the essence of life means understanding cellular life. Living cells display a set of remarkable properties, such as self-maintenance, self-healing, self-reproduction, homeostasis, and several others. Even in the simplest bacteria, all these properties are given by a structural–functional organization that is currently unsurpassed by any manmade device. This huge complexity is fascinating, but limits the extent of our understanding of the basic principles of living cells. One may therefore ask whether it is possible to *reduce* the complexity of such organization and still maintain the essential (minimal) aforementioned features. Is it possible to assemble a structure with a *minimal* number of molecular components and still see some key living properties, although reasonably dumped down to their minimal efficiency (Luisi et al., 2006)? This general idea has been translated into an experimental project that aims at understanding the principle of cellular organization and therefore of the earlier cited list of properties, by *constructing* cell models. Note that this approach radically differs from the classical *analytical* approach, where a system is decomposed into its parts with the aim of understanding it. The researcher takes inspiration from living cell and develops biomimetic or bioinspired molecular systems capable, for example, of self-healing or self-maintaining, competing for nutrients, and so on.

At this aim, the first step consists in framing the constructive approach into a theoretical landscape, which in this context is given by the autopoietic theory of Maturana and Varela (Varela et al., 1974). The autopoiesis offers us a strong and fruitful link between the concepts of cell organization, synthetic cells, and the topic of self-healing. In fact, self-healing in dynamical and structured chemical systems can be seen as the self-maintenance of internal organization by means of regeneration of "self" components, and against the occasional appearance of damages of "non-self" components that do not fit into the autopoietic organization. This viewpoint differs from the classical discussion of self-healing aspects of lipid-based systems, which are shortly commented in Section 14.5. After introducing autopoiesis, we will introduce the self-reproduction of micelles, reverse micelles, and lipid vesicles, and then we will shortly discuss the evolution of such studies into more complex systems, namely, the construction of cell-like structures, always based on lipid compartment, named "minimal cells" (MCs) (Luisi, 2002; Luisi et al., 2006). A critical analysis of next developments is also presented. Constantly, we will link self-healing features with the issue of structural/dynamical organization in microcompartmentalized systems.

14.2 AUTOPOIESIS

Autopoiesis (self-production) is a theory that describes the organization of living cells. Developed in the 1970s by two Cilean neurobiologists, Humberto Maturana and Francisco Varela (Varela et al., 1974), autopoiesis is classically considered a system theory. Looking at a functioning cell, we see the next essential facts of life (see also Figure 14.1):

1. There is a *boundary* that defines the individuality of the living entity. The boundary determines a microenvironment which is physically or chemically different from the environment.
2. The boundary permits the *traffic* of nutrients and wastes, a necessary dynamical aspect of life (life as open and nonequilibrium system).
3. Inside the compartment, there is a reaction *network* that continuously transforms the internal components (included those forming the boundary), by replacing those components that disappear being destructed by side reactions or due to their transformation in other components of the network.
4. In this way, the cell maintains its own *identity* by a self-regeneration process despite the continuous transformation of material parts by anabolic and catabolic processes. The cell exists as a *dynamical* set of material parts, but it stays at a higher hierarchical level.

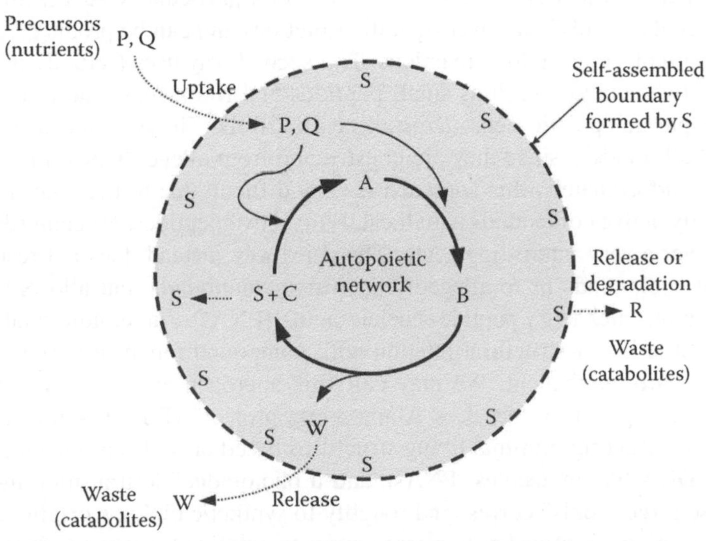

FIGURE 14.1 A cartoon representing a simple autopoietic system. It consists in a series of processes that produce all the cell's components that in turn produce the processes that produce such components (the "autopoietic network," here composed by A, B, C, and S molecules). The network is physically embodied into a self-bounded physical system (here constituted by S molecules), here represented as gray circle. In order to feed the cycle, precursors P and Q are taken up from the environment; waste products W and R are also generated and released. There is no central organizing unit in autopoiesis. Further discussion on chemical autopoiesis can be found in Luisi (2003).

It is remarkable that internal organization tends to maintain its own existence in recursive manner. More specifically, the circular logic is expressed as it follows: an autopoietic reaction network is a set of reactions that produce molecular components, which in turn structurally determine a self-bounded system wherein a reaction network operates, which in turn produces molecular components, and so on.

Clearly, the main consequence of the autopoietic organization is the maintenance of its identity. The autopoietic unit processes nutrients from the environment and transforms them into its own components within the network of coupled reactions. The aspect of self-maintenance is that one more closely related to self-healing properties of material. Here, we see very clearly the implications of self-maintenance for self-healing. Self-healing can be considered as the consequence of the continuous internal regeneration of the system's parts, replacing the missing or damaged ones. The information stored in autopoietic organization allows self-healing dynamics because each autopoietic unit, in terms of the reciprocal relations among the parts, contains the information for the development of properly organized structure, as a normal turnover of components.

Very importantly, autopoiesis acts as a theoretical framework for the development of an experimental plan about the construction of minimal living cells.

To date, there are essentially three directions for the construction of minimal autopoietic cells in the laboratory. The first way uses modern macromolecules such as enzymes, and nucleic acids, and construct MCs that are called consequently "semisynthetic" (Luisi et al., 2006). This route is particularly successful because these molecular "tools" have well-specific functions and can be produced according to classical molecular biology reactions. The second way uses (actually it aspires to use) primitive catalysts such as small peptides and ribozymes, and aims at developing simple autopoietic cells (Szostak et al., 2001). These could be relevant as primitive cell models, since they are constructed from allegedly primitive chemical structure. Unfortunately, this approach is very difficult due to the unavailability of catalytically active compounds (small catalytic active peptides, efficient ribozymes). It is however a very interesting route. The third way instead does not restrict itself to natural compounds or to allegedly primitive compounds, but allows the use of synthetic molecules (e.g., peptide–nucleic acids [PNAs], ruthenium catalysts, etc.) that have little if any structural relation with compound found in living forms, but are functionally equivalent. We may call this approach as a fully synthetic one. An example is given by the "Los Alamos bug project" (Rasmussen et al., 2004), aimed at constructing minimal living structures based on self-reproducing micelles, adsorbed DNA-like molecules (PNAs), and a photoinducible transition metal catalyst. These three "souls" correspond roughly to synthetic biology, origin of life, and systems chemistry approaches to microcompartmentalized systems (a more detailed comment on systems chemistry is given in the concluding remarks). The distinction is however very elusive, since cross-fertilization among these fields is often more advantageous than a strict separation. Instead of emphasizing differences, in fact, it is useful to focus on common aspects. These are indeed all reducible to the dynamics illustrated in Figure 14.1, but using different components and compartments. Several researchers, especially in systems chemistry, refer to another theoretical framework for the construction of MC: this is the *chemoton*, introduced independently by Tibor

Ganti (1975) approximately at the same time of autopoiesis. The two approaches do not differ in the essence of mechanistic description of living cells, but some important conceptual differences remain, in particular the concepts of organizational closure, emergence, biological autonomy, and minimal cognition, which are elegantly expressed in autopoiesis. A comparison between autopoiesis, chemoton, and (M,R)-systems (Rosen, 1991) has been recently published; the interested reader can refer to the work of Letelier et al. (2006) and Cárdenas et al. (2010).

In Sections 14.3 and 14.4, we will see how simple lipid-based compartments such as micelles, reverse micelles, and vesicles have been used to build minimal autopoietic systems, and later how these initial studies have been advanced according to the semisynthetic route, that is, combining molecular biology reactions and lipid vesicles (liposomes).

14.3 LIPID-BASED SUPRAMOLECULAR SYSTEMS

The first experimental implementation of autopoiesis was carried out by using lipid-based supramolecular systems, such as micelles, reverse micelles, and vesicles as self-bounded compartments, and chemical or biochemical reaction compartmentalized inside them (Figure 14.2) (for a review, see Stano and Luisi, 2010). In the early 1990s, two important cases were first studied. These are the autopoietic self-reproduction of micelles and reverse micelles. The rationale underlying such approaches is the following. In the simplest possible autopoietic system, the boundary is composed by only one component, whose synthesis can be achieved inside the compartment itself due to the presence of a chemical reaction, catalyzed or stoichiometrically supported by other compounds. If the precursor of the boundary-forming compound and the helper compounds can be taken up by the compartment, they will react internally and produce the desired boundary-forming compound. This dynamics constitute

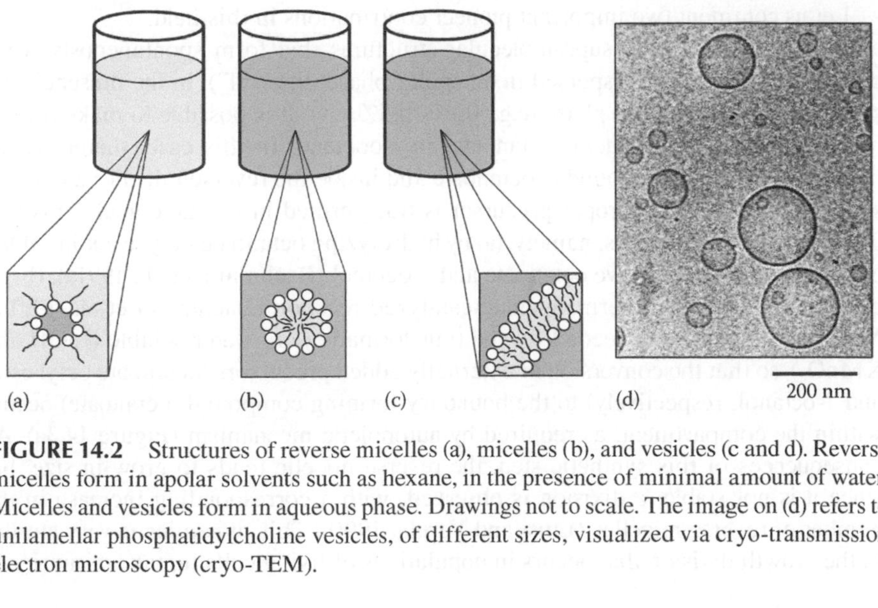

FIGURE 14.2 Structures of reverse micelles (a), micelles (b), and vesicles (c and d). Reverse micelles form in apolar solvents such as hexane, in the presence of minimal amount of water. Micelles and vesicles form in aqueous phase. Drawings not to scale. The image on (d) refers to unilamellar phosphatidylcholine vesicles, of different sizes, visualized via cryo-transmission electron microscopy (cryo-TEM).

the *self-production* of parts from within a system. Accordingly, the compartmentalized system must grow (there are no catabolic/destructive routes to compete with synthesis, so that homeostasis cannot be reached). However, this is the interesting aspect; in some cases the self-assembled compartment cannot grow indefinitely. Due to internal constrains, its stability is size-dependent. As consequence, a division can follow the initial growth, giving rise to a coupled growth division mechanism that is equivalent to *self-reproduction*.

In the cases of micelles and reverse micelles, the fact that only a certain size is generally stable directly derives from the balance of forces that hold these structures. Aqueous micelles cannot grow spherically because this means that the internal space has to be filled by other micelle-forming compound that will bring a polar or charge head group in the hydrophobic micelle core, missing the opportunity of interacting with water. Therefore, a cylinder-like shape becomes favored, but this also means that it can be segmented to give again smaller micelles. Reverse micelles' size, on the other hand, is often determined by the water/surfactant ratio. Note that the water/surfactant ratio can be ultimately seen as a volume-to-surface ratio of each micelle. When the amount of water phase is fixed, and the amount of reverse micelle–forming compound increases, a spherical micelle should turn into cylindrical one, in order to decrease its volume-to-surface ratio, and again this structure, if unstable, can be a precursor of micelle division. In contrary, vesicles can indeed exist in a very wide range of size and their spontaneous division probably derives by a specific growth mechanism that is still not well known (see the following). To summarize, micelle and reverse micelle self-reproduction is based on thermodynamic reasons, whereas vesicle self-reproduction is probably based on kinetic constrains.

It is important to remark that since only some of the components of the autopoietic unit are produced, these systems are just an approximation to the true autopoietic mechanism. Nevertheless, they represent very interesting cases for investigating this interesting coupling between structure and organization.

Let us comment two important pioneer contributions in this field.

Reverse micelles are supramolecular structures that form spontaneously when certain surfactants are dispersed in an apolar phase (the "oil"), in the presence of a small amount of aqueous phase (e.g., 0.1%–0.5% v/v). It is possible to make reverse micelles from octanoic acid (octanoate) in isooctane. In this case, therefore, the boundary-forming compound is octanoate and inside the reverse micelles a reaction should occur so that a proper precursor is transformed in octanoic acid. This was done in two different ways, namely, (a) by hydrolyzing octanoate octyl ester in LiOH-catalyzed reactions, to give octanoate and 1-octanol (Bachmann et al., 1991a); (b) by oxidizing 1-octanol in permanganate-catalyzed reaction (Bachmann et al., 1991b). Note that the reactants needed for the transformation are water soluble (LiOH and $KMnO_4$), so that the conversion of externally added precursors (octanoate octyl ester and 1-octanol, respectively) to the boundary-forming compound (octanoate) occurs within the compartment, as required by autopoietic mechanism (Figure 14.3a). As consequences of this synthetic step, the reverse micelle tends to grow in size, but since it is not stable, a division is observed, with a corresponding increase in the number of reverse micelles (Luisi and Varela, 1990). This dynamics is very similar to the growth division that occurs in populations of living cells (uptake of nutrients,

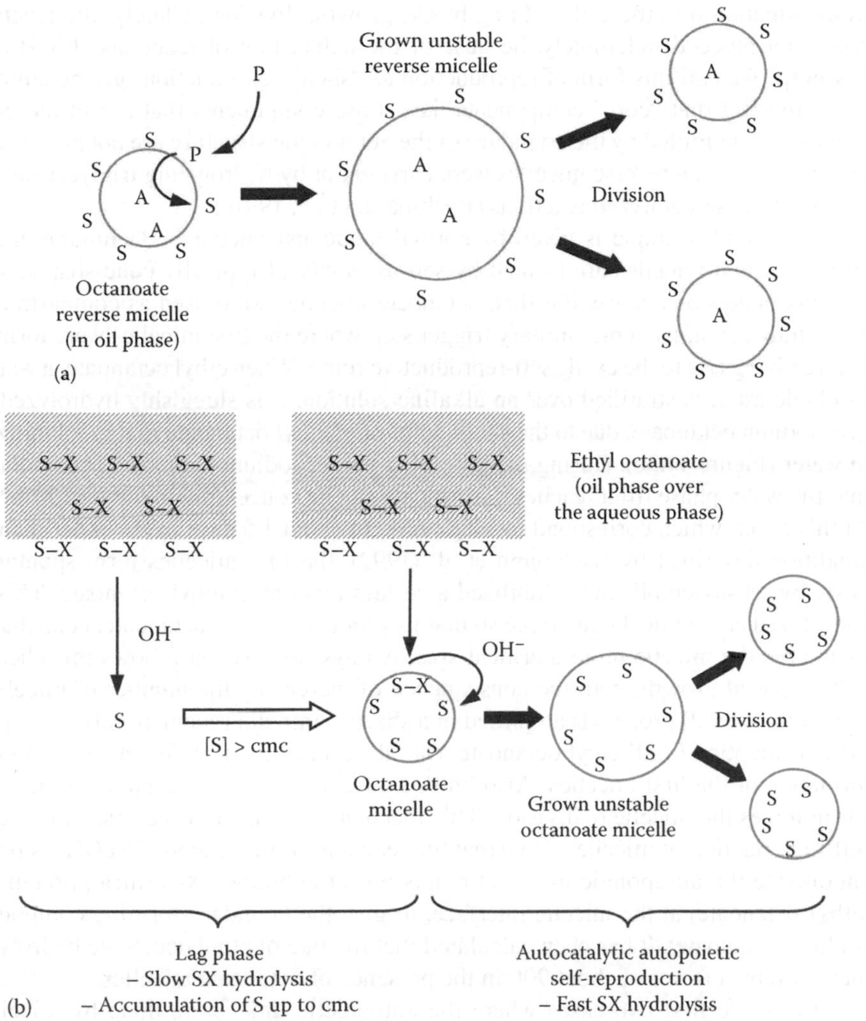

FIGURE 14.3 Self-reproduction of micelles and reverse micelles. (a) The case of reverse micelles. Octanoate reverse micelles in isooctane/1-octanol 90/10 are prepared in the presence of LiOH or $KMnO_4$ (here indicated as "A" entrapped inside the reverse micelle). 1-Octanol actually acts as cosurfactant. In the first case (LiOH) (Bachmann et al., 1991a), octyl octanoate was added in oil phase, so that it is adsorbed on the reverse micelle shell and hydrolyzed to octanoate and 1-octanol, giving rise to micelle growth, destabilization, and division. In the case of $KMnO_4$, 1-octanol was oxidized and octanoate was produced, which induces micelle growth and division (Bachmann et al., 1991b). (b) The case of "normal" micelles. Ethyl octanoate (water insoluble) is first stratified over an alkaline solution. The hydrolysis of the ester is initially very slow, due to its low solubility in aqueous phase. However, the sluggish reaction slowly accumulates octanoate in the water phase, until the critical micelle concentration (cmc) is reached. The formation of first micelle helps to solubilize ethyl octanoate, by taking up this hydrophobic molecule in the micelle, where it is hydrolyzed. Consequently, the micelle grows and divides to give more micelles, which further amplify the process in autocatalytic way. The kinetic trace of ethyl octanoate hydrolysis follows a sharp sigmoidal pattern. (From Bachmann, P.A. et al., *Nature*, 357, 57, 1992.)

transformation into the cell building block, growth, division). Clearly, the reaction could not proceed indefinitely, because of the exhaustion of reactants (LiOH and $KMnO_4$). We call this form of reproduction as "shell" reproduction (giving emphasis to the fact that "core" components, i.e., those components that are in the confined space delimited by the boundary of the autopoietic structure are not produced). Further studies on reverse micelles were carried out by hydrolyzing triglycerides by means of lipase-catalyzed reactions (Bachmann et al., 1991b).

The second example is given by normal (aqueous) micelles (Bachmann et al., 1992). Normal micelles are formed by self-assembly of typically cone-shaped surfactants in aqueous phase. For this, octanoate micelles were used as compartment. This study contained a preliminary trigger step where the first micelles were formed before giving rise to the cyclic self-reproductive route. When ethyl octanoate, a water-insoluble ester, is stratified over an alkaline solution, it is sluggishly hydrolyzed to give sodium octanoate, due to the small amount of ethyl octanoate that is solubilized in water (Figure 14.3b). During the initial lag phase, sodium octanoate accumulates into the water phase till it reaches the critical micelle concentration (about 100 mM). At this point, which correspond to a lag phase of about 1.5 days in the experimental condition described by Bachmann et al. (1992), the first micelles form spontaneously by self-assembly and solubilized a certain amount of ethyl octanoate. This is hydrolyzed at the micelle interface so that its product, sodium octanoate, contributes to enlarge the micelle up to a critical size. A physical division follows this chemically induced growth, with the consequence of increasing the number of micelles, so that the overall process is amplified in a sharp sigmoidal (geometrical) way, up to the consumption of all ethyl octanoate available, that is, within few hours from the formation of the first micelles. Also, in this case, therefore, a thermodynamic constrain forces the micelle to division. This mechanism is an autocatalytic autopoietic self-reproduction of micelles. Note that the reactant for the reaction (NaOH) is present outside the autopoietic unit, but reacts with the boundary-forming precursor (ethyl octanoate) at the micelle interface, to give the boundary-forming compound (sodium octanoate). It has been calculated that the rate of ethyl octanoate hydrolysis increased by a factor of about 900 in the presence of octanoate micelles.

After these first two cases where the autopoietic units were done by micelles and reverse micelles (two "equilibrium" structures, where the limit to their size caused a spontaneous division after growth), the interest was shifted toward the more complicated (but also more interesting) case of vesicles (Walde et al., 1994). In contrast to micelles and reverse micelles, vesicles exist in all possible size (except, perhaps, very small size (diameter <30–40 nm) due to the increase of the bending energy of the bilayer). Most of the work has been done with fatty acid vesicles, for example, oleic acid/oleate vesicles (shortly, oleate vesicles). This is due to two convergent reasons: (1) fatty acid vesicles have the distinctive properties of being quite "dynamical" systems when compared with classical phospholipid vesicles (liposomes); this is due to the higher solubility of fatty acids in water that translate in higher critical aggregation concentration, rapid uptake and release by and from the membrane, higher membrane permeability (Walde et al., 2006; Chen and Walde, 2010); (2) fatty acid vesicles are, to date, the most plausible candidate to play the role of prebiotic compartments (Hargreaves and Deamer, 1978; Monnard and

Deamer, 2002; Deamer and Dworkin, 2005; Walde, 2006), and therefore, their use as autopoietic structures is very attractive.

The physicochemistry of fatty acid assemblies is complicated mainly by the pH dependence (but also by cation compatibility), and the interested reader should refer to already published treatment for a deep understanding (Gebicki and Hicks, 1976; Haines, 1983; Cistola et al. 1986, 1988). The essential fact is that fatty acid vesicles exist only in a narrow pH region; for example, oleate vesicles are maximally stable at pH 8.5, where the amount of carboxylic and carboxylate forms are approximately equal (note that this value is about 3.5 pH units above the p*K* of free carboxylic acid in aqueous phase).

The autopoietic self-reproduction of oleate vesicles has been established in the following manner (Walde et al., 1994) (see Figure 14.4). A determined amount of oleic anhydride (water insoluble) was stratified over an alkaline (pH 8.5) solution. As in the case of ethyl octanoate (see the preceding text), the spontaneous hydrolysis of oleic anhydride proceeded sluggishly in the absence of oleate vesicles. After the initial lag phase, where oleate molecules accumulate in the aqueous phase, a critical aggregation concentration is reached (about 0.02 mM), so that vesicles form. At this point, oleic anhydride is taken up by oleate vesicles, hydrolyzed at the vesicle interface, and a strong acceleration of anhydride hydrolysis is observed. Clearly, the

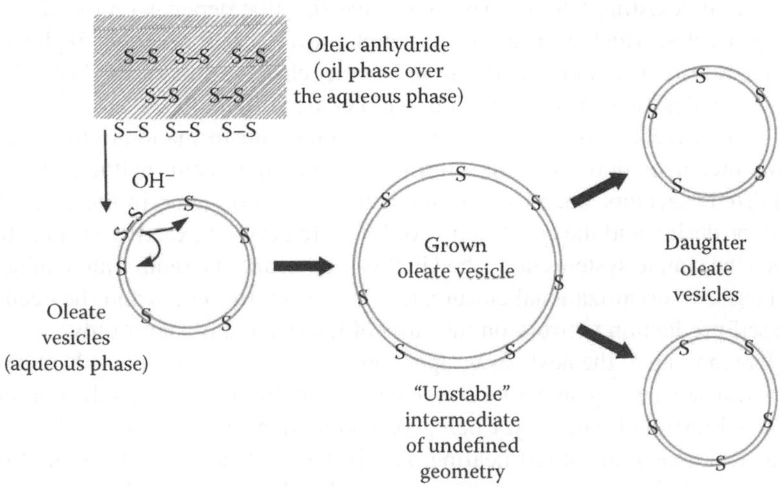

FIGURE 14.4 Self-reproduction of oleate vesicles. Oleic anhydride (water insoluble) is first stratified over an alkaline solution, containing or not containing preformed oleate vesicles. The hydrolysis of the ester is slow when oleate vesicles are absent, but it is accelerated by the presence of oleate vesicles that uptake oleic anhydride on the membrane allowing its hydrolysis. The products of the reaction, two oleate molecules, are incorporated into the membrane, so that the vesicle grows to an unstable state and eventually divides. The kinetic trace of oleic anhydride hydrolysis follows a sharp sigmoidal pattern (without a lag phase when preformed oleate vesicles are present) (Walde et al., 1994). Oleic anhydride can be substituted by oleate micelles that act as an oleate reservoir for vesicle growth and division, with the advantage of being water soluble, so that spectroscopic methods can be utilized for investigating the course of the reaction.

in situ formed oleate molecules (derived from the hydrolysis of oleic anhydride) are incorporated in the oleate vesicles and their growth is expected. It was instead not expected to find evidences, via cryo-transmission electron microscopy (Berclaz et al., 2001a,b) of vesicle growth and *division*, as in the case of micelles. The reasons and the mechanism for such observed behavior are—after about 20 years—still under investigation, but there are suggestions that this can be due to unpaired surface-to-volume growth (vesicle surface grows faster than vesicle volume, resulting in a flaccid vesicle that can divide due to shearing forces).

Several studies have been devoted to the study of vesicle self-reproduction, both using heterogeneous systems (precursor: water-insoluble anhydride) and homogeneous systems (precursor: fatty acid micelles; Bloechliger et al., 1998), and employing a variety of techniques such as spettroturbidimetry, fluorimetry (Chen and Szostak, 2004), stopped flow (Rogerson et al., 2006), dynamic light scattering (Rasi et al., 2003), size exclusion chromatography (Chungcharoenwattana and Ueno, 2004), cryo-transmission (Berclaz et al., 2001a,b) and freeze-fracture electron microscopy (Stano et al., 2006). Among others, it is noteworthy to mention the so-called "matrix effect" (Bloechliger et al., 1998; Lonchin et al., 1999; Rasi et al., 2003; Stano et al., 2006). By this term, we indicate the effect of preexisting vesicle size on the size of the newly formed vesicles, for example, those derived by feeding oleate vesicles with oleate micelles. It was observed that vesicle population resulted from the growth division of preexisting vesicles has an average size that depends on the size of preexisting vesicles, which exert a kind of templating effect of unknown explanation.

In addition to fatty acid–based system, vesicle self-reproduction has been observed with ad hoc–designed surfactants (Takakura et al., 2003).

To summarize, it is important to remark that the transformation of the boundary-forming precursor into a boundary-forming compound occurs without the help of internalized reactants, such as in living cell. The efforts done in the last 10 years toward the design and the construction of MCs are devoted exactly to bridge the gap between the simple systems described in this section and the requirement of autopoiesis in terms of organizational closure, that is, the determination and the occurrence of the self-production network on the basis of internal components only.

Before moving to the next paragraph, however, it is useful to report here a special case of vesicle usage as autopoietic systems, which differ from the self-reproduction dynamics illustrated earlier, yet being very compliant with the autopoietic framework and with the issue of self-healing. This is the first (and unique) attempt to establish a homeostatic dynamic in fatty acid vesicles (Figure 14.5). Zepik et al. (2001) reported on a homeostatic vesicle system based on the simultaneous occurrence—within the same vesicle—of anabolic and catabolic processes. In particular, oleate vesicles were designed to support to concurrent paths: (1) uptake of oleic anhydride that, as shown earlier, is hydrolyzed at the vesicle interface, producing membrane-forming oleate molecules; and (2) oxidation of oleate molecules by OsO_4 to give the diol 9,10-dihydroxystearate (which does not form stable vesicles, therefore inducing the breakdown of vesicles). By tuning the rate of these two concurrent reactions, it was possible to show that a homeostatic pattern could be generated experimentally by balancing the two opposite reactions. Very recently, this pattern has been modeled stochastically (Mavelli and Stano, 2010).

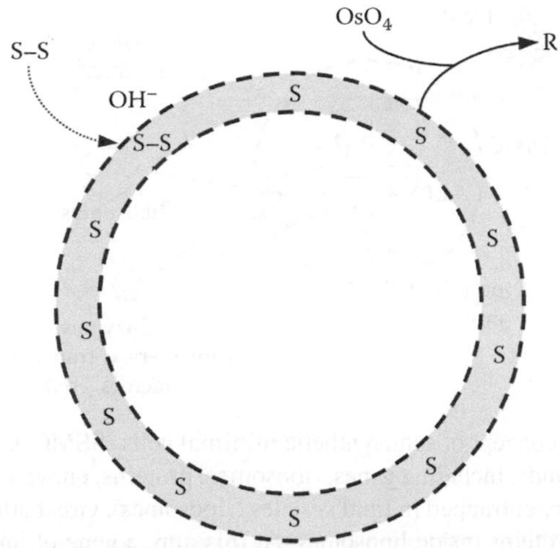

FIGURE 14.5 Homeostatic autopoietic vesicles. Oleic acid/oleate vesicles uptake oleic anhydride that is hydrolyzed at the vesicle interface, producing two membrane-forming molecules. At the same time, OsO_4 oxidizes oleate molecules to give the diol 9,10-dihydroxystearate. By tuning the rate of these two concurrent reactions, Zepik et al. (2001) were able to show that a vesicle-based chemical homeostatic system follows the general autopoietic principles described in Figure 14.1.

Although very simple in its realization, we can ascribe to homeostatic vesicles a kind of self-healing capacity. In fact, the continuous regeneration of vesicle components "heals" the damage occurred by the destruction of fatty acids by the catabolic reaction.

14.4 TOWARD MORE COMPLEX CELL-LIKE MODELS: THE CASE OF SEMISYNTHETIC MINIMAL CELLS

14.4.1 Minimal Cell Concept: The Current Experimental Approaches

From Figure 14.1, considering the minimal autopoietic scheme, it is clear that in order to achieve a compartmentalized system capable of self-producing all its components, it is necessary to design and build an operationally closed reaction network, capable of giving, as output of its operation, the reaction network itself. As we have anticipated, one of the most fruitful approaches to design a minimal autopoietic living cell is called "semisynthetic" (Luisi et al., 2006), and it indicates a research line where enzymes, DNAs, and RNAs are encapsulated inside liposomes at the aim of reconstituting the minimal autopoietic dynamics (Figure 14.6). Semisynthetic MCs (SSMCs) are lab-made cells containing the minimal and sufficient number of components to be defined as alive. Although the definition of life is certainly a complicated task (for a discussion, see Luisi, 1998), we argue that many experts in the field would be convinced by the simultaneous occurrence of (1) self-maintenance, (2) self-reproduction, and (3) possibility to evolve.

The experimental strategy for constructing SSMCs consists in the insertion of the minimal number of genes and enzymes inside a synthetic vesicle (liposome), so as

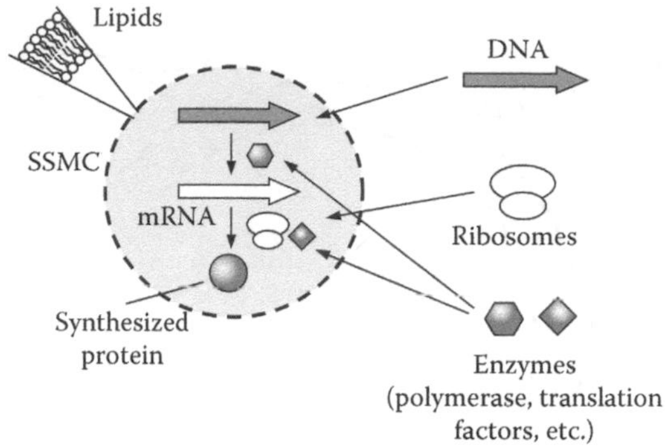

FIGURE 14.6 The concept of semisynthetic minimal cells (SSMCs). The minimal number of molecular compounds, including genes, ribosomes, proteins, enzymes, and low-molecular-weight compounds are entrapped in lipid vesicles (liposomes). Great attention has been given to the synthesis of proteins inside liposomes. At this aim, a gene of interest is co-entrapped with RNA polymerase, ribosomes, translation factors, tRNAs, amino acyl-tRNA synthetases, and energy-recycling enzymes into vesicles, and the production of the gene-encoded functional protein is followed in time.

to have an operational viable cell. This is schematically represented in Figure 14.6. How many genes are needed for making SSMCs? The question of minimal genome is a concept strictly connected to the issue of SSMCs, and it has been recently discussed in the literature (Gil et al., 2004; Forster and Church, 2006; Luisi et al., 2006; Moya et al., 2009; Henry et al., 2010). It turns out that the minimal genome consists in about 200 genes; most of them essentially related to the synthesis of proteins. This important and essential process is indeed the key module in living cells and SSMCs, because the production of well-folded proteins signifies to develop the possibility of carrying out all functions due to the presence of protein (enzyme) catalysis.

The current stage of experimental investigation, unfortunately, does not allow the insertion of 200 genes in synthetic lipid compartments, but several interesting advancements have been reported in the last 10 years.

Currently, the synthesis of well-folded proteins inside vesicles is the key aspect of the semisynthetic approach to minimal autopoietic cell. This step can be seen as a preliminary step for achieving more complex functions, such as the establishment of internal circular organization, including the self-production of "core" components and the self-production of "shell" components. Moreover, protein synthesis allows additional performance, as it will be seen later.

The model system that has been studied in great detail is the synthesis of green fluorescent protein (GFP) inside liposome (Figure 14.6). Actually the first report on this reaction dates back to 2001, when the group of Yomo in Osaka describes for the first time the observation of fluorescent micrometric vesicles after co-encapsulation of a *gfp* gene carrying plasmid, RNA polymerase, and the whole translation machinery, composed by ribosomes, tRNAs, amino acyl-tRNA synthases, and low-molecular-weight compounds (Yu et al., 2001). This report was preceded only by a study on the expression of poly(Phe) inside liposomes, using poly(U) as messenger (Oberholzer et al., 1999).

Starting from these two reports, several studies have been published in order to understand in great detail all aspects of protein synthesis inside liposomes. These results have been reviewed in details (Luisi et al., 2006; Chiarabelli et al., 2009; Stano, 2010a,b).

Protein synthesis studies were typically carried out by using not well-characterized mixtures for the occurrence of transcription–translation-coupled reactions, for example, homemade or commercial cell extracts were used. Since 2006, this approach was substituted by using a fully reconstituted system for protein synthesis, called PURE system, that is, protein synthesis using recombinant elements (Shimizu et al., 2001). The PURE system is composed of 36 individually purified enzymes, tRNAs, ribosomes, and low-molecular-weight compounds. From the viewpoint of functionality, it is composed by four modules: (1) DNA-to-RNA transcription, (2) RNA-to-protein translation; (3) tRNA charging; and (4) energy regeneration. The first module consists into one enzyme, namely, the T7 RNA polymerase, and it is specific for genes to be expressed under T7 promoter. The second module consists into 11 translation factors and ribosomes. The third module consists into 20 amino acyl-tRNA synthases. The fourth module consists into four enzymes dedicated to chemical energy conversion. To date, the encapsulation of the PURE system for synthesizing proteins inside lipid vesicles has become the standard for synthetic biology applications of SSMCs. This approach highly resonates with the concept of standard biological parts introduced by the pioneers of synthetic biology (see http://partsregistry.org/Main_Page).

Among the most advanced studies, there are three important results that deserve special mention.

The first one refers to the first attempt to replicate genetic information by an *in situ*–synthesized replicase (Kita et al., 2008) (see Figure 14.7a). The system consists into RNA replication by Qβ-replicase (the "Spiegelman enzyme"; Haruna et al., 1963). This work gives an important contribution for the construction of minimal autopoietic cell because it addresses the question of genetic information replication. In their design, Kita et al. (2008) conceive a molecular toolbox that consists into the reciprocal formation (self-production) of two species, by means of a third one, encoded in their sequence. Note that the reciprocal RNA production can be seen as a self-healing dynamics, because a damaged (+)- or (–)-RNA strand can be substituted by a newly synthesized one on the next replication cycle.

The second example deals instead on development of a "transport" functionality codified by encapsulated molecular network. At this aim, it used a DNA sequence encoding α-hemolysin sequence (Noireaux and Libchaber, 2004) (see Figure 14.7b). This small protein (36 kDa) is able to self-assemble as a heptamer pore on the membrane, allowing an effective nutrient/waste transport across the otherwise impermeable membrane.

As a third example of the construction of complex reaction networks inside liposomes, it is useful to consider here the work of Tetsuya Yomo and coworkers on the two-stage genetic cascade reaction (Ishikawa et al., 2004) (see Figure 14.7c). In particular, in this work it has been shown that the expression of one gene can be regulated in simple manner by producing the corresponding RNA polymerase. The process is equivalent to switching-on a desired function.

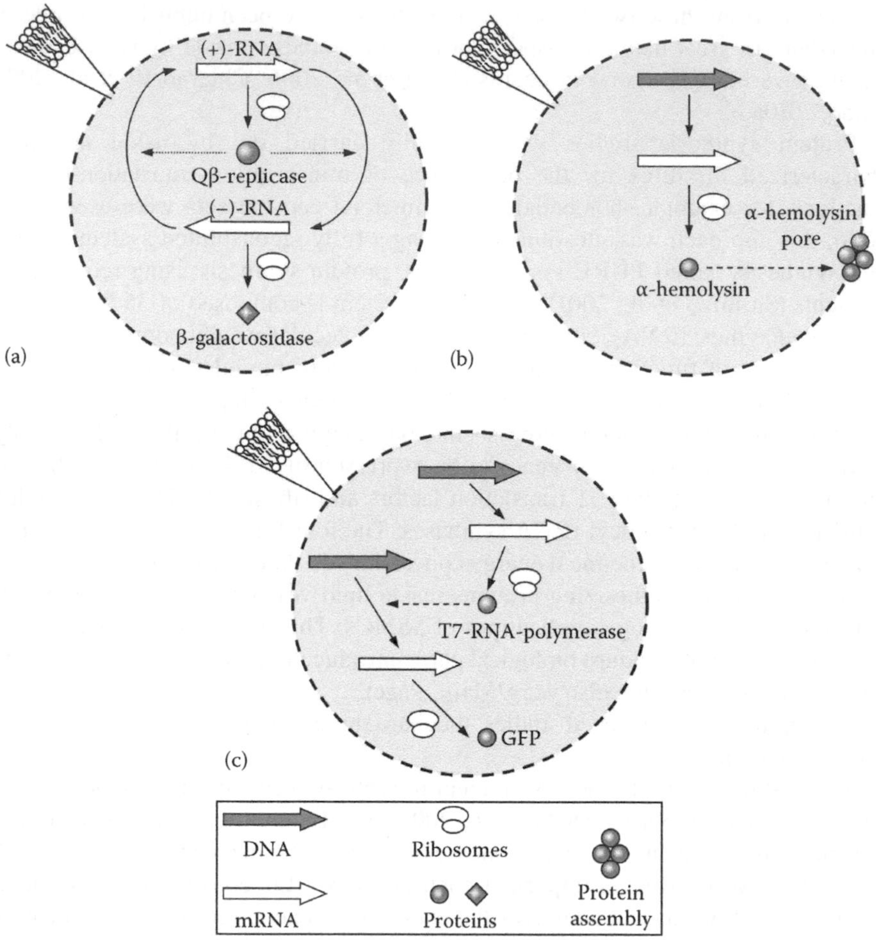

FIGURE 14.7 Advanced experimental models in minimal cell research. (a) Self-replication of RNA by encoded Qβ-replicase inside liposomes. This *in situ*–synthesized enzyme replicates both RNA strands by reciprocal template-directed synthesis. The additional enzyme β-galactosidase is also produced (Kita et al., 2008); (b) production of α-hemolysin inside giant lipid vesicles, with the formation of a heptamer pore on the membrane (cutoff: 3 kDa), which gives the possibility of feeding lipid vesicles by externally added "nutrients" (Noireaux and Libchaber, 2004); (c) two-stage genetic cascade reaction inside liposomes: the first gene is transcripted and translated to give T7-RNA-polymerase, which in turn transcribe the second gene, to give the green fluorescent protein (GFP). Notice that the first mRNA is synthesized by SP6RNA polymerase (Ishikawa et al., 2004). In all cases, vesicles are composed by phosphatidylcholine, occasionally admixed with helper lipids. Further details are given in the text.

Future developments have to face some current technical difficulties. First of all, there is the need to control vesicle formation and solute encapsulation. Despite the fact that vesicles have been used for decades as drug delivery vectors, so that the methods of vesicle preparation are well established, the combination of cell-free systems and liposomology is still a challenge for the researchers. In fact, the presence of macromolecules often implies tailoring each preparation step in agreement with the specific case under study. Very recent studies have reported a new method of vesicle formation (the "droplet transfer" method, reviewed by Walde et al., 2010) that allows the control of solute entrapment (Pautot et al., 2003; Noireaux and Libchaber, 2004; Yamada et al., 2007; Pontani et al., 2009; Abkarian et al., 2011), and interestingly this method has inspired further research in microfluidic technology (Ota et al., 2009; Matosevic and Paegel, 2011).

14.4.2 Toward Shell Production

Inspired by the work on the self-reproduction of micelles, reverse micelles, and vesicles, great attention has been devoted to the issue of MC growth and division as a result of lipid synthesis (rather than addition of lipids or lipid precursors from outside). This kind of investigation represents an essential step toward the problem of full ("core-and-shell") self-reproduction of SSMCs.

The first level that has been proposed is the construction of an enzyme-based compartment capable of synthesizing lipids (or fatty acids) from water-soluble precursors. An early attempt was done by Schmidli et al. (1991) by exploiting the so-called "lipid-salvage pathway" (Figure 14.8a). More recently, a similar approach was attempted, but focusing on fatty acid synthesis inside phosphatidylcholine vesicles (Murtas, 2010). At this aim, fatty acid synthase I from *Brevibacterium ammoniagenes* was entrapped into liposomes, together with malonyl- and acetyl coenzymes A, the two reactants needed for palmitate synthesis (Figure 14.8b). Also in this case, the production of the bilayer-forming compound (palmitate) was insignificant with respect to observing vesicle growth.

The second level is instead focused on the encapsulation of the genes encoding for the enzymes that catalyze the lipid-production reactions. Kuruma et al. (2009) designed a liposome system that had to endogenously biosynthesize of lipid after the *in situ* expression of the enzymes indicated in Figure 14.8c (Kuruma et al., 2009). It was shown that the choice of the lipid composition is particularly important for determining the yield of protein synthesis, the encapsulation efficiency, and the catalytic activity of the *in situ*–synthesized enzymes. Clearly, additional work is required for achieving the first lipid-producing vesicle and for observing a macroscopic behavior (growth division) that can be comparable with what has been shown in the case of fatty acid self-reproduction.

14.5 CLASSICAL VIEWS ON SELF-HEALING OF LIPID-BASED SYSTEMS. WHAT CAN WE LEARN?

It is interesting to shortly resume some classical aspects of self-healing in lipid-based systems and try to find some contact points with the phenomenology and technology of SSMCs. Classically, the self-healing properties of lipid-based systems are described in terms of their self-assembly capability. Self-assembly occurs when an ensemble of objects, the building blocks of the system, spontaneously organize in an ordered structure by means of physical interactions. Lipid bilayers in form of flat membranes or closed vesicles have been typically studied.

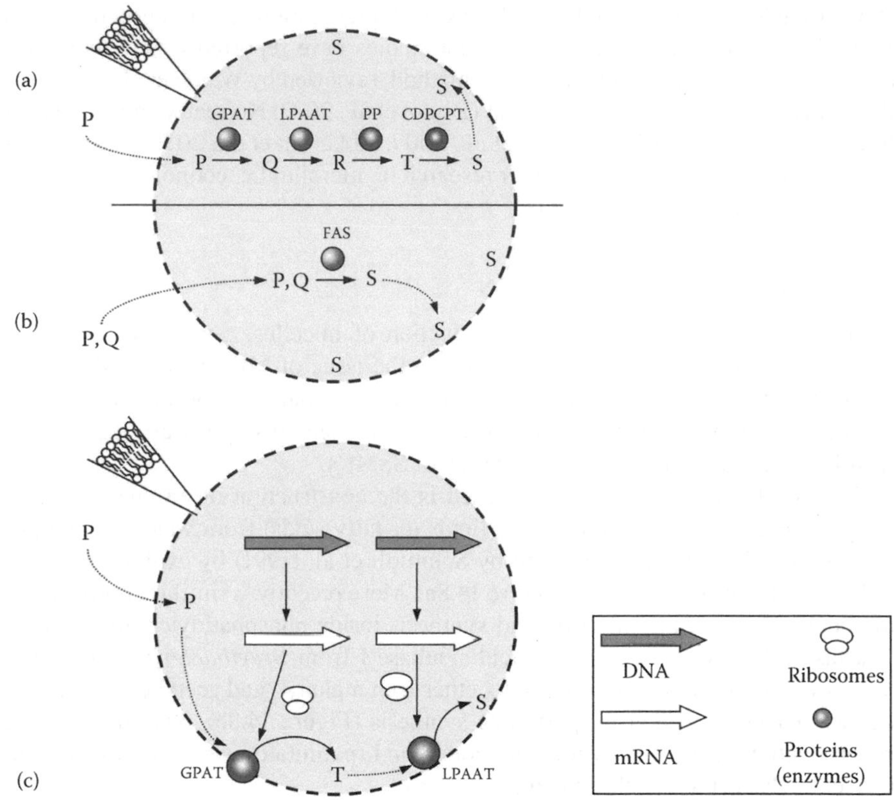

FIGURE 14.8 Lipid-producing minimal cells, comparison between two strategies: enzymes entrapped inside vesicles (a and b), and enzyme synthesized inside vesicles starting from the corresponding genes (c). (a) Reconstitution of the lipid-salvage pathway in phosphatidylcholine vesicles (Schmidli et al., 1991). The four enzyme glycerol-3-phosphate acyl transferase (GPAT), lysophophatidic acid acyl transferase (LPAAT), phosphatidate phosphorylase (PP), cytidine diphosphocholine phosphocholine transferase (CDPCPT) have been co-entrapped into lipid vesicles together with the reactants (P) from synthesizing lecithins (glycerol-3-phosphate, oleoyl-CoA, palmitoyl-CoA, CDP-choline), through the intermediates Q, R, T. The final product (S) is a phosphatidylcholine (b) Reconstitution of fatty acid synthesis inside phosphatidylcholine vesicles (Murtas, 2010). Fatty acid synthase (FAS) is co-encapsulated with the reactants (P, Q) needed for palmitate synthesis (acetyl-CoA, malonyl-CoA, NADPH). Palmitate is the product (S) of the reaction. (c) Expression, inside lipid vesicles of the first two enzymes (GPAT, LPAAT) of the lipid-salvage pathway, starting from the corresponding genes *plsB* and *plsC*, and the PURE system (Kuruma et al., 2009). The first protein is a membrane-integral enzyme, whereas the second is an associated membrane one. The substrates P are first transformed into the lysophosphatidic acid T, which is in turn converted into the product (phosphatidic acid) S.

Flat adsorbed lipid membranes and black membranes (Ou-Yang et al., 1998) are the first representative case. Bilayer lipid membranes, prepared by sequentially depositing two lipid monolayers onto chitosan films, were shown to spread (from a lipid reservoir) on the support with a $t^{-1/2}$ time dependence. This was interpreted as self-healing of bilayers on chitosan (Baumgart and Hoffenhäusser, 2003). Similarly,

Nishimura et al. (2006) reported self-healing of supported lipid monolayer on alkane thiol–gold layer, demonstrating a fast self-healing dynamics after application of an electric pulse. The concept of self-healing of membranes is further discussed by Nissen et al. (1999), with emphasis on the wetting behavior of phospholipid membranes on solid surfaces immersed in aqueous solution.

Lipid vesicles have been shown to have self-healing properties. For example, macroscopic transient pores have been imaged in mechanically stretched giant vesicles (Sandre et al., 1999). It was observed that holes open above a critical radius, first grow up, and then close according to the self-healing capabilities (Figure 14.9). In similar way, it was shown that a small vesicle could exit from the internal space of a larger vesicle (the so-called "translocation" mechanism; Menger and Gabrielson, 1994), with a consequent self-healing of the membrane after transient rupture (Wick et al., 1995; Moroz et al., 1997; Takakura et al., 2003). In particular, it was shown that the inner vesicles are expelled from the larger mother vesicle via the accompanying decrease in the vesicle area by ca. 25%, which forces a pore to open in the mother vesicle (Leirer et al., 2009).

Some self-healing aspects of lipid assemblies can be found in the review of Amendola and Meneghetti (2009).

What can we learn from these more classical aspects of self-healing for improving our capacity of building MCs? Quite probably, the strongest link between membrane self-healing properties and MCs lies in self-reproduction dynamics. Mechanically, the self-reproduction of vesicles and micelles consists in a dynamical balance between instability (breaking) forces counterbalanced, also at the proper time scale, by self-assembling (healing) tendencies. For example, vesicle growth and division, one of the pillars of MC research, is a particularly rich example. During vesicle

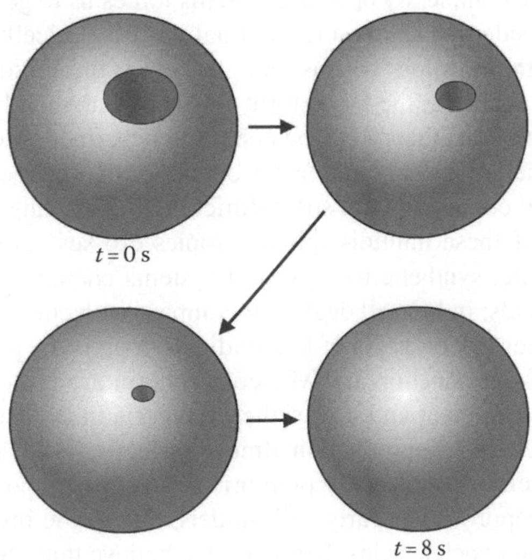

FIGURE 14.9 Cartoon representing the time sequences of a pore closure in a vesicle tensed by intense illumination. Addition of glycerol raises the viscosity and slows down the leaking of the inner liquid, which enables to observe macroscopic transient pores. (From Sandre, O. et al., *Proc. Natl. Acad. Sci. U.S.A.*, 96, 10591, 1999.)

growth, a membrane can grow evenly only if efficient flip-flop mechanisms avoid a too strong enrichment of the outer membrane leaflet and if the vesicle volume grows accordingly. Both mechanisms cannot be really effective, due to kinetic reason and due to the onset of osmotic gradients, respectively, Accordingly, nonspherical growth has been proposed as the main trigger of vesicle division (Luisi et al., 2008). In other words, vesicles are not able to cope with a fast addition of excess membrane-forming compound without changing their structure. As consequence of their deformation, division becomes possible. In order to divide, two facing membranes have to reach a close reacting distance, where they have to break and reseal (also in opposite order). The resealing (fusion) process is topologically equivalent to the aforementioned examples of membrane hole self-closure. It is therefore clear that for the establishment of complex dynamics, such as the compartment transformations depicted in Section 14.3 and future goals illustrated in Section 14.4, the self-assembly and self-repairing properties of lipid bilayers again become the central issue, together with the earlier discussed containment function.

14.6 CONCLUSIONS AND PERSPECTIVES FOR FUTURE RESEARCH AND APPLICATIONS

In this chapter, we have discussed some of the most relevant studies on the issue of construction of compartment-based systems at the aim of displaying minimal living properties, in particular self-maintenance and autopoietic behavior. The starting consideration refers to the idea of constructing simple forms of life in the laboratory. This means to demonstrate that life is an *emergent property*, deriving from the organization of a complex molecular network, rather than deriving from vitalistic forces.

Clearly, the huge complexity of living systems forces us to go back to the roots of simplicity, and consider the simplest level of natural life, unicellular organisms. The first assessment to the question "what is life?" as well as the approach to the construction of simple life forms in the laboratories, starts from such basic considerations and due to the autopoietic theory can be concisely described. Moreover, autopoietic systems provide the correct inspiration for constructing chemical and biochemical systems of minimal complexity, yet still mimicking basic living properties.

The relevance of these multidisciplinary topics crosses several different fields, such as origin of life, synthetic biology, and systems chemistry (for some aspects, these are related fields; in fact, all deal with complex molecular systems).

Within the framework of origin of life studies, which is the place where the studies on vesicles self-reproduction and MC construction are rooted in, the interest is focused on the development and use of these constructions as primitive cell models. Fatty acid vesicles self-reproduction almost perfectly fulfills these requirements, due to their simplicity and allegedly prebiotic chemical composition. SSMCs have been sometime disapproved as early cell models, due to the involvement of highly evolved enzymes and nucleic acids. However, we believe that SSMCs are useful not only because they are the only experimentally accessible simple cell model, but also because, despite the use of complex molecules, the resulting compartmentalized system still has the minimal number of essential functions. These essential functions are a model of primitive functions, but they are carried out by different molecules.

In other words, if one looks only at the organization and not at the structure, an autopoietic MC created with modern molecules can be considered as functionally equivalent to an autopoietic MC created with ancient molecules.

In synthetic biology, a new emerging biology discipline, the issue of construction of synthetic cells is well recognized as one of the pillar projects (De Lorenzo and Danchin, 2008). Here, we distinguish between two approaches. The top-down approach starts from complex cells and removes unessential "modules" to reduce it at the minimal complexity, often aiming at use the resulting "engineered" bacteria for producing biochemicals of human interests. System-level genetic and metabolic engineering aspires to design a synthetic cell with practical applications beyond basic science. For example, specific goals may involve the removal of mobile DNA elements, and/or of competing metabolic pathways that drive raw materials away from desired end products and toward useless or even toxic by-products, as well as the complex layer of transcriptional regulatory interactions that make natural microorganisms resistant to engineering efforts (Henry et al., 2010). Therefore, synthetic cells might find applications in medicine, medicine/pharmacology, fuel production, bioremediation, and so on. The announcement of the construction of a synthetic cell from the transplantation of a synthetic genome into a host cell recently released by the Craig Venter Institute had a worldwide resonance (Gibson et al., 2010).

On the contrary, the SSMCs approach markedly differs from the top-down approach. In fact, in SSMCs studies, that we may call here "bottom-up," the interest is focused on the self-organization aspect of cell construction, and starts from *separated* components, rather than from preexisting cells. The very simple cell-like structures created by the bottom-up approach can have biotechnological relevance. For example, it has been recently proposed that ad hoc–designed SSMCs might be considered intelligent nano-biofactories for injection in the body (similar to the case of liposomal drugs), tissue recognition, and *in situ* drug production (LeDuc et al., 2007).

Systems chemistry is a recently coined expression that embraces a broad collection of interests on complex and dynamic molecular systems such as prebiotic and supramolecular chemistry, autocatalysis, oscillations, symmetry breaking, computer modeling, and complex systems dynamics. Born within the community of origin of life research, systems chemistry is now attracting the interest of a wide audience (Lodlow and Otto, 2008; Nitschke, 2009; Von Kiedrowski et al., 2010). With respect to the construction of MCs, the systems chemistry approach explicitly follows the Ganti chemoton approach (Ganti, 1975), by designing pairwise integration of three modular functions (containment, metabolism, information) into higher organized supersystems, up to the final goal of triple integration into minimal chemical cells (Figure 14.10). Somehow related to the interest of systems chemistry for the vital problem of cell assembly, there are the theoretical studies on complex systems, which have often supported debates on experimentally difficult problems in origin of life.

From all these considerations, it is evident how the effort for understanding cellular life via the construction of cell models (learning by constructing) is a central issue in science, which can be approached and discussed from several viewpoints, from basic science to technology. At the core of the phenomena, we are investigating; we do not find any special chemical species, but we realize that the prime role is for self-organization. The capacity of a coordinate self-regulating behavior of a collection of molecules is indeed a remarkable property of living assembly, determining

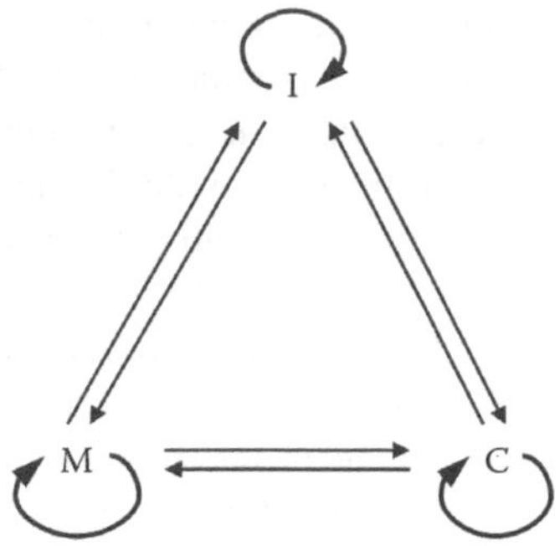

FIGURE 14.10 The minimal cell concept in systems chemistry. "Systems chemistry is the joint effort of prebiotic and supramolecular chemistry assisted by computer science from theoretical chemistry, biology, and complex systems research to tackle dynamic supersystem integration embedding at least one autocatalytic subsystem. Subsystems may be classified as genetic, metabolic, or compartment-building. Pairwise integration into higher organized supersystems is expected to yield the knowledge enabling later the triple integration into minimal chemical cells. The integration approach will necessarily link to the question of asymmetric autocatalysis and chiral symmetry breaking, while the key challenge is to find the roots of Darwinian evolvability in chemical systems." (From the COST CM0703 Action website: http://www.cost.esf.org/domains_actions/cmst/Actions/Systems_Chemistry).

their self-maintenance of structural and functional identity, including the self-healing functions to cope with environmental stresses or internal "natural" molecular turnover. As a result of the contemplation of these dynamical systems, we aim at designing manmade material inspired from them. Several targets have been already reached along the roadmap to the construction of MC models, but many other are still missing. As we have seen, the complete set of transformation for core-and-shell self-reproduction is still an experimentally complex task, but we are confident that the efforts carried out in the last years will help to reach this goal. In fact, the rise of interest of the scientific community toward MCs witnesses the relevance of this research for basic science and biotechnology.

ACKNOWLEDGMENTS

I greatly thank Pier Luigi Luisi for inspiring this research and for stimulating discussions. Funding/networking agencies: SYNTHCELLS project (Approaches to the Bioengineering of Synthetic Minimal Cells, EU-FP6 043359), HFSP (RGP0033/2007-C), ASI (I/015/07/0), PRIN2008 (2008FY7RJ4), SynBioNT, and COST Systems Chemistry action (CM0703).

REFERENCES

Abkarian, M., Loiseau, E., and Massiera, G. 2011. Continuous droplet interface crossing encapsulation (cDICE) for high throughput monodisperse vesicle design. *Soft Matter* 7:4610–4614.

Amendola, V. and Meneghetti, M. 2009. Self-healing at the nanoscale. *Nanoscale* 1:74–88.

Bachmann, P. A., Luisi, P. L., and Lang, J. 1991a. Self-replication of reverse micelles. *Chimia* 45:266–268.

Bachmann, P. A., Luisi, P. L., and Lang, J. 1992. Autocatalytic self-replicating micelles as models for prebiotic structures. *Nature* 357:57–59.

Bachmann, P. A., Walde, P., Luisi, P. L., and Lang, J. 1991b. Self-replicating micelles: Aqueous micelles and enzymatically driven reactions in reverse micelles. *J. Am. Chem. Soc.* 113:8204–8209.

Baumgart, T. and Hoffenhäusser, A. 2003. Polysaccharide-supported planar bilayer lipid model membranes. *Langmuir* 19, 1730–1737.

Berclaz, N., Blochliger, E., Muller, M., and Luisi, P. L. 2001b. Matrix effect of vesicle formation as investigated by cryotransmission electron microscopy. *J. Phys. Chem. B* 105:1065–1071.

Berclaz, N., Muller, M., Walde, P., and Luisi, P. L. 2001a. Growth and transformation of vesicles studied by ferritin labeling and cryotransmission electron microscopy. *J. Phys. Chem. B* 105:1056–1064.

Bloechliger, E., Blocher, M., Walde, P., and Luisi, P. L. 1998. Matrix effect in the size distribution of fatty acid vesicles. *J. Phys. Chem.* 102:10383–10390.

Cárdenas, M. L., Letelier, J. C., Gutierrez, C., Cornish-Bowden, A., and Soto-Andrade, J. 2010. Closure to efficient causation, computability and artificial life. *J. Theor. Biol.* 263:79–92.

Chen, I. A. and Szostak, J. W. 2004. A kinetic study of the growth of fatty acid vesicles. *Biophys. J.* 87:988–998.

Chen, I. A. and Walde, P. 2010. From self-assembled vesicles to protocells. *Cold Spring Harb. Perspect. Biol.* DOI: 10.1101/cshperspect.a002170.

Chiarabelli, C., Stano, P., and Luisi, P. L. 2009. Chemical approaches to synthetic biology. *Curr. Opin. Biotechnol.* 20:492–497.

Chungcharoenwattana, S. and Ueno, M. 2004. Size control of mixed egg yolk phosphatidylcholine (EggPC)/oleate vesicles. *Chem. Pharm. Bull.* 52:1058–1062.

Cistola, D. P., Atkinson, D., Hamilton, J. A., and Small, D. M. 1986. Phase behavior and bilayer properties of fatty acids: Hydrated 1:1 acid-soaps. *Biochemistry* 25:2804–2812.

Cistola, D. P., Hamilton, J. A., Jackson, D., and Small, D. M. 1988. Ionization and phase behavior of fatty acids in water: Application of the Gibbs phase rule. *Biochemistry* 27:1881–1888.

De Lorenzo, V. and Danchin, A. 2008. Synthetic biology: Discovering new worlds and new words. *EMBO Rep.* 9:9.

Deamer, D. and Dworkin, J. P. 2005. Chemistry and physics of primitive membranes. *Top. Curr. Chem.* 259:1–27.

Forster, A. C. and Church, G. M. 2006. Towards synthesis of a minimal cell. *Mol. Syst. Biol.* 2:45.

Ganti, T. 1975. Organization of chemical reactions into dividing and metabolizing units: The chemotons. *BioSystems* 7:15–21.

Gebicki, J. M. and Hicks, M. 1976. Preparation and properties of vesicles enclosed by fatty acid membranes. *Chem. Phys. Lipids* 16:142–160.

Gibson, D. G., Glass, J. I., Lartigue, C. et al. 2010. Creation of a bacterial cell controlled by a chemically synthesized genome. *Science* 329:52–56.

Gil, R., Silva, F. J., Peretó, J., and Moya, A. 2004. Determination of the core of a minimal bacteria gene set. *Microbiol. Mol. Biol. Rev.* 68:518–537.

Haines, T. H. 1983. Anionic lipid headgroups as a proton-conducting pathway along the surface of membranes: A hypothesis. *Proc. Natl. Acad. Sci. U.S.A.* 80:160–164.

Hargreaves, W. R. and Deamer, D. W. 1978. Liposomes from ionic, single-chain amphiphiles. *Biochemistry* 17:3759–3768.

Haruna, I., Nozu, K., Ohtaka, Y., and Spiegelman, S. 1963. A RNA "replicase" induced by and selective for a viral RNA: Isolation and properties. *Proc. Natl. Acad. Sci. U.S.A.* 50:905–911.

Henry, C. S., Overbeek, R., and Stevens, R. L. 2010. Building the blueprint of life. *Biotechnol. J.* 5:695–704.

Ishikawa, K., Sato, K., Shima, Y., Urabe, I., and Yomo, T. 2004. Expression of a cascading genetic network within liposomes. *FEBS Lett.* 576:387–390.

Kita, H., Matsuura, T., Sunami, T., Hosoda, K., Ichihashi, N., Tsukada, K., Urabe, I., and Yomo, T. 2008. Replication of genetic information with self-encoded replicase in liposomes. *ChemBioChem* 9:2403–2410.

Kuruma, Y., Stano, P., Ueda, T., and Luisi, P. L. 2009. A synthetic biology approach to the construction of membrane proteins in semi-synthetic minimal cells. *Biochim. Biophys. Acta* 1788:567–574.

LeDuc, P., Wong, M. S., Ferreira, P. M. et al. 2007. Towards an in vivo biologically inspired nanofactory. *Nat. Nanotechnol.* 2:3–7.

Leirer, C. T., Wunderlich, B., Wixforth, A., and Schneider, M. F. 2009. Thermodynamic relaxation drives expulsion in giant unilamellar vesicles. *Phys. Biol.* 6:016011.

Letelier, J. C., Soto-Andreade, J., Abarzuab, F. G., Cornish-Bowden, A., and Cárdenas, M. L. 2006. Organizational invariance and metabolic closure: Analysis in terms of (M,R) systems. *J. Theor. Biol.* 238:949–961.

Lodlow, R. F. and Otto, S. 2008. Systems chemistry. *Chem. Soc. Rev.* 37:101–108.

Lonchin, S., Luisi, P. L., Walde, P., and Robinson, B. H. 1999. A matrix effect in mixed phospholipid/fatty acid vesicle formation. *J. Phys. Chem. B* 103:10910–10916.

Luisi, P. L. 1998. About various definitions of life. *Orig. Life Evol. Biosph.* 28:613–622.

Luisi, P. L. 2002. Toward the engineering of minimal living cells. *Anat. Rec.* 268:208–214.

Luisi, P. L. 2003. Autopoiesis: A review and a reappraisal. *Naturwissenschaften* 90:49–59.

Luisi, P. L., Ferri, F., and Stano, P. 2006. Approaches to semi-synthetic minimal cells: A review. *Naturwissenschaften* 93:1–13.

Luisi, P. L., Souza, T., and Stano, P. 2008. Vesicle behavior: In search of explanations. *J. Phys. Chem. B* 112:14655–14664.

Luisi, P. L. and Stano, P. (Eds.). 2011. *The Minimal Cell. The Biophysics of Cell Compartment and the Origin of Cell Functionality*. Dordrecht, the Netherlands: Springer.

Luisi, P. L. and Varela, F. J. 1990. Self-replicating micelles. A chemical version of minimal autopoietic systems. *Orig. Life Evol. Biosph.* 19:633–643.

Mansy, S. S. and Szostak, J. W. 2009. Reconstructing the emergence of cellular life through the synthesis of model protocells. *Cold Spring Harb. Symp. Quant. Biol.* 74:1–8.

Matosevic, S. and Paegel, B. M. 2011. Stepwise synthesis of giant unilamellar vesicles on a microfluidic assembly line. *J. Am. Chem. Soc.* 133:2798–2800.

Mavelli, F. and Stano, P. 2010. Kinetic models for autopoietic chemical systems: Role of fluctuations in homeostatic regime. *Phys. Biol.* 7:016010.

Menger, F. M. and Gabrielson, K. 1994. Chemically-induced birthing and foraging in vesicle systems. *J. Am. Chem. Soc.* 116:1567–1568.

Monnard, P. A. and Deamer, D. W. 2002. Membrane self-assembly processes: Steps toward the first cellular life. *Anat. Rec.* 268:196–207.

Morowitz, H. 1992. *Beginnings of Cellular Life. Metabolism Recapitulates Biogenesis*. New Haven, CT: Yale University Press.

Moroz, J. D. and Nelson, P. 1997. Dynamically stabilized pores in bilayer membranes. *Biophys J.* 72:2211–2216.

Moya, A., Gil, R., Latorre, A., Peretó, J., Garcillán-Barcia, M. P., and de la Cruz, F. 2009. Toward minimal bacterial cells: Evolution vs. design. *FEMS Microbiol. Rev.* 33:225–235.

Murtas, G. 2010. Internal lipid synthesis and vesicle growth as a step toward self-reproduction of the minimal cell. *Syst. Synth. Biol.* 4:85–93.

Nishimura, Y., Wagner, R., Manaka, T., and Iwamoto, M. 2006. Preparation and surface morphology change observation of hybrid bilayer membranes. *Thin Solid Films* 499:40–43.

Nissen, J., Gritsch, S., Wiegand, G., and Raedler, J. O. 1999. Wetling of phospholipid membranes on hydrophilic surfaces-concepts towards self-healing membranes. *Eur Phys. JB* 10:335–344.

Nitschke, J. R. 2009. Molecular networks come of age. *Nature* 462:736–738.

Noireaux, V. and Libchaber, A. 2004. A vesicle bioreactor as a step toward an artificial cell assembly. *Proc. Natl. Acad. Sci. U.S.A.* 101:17669–17674.

Oberholzer, T., Nierhaus, K. H., and Luisi, P. L. 1999. Protein expression in liposomes. *Biochem. Biophys. Res. Commun.* 261:238–241.

Ota, S., Yoshizawa, S., and Takeuchi, S. 2009. Microfluidic formation of monodisperse, cell-sized, and unilamellar vesicles. *Angew. Chem. Int. Ed.* 48:6533–6537.

Ou-Yang, Z. C., Liu, J. X., and Zhang, X. Y. 1998. *Geometric Methods in the Elastic Theory of Membranes in Liquid Crystal Phases*. Singapore: World Scientific.

Pautot, S., Frisken, B. J., and Weitz, D. A. 2003. Production of unilamellar vesicles using an inverted emulsion. *Langmuir* 19:2870–2879.

Pontani, L. L., van der Gucht, J., Salbreux, G., Heuvingh, J., Joanny, J.-F., and Sykes, C. 2009. Reconstitution of an actin cortex inside a liposome. *Biophys. J.* 96:192–198.

Rasi, S., Mavelli, F., and Luisi, P. L. 2003. Cooperative micelle binding and matrix effect in oleate vesicle formation. *J. Phys. Chem. B* 107:14068–14076.

Rasmussen, S., Chen, L., Deamer, D., Krakauer, D. C., Packard, N. H., Stadler, P. F., and Bedau, M. A. 2004. Transitions from nonliving to living matter. *Science* 303:963–965.

Rogerson, M. L., Robinson, B. H., Bucak, S., and Walde, P. 2006. Kinetic studies of the interaction of fatty acids with phosphatidylcholine vesicles (liposomes). *Coll. Surf. B Bioint.* 48:24–34.

Rosen, R. 1991. *Life Itself.* New York: Colombia University Press.

Sandre, O., Moreaux, L., and Brochard-Wyart, F. 1999. Dynamics of transient pores in stretched vesicles. *Proc. Natl. Acad. Sci. U.S.A.* 96:10591–10596.

Schmidli, P. K., Schurtenberger, P., and Luisi, P. L. 1991. Liposome-mediated enzymatic synthesis of phosphatidylcholine as an approach to self-replicating liposomes. *J. Am. Chem. Soc.* 113:8127–8130.

Shimizu, Y., Inoue, A., Tomari, Y., Suzuki, T., Yokogawa, T., Nishikawa, K., and Ueda, T. 2001. Cell free translation reconstituted with purified components. *Nat. Biotechnol.* 19:751–755.

Stano, P. 2010a. Minimal and primitive cells in origins of life, systems chemistry and synthetic biology. *Orig. Life Evol. Biosph.* 40:452–457.

Stano, P. 2010b. Synthetic biology of minimal living cells: Primitive cell models and semi-synthetic cells. *Syst. Synth. Biol.* 4:149–156.

Stano, P. and Luisi, P. L. 2010. Achievements and open questions in the self-reproduction of vesicles and synthetic minimal cells. *Chem. Comm.* 46:3639–3653.

Stano, P., Wehrli, E., and Luisi, P. L. 2006. Insights on the oleate vesicles self-reproduction. *J. Physics: Condensed Matter* 18:S2231–S2238.

Szostak, J. W., Bartel, D. P., and Luisi, P. L. 2001. Synthesizing life. *Nature* 409:387–390.

Takakura, K., Toyota, T., and Sugawara, T. 2003. A novel system of self-reproducing giant vesicles. *J. Am. Chem. Soc.* 125:8134–8140.

Varela, F. J., Maturana, H. R., and Uribe, R. 1974. Autopoiesis: The organization of living systems, its characterization and a model. *Biosystems* 5:187–196.

Von Kiedrowski, G., Otto, S., and Herdewijn, P. 2010. Welcome home, systems chemists! *J. Syst. Chem.* 1:1.

Walde, P. 2006. Surfactant assemblies and their various possible roles for the origin(s) of life. *Orig. Life Evol. Biosph.* 36:109–150.

Walde, P., Cosentino, K., Hengel, H., and Stano, P. 2010. Giant vesicles: Preparations and applications. *ChemBioChem* 11:848–865.

Walde, P., Namani, T., Morigaki, K., and Hauser, H. 2006. Formation and properties of fatty acid vesicles (liposomes). In *Liposome Technology* (Vol. I, 3rd edn.), G. Gregoriadis (Ed.). New York: Informa Healthcare, pp. 1–19.

Walde, P., Wick, R., Fresta, M., Mangone, A., and Luisi, P. L. 1994. Autopoietic self-reproduction of fatty acid vesicles. *J. Am. Chem. Soc.* 116:11649–11654.

Wick, R., Walde, P., and Luisi, P. L. 1995. Autocatalytic self-reproduction of giant vesicles. *J. Am. Chem. Soc.* 117:1435–1436.

Yamada, A., Le Berre, M., Yoshikawa, K., and Baigl, D. 2007. Spontaneous generation of giant liposomes from an oil/water interface. *ChemBioChem* 8:2215–2218.

Yu, W., Sato, K., Wakabayashi, M., Nakatshi, T., Ko-Mitamura, E. P., Shima, Y., Urabe, I., and Yomo, T. 2001. Synthesis of functional protein in liposome. *J. Biosci. Bioeng.* 92:590–593.

Zepik, H. H., Blochliger, E., and Luisi, P. L. 2001. A chemical model of homeostasis. *Angew. Chemie Int. Ed. Eng.* 40:199–202.

15 Molecular Motors and Self-Healing at the Neuromuscular Synapse

Alf Månsson, Marlene Norrby, and Sven Tågerud

CONTENTS

15.1 GENERAL INTRODUCTION

The ability of autonomous damage recovery, characteristic of any self-healing material (Amendola and Meneghetti, 2009), is an inherent property of living organisms. Many aspects of this self-healing occur on the nanoscale and depend on the coordinated action of a range of cellular nanomachines powered by high-energy phosphate compounds such as adenosine triphosphate (ATP). A prerequisite for this capability of living organisms is the fact that they are kept in a state far from equilibrium by the continuous supply of high-energy phosphate compounds as a result of the metabolism of carbohydrates, lipids, and proteins. Since the self-healing processes in living systems have been fine-tuned by evolution, they should provide many important insights of value for the design and production of artificial nanoscale self-healing systems. As for many other aspects of life, self-healing occurs at different levels of organization, from the molecular level to the levels of tissues and body parts. Self-healing at the molecular level includes the continuously ongoing DNA repair (Lindahl, 1993; Sancar et al., 2004), protein refolding (Saibil and Ranson, 2002; Kinbara and Aida, 2005), and other types of repair or regeneration of protein function (Barber and Andersson, 1992; Rando, 2001). Another continuously ongoing repair process within living cells is the turnover of most proteins on a daily to monthly basis. This is made possible by the tagging of nonfunctional proteins for destruction (Hershko and Ciechanover, 1998) and the existence of a DNA template and protein synthesis machinery (ribosomes) for producing new protein components on demand. These processes may be strongly speeded up or modulated by damage to the cell. On a higher level, there is continuous turnover of entire cells and tissues (e.g., in the gastrointestinal tract, the blood, and the skin). This is enabled by the presence of certain stem cells that continuously divide, and whose daughter cells, differentiate and distribute in a manner carefully regulated by various extracellular cues. Upon acute damage, for example, a skin wound, several processes start to counteract deleterious effects (e.g., infection, fluid and electrolyte imbalances), remove the damaged tissue, and finally restore the tissue to its original form and function. The potency of this process is clearly demonstrated in lower vertebrates, for example, lizards that may lose entire body parts followed by ordered regeneration.

A characteristic of all self-healing processes in living organisms is the regulation by complex signaling processes within cells and/or between cells and the coordinated activation of a range of transport and motion-generating processes. Signaling within cells may be initiated by such diverse factors as binding of extracellular signaling molecules to cell surface receptors, entry of calcium ions into the cell interior, or tension between the cell and the extracellular matrix sensed by processes involving the internal cytoskeleton. These events activate a range of interacting signaling systems composed of, for example, various enzymes, often kinases (with phosphorylating capability) or G-proteins (GTP-hydrolyzing molecular switches). These activate or inactivate other signaling or effector proteins in complex serial and parallel networks along a number of strongly interconnected paths. The outcome is a cell response, for example, in the form of increased synthesis of certain proteins, reorganization of the actin and microtubule cytoskeleton, and activation of transport processes. These include the transport of ions through the plasma membrane via protein channels and transportation of intracellular components and entire cells due to cytoskeletal reorganization and the action of a plethora of molecular motors. Indeed, it is self-evident that motion is a key process in biological self-healing, from refolding of proteins by the action of cellular chaperones (Saibil and Ranson, 2002), over the replacement of proteins or plasma membranes in a certain part of a cell, to wound healing on the tissue and organ levels. Here, we will focus on aspects of the fascinating self-healing processes that occur following destruction (by crush injury) of the axon of an adult motor neuron. The latter structure is the extension of the neuron that transmits electric signals (action potentials) to activate contraction of the muscle fibers innervated by the neuron. As a result of the destruction of the axon, a number of processes start both in the axon stump and in the motor end plate, the region where the motor neuron contacts a muscle fiber. These processes are well-defined examples of self-healing, which have been extensively studied both with regard to cellular sensing and signaling events and with regard to the cytoskeletal rearrangements and cellular motion-generating machinery responsible for the actual morphological changes during self-healing. However, the number of integrated studies that consider *both* signaling and cytoskeletal responses at the same time as well as changes in both muscle and nerve is more limited. The focus of the present review will be this overall picture. Particularly, we will consider how the cell signaling machinery, in a coordinated way, regulates the cytoskeleton and molecular motors to produce the ultrastructurally observable morphological changes in the nerve axon and at the motor end plate following injury. With regard to motors and the cytoskeleton, the focus will be slightly more on the actin filaments (microfilaments) and actin-based motors (myosins) than on microtubules and microtubule-based motors (kinesins and dyneins). Since the present review spans over a wide field, we refer to more specialized reviews for details, for example, on molecular motors and the cytoskeleton (Howard, 2001; Pollard and Borisy, 2003; Vale, 2003; Geeves et al., 2005; Gennerich and Vale, 2009; Etienne-Manneville, 2010; Lee and Dominguez, 2010; Spudich, 2011), axonal transport (Gunawardena and Goldstein, 2004; Hirokawa et al., 2010), growth-cone mobility (Geraldo and Gordon-Weeks, 2009; Lowery and Van Vactor, 2009), formation of the neuromuscular synapse (Sanes and Lichtman, 1999; Hughes et al., 2006; Wu et al., 2010) as well as on Wallerian degeneration, subsequent regeneration and Schwann cell responses (Stoll et al., 2002; Kang et al., 2003; Fenrich and Gordon, 2004; Navarro et al., 2007; Campbell, 2008; Radtke and Vogt, 2009). We also relate the account of the self-healing processes upon denervation to recent efforts in bionanotechnology and materials science (e.g., Soong et al., 2000; Månsson et al., 2005; Bakewell and Nicolau, 2007; Fischer and Hess, 2007; van den Heuvel and Dekker, 2007; Goel and Vogel, 2008; Agarwal and Hess, 2010; Korten et al., 2010).

15.2 MOLECULAR MOTORS AND THE CYTOSKELETON

15.2.1 General

In travesty of Sherrington (1924; cf. Kandel et al., 1995), to move is all that cells and cell components can do whether dividing, migrating to develop organ systems, taking part in self-healing, and so on, for such the sole executants are molecular motors and the cytoskeleton. With regard to the molecular motors, these are specialized enzymes (catalytic proteins). Although enzyme catalysis, in general, is associated with structural changes in the enzyme, the motion produced by a molecular motor is more extensive. Thus, a molecular motor may be defined (Cross, 1997; Vale and Milligan, 2000; Howard, 2001) as an enzyme where a local chemical change around, for example, an active site for high-energy phosphate turnover, is amplified through the protein structure leading to swing of a lever arm–like component. This enables typically several nanometers of movement upon turnover of one high-energy phosphate compound, for example, ATP. The movement may be transformed into rotation as in the bacterial flagellar motor. However, a majority of the motors are termed "linear," and move along linear tracks. Thus, by synchronizing the structural changes in lever arm position during different steps of high-energy phosphate turnover to varying affinity of another part of the motor to its track, the motor will be able to move one or several steps along this track. If the motor generally takes several steps before detaching it is regarded as processive, otherwise, nonprocessive. With the earlier definition, enzymes moving along nucleic acids such as DNA and RNA polymerases can be included among the linear motors as well as the ribosomal components that move mRNA and the newly formed polypeptide chain. However, the linear motors in focus here are the motors that underlie cell motility, intracellular transport, tension generation (e.g., cytokinesis, muscle contraction), and so on. These motors can be divided into the actin-based myosins and microtubule-based kinesin and dyneins moving along "tracks" of the actin and microtubule cytoskeleton (Figure 15.1). Although the name implies that this cytoskeleton mainly has a role in maintaining the structural integrity and shape of the cell, this is true only to a limited extent. Although a third cytoskeletal component, in addition to actin filaments and microtubules, the intermediate filaments, might have this role to a larger extent, the actin filaments and microtubules are highly dynamic. They thus react, for example, by changing their stability and/or the rate of addition/exchange of monomers in

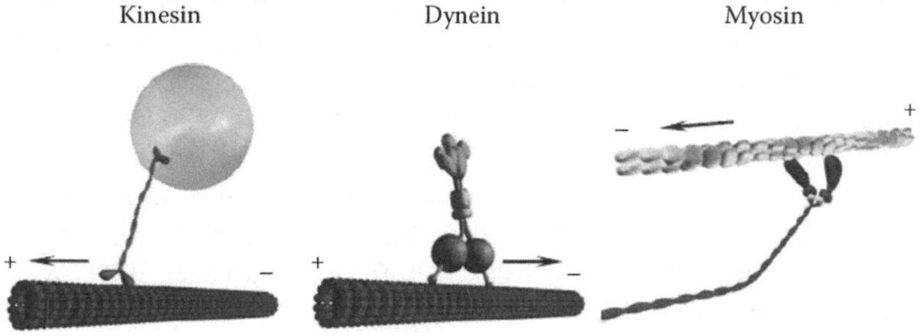

FIGURE 15.1 Key players in the motion-generating machinery of the cell. From the left to the right: kinesin motor (with vesicular cargo) moving toward the microtubule plus end (fast polymerizing end), dynein motor moving toward the microtubule minus end, and myosin motor moving toward actin filament plus end. If the myosin motors are immobilized in filaments as for the illustrated myosin II motor, this would lead to translocation of the actin filament as indicated by the arrow. (Adapted from Balaz, M., Interaction of actomyosin with synthetic materials—Effects on motor function and potential for exploitation, PhD thesis, University of Kalmar, Kalmar, Sweden, p. 202, 2008.)

response to intracellular chemical changes (e.g., related to cell signaling networks) or upon sensing changes in the extracellular environment (e.g., via integrins or other cell-adhesion molecules). Related to these responses, they also play important roles in cell motility independent of their roles as "tracks" for molecular motors (Pollard and Borisy, 2003; Lindberg et al., 2008).

15.2.2 Actin Cytoskeleton

Actin filaments (F-actin; 10 nm diameter) are formed by a large number of identical G-actin monomers polymerized into a right-handed double helical arrangement with a half-pitch of about 36 nm and a distance between the monomers along a protofilament of ~5.5 nm (Figure 15.2) (Prochniewicz, 1979; Egelman et al., 1982;

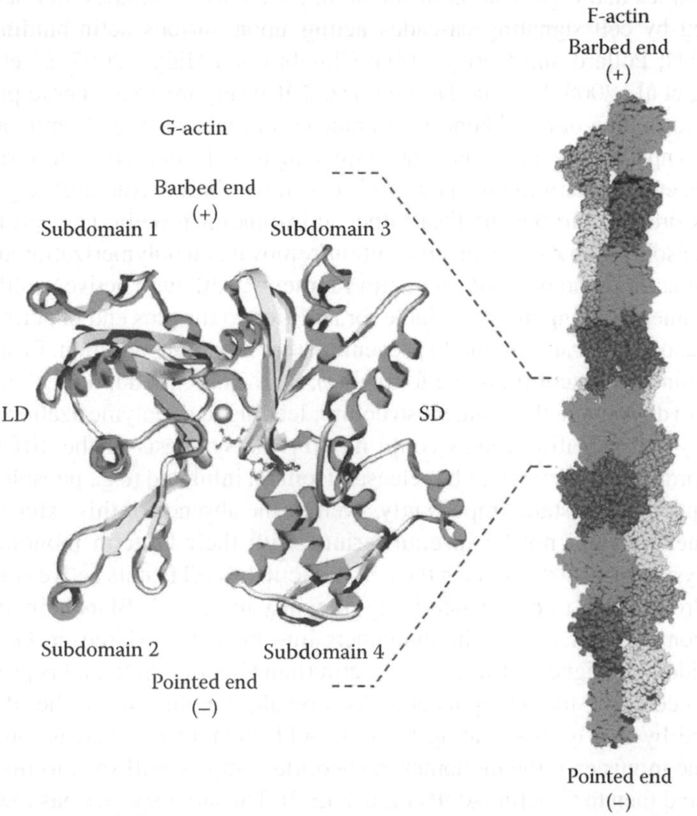

FIGURE 15.2 (See color insert.) The actin monomer, G-actin to the left with "large" (LD) and "small" (SD) domains and nucleotide and divalent metal ion (light gray sphere) indicated in the catalytic cleft between these domains. The figure to the right indicates how the monomers are incorporated into F-actin in two right-handed helical protofilaments. The dimensions of the actin monomer is approximately 5.5 × 5.0 × 3.5 nm^3. (Reproduced from Balaz, M., Interaction of actomyosin with synthetic materials—Effects on motor function and potential for exploitation, PhD thesis, University of Kalmar, Kalmar, Sweden, p. 202, 2008.)

Yanagida et al., 1984; Kabsch et al., 1990; Schutt et al., 1993; Egelman and Orlova, 1995; McGough, 1998; Oda et al., 2009; Bugyi and Carlier, 2010; Fujii et al., 2010; Murakami et al., 2010). The actin filaments are semiflexible (persistence length 10–20 μm) (Isambert et al., 1995) and exhibit structural polarity as a result of the asymmetric structure of each monomer. This polarity is reflected in faster addition of monomers at one end of the filament (termed the plus end or the barbed end) than the other (the minus end or pointed end). Moreover, each type of myosin motor also moves either toward the plus end (the usual case) or the minus end (myosin VI and IX) (Wells et al., 1999; Inoue et al., 2002) of the filament. The actin filaments are highly dynamic both in that they can effectuate length changes by polymerization–depolymerization events in different patterns (Chhabra and Higgs, 2007; Bugyi and Carlier, 2010) and in that the filaments can fluctuate between different structural and functional states without polymerization/depolymerization (Kozuka et al., 2006). The dynamics and organization of the actin filaments in bundles and networks are controlled by cell signaling cascades acting upon various actin-binding proteins (Otto, 1994; Pollard and Borisy, 2003; Chhabra and Higgs, 2007; Li et al., 2008; Lindberg et al., 2008; Lee and Dominguez, 2010) (Figure 15.3). These proteins can nucleate formation of new filaments or branches of filaments (e.g., formin proteins, the Arp 2/3 complex, spire) and lead to cross-linking (e.g., fascin, villin, fimbrin, filamin, α-actinin, etc.) of actin filaments in different higher-order structures (e.g., branched networks, ordered parallel bundles). Other actin-binding proteins (e.g., profilin, ADF/cofilin, gelsolin, Cap Z) control the addition/removal and polymerization/depolymerization of actin monomers to/from actin filaments. Although active profilin sequesters monomers making them available for addition to the plus end of actin filaments, gelsolin and Cap Z cap this end to prevent further monomer addition. Proteins of the ADF/cofilin family (cofilin in the following), on the other hand, bind along the actin filament to destabilize the filament structure, leading to depolymerization. Either the regulatory cell signaling events could modify the synthesis of the different actin-binding proteins or they might be released from an inhibited (e.g., phosphorylated or nonphosphorylated) state. Importantly, even in the absence of this external control, the filaments would not be in equilibrium with their G-actin monomers. Thus, under physiological conditions in the cell, G-actin has ATP at its active site with very slow hydrolysis, the rate of which is significantly increased (Blanchoin and Pollard, 2002) upon incorporation of the monomers into filaments. Moreover, G-actin–ATP has considerably higher "affinity" to F-actin than G-actin–ADP and is preferentially incorporated at the filament plus end. As a result, the plus end of the filament will be capped by slowly dissociating G-actin–ATP monomers whereas, progressively toward the minus end, the monomer–nucleotide complex will shift to first G-actin–ADP-P_i and then to G-actin–ADP (Figure 15.3). The latter species has lower affinity to the filament form and will be the one that dissociates at the minus end. The rate of the different steps in the ATP turnover and the rate of monomer addition (e.g., due to availability) are of considerable importance for actin dynamics during cytoskeletal remodeling, for example, upon growth-cone propulsion (see the following). In this connection, it is worth mentioning that the arguments for filament growth and shrinkage by monomer addition/removal rather than by annealing/breaking off of filament fragments are quite interesting (Howard, 2001). In addition to the control of

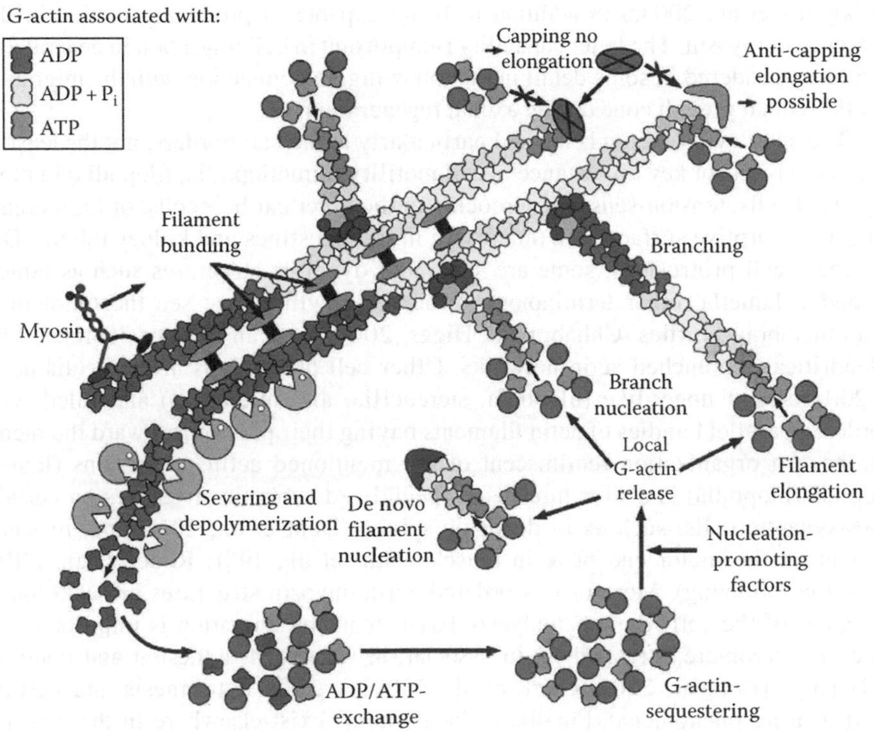

FIGURE 15.3 **(See color insert.)** Schematic illustration of actin dynamics showing the incorporation of G-actin–ATP at the plus end and subsequent shift of these monomers while they are first hydrolyzing ATP followed by release of inorganic phosphate (P_i). The figure also schematically illustrates the action of several actin-binding proteins that (1) nucleate formation of new filaments (e.g., formins), (2) nucleate branching of filaments (e.g., the ARP 2/3 complex), (3) sever (cut, e.g., gelsolin) and depolymerize (e.g., cofilin) filaments, (4) sequester G-actin monomers (e.g., profilin), (5) cap filaments to block further monomer incorporation (e.g., Cap Z and gelsolin), and (6) bundle filaments in different geometric patterns (e.g., filamin, fascin, α-actinin). (Reprinted from *Curr. Opin. Neurobiol.*, 18, Witte, H. and Bradke, F., The role of the cytoskeleton during neuronal polarization, 479–487, Copyright 2008, with permission from Elsevier.)

actin dynamics by ATP turnover, and various actin-binding proteins, actin dynamics may also be affected through mechanical forces applied to the cell membrane (see the following). This is possible since the actin cytoskeleton is coupled to the extracellular matrix and the cell membrane, for example, via cell-adhesion molecules and various adaptor proteins. The actin cytoskeleton is therefore important for sensing external forces, for example, related to altered topography and mechanical properties of the surrounding extracellular environment.

In studies of muscle contraction and some aspects of other cellular functions of myosins, actin filaments are often viewed simply as tracks for myosins. However, as should be clear from the preceding text, this is a gross oversimplification. Thus, the structural dynamics of actin plays important roles in modulating actomyosin function (Yanagida et al., 1984; Orlova and Egelman, 1993; Kozuka et al., 2006;

Vikhorev et al., 2008a) in addition to being capable of producing motion in the absence of myosin. The latter capability is important in cell migration in general and will be considered in some detail in the following, in connection with the migration of the axonal growth cone during axonal regeneration.

The actin cytoskeleton is located particularly at the cell borders, not the least in cell extensions of key importance in cell motility (lamellopodia, filopodia) in most types of cells, tension-sensing (stereocilia in the inner ear hair cells) or for increasing an absorptive surface area (microvilli in the intestines and kidney tubuli). Out of these cell protrusions, some are sheet-like dynamic structures such as lamellopodia, lamella (other terminology in nerve growth cones; see the following), and membrane ruffles (Chhabra and Higgs, 2007) with an interior dominated by dendritically branched actin networks. Other cell protrusions are thin (diameter <200 nm) and finger-like (filopodia, stereocilia, and microvilli) and filled with ordered parallel bundles of actin filaments having their plus ends toward the membrane. An organization reminiscent of the mentioned actin protrusions (lamellopodia/filopodia) is also exhibited by specialized synaptic structures on certain postsynaptic cells, such as in dendritic spines (Dent et al., 2011) and, to some extent, at the motor end plate in muscle (Hall et al., 1981; Roder et al., 2010) (see the following). More or less ordered actin–myosin structures are also found in areas of the cell where actomyosin-based tension generation is important (the muscle sarcomere, stress fibers in association with focal adhesion and contractile rings [Bement, 2002; Clark et al., 2009], e.g., in cytokinesis and certain self-healing phenomena). Finally, actin filaments exist elsewhere in the cell, for example, around intracellular organelles such as the Golgi apparatus and even in the nucleus (Chhabra and Higgs, 2007).

15.2.3 Microtubule Cytoskeleton

Microtubules are hollow tubes, ~25 nm in diameter, formed by polymerization of α-tubulin and β-tubulin heterodimers (Nogales et al., 1998; Howard, 2001; Li et al., 2002; Witte and Bradke, 2008) (Figure 15.4). These dimers that cannot be split into monomers without denaturing the protein (Howard, 2001) are arranged axially (with about 8 nm interdimer distance) along protofilaments. Eleven to fifteen (13 on average) (Tilney et al., 1973) of such protofilaments in parallel make up the wall of the hollow microtubules (Erickson, 1974; Howard, 2001). Their structure causes the microtubules to exhibit substantially higher bending rigidity than actin filaments with a persistence length >1 mm (Gittes et al., 1993). However, there are also similarities between actin filaments and microtubules. Thus, similar to the actin filaments, the microtubules are polar structures with a fast polymerizing plus end and a slowly polymerizing minus end (Howard, 2001; Etienne-Manneville, 2010). In further similarity to actin filaments, the tubulin dimers have a nucleotide triphosphate (in this case GTP) at the active site and this is hydrolyzed upon addition of the dimers to the filament (Etienne-Manneville, 2010). In similarity with the actin-based motors, microtubule-based motors move either toward the microtubule plus or minus

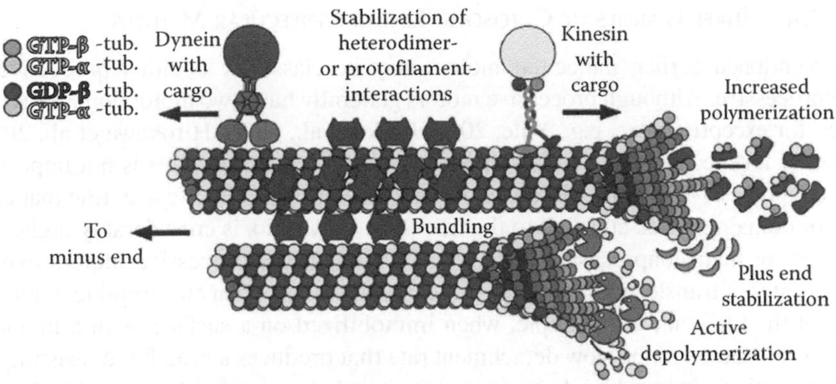

FIGURE 15.4 **(See color insert.)** Microtubule dynamics showing the plus end incorporation of α/β-tubulin heterodimers with GTP at their active site and subsequent shift of these dimers along the filaments with hydrolysis of GTP, leaving GDP at the active site. The figure also schematically illustrates the action of several microtubule-associated proteins (MAPs) that (1) stabilize dimer interactions within the filament (e.g., tau), (2) link neighboring filaments or facilitate bundling (e.g., MAP2), (3) facilitate polymerization by binding to α/β-tubulin heterodimers (e.g., CRMP-2), and (4) stabilize the dynamic plus end (e.g., EB1). (Reprinted from *Curr. Opin. Neurobiol.*, 18, Witte, H. and Bradke, F., The role of the cytoskeleton during neuronal polarization, 479–487, Copyright 2008, with permission from Elsevier.)

end (Hirokawa et al., 2010). Although most kinesin motors move toward the plus end, dyneins and some kinesin motors move toward the minus end (Vale, 2003; Hirokawa et al., 2010). In similarity to the actin filaments, the microtubules are also highly dynamic structures with polymerization and depolymerization particularly at the plus end. The so-called "dynamic instability" at this site, with phases of rapid growth intervened with phases of rapid shrinkage ("catastrophes"), is regulated by microtubule-binding proteins, in turn controlled by cell signaling systems (Witte and Bradke, 2008; Akhmanova and Steinmetz, 2010; Etienne-Manneville, 2010). Not the least are a number microtubule plus end tracking proteins (+TIPs) of key importance in this regard. Some of these proteins also seem to be central in connecting the microtubules to the actin cytoskeleton, for example, in nerve growth cones (Lowery and Van Vactor, 2009; Akhmanova and Steinmetz, 2010).

The microtubules run from near the cell center (organized close to centrosomes) and extend all the way to the cell periphery where the microtubule plus end is generally located. Due to their high bending rigidity, the microtubules provide structural stability to cells but they also form suitably positioned "highways" for transport of various intracellular cargoes (mitochondria, lipid vesicles with proteins, etc.) from the cell center to the periphery or vice versa using either plus end–directed or minus end–directed microtubule-based molecular motors (as in axonal transport, see the following). The microtubule cytoskeleton is also of critical importance during mitosis when it undergoes dynamic changes in interaction with specific kinesin motors to organize the mitotic spindle and move the two sets of chromosomes after DNA replication to the opposite cell poles (Zhu et al., 2005).

15.2.4 Three Families of Cytosolic Linear Molecular Motors

As mentioned earlier, molecular motors may be classified as either processive or nonprocessive. Although processive motors generally have two motor domains (however, for exceptions see, e.g., Vale, 2003; Iwaki et al., 2006; Hirokawa et al., 2010), in order for the two domains to step sequentially on the track, this is not important for nonprocessive motors. Moreover, the fraction of the ATPase cycle time that each motor domain spends attached to the track, the duty ratio, is considerably higher for processive than nonprocessive motors and, related, the processive motors exhibit lower rates of translocation of the cytoskeletal filaments when an ensemble of motors propel the filament, for example, when immobilized on a surface or in a filament. This is attributed to the slow detachment rate that produces a drag force resisting the sliding action produced by the power stroke of other motors in the ensemble. On the other hand, the slow detachment rate is important as a basis for processivity since it prevents the first head from detaching until the second head has had a chance to attach to the track.

Clearly, the processive and nonprocessive motors are adapted for different tasks. On the one hand, slow, high-duty ratio motors with strain-dependent kinetic steps are useful for, for example, cargo transportation and tension-sensing. On the other hand, the low-duty nonprocessive motors that work in teams are well suited for high force development, power output, and high-speed motion.

Most kinesins and cytoplasmic dyneins are processive, high-duty ratio motors. Several myosins (e.g., myosins V, VII, X) are also generally classified as processive, whereas the so-called conventional muscular and nonmuscular myosin II, as well as myosin I, are classified as nonprocessive. However, importantly, also the latter motors may be processive (i.e., taking several steps along their tracks before detaching) under certain conditions (Albet-Torres et al., 2009; Mansson, 2010; McConnell and Tyska, 2010).

15.2.4.1 Myosins

The myosin motors belong to a large superfamily of 24 classes (in humans 11 classes from 40 different genes) (Foth et al., 2006). They move along actin filaments or propel these in a process that is driven by ATP turnover at the myosin active site.

The myosins are heteromultimeric proteins with heavy chains, light chains and, in some cases, additional subunits (e.g., Figure 15.5). The heavy chain has an N-terminal motor domain with ATPase site and actin-binding site. The amino acid sequence of the motor domain is well conserved resulting in similar folding and characteristic motifs that are common to all myosins (Sellers, 1999). However, subtle differences in amino acid sequence of the motor domain between classes and between isoforms within classes lead to varying kinetic and mechanical properties (see the following). Important structural motifs in the myosin motor domain (Rayment et al., 1993, 1996; Smith and Rayment, 1996) include the switch-I, switch-II, and P-loops (in close proximity to the nucleotide-binding pocket). These have roles in coordinating nucleotide binding and in relaying different steps in the ATP turnover, for example, the release of phosphate, to distant structural changes leading to modification of actin affinity and swing of the myosin neck region

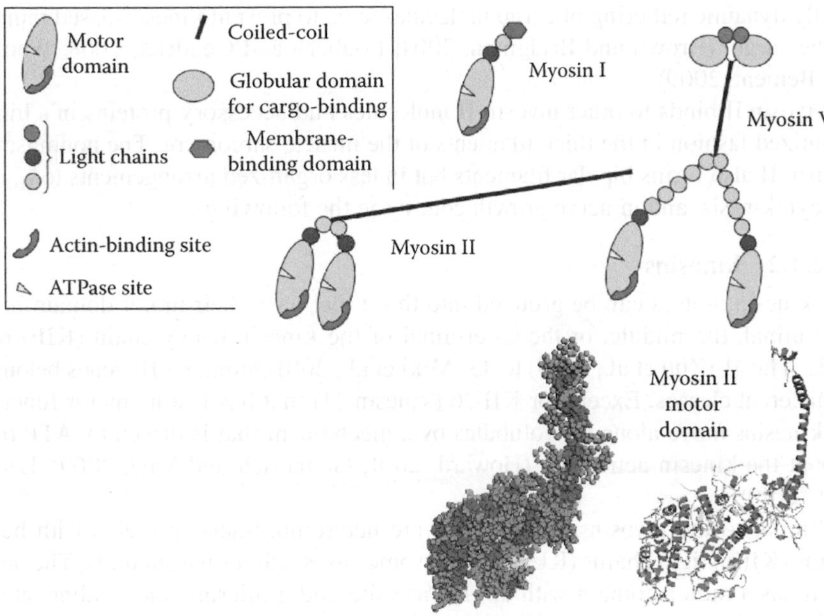

FIGURE 15.5 **(See color insert.)** Schematic illustration of motors from three myosin classes (I, II, and V) and, to the lower right the atomic model of the myosin II motor and neck domain with (left model) and without (right) the light chains. Color code: secondary structure. The atomic coordinates from the Brookhaven Data Bank (MYS2) illustrated using PyMOL software.

(the lever arm). The latter consists of an α-helix that is stabilized by a number of myosin light chains (often calmodulins or calmodulin-like proteins, e.g., with regulatory roles) and it connects the motor domain to the C-terminal tail domain. The length of the neck α-helix varies substantially between myosins having binding sites (IQ domains) for 1 (as in certain myosin I isoforms) to 6 (as in myosin V) light chains. The tail domain, finally, is the most variable among the different myosins but it generally has a coiled-coil region that allows dimerization of two heavy chains. Moreover, in most myosins the most C-terminal part of the tail has roles in anchorage, regulation, and cargo-binding (Sellers, 1999; Redowicz, 2001; Li and Nebenfuhr, 2008). In myosin II, this coiled-coil region extends for >150 nm and it contains, for example, charged regions of importance for polymerization of several myosin molecules into bipolar filaments with the N-terminal heads extending in bipolar directions. In myosin I, on the other hand, the tail is highly truncated and the motor domain and the neck domain form a monomer that can bind directly to lipid membranes (McConnell and Tyska, 2010).

Motors from the myosin family underlie muscle contraction, cytokinesis (cell division), and most other aspects of cell motility (together with the actin polymerization–depolymerization motor considered earlier). Some of the myosins also seem to have important roles in short-distance transportation of intracellular cargoes, endocytosis/exocytosis, tension-sensing, organization of the actin cytoskeleton, and finally dynamic tethering of cargo molecules (e.g., to prevent kinesin-based removal of the cargo) (Brown and Bridgman, 2004; Loubery and Coudrier, 2008; Woolner and Bement, 2009).

Myosin II binds to other myosin II molecules and accessory proteins in a highly organized fashion in the thick filaments of the muscle sarcomere. The nonmuscular myosin II also forms bipolar filaments but in less organized arrangements (e.g., during cytokinesis, and in nerve growth cones; see the following).

15.2.4.2 Kinesins

The kinesin motors can be grouped into those that have their motor domain in the N-terminal, the middle, or the C-terminal of the kinesin heavy chain (KIF) molecule. The 41 (Zhu et al., 2005) to 45 (Miki et al., 2001) human KIF genes belong to 14 different classes. Except for KIF26 (kinesin 11) that has lost its motor function, the kinesins move along microtubules by a mechanism that is driven by ATP turnover at the kinesin active site (Howard, 2001; Gennerich and Vale, 2009;, Endow et al., 2010).

Similar to the myosins, the kinesins are heteromultimeric proteins with heavy chains (KIFs), light chains (KLC) and, in some cases, additional subunits. The heavy chain has a motor domain with an ATPase site and a microtubule-binding site. In further analogy with myosin, the motor domain is well conserved and even if it is smaller than that of myosin it contains similar switch-I, switch-II, and P-loop motifs (Kull et al., 1996) (also existing in G-proteins). In kinesin, structural changes around the ATP-binding pocket are coupled to the swing of an extension from the motor domain, in this case, the so-called neck linker. The latter, that connects the motor domain to the stalk, positions, upon its swing during the kinesin motor stroke, the partner head of dimeric kinesins for binding to the next tubulin-heterodimer, about 8 nm along the filament. Other mechanisms could apply for monomeric kinesins (although these might dimerize *in vivo*) (Vale, 2003). The two heavy chains of dimeric kinesins are connected to each other by a coiled-coil region in the stalk of the molecule, followed by a globular tail at the end opposite to the motor domain. The tail region also contains light chains and cargo-binding sites and, as for the myosins, the variability in amino acid sequence between the different KIFs is considerably greater in this region than in the motor domain.

The kinesin motors play important roles in mitosis and cytokinesis (Zhu et al., 2005) and in various forms of intracellular transport, not the least long-distance transport (Vale, 2003; Hirokawa et al., 2010). For example, of major importance here, a range of KIFs (Figures 15.1 and 15.6) underlie the anterograde axonal transport (toward the periphery). Thus, these motors carry their cargoes (synaptic vesicle precursors, growth factor receptors, mitochondria, other motors and cytoskeletal components, etc.; Figure 15.7) while moving toward the microtubule plus end along the axon shaft. These motors are responsible for both fast axonal transport (rate 400 mm/day) and slow axonal transport (<10 mm/day). The former kind of transport involves nearly uninterrupted movement of kinesins along their straight microtubule tracks within the axons. Although slow axonal transport is less well characterized, it is believed to involve fast kinesin and dynein-driven transport interrupted by changes in direction and pauses of substantial duration (Shah and Cleveland, 2002;

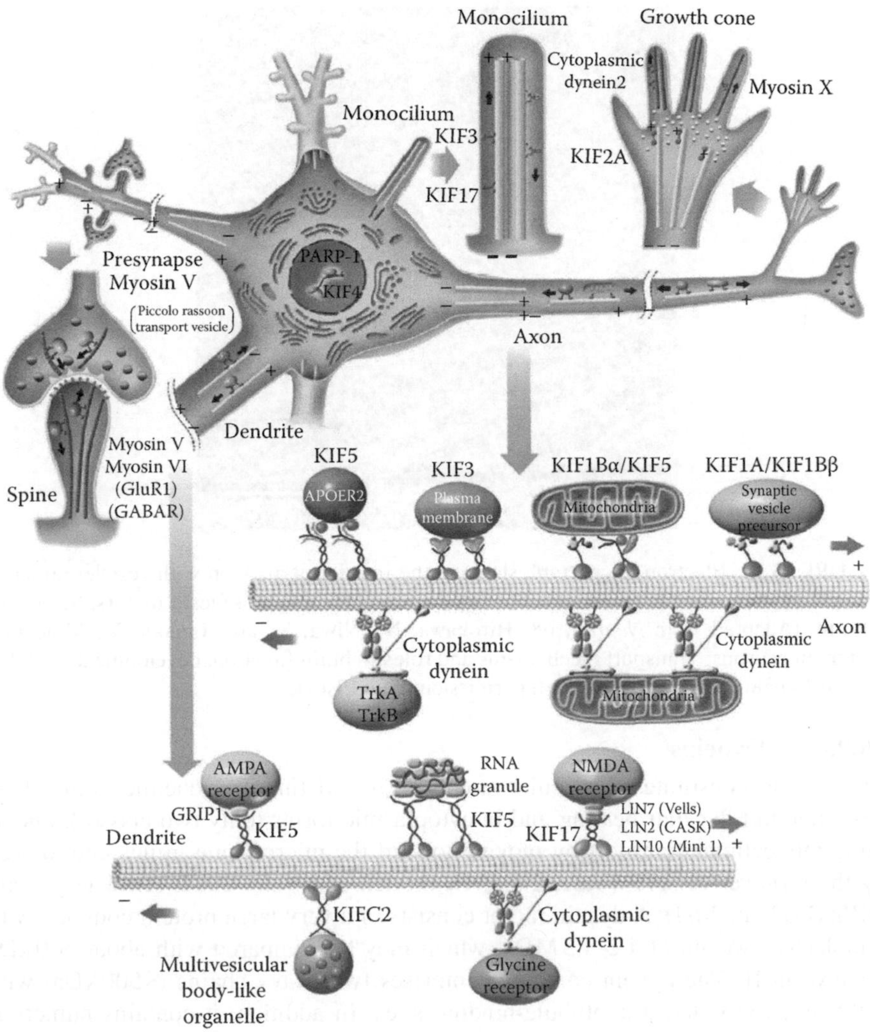

FIGURE 15.6 (See color insert.) The role of cytoplasmic dynein and different myosin and kinesin (KIF) classes of motors in a neuron (e.g., in axonal/dendritic transport, synaptic plasticity/receptor recycling, and growth-cone motility). Actin filaments (red) and microtubules (yellow). (Reprinted from *Neuron*, 68, Hirokawa, N., Niwa, S., and Tanaka, Y., Molecular motors in neurons: Transport mechanisms and roles in brain function, development, and disease, 610–638, Copyright 2010, with permission from Elsevier.)

Hirokawa et al., 2010). Not unexpectedly, several severe neurological diseases have been confirmed and others have been proposed to be associated with disturbances in these motors (Hirokawa et al., 2010). As will be seen in the following, the motors also play pivotal roles in nerve regeneration after damage. The kinesins also take part in both anterograde and retrograde transport in the dendrites and play roles in growth-cone propulsion.

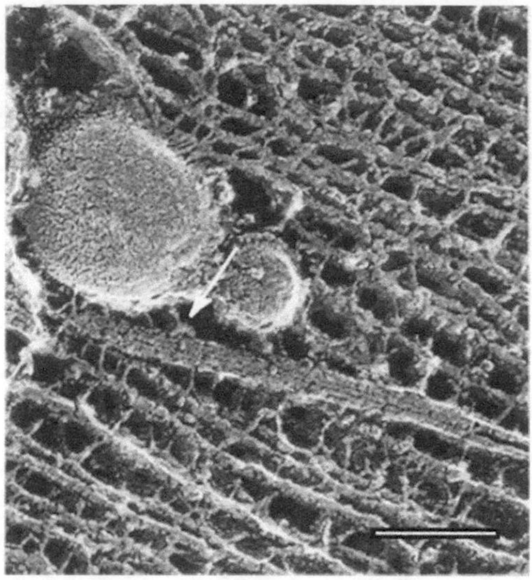

FIGURE 15.7 Electron micrograph showing the interior of an axon with vesicles attached to a microtubule via feet-like extensions (e.g., arrow) presumably kinesin motors. Scale bar: 100 nm. (Adapted from *Neuron*, 68, Hirokawa, N., Niwa, S., and Tanaka, Y., Molecular motors in neurons: Transport mechanisms and roles in brain function, development, and disease, 610–638, Copyright 2010, with permission from Elsevier.)

15.2.4.3 Dyneins

The dyneins constitute a molecular motor family existing in axonemic forms, driving ciliar and flagellar beating and in cytoplasmic forms (only two classes), underlying intracellular transport by moving toward the microtubule minus end, driven by the turnover of ATP (Vale, 2003; Gennerich and Vale, 2009; Hirokawa et al., 2010; Spudich, 2011). A dynein motor consists of a very large protein complex with a molecular weight of 1.2–1.5 MDa, which may be compared with about 500 kDa for myosin II. The dynein complex comprises two heavy chains (~500 kDa) with ATPase activity and microtubule-binding sites. In addition, it contains numerous subunits of lower molecular weight (~10–70 kDa). Moreover, cytoplasmic dynein has an associated protein complex, dynactin, of importance for the regulation of motor activity and cargo-binding.

In addition to the large size, the dynein motors also have a very different evolutionary origin than the kinesin and myosin motors and also the mechanisms of chemomechanical transduction differ substantially from myosins and kinesins (that are phylogenetically related). Thus, in contrast to the latter motor families, the dynein motor belongs to the AAA^+ ATPases superfamily. This family, unlike the kinesins and myosins, has members both in eukaryotes and prokaryotes and with varying roles, for example, as unfoldases, targeting proteins for degradation or as DNA and RNA helicases (Spudich, 2011). These roles are fulfilled, using ATP and feeding the polypeptide or nucleic acid chains through a pore formed by six domains or subunits in ring-like structures. A corresponding ring-like hexameric structure forms the core of each of the globular motor domains in the dynein dimer where four domains bind

ATP and one is the main ATPase site. On the opposite side of the molecule (several nm) from the latter site (Carter et al., 2011), a long (~15 nm) coiled-coil domain (denoted the stalk) extends, ending in a globular microtubule-binding site. Both the angles that this extended domain forms with the AAA ring structure and the fine structure and microtubule affinity of the globular end domain vary allosterically during ATP turnover. This leads to appropriately synchronized changes in angle and microtubule affinity to allow processive motion along the microtubules. By this action and the varying cargo-binding capacity, the cytoplasmic dyneins are the dominating motors for retrograde axonal transport with transport rates of >100 mm/day, similar to that for fast anterograde transport. In spite of only two heavy chain genes of cytoplasmic dynein, different cargoes may be transported by the existence of a plethora of complex dynein-binding proteins that may act as adaptors for the attachment of the cargoes. In the axon, the latter include mitochondria, myosin V, and activated neurotrophin receptors (Hirokawa et al., 2010, and references therein).

15.2.5 Summary

The molecular motors and the actin and microtubule cytoskeleton, with all accessory proteins, form a highly versatile toolbox that allows execution of a range of transportation and motion-generating tasks extensively used by all cellular players (e.g., neurons, muscle fibers, Schwann cells, etc.) to achieve the self-healing following axonal injury. The morphological and functional consequences of these motor actions are described below.

15.3 GENERAL ASPECTS OF NEUROMUSCULAR COMMUNICATION AND THE OVERALL EVENTS FOLLOWING DENERVATION AND REINNERVATION

The three main cellular elements in neuromuscular communication are the motor neuron, the Schwann cell, and the muscle fiber (Figure 15.8). Motor neurons have their cell bodies and dendrites in the spinal cord where they receive information from other neurons. They have very long axons that extend uninterrupted all the way from the spinal cord to the muscle fibers they innervate in a skeletal muscle. Schwann cells are the glia (supportive) cells of the peripheral nervous system and are either myelinating or nonmyelinating (see Woodhoo and Sommer, 2008). Myelinating Schwann cells form a myelin sheath around the motor axon, which increases the speed with which an electrical impulse, an action potential, can travel along the axon. Specialized Schwann cells, known as terminal Schwann cells, cover the neuromuscular junction (see Sanes and Lichtman, 1999), the point of contact between a motor neuron and a muscle fiber.

Muscle fibers are very large elongated cells (up to 100 μm and tens of cm diameter and length, respectively) that are specialized for generation of force and movement. Each muscle fiber typically contains thousands of nuclei that originally arise from individual mononuclear cells, myoblasts, that fuse to form myotubes during embryonic muscle formation (Hall and Sanes, 1993; Sanes and Lichtman, 1999).

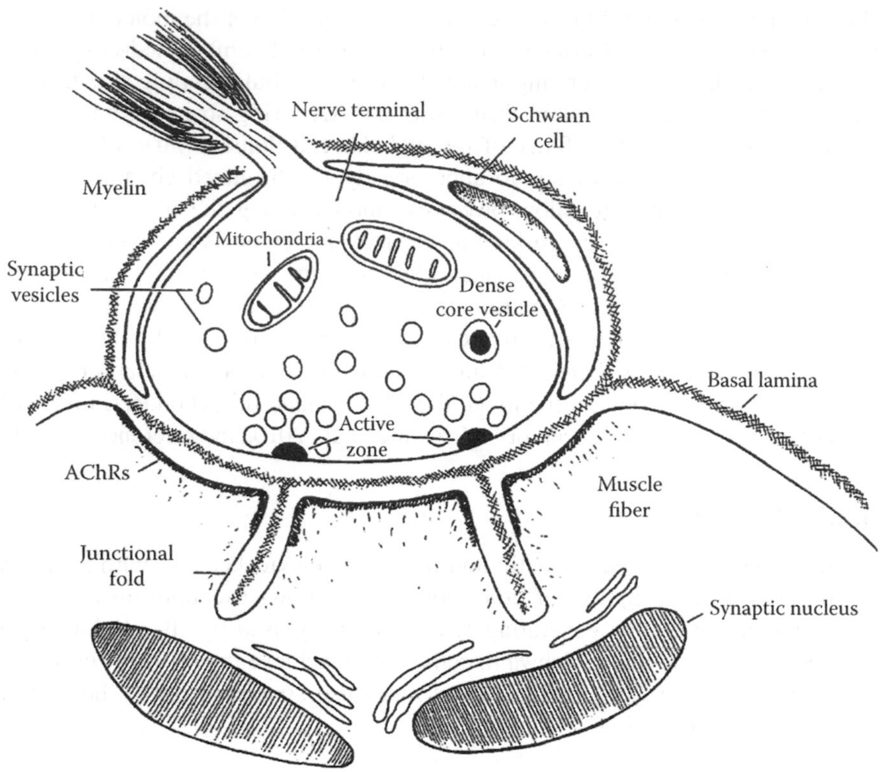

FIGURE 15.8 Schematic illustration of the neuromuscular junction showing the three main cellular elements in neuromuscular communication, the motor neuron (axon terminal), the Schwann cell, and the muscle fiber. The junctional folds with acetylcholine receptors located at the crests are aligned with active zones where acetylcholine is released from the nerve terminal. The basal lamina that surrounds all cell components also extends into the junctional folds. (Reproduced from *Cell*, 72(Suppl), Hall, Z.W. and Sanes, J.R., Synaptic structure and development: The neuromuscular junction, 99–121, Copyright 1993, with permission from Elsevier.)

The myotubes, when contacted by an innervating nerve, differentiate into mature muscle fibers. The neuromuscular junction is generally situated close to the middle of the muscle fiber and occupies a small fraction, in the range of 0.1% (Sanes, 2003), of the total muscle fiber surface area.

15.3.1 Neuromuscular Transmission

The neuromuscular junction is a chemical synapse in which information from the nerve is transmitted to the muscle fiber by means of the chemical transmitter substance acetylcholine that is stored in vesicles in the nerve terminal. When an action potential arrives to the nerve terminal of the axon, voltage-sensitive calcium channels open and the inflow of calcium into the nerve terminal triggers release of acetylcholine into the tiny space (about 50 nm, see Hughes et al., 2006) between the axon terminal and the muscle fiber, the synaptic cleft. By diffusion, acetylcholine reaches the muscle fiber surface, which at the motor end plate (the muscle surface part of the neuromuscular junction) is covered with receptor proteins for acetylcholine. The receptor protein, known as the skeletal muscle type of nicotinic acetylcholine receptor, is a ligand-operated ion channel (opened by binding of, e.g., acetylcholine). Upon opening, it is permeable primarily to sodium and potassium ions. As the ions start to flow through the open receptor channels, the end-plate region of the muscle fiber depolarizes (reversing the electric potential on the inside and the outside of the membrane). As a result, voltage-sensitive sodium channels in the muscle fiber membrane open leading to the generation of an action potential in the muscle fiber. The action potential rapidly spreads from the end-plate region in both directions of the muscle fiber and triggers further events that rapidly lead to release of calcium in the muscle fiber and activation of the contractile proteins (actin and muscular myosin II) that are responsible for the contraction or force generation. Acetylcholine, once released into the synaptic cleft, is rapidly inactivated by the enzyme acetylcholine esterase to ensure a brief activation of the muscle fiber and in order to restore the system for further activation from the motor neuron.

15.3.2 Muscle Fiber End Plate

The neuromuscular junction is highly specialized for efficient transmission of signals from the motor neuron to the muscle fiber. In the end-plate region, the muscle fiber membrane contains folds (Figure 15.8) about 0.1 μm wide and 1 μm deep, spaced at 1–3 μm intervals and aligned with active zones in the presynaptic nerve terminal where acetylcholine release occurs (Sanes and Lichtman, 2001). These folds show certain resemblance to actin protrusions in the nerve growth cone with a well-developed actin cytoskeleton. Acetylcholine receptors are present at high concentrations (>10,000/μm^2) (Fertuck and Salpeter, 1974) on the crests of the folds, whereas voltage-gated sodium channels, for generation of action potentials in the muscle fiber, are located deeper in the folds. The sodium channels are present at 10–20 times higher concentration in the end-plate region (about 2000/μm^2) than in extrasynaptic regions (Caldwell, 2000). Acetylcholine receptors on the crests of folds are clustered by the protein rapsyn and are associated with the actin cytoskeleton via adaptor scaffolds including, for example, α-actinin, and utrophin, a synapse-specific homologue of dystrophin (Apel and Merlie, 1995; Berthier and Blaineau, 1997; Mitsui et al., 2000; Hughes et al., 2006; Dobbins et al., 2008) (the important membrane protein related to Duchenne's muscular dystrophy). Voltage-gated sodium channels in the depth of the folds are instead associated with ankyrin and dystrophin. Utrophin, at the crests of folds (Bewick et al., 1992), and dystrophin, in the depth of the folds (Sealock et al., 1991), associate with a number of other proteins in a complex that links the cytoskeleton of the cell with the extracellular basal lamina (Banks et al., 2003). Neural cell-adhesion molecule (NCAM) is another membrane-bound glycoprotein localized to the depth of folds (Covault and Sanes, 1986). In addition to anchoring proteins, actin is also likely to direct vesicles with acetylcholine receptors to the end plate by actin-depolymerizing factor cofilin-directed trafficking (Lee et al., 2009) and/or myosin Va-dependent mechanisms (Roder et al., 2008, 2010; Rudolf et al., 2011).

15.3.3 Extracellular Matrix

Another important constituent of the neuromuscular junction is the extracellular matrix component known as the basal lamina that surrounds all the cellular components (Figure 15.8). The basal lamina is also present in the synaptic cleft where it even extends into the junctional folds (Sanes and Lichtman, 1999; Sanes, 2003). In comparison with extrasynaptic regions, the synaptic basal lamina contains a number of special features. Acetylcholine esterase, the enzyme that hydrolyzes acetylcholine to terminate its transmitter action, is attached to the synaptic basal lamina. Moreover, certain subunits of collagen IV (chains α3, α4, and α5) and laminins (α4, α5, and β2) occur specifically in the synaptic basal lamina, whereas other collagens (chains α1 and α2) as well as laminin β1 are excluded but occur extrasynaptically. Laminins containing α2 and γ1 chains occur at both sites (Sanes, 2003).

The importance of the synaptic basal lamina for structuring components of the neuromuscular junction has been illustrated in a number of self-healing experiments in which cellular components of the neuromuscular junction are allowed to regenerate after damage. In this regard, it is of interest to consider in some detail damages that lead to death of muscle fibers and removal of tissue debris followed by regeneration of new muscle fibers from precursor cells (satellite cells). Following certain types of damage, such regeneration of muscle fibers occurs within empty basal lamina sheaths that remain in the tissue and that have previously surrounded the original muscle fibers. If the muscle is damaged in such a manner and the nerve is also damaged and prevented from regeneration, new muscle fibers will regenerate without contact with a nerve. Despite the absence of a nerve, the new muscle fibers develop an end-plate region with folds and clusters of acetylcholine receptors exactly at the position where the original end plate was located (McMahan et al., 1980; McMahan and Slater, 1984). Conversely, if new muscle fibers are prevented from forming (by X-irradiation) but a damaged nerve is allowed to regenerate, the nerve will grow toward the previous synaptic region of the empty basal lamina sheaths and acquire concentrations of synaptic vesicles and active zones typical of differentiated nerve terminals (McMahan et al., 1980). In similar experiments, voltage-gated sodium channels have also been shown to accumulate in regenerating end plates without neural influence (Lupa and Caldwell, 1994). Presumably, in all these situations, information remaining in the basal lamina after damage is enough to direct differentiation of presynaptic as well as postsynaptic elements of the neuromuscular junction. Possible stop signals for growing axons include laminin 11 (α5β2γ1) and certain semaphorins such as semaphorin 6C (Patton et al., 1997; Svensson et al., 2008). The previous results demonstrate the pivotal role of communication between the extracellular matrix and the motion-generating machinery (cytoskeleton and molecular motors) that drives the outgrowth of a new axon and the formation of a new synaptic terminal (see the following).

One component of the basal lamina that is important for the clustering of acetylcholine receptors is the heparan sulfate proteoglycan agrin (Sanes and Lichtman, 2001). Agrin is synthesized by motor neurons and is transported in the axons and released at nerve terminals (see Sanes and Lichtman, 1999). Agrin is suggested to induce a tyrosine phosphorylation of the β subunit of acetylcholine receptors, which has been demonstrated to be important for the linking of acetylcholine receptors to the cytoskeleton (Borges and Ferns, 2001). The receptor for agrin appears to be Lrp4 (low-density lipoprotein receptor-related protein 4), which forms a complex with MuSK (a muscle-specific kinase) and mediates MuSK activation by agrin (Kim et al., 2008; Zhang et al., 2008). Acetylcholine receptor clustering induced by agrin, as well as by laminin, is mediated by activation of the Rho-GTPases Rho, Rac, and Cdc42 (Weston et al., 2000, 2003, 2007). Rho-GTPases are activated by upstream guanine nucleotide exchange factors (GEF), and one such RhoGEF, ephexin1, has recently been shown to regulate the stability of acetylcholine receptor clusters. Interestingly, the most downstream parts of these signaling networks exhibit great similarities to those existing in the nerve growth cone (Shi et al., 2010a,b) (see the following; Figure 15.9).

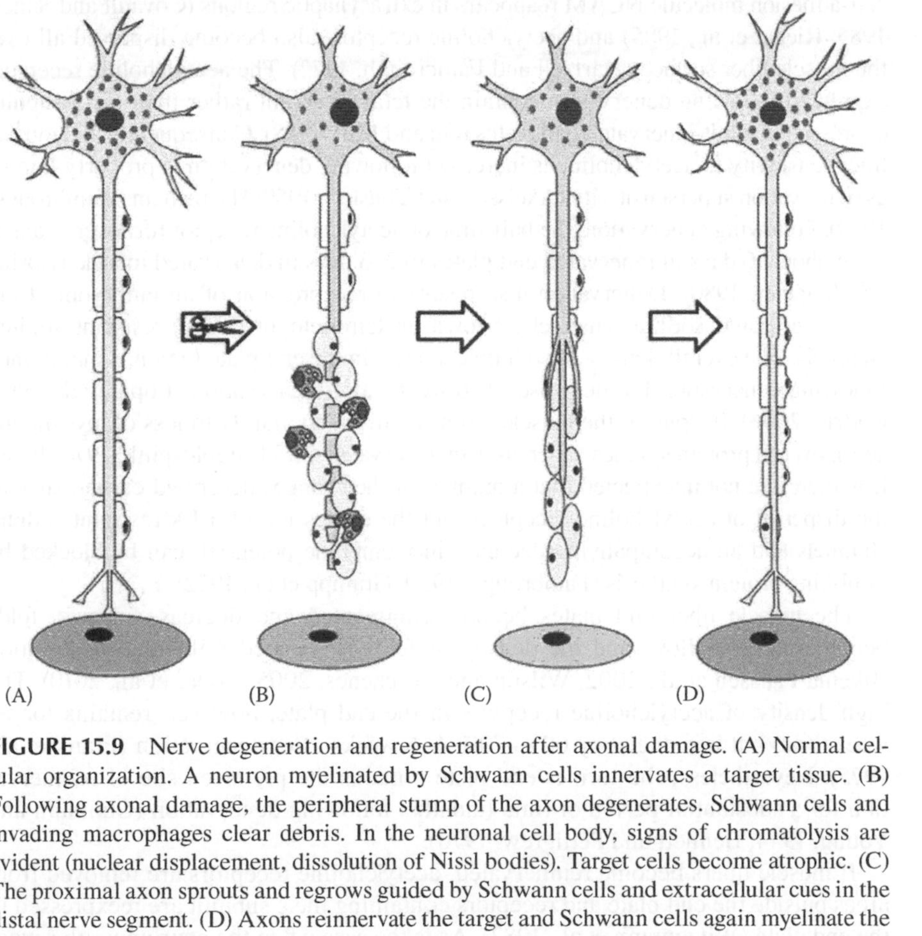

FIGURE 15.9 Nerve degeneration and regeneration after axonal damage. (A) Normal cellular organization. A neuron myelinated by Schwann cells innervates a target tissue. (B) Following axonal damage, the peripheral stump of the axon degenerates. Schwann cells and invading macrophages clear debris. In the neuronal cell body, signs of chromatolysis are evident (nuclear displacement, dissolution of Nissl bodies). Target cells become atrophic. (C) The proximal axon sprouts and regrows guided by Schwann cells and extracellular cues in the distal nerve segment. (D) Axons reinnervate the target and Schwann cells again myelinate the axon. The neuronal cell body and the target tissue return to their initial states. (Reproduced from *Prog. Neurobiol.*, 82, Navarro, X., Vivo, M., and Valero-Cabre, A., Neural plasticity after peripheral nerve injury and regeneration, 163–201, Copyright 2007, with permission from Elsevier.)

15.3.4 Denervation Changes in Muscle Fibers

When a skeletal muscle loses contact with its innervating nerve (denervation), it loses its ability to contract in response to nerve stimulation and consequently becomes inactive although some spontaneous electrical activity and associated contractions (fibrillations) occur in denervated muscle fibers (Purves and Sakmann, 1974; Smith and Thesleff, 1976). Denervation leads to a number of changes in all cellular components of the neuromuscular junction, for example, atrophy of muscle fibers, degeneration and regeneration of the damaged nerve, and activation of Schwann cells. Many of the alterations in the properties of muscle fibers that occur following denervation are likely aimed at establishing optimal conditions for the self-healing process that leads to reinnervation of the muscle fibers. In several ways, the muscle fibers following denervation revert to a protein expression pattern resembling that of developing muscle fibers. Thus, the cell-adhesion molecule NCAM reappears in extrasynaptic regions (Covault and Sanes, 1985; Rieger et al., 1985) and acetylcholine receptors also become dispersed all over the muscle fiber surface (Hartzell and Fambrough, 1972). The acetylcholine receptors expressed following denervation contain the fetal γ subunit rather than the ε subunit expressed in adult-innervated end plates (Gu and Hall, 1988). Consequently, the muscle fiber sensitivity to acetylcholine is increased following denervation, a property known as denervation supersensitivity (Axelsson and Thesleff, 1959; Hartzell and Fambrough, 1972). Following denervation, the half-time of acetylcholine receptor turnover changes from about 10 days in innervated end plates to 2–3 days in denervated muscle (Loring and Salpeter, 1980). Denervation also results in reexpression of an embryonic form of voltage-gated sodium channels, known as tetrodotoxin (TTX)-resistant sodium channels. The overall density of sodium channels in the end-plate region is maintained following denervation but decreases 4–6 weeks after denervation (Lupa et al., 1995; Midrio, 2006). In spite of the muscle atrophy, attributed mainly to loss of myosin and actin, overall protein synthesis increases in denervated muscle (Goldspink, 1976, 1978). It is therefore not unexpected that a number of the changes described earlier, such as the dispersal of acetylcholine receptors and the expression of TTX-resistant sodium channels and an accompanying decrease in membrane potential, can be blocked by inhibiting protein synthesis (Fambrough, 1970; Grampp et al., 1972).

The muscle fiber end plates become fragmented and decrease in size, folds become more shallow, and the density of folds is reduced following denervation (Ijkema-Paassen et al., 2002; Wilson and Deschenes, 2005; Roder et al., 2010). The high density of acetylcholine receptors in the end plate, however, remains for >2 weeks (rat) but is reduced by 60%–70% 4–6 weeks after denervation (Frank et al., 1975). Nevertheless, the former end plate remains the preferred site of reinnervation for a substantial period of time (months) following denervation (Gutmann and Young, 1944; Bennett and Pettigrew, 1976).

If muscle fibers become reinnervated, acetylcholine receptors are removed from areas outside the end plate and receptors containing the ε subunit are reexpressed in the end plate (Witzemann et al., 1987). As is the case when the neuromuscular junction develops, this is likely to be regulated by a number of neurally derived factors including neuregulin1/ARIA acting on its receptors Erbb2, Erbb3, and Erbb4 as well as calcitonin gene-related peptide (CGRP) and its receptor. ARIA, released from the nerve, is believed to be at least partly responsible for transcription of synapse-specific genes (e.g., acetylcholine receptor subunits and utrophin) in synaptic nuclei (Chu et al., 1995; Sandrock et al., 1997; Sanes and Lichtman, 1999). The peptide, CGRP, which is also released from motor neurons, has recently been shown to block the fragmentation of end plates following denervation, an action dependent on myosin Va (Roder et al., 2010). While the expression of synapse-specific genes and the maintenance of the end plate appear to depend on neural factors such as agrin, ARIA, and CGRP, the repression of transcription of, for example, acetylcholine receptor subunit genes in extrasynaptic nuclei is believed to result from the muscle activity induced by acetylcholine signaling from the nerve (Sanes and Lichtman, 1999).

15.3.5 Peripheral Nerve Regeneration

Injury to peripheral nerves (Figure 15.9) leads to degeneration of the distal parts of the injured axons, a process known as Wallerian degeneration (see Stoll et al., 2002). Although the distal part of the axon degenerates, the basement membrane together with connective tissue remains as an endoneurial tube (see Campbell, 2008). In response to injury, Schwann cells are modulated and reacquire the ability to proliferate and to migrate but also gain ability of phagocytosis and may, together with invading macrophages, be involved in removing debris from the degenerating axons (Fenrich and Gordon, 2004). Proliferation of Schwann cells as a response to denervation leads to the formation of the so-called band of Büngner (Schwann cell tubes), a line up of Schwann cells within the endoneurial tube (see Campbell, 2008).

Early after injury, the proximal end of the neuron is sealed. This is a process that depends on molecular motors and the actin cytoskeleton as this type of sealing could occur by exocytosis with addition of new membrane. However, this type of cellular wound healing has also, in some cases, been associated with contraction of an actomyosin contractile ring (Bement, 2002). A number of retrograde signals are generated in axons following injury including rapid signals such as high-frequency action potential discharges as well as slower signals dependent on molecular motors mediating retrograde axonal transport (Fenrich and Gordon, 2004; Navarro et al., 2007; Abe and Cavalli, 2008; Rishal and Fainzilber, 2010). It is of interest to note, in this connection, that the neuronal cell body displays a number of characteristic changes following axotomy. These are collectively referred to as chromatolysis and include movement of the nucleus from the center of the cell, dispersion of Nissl bodies (ribosome clusters), and nuclear and nucleolar enlargement. The changes indicate a transformation from a state of transmitting electrical signals into a state of regeneration and are associated with increased protein synthesis and an up-regulation of regeneration-associated genes such as tubulin and actin. Retrograde transport of injury signals, for example, extracellular signal-regulated kinase (ERK) and c-jun N-terminal kinase (JNK), via interactions with the retrograde molecular motors (dynein/dynactin) in the axon is believed to regulate the events leading to regeneration (Abe and Cavalli, 2008).

Following these responses, the proximal part of the axon extends processes (sprouts) within the band of Büngner or along Schwann cell processes to reinnervate target tissue (Figure 15.9). Axons regenerate at 1–3 mm/day, which is the rate of migration of the leading edge the growth cone. Interestingly, this rate is similar to that of slow axonal transport (Fenrich and Gordon, 2004) consistent with the idea that this process is rate limiting for axonal regeneration. In the neuromuscular system, the new growth cone will eventually target the axon to, in most cases, the original synaptic site on muscle fibers (see Sanes and Lichtman, 1999; Sanes, 2003).

Following denervation of skeletal muscle, terminal Schwann cells also give rise to sprouts from the end plate. Subsequent to partial denervation, where not all muscle fibers are denervated, the Schwann cells may extend their sprouts to end plates on adjacent innervated muscle fibers. When Schwann cell processes reach an innervated end plate, sprouting is induced from the innervating axon and the sprouts grow along the Schwann cell processes to reinnervate the denervated end plate (Figure 15.10) (Son and Thompson, 1995, see also Kang et al., 2003).

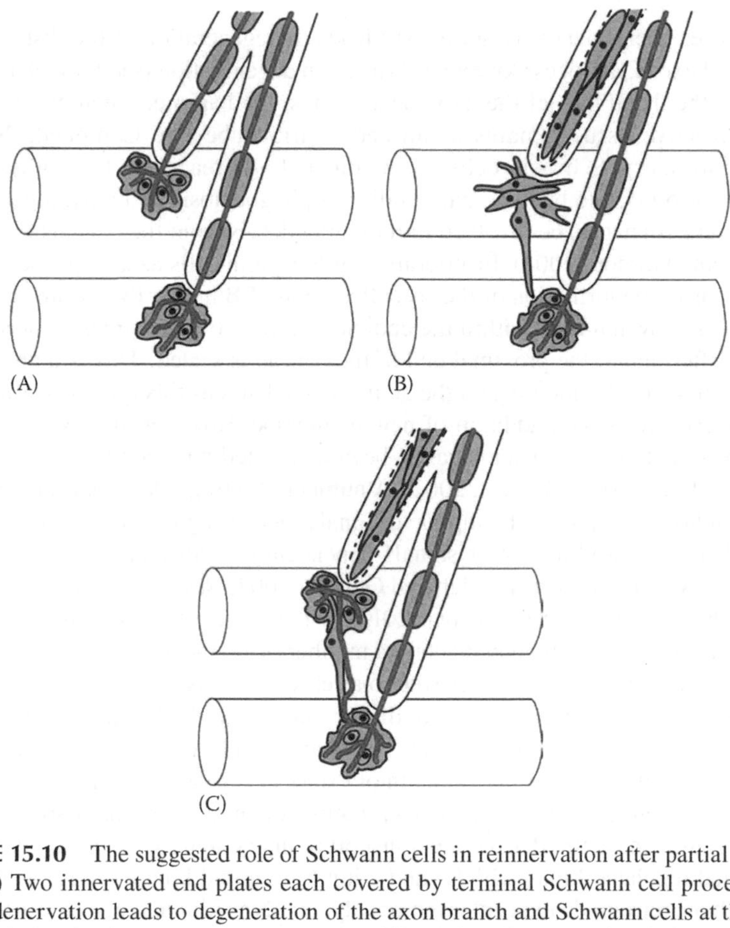

FIGURE 15.10 The suggested role of Schwann cells in reinnervation after partial denervation. (A) Two innervated end plates each covered by terminal Schwann cell processes. (B) Partial denervation leads to degeneration of the axon branch and Schwann cells at the denervated end plate begin to extend processes that (C) reach an innervated end plate on an adjacent muscle fiber. The innervating axonal branch also gives rise to sprouts that grow along the Schwann cell processes to innervate the denervated end plate, thus restoring functional neuromuscular communication. (With kind permission from Springer Science+Business Media: *J. Neurocytol.*, Terminal Schwann cells guide the reinnervation of muscle after nerve injury, 32, 2003, 975–985, Kang, H., Tian, L., and Thompson, W.)

15.3.6 Summary

The self-healing processes following nerve injury are, to a large extent, effectuated by molecular motors and the cytoskeleton, for example, (1) sealing of the proximal end of the axon at the site of injury, (2) formation of a growth cone, (3) outgrowth of new axons with the axon growth cone finding its way back toward the motor end plate, and finally (4) the formation of a new synapse with specialized organization both presynaptically (at the nerve terminal) and postsynaptically (at the motor end plate). It is clear that such complex events must be under detailed and synchronized control. In the following, we consider this control with focus on how it converges on the cytoskeleton and molecular motors. Moreover, we describe in detail how the concerted action of these motors and the cytoskeleton effectuate growth-cone motility while also, themselves, taking part in complex signaling networks with sensing of the extracellular matrix and transportation of growth-related factors to the cell body.

15.4 ROLE OF MOLECULAR MOTORS AND THE CYTOSKELETON DURING GROWTH OF A NERVE AXON

The following account describes the function of a generic axonal growth cone based on experimental evidence from a range of different neurons and species as diverse as neurons from the sea slug Aplysia to human neural cell lines. We thus make the simplifying assumption that these different neurons are similar. Naturally, this introduces uncertainties but we feel that these are warranted considering our aim to describe the principles of the complex sensory–migratory machine of the growth cone.

The motile and sensory leading edge of a growing axon, the growth cone, is of pivotal importance during development and regeneration after injury. However, one also needs to consider the, usually, very long (up to 1 m) and thin (in the order of 1–10 μm) major part of the axon (the axon shaft) that separates the growth cone from the cell body. In spite of possible axonal protein synthesis, this part is critical for signaling about peripheral damage to the cell body in order to initiate increased rate of protein synthesis supplying the growth cone with new organelles and macromolecular components for sustained growth (Fenrich and Gordon, 2004; Navarro et al., 2007; Abe and Cavalli, 2008; Cosker et al., 2008; Hirokawa et al., 2010; Rishal and Fainzilber, 2010). Since this must occur via active axonal transport, due to the appreciable length of the axon, it is not surprising that the center of the axon shaft is dominated by ordered bundles (Figure 15.11) of microtubules with their plus end at the axon terminal. This ordered arrangement depends on cross-linking by microtubule-associated proteins (MAP; Figure 15.4) and it makes the microtubules excellent ordered highways for anterograde and retrograde transport by plus end– and minus end–directed microtubule-based motors, respectively. In addition to the microtubules, the axon shaft also contains structurally important neurofilaments (a type of intermediate filaments) and mainly peripherally located actin filaments. The microtubule bundles diverge at the border between the axon shaft and the most proximal parts of the growth cone, the wrist and the central compartment (Figure 15.11). The latter is also filled with mitochondria. Moreover, the microtubule bundles, becoming

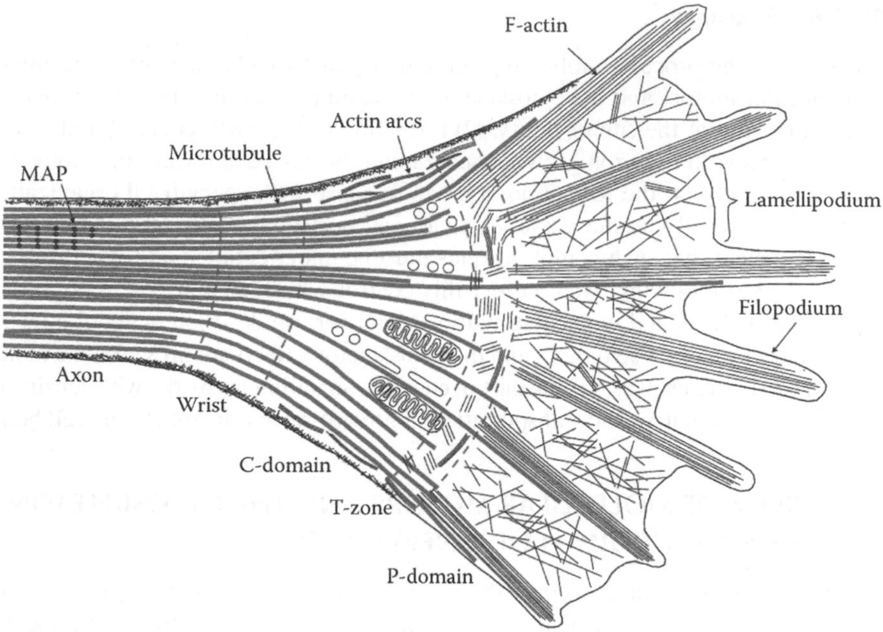

FIGURE 15.11 Nerve growth cone with the actin and microtubule cytoskeletons in focus. The width of the axon shaft to the left is on the order of 1–10 μm depending on the exact neuron type. (Reproduced from Geraldo, S. and Gordon-Weeks, P.R., *J. Cell Sci.*, 122, 3595, 2009. With permission from The Company of Biologists.)

sparse toward the distal end of the central compartment are, in this location, surrounded by arcs of actomyosin contractile bundles, presumably being composed of bipolar filaments containing nonmuscular myosin IIa/b. These probably play a role in compressing the microtubule lattice, thereby facilitating bundle formation when this lattice enters the axon shaft (Geraldo and Gordon-Weeks, 2009; Lowery and Van Vactor, 2009). This is particularly important when the axon shaft extends following propulsion of the leading edge of the growth cone that contains the actual motile machinery. At the very front are finger-like filopodia with parallel bundles of actin filaments that extend into the more proximal lamellopodium (Geraldo and Gordon-Weeks, 2009; Lowery and Van Vactor, 2009). These parallel actin bundles seem to interact with a few microtubules (Geraldo and Gordon-Weeks, 2009) extending the entire way into the P-domain (Figure 15.11) of the growth cone. The filopodia are the main sensory structures of the growth cone with cell surface receptors for growth factors (Tani et al., 2005) and other, facilitating or inhibitory cues (Dickson, 2002; Bonanomi and Pfaff, 2010). The latter may be diffusible (e.g., netrins and certain semaphorins) forming gradients that can be either inhibitory or attractive. Moreover, the cytoskeleton is coupled to the underlying matrix via components (e.g., fibronectin, and certain laminins) that bind to cell-adhesion molecules (integrins, *N*-cadherins) or the matrix may contain molecules (e.g., slits or ephrins) that are nonadhesive (Lowery and Van Vactor, 2009). Here, the density and types of surface molecules determine the degree of adhesion of filopodia/lamellopodia to the matrix and propulsion of the growth cone (Aratyn-Schaus and Gardel, 2008; Bard et al., 2008; Gardel et al., 2010). Even the mechanical properties of the matrix itself may have effects on the mode of migration of the growth cone (Bard et al., 2008; Chan and Odde, 2008).

Depending on the guiding/inhibiting cues, the cytoskeleton and molecular motor function will modify to (1) retract the growth cone and stop movement, (2) stabilize (with increased adhesion to underlying surface) and advance the growth cone straight ahead, or (3) make the growth cone turn in one direction or the other. The process of retraction is associated with loss of adhesion of the filopodia to the underlying surface, retraction of the growth-cone tip by the action of myosin II motors in zone T (Figure 15.11) of the growth cone, and depolymerization of actin filaments at the minus end (Brown and Bridgman, 2009; Lowery and Van Vactor, 2009). The retraction of the growth cone is also associated with disassembly of the microtubules in the central zone of the growth cone and retrieval of membrane by endocytosis (Prager-Khoutorsky and Spira, 2009). If the cues are favorable for growth-cone progression, the cell signaling leads to maintenance and further strengthening of the surface adhesions while simultaneously leading to extension of the filopodia and progression of the leading edges of the lamellopodia. This process can be described on basis of the continuous ongoing addition of actin monomers by the action of formins and profilin at the leading end of the filopodia and, for example, by nucleating new branches in the lamellopodia actin networks (relying on the Arp 2/3 complex) (Chhabra and Higgs, 2007). Under stable, resting conditions without filopodia and lamellopodia protrusion, this addition of monomers is associated with continuous retrograde flow of actin monomers along the actin filament and depolymerization at the minus end (treadmilling) (Lowery and Van Vactor, 2009, and references therein). In addition to the push by the plasma membrane, an important role in producing the retrograde flow is played by the action of myosin II (Aratyn-Schaus and Gardel, 2008; Chan and Odde, 2008; Lowery and Van Vactor, 2009) or myosin I (Diefenbach et al., 2002) motors in the T-zone of the growth cone. Upon appropriate cues for filopodia protrusion, the filopodia adhere to the underlying surface, depending on a suitable presence of extracellular receptors (e.g., fibronectins, laminins) and cell-adhesion molecules (integrins, *N*-cadherin, etc.) in the cell membrane. This is believed to lead to the engagement of a clutch that via the cell-adhesion molecules and additional linkage proteins couples the actin cytoskeleton to the extracellular matrix (Aratyn-Schaus and Gardel, 2008; Lowery and Van Vactor, 2009). If this link is strong enough, the retrograde actin flow will be reduced and the elongating actin filament will push on the leading membrane at the same time as traction forces are applied to the extracellular matrix with propulsion of the filopodium.

If the external cues signal that turning in one direction is favorable, then the filopodia stabilize with increased adhesion and propulsion of their tip at this side while retracting at the other, causing turning of the growth cone toward the stimulating cues (Geraldo and Gordon-Weeks, 2009). For effective turning, it seems that interactions between the actin bundles from filopodia and microtubules extending into the growth-cone P-zone are important (Geraldo and Gordon-Weeks, 2009). However, the detailed mechanisms remain to be elucidated as well as several other aspects of the actin–microtubule interaction in growth-cone mobility. In addition to the turning of the growth cone toward certain extracellular cues, the filopodia also turn autonomously in order to probe the environment. This autonomous turning seems to rely on the activity of myosin V, illustrating one example where nonconventional myosins (other myosin motors than myosin II) play a role in growth-cone function (Tamada et al., 2010). It is also clear that myosin X exists at the front of filopodia with roles in their maintenance and formation (Zhang et al., 2004; Sousa and Cheney, 2005) possibly related to the accumulation of integrins. Other myosins that may be of importance in growth-cone function are myosin I and myosin VI. Although these myosins have been found in growth cones (Suter et al., 2000; Diefenbach et al., 2002) of some neurons, their functions have been studied only to a limited extent. However, evidence from dorsal root ganglion axonal growth cones (Diefenbach et al., 2002) suggest that myosin Ic may play a role in the retrograde flow of actin at the leading edge of lamellopodia.

Opposite to growth-cone retraction, growth-cone progression is associated with addition of new membrane by exocytosis and extension of the microtubule bundles with a similar rate as the progression of the growth-cone tip. This occurs both by addition of new monomers at the microtubule plus end and by transportation of microtubules from more proximal regions of the axon. In some neurons, it is important with transport of microtubules anterogradely along the axon following injury. This transport may actually be inhibited by certain kinesins and it has been found that pharmacological inhibition of these motors may promote axonal outgrowth after injury (Lin et al., 2011). Another pharmacological intervention to promote axonal outgrowth in the spinal cord was inhibition of myosin II to prevent growth-cone retraction under certain conditions (Hur et al., 2011).

Importantly, growth-cone progression requires continuous addition of new components, particularly cytoskeletal and membrane components. Thus, damage to the axon with degeneration of the terminal stump will require increased protein synthesis, mainly in the cell body (which can be a meter away from the site of damage) and increased axonal transport of the synthesized components. Upon synthesis, some of the proteins are likely to be transported in lipid vesicles (also contributing new material to the plasma membrane), whereas cytoskeletal components may be transported in filament form. Transport of lipid vesicles and mitochondria along axons has been associated with the fast axonal transport whereas, for example, cytoskeletal filaments are believed to be generally transported with the slow component. The availability of cytoskeletal components may thus be rate limiting on the rate of growth-cone migration (see the preceding text) since this appeared to be limited by the rate of the slow component of axonal transport. Anyway, it is clear that growth-cone progression cannot start until newly synthesized components have reached the site of the damage and then the rate of progression cannot be faster than the rate of delivery of protein and lipid components by anterograde axonal transport.

15.4.1 Summary

The preceding account describes the important motile and transportation events that are involved in regeneration and outgrowth of an injured axon, as in the case with the motor neuron axon. We have also summarized evidence (with external diffusible and matrix cues) that explain why the growth cone follows the path back toward the motor end plate that it previously innervated. In the following, we describe briefly how these cues, via cell signaling events, can influence the motion-generating machinery of the growth cone. We also relate the control of this motion-generating machinery to the control of motor and cytoskeletal activities (see also Section 15.2) responsible for the detailed organization of the neuromuscular synapse.

15.5 COORDINATION OF CYTOSKELETAL CHANGES AND MOLECULAR MOTORS BY CELL SIGNALING

The cell signaling events that start upon denervation are extremely complex and difficult to interpret without taking a system approach. The signaling thus involves intracellular signaling within several cell types (motor neuron, muscle fiber, Schwann cells, etc.) as well as signaling between these cells (by paracrine substances) and between the cells and the underlying extracellular matrix (e.g., laminins, collagen, fibronectin). Although the preceding text has considered some aspects of signaling both between and within cells, the following account indicates how cell signaling events converge to affect the cytoskeleton and molecular motors.

The upstream stimuli that affect the cytoskeletal organization are somewhat different in the growth cone and in the synaptic region. As outlined earlier they are either soluble diffusible cues that affect the cell by binding to cell surface receptors or cues present in the extracellular matrix that may affect the cytoskeleton via binding to cell-adhesion molecules such as integrins and *N*-cadherins. Interestingly, independent of whether the extracellular cue is diffusible or an integral constituent of the extracellular matrix or whether it acts on the growth cone or the motor end plate, there appears to be considerable overlap in the downstream signaling paths and effectors (Lowery and Van Vactor, 2009; Wu et al., 2010). This is illustrated schematically in Figure 15.12 where it can be seen that signaling via the Rho-GTPases (Rho, Rac, and Cdc42) is of central importance (although the details may vary) both at the neuromuscular junction and at the nerve growth cone. Moreover, altered Ca^{2+} concentrations play important roles (Henley and Poo, 2004; Wu et al., 2010) in the signaling. The different signaling paths affect cytoskeletal organization and force development (causing actin polymerization, depolymerization, and/or branching) via downstream influence over actin-binding proteins and myosin light-chain kinase that phosphorylates the myosin II light chains, thereby activating myosin II function. The coupling between the cytoskeleton and the extracellular matrix via cell-adhesion molecules can initiate cell signaling directly. For example, high forces may affect the conformation of the proteins that link actin filaments to the cell-adhesion molecules (e.g., integrins, *N*-cadherins) and cause these to activate, for example, focal adhesion kinase (FAK) with secondary effects on Rho-GTPases (Parsons et al., 2010) (Figure 15.12). However, also the dynamics of the cytoskeletal extracellular matrix clutch may be varied, for example, simply due to altered stiffness of the extracellular matrix without any alterations in cell-signaling cascades and phosphorylation status (Chan and Odde, 2008). Finally, the density and type of extracellular binding partners for the cell-adhesion molecules can act as a very direct signaling mechanism promoting growth or retraction of the growth

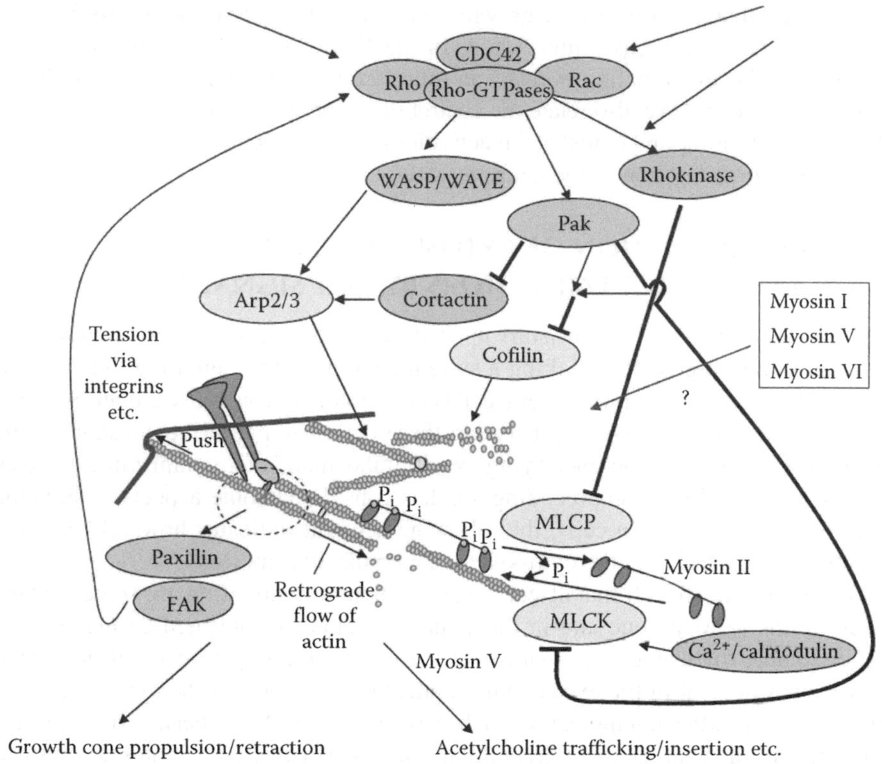

FIGURE 15.12 Generic scheme indicating the approximate organization in both growth cone, presynaptic terminal and motor end plate of key downstream elements in signaling networks that control myosin motors and cytoskeletal networks.

cone (Bard et al., 2008). Thus, this density needs to be sufficiently high to allow production of the traction forces on the extracellular matrix, the reduced rate of retrograde actin flow, and the propulsion of the leading edges of the filopodia and lamellopodia that are required for growth-cone propulsion.

It is important to emphasize that the preceding account only gives an overall simplified view of the intracellular mechanisms whereby various soluble factors and extracellular matrix–based ligands could control growth-cone motility and neuromuscular plasticity. This simplification is in line with the purpose of this review and further complexities can be found elsewhere. However, there are also several detailed mechanisms that remain to be fully elucidated. For example, it may be of interest in the future to take a system's view that incorporates the considerably larger and more strongly interconnected networks of signaling molecules as well as spatiotemporal differences in the signaling, for example, with compartmentalization of signaling paths and/or time of activation in different parts of the growth cone. Additionally, the actin–microtubule interactions (Geraldo and Gordon-Weeks, 2009; Lowery and Van Vactor, 2009) as well as the possible role of some unconventional myosins also need to be explored.

15.6 MOLECULAR MOTORS IN NANOBIOTECHNOLOGY

A striking feature of a process such as nerve regeneration is the enormous complexity involving a large range of protein effectors and, for example, proteins that bind to motors and cytoskeletal filament and that are regulated by a complex network of signaling cascades. To achieve something similar in nanotechnological applications would be an enormous challenge. However, there are ongoing efforts to explore the self-organizing properties of molecular motors and cytoskeletal components *in vitro* to create gradients (Vikhorev et al., 2008b), ordered surface patterns (Interliggi et al., 2007; Vikhorev et al., 2008b; Butt et al., 2010; Schaller et al., 2010), biosensing and sorting devices, and so on (Ramachandran et al., 2006; Sundberg et al., 2006; van den Heuvel and Dekker, 2007; Goel and Vogel, 2008; Lin et al., 2008; Fischer et al., 2009; Agarwal and Hess, 2010; Korten et al., 2010). Moreover, instead of external guidance cues the motors or their cytoskeletal filaments have been organized on the nanoscale *in vitro* by combining with appropriate nanostructured surfaces. Also, cell growth and neurite (such as axon) extension have been organized by allowing the cells to grow on various nanostructured surfaces (Johansson et al., 2006; Gustavsson et al., 2007). In the cell-free systems, the devices have generally been developed, for example using pure actin–myosin or microtubule–kinesin systems. It may be interesting in the future to integrate more actin-binding proteins and cell signaling molecules to reconstruct a dynamic actin–myosin network *in vitro.* However, one problem with such cell-free systems is actually that the components themselves have considerably reduced capability to self-heal.

For example, in applications involving myosin motors, the latter may denature without the access to cellular chaperones to refold them or the motors may oxidize due to poor antioxidation protection. The actin filaments, on the other hand, are sensitive to proteolysis and may lose their capability to form filaments or decay irreversibly into monomers due to insufficient stability and low concentration of free monomers in the solution. Ideally, a molecular motors–based nanodevice should have self-healing capability. How to achieve this in a controlled way is extremely challenging. However, evidence exists that certain drugs (molecular chaperones) may rescue proteins from unfolding, thus prolonging their active life. Moreover, oxidation that is presently prevented by adding reducing agents such as dithiothreitol (DTT) to molecular motors–based nanodevices may be more effectively prevented by exploration of cellular components that protect against oxidative damage. Another potential way to stabilize the molecular motors and their cytoskeletal filaments may be by adding some of the protein components that have stabilizing actions in the living cell. Thus, some of the proteins as well as small molecular substances that stabilize actin filaments against depolymerization also appear to stabilize actin against thermal unfolding (Levitsky et al., 2008). We showed quite recently that phosphorylation of the regulatory light chains of skeletal muscle myosin might reduce the interaction of the myosin motor domain with the underlying surface (Vikhoreva and Mansson, 2010). While possibly not causing healing of damaging structural changes, it may reduce the risk that such changes occur in the first place by reducing surface-induced denaturation.

In conclusion, although some aspects of the cellular self-healing machinery composed of molecular motors and cytoskeletal filaments are being exploited for nanotechnological applications today, the complexity is much less than for the cellular systems and these devices are themselves actually lacking a self-healing capacity. It would be of interest in the future to develop more complex devices *in vitro* that could be controlled more extensively, for example, by chemical gradients and where cellular strategies for self-healing could be implemented to increase the lifetime of the motor-based devices.

15.7 CONCLUSIONS

Neuromuscular communication is of vital importance for all kinds of voluntary body movements as well as for breathing and speaking. The neuromuscular system shows a remarkable ability for self-healing at least after certain types of damage. This requires the processing of large amounts of information by a number of different cell types that need to cooperate for successful self-healing. The required information is largely conveyed by molecular components present in the extracellular environment of the cells, either freely diffusible molecules or molecules bound to or making up the extracellular matrix. The signaling events are conveyed to the cell interior in various ways and often result in cytoskeletal rearrangements in the different cell types and in reactions that require the participation of a number of different types of molecular motors. After a crush damage to motor axons, the result of this self-healing process will be regrowth of the axons back to the denervated skeletal muscle fibers. The reinnervation of muscle fibers, which usually occurs at exactly the same spots (the motor end plates) where motor axons previously contacted the muscle fibers, demonstrates the remarkable precision of the self-healing process.

The most downstream cell signaling machinery that controls the cytoskeleton and molecular motor dynamics upon nerve regeneration and synapse formation has appreciable similarities in growth cones and in the presynaptic and postsynaptic parts of the synapse revealing a certain modularity of the control–effector machinery. The transport strategies also appear similar with long-distance transport (from nerve cell body or muscle fiber nuclei) by kinesin and dynein motors but with short-distance transport by myosin, either by processive transport or indirectly through rearrangements of the actin cytoskeleton.

To mimic the cellular self-healing processes in artificial systems would be a formidable task but limited aspects may be exploited or, indeed, mimicked. The challenges are attributed to several different facts. First, the processes take place far from equilibrium with constant supply of high-energy phosphates. Second, they rely on the action of a range of complex nanomachines, most notably the cytoskeleton and molecular motors that are under coordinated control by the cell signaling machinery and, on a higher level, by signaling between cells and between cells and the extracellular matrix.

ACKNOWLEDGMENTS

This work was funded by contributions from The Swedish Research Council (Project # 621-2010-5146) and the Faculty of Natural Sciences and Engineering at Linnaeus University. We apologize to those authors whose original work has not been cited.

Due to the broad scientific field covered, the references are dominated by recent review articles or books where original citations can be found. Limited citations to key original references are given, partly biased by our own main research interests.

REFERENCES

Abe, N. and Cavalli, V., Nerve injury signaling. *Curr Opin Neurobiol* 2008, 18, 276–283.

Agarwal, A. and Hess, H., Biomolecular motors at the intersection of nanotechnology and polymer science. *Prog Polymer Sci* 2010, 35, 252–277.

Akhmanova, A. and Steinmetz, M. O., Microtubule +TIPs at a glance. *J Cell Sci* 2010, 123, 3415–3419.

Albet-Torres, N., Bloemink, M. J., Barman, T. et al., Drug effect unveils inter-head cooperativity and strain-dependent ADP release in fast skeletal actomyosin. *J Biol Chem* 2009, 284, 22926–22937.

Amendola, V. and Meneghetti, M., Self-healing at the nanoscale. *Nanoscale* 2009, 1, 74–88.

Apel, E. D. and Merlie, J. P., Assembly of the postsynaptic apparatus. *Curr Opin Neurobiol* 1995, 5, 62–67.

Aratyn-Schaus, Y. and Gardel, M. L., Biophysics clutch dynamics. *Science* 2008, 322, 1646–1647.

Axelsson, J. and Thesleff, S., A study of supersensitivity in denervated mammalian skeletal muscle. *J Physiol* 1959, 147, 178–193.

Bakewell, D. J. G. and Nicolau, D. V., Protein linear molecular motor-powered nanodevices. *Aust J Chem* 2007, 60, 314–332.

Balaz, M., Interaction of actomyosin with synthetic materials—Effects on motor function and potential for exploitation. University of Kalmar, Kalmar, Sweden, p. 202, 2008.

Banks, G. B., Fuhrer, C., Adams, M. E., and Froehner, S. C., The postsynaptic submembrane machinery at the neuromuscular junction: Requirement for rapsyn and the utrophin/dystrophin-associated complex. *J Neurocytol* 2003, 32, 709–726.

Barber, J. and Andersson, B., Too much of a good thing: Light can be bad for photosynthesis. *Trends Biochem Sci* 1992, 17, 61–66.

Bard, L., Boscher, C., Lambert, M., Mege, R. M., Choquet, D., and Thoumine, O., A molecular clutch between the actin flow and *N*-cadherin adhesions drives growth cone migration. *J Neurosci* 2008, 28, 5879–5890.

Bement, W. M., Actomyosin rings: The riddle of the sphincter. *Curr Biol* 2002, 12, R12–R14.

Bennett, M. R. and Pettigrew, A. G., The formation of neuromuscular synapses. *Cold Spring Harb Symp Quant Biol* 1976, 40, 409–424.

Berthier, C. and Blaineau, S., Supramolecular organization of the subsarcolemmal cytoskeleton of adult skeletal muscle fibers. A review. *Biol Cell* 1997, 89, 413–434.

Bewick, G. S., Nicholson, L. V., Young, C., O'Donnell, E., and Slater, C. R., Different distributions of dystrophin and related proteins at nerve-muscle junctions. *Neuroreport* 1992, 3, 857–860.

Blanchoin, L. and Pollard, T. D., Hydrolysis of ATP by polymerized actin depends on the bound divalent cation but not profilin. *Biochemistry* 2002, 41, 597–602.

Bonanomi, D. and Pfaff, S. L., Motor axon pathfinding. *Cold Spring Harb Perspect Biol* 2010, 2, a001735.

Borges, L. S. and Ferns, M., Agrin-induced phosphorylation of the acetylcholine receptor regulates cytoskeletal anchoring and clustering. *J Cell Biol* 2001, 153, 1–12.

Brown, J. A. and Bridgman, P. C., Disruption of the cytoskeleton during semaphorin 3A induced growth cone collapse correlates with differences in actin organization and associated binding proteins. *Dev Neurobiol* 2009, 69, 633–646.

Brown, M. E. and Bridgman, P. C., Myosin function in nervous and sensory systems. *J Neurobiol* 2004, 58, 118–130.

Bugyi, B. and Carlier, M. F., Control of actin filament treadmilling in cell motility. *Annu Rev Biophys* 2010, 39, 449–470.

Butt, T., Mufti, T., Humayun, A., Rosenthal, P. B., Khan, S., and Molloy, J. E., Myosin motors drive long range alignment of actin filaments. *J Biol Chem* 2010, 285, 4964–4974.

Caldwell, J. H., Clustering of sodium channels at the neuromuscular junction. *Microsc Res Tech* 2000, 49, 84–89.

Campbell, W. W., Evaluation and management of peripheral nerve injury. *Clin Neurophysiol* 2008, 119, 1951–1965.

Carter, A. P., Cho, C., Jin, L., and Vale, R. D., Crystal structure of the dynein motor domain. *Science* 2011, 331, 1159–1165.

Chan, C. E. and Odde, D. J., Traction dynamics of filopodia on compliant substrates. *Science* 2008, 322, 1687–1691.

Chhabra, E. S. and Higgs, H. N., The many faces of actin: Matching assembly factors with cellular structures. *Nat Cell Biol* 2007, 9, 1110–1121.

Chu, G. C., Moscoso, L. M., Sliwkowski, M. X., and Merlie, J. P., Regulation of the acetylcholine receptor epsilon subunit gene by recombinant ARIA: An in vitro model for transynaptic gene regulation. *Neuron* 1995, 14, 329–339.

Clark, A. G., Miller, A. L., Vaughan, E., Yu, H. Y. E., Penkert, R., and Bement, W. M., Integration of single and multicellular wound responses. *Curr Biol* 2009, 19, 1389–1395.

Cosker, K. E., Courchesne, S. L., and Segal, R. A., Action in the axon: Generation and transport of signaling endosomes. *Curr Opin Neurobiol* 2008, 18, 270–275.

Covault, J. and Sanes, J. R., Distribution of N-CAM in synaptic and extrasynaptic portions of developing and adult skeletal muscle. *J Cell Biol* 1986, 102, 716–730.

Covault, J. and Sanes, J. R., Neural cell adhesion molecule (N-CAM) accumulates in denervated and paralyzed skeletal muscles. *Proc Natl Acad Sci USA* 1985, 82, 4544–4548.

Cross, R. A., Molecular motors: The natural economy of kinesin. *Curr Biol* 1997, 7, R631–R633.

Dent, E. W., Merriam, E. B., and Hu, X., The dynamic cytoskeleton: Backbone of dendritic spine plasticity. *Curr Opin Neurobiol* 2011, 21, 175–181.

Dickson, B. J., Molecular mechanisms of axon guidance. *Science* 2002, 298, 1959–1964.

Diefenbach, T. J., Latham, V. M., Yimlamai, D., Liu, C. A., Herman, I. M., and Jay, D. G., Myosin 1c and myosin IIB serve opposing roles in lamellipodial dynamics of the neuronal growth cone. *J Cell Biol* 2002, 158, 1207–1217.

Dobbins, G. C., Luo, S., Yang, Z., Xiong, W. C., and Mei, L., Alpha-actinin interacts with rapsyn in agrin-stimulated AChR clustering. *Mol Brain* 2008, 1, 18.

Egelman, E. H., Francis, N., and DeRosier, D. J., F-Actin is a helix with a random variable twist. *Nature* 1982, 298, 131–135.

Egelman, E. H. and Orlova, A., Allostery, cooperativity, and different structural states in F-actin. *J Struct Biol* 1995, 115, 159–162.

Endow, S. A., Kull, F. J., and Liu, H., Kinesins at a glance. *J Cell Sci* 2010, 123, 3420–3424.

Erickson, H. P., Microtubule surface lattice and subunit structure and observations on reassembly. *J Cell Biol* 1974, 60, 153–167.

Etienne-Manneville, S., From signaling pathways to microtubule dynamics: The key players. *Curr Opin Cell Biol* 2010, 22, 104–111.

Fambrough, D. M., Acetylcholine sensitivity of muscle fiber membranes: Mechanism of regulation by motoneurons. *Science* 1970, 168, 372–373.

Fenrich, K. and Gordon, T., Canadian Association of Neuroscience review: Axonal regeneration in the peripheral and central nervous systems—Current issues and advances. *Can J Neurol Sci* 2004, 31, 142–156.

Fertuck, H. C. and Salpeter, M. M., Localization of acetylcholine receptor by ^{125}I-labeled alpha-bungarotoxin binding at mouse motor endplates. *Proc Natl Acad Sci USA* 1974, 71, 1376–1378.

Fischer, T., Agarwal, A., and Hess, H., A smart dust biosensor powered by kinesin motors. *Nat Nanotechnol* 2009, 4, 162–166.

Fischer, T. and Hess, H., Materials chemistry challenges in the design of hybrid bionanodevices: Supporting protein function within artificial environments. *J Mater Chem* 2007, 17, 943–951.

Foth, B. J., Goedecke, M. C., and Soldati, D., New insights into myosin evolution and classification. *Proc Natl Acad Sci USA* 2006, 103, 3681–3686.

Frank, E., Gautvik, K., and Sommerschild, H., Cholinergic receptors at denervated mammalian motor end-plates. *Acta Physiol Scand* 1975, 95, 66–76.

Fujii, T., Iwane, A. H., Yanagida, T., and Namba, K., Direct visualization of secondary structures of F-actin by electron cryomicroscopy. *Nature* 2010, 467, 724–728.

Gardel, M. L., Schneider, I. C., Aratyn-Schaus, Y., and Waterman, C. M., Mechanical integration of actin and adhesion dynamics in cell migration. *Annu Rev Cell Dev Biol* 2010, 26, 315–333.

Geeves, M. A., Fedorov, R., and Manstein, D. J., Molecular mechanism of actomyosin-based motility. *Cell Mol Life Sci* 2005, 62, 1462–1477.

Gennerich, A. and Vale, R. D., Walking the walk: How kinesin and dynein coordinate their steps. *Curr Opin Cell Biol* 2009, 21, 59–67.

Geraldo, S. and Gordon-Weeks, P. R., Cytoskeletal dynamics in growth-cone steering. *J Cell Sci* 2009, 122, 3595–3604.

Gittes, F., Mickey, B., Nettleton, J., and Howard, J., Flexural rigidity of microtubules and actin-filaments measured from thermal fluctuations in shape. *J Cell Biol* 1993, 120, 923–934.

Goel, A. and Vogel, V., Harnessing biological motors to engineer systems for nanoscale transport and assembly. *Nat Nanotechnol* 2008, 3, 465–475.

Goldspink, D. F., The effects of denervation on protein turnover of rat skeletal muscle. *Biochem J* 1976, 156, 71–80.

Goldspink, D. F., The effects of denervation on protein turnover of the soleus and extensor digitorum longus muscles of adult mice. *Comp Biochem Physiol B* 1978, 61, 37–41.

Grampp, W., Harris, J. B., and Thesleff, S., Inhibition of denervation changes in skeletal muscle by blockers of protein synthesis. *J Physiol* 1972, 221, 743–754.

Gu, Y. and Hall, Z. W., Characterization of acetylcholine receptor subunits in developing and in denervated mammalian muscle. *J Biol Chem* 1988, 263, 12878–12885.

Gunawardena, S. and Goldstein, L. S. B., Cargo-carrying motor vehicles on the neuronal highway: Transport pathways and neurodegenerative disease. *J Neurobiol* 2004, 58, 258–271.

Gustavsson, P., Johansson, F., Kanje, M., Wallman, L., and Linsmeier, C. E., Neurite guidance on protein micropatterns generated by a piezoelectric microdispenser. *Biomaterials* 2007, 28, 1141–1151.

Gutmann, E. and Young, J. Z., The re-innervation of muscle after various periods of atrophy. *J Anat* 1944, 78, 15–43.

Hall, Z. W., Lubit, B. W., and Schwartz, J. H., Cytoplasmic actin in postsynaptic structures at the neuromuscular junction. *J Cell Biol* 1981, 90, 789–792.

Hall, Z. W. and Sanes, J. R., Synaptic structure and development: The neuromuscular junction. *Cell* 1993, 72(Suppl), 99–121.

Hartzell, H. C. and Fambrough, D. M., Acetylcholine receptors. Distribution and extrajunctional density in rat diaphragm after denervation correlated with acetylcholine sensitivity. *J Gen Physiol* 1972, 60, 248–262.

Henley, J. and Poo, M. M., Guiding neuronal growth cones using Ca^{2+} signals. *Trends Cell Biol* 2004, 14, 320–330.

Hershko, A. and Ciechanover, A., The ubiquitin system. *Annu Rev Biochem* 1998, 67, 425–479.

Hirokawa, N., Niwa, S., and Tanaka, Y., Molecular motors in neurons: Transport mechanisms and roles in brain function, development, and disease. *Neuron* 2010, 68, 610–638.

Howard, J., *Mechanics of Motor Proteins and the Cytoskeleton*. Sinauer Associates Inc.: Sunderland, MA, 2001.
Hughes, B. W., Kusner, L. L., and Kaminski, H. J., Molecular architecture of the neuromuscular junction. *Muscle Nerve* 2006, 33, 445–461.
Hur, E. M., Yang, I. H., Kim, D. H. et al., Engineering neuronal growth cones to promote axon regeneration over inhibitory molecules. *Proc Natl Acad Sci USA* 2011, 108, 5057–5062.
Ijkema-Paassen, J., Meek, M. F., and Gramsbergen, A., Reinnervation of muscles after transection of the sciatic nerve in adult rats. *Muscle Nerve* 2002, 25, 891–897.
Inoue, A., Saito, J., Ikebe, R., and Ikebe, M., Myosin IXb is a single-headed minus-end-directed processive motor. *Nat Cell Biol* 2002, 4, 302–306.
Interliggi, K. A., Zeile, W. L., Ciftan-Hens, S. A., McGuire, G. E., Purich, D. L., and Dickinson, R. B., Guidance of actin filament elongation on filament-binding tracks. *Langmuir* 2007, 23, 11911–11916.
Isambert, H., Venier, P., Maggs, A. C. et al., Flexibility of actin filaments derived from thermal fluctuations. Effect of bound nucleotide, phalloidin, and muscle regulatory proteins. *J Biol Chem* 1995, 270, 11437–11444.
Iwaki, M., Tanaka, H., Iwane, A. H., Katayama, E., Ikebe, M., and Yanagida, T., Cargo-binding makes a wild-type single-headed myosin-VI move processively. *Biophys J* 2006, 90, 3643–3652.
Johansson, F., Carlberg, P., Danielsen, N., Montelius, L., and Kanje, M., Axonal outgrowth on nano-imprinted patterns. *Biomaterials* 2006, 27, 1251–1258.
Kabsch, W., Mannherz, H. G., Suck, D., Pai, E. F., and Holmes, K. C., Atomic structure of the actin: DNase I complex [see comments]. *Nature* 1990, 347, 37–44.
Kandel, E. R., Schwartz, J. H., and Jessell, T. H., *Essentials of Neural Science and Behavior*. Appleton & Lange: Stamford, CT, 1995.
Kang, H., Tian, L., and Thompson, W., Terminal Schwann cells guide the reinnervation of muscle after nerve injury. *J Neurocytol* 2003, 32, 975–985.
Kim, N., Stiegler, A. L., Cameron, T. O. et al., Lrp4 is a receptor for agrin and forms a complex with MuSK. *Cell* 2008, 135, 334–342.
Kinbara, K. and Aida, T., Toward intelligent molecular machines: Directed motions of biological and artificial molecules and assemblies. *Chem Rev* 2005, 105, 1377–1400.
Korten, T., Månsson, A., and Diez, S., Towards the application of cytoskeletal motor proteins in molecular detection and diagnostic devices. *Curr Opin Biotechnol* 2010, 21, 477–488.
Kozuka, J., Yokota, H., Arai, Y., Ishii, Y., and Yanagida, T., Dynamic polymorphism of single actin molecules in the actin filament. *Nat Chem Biol* 2006, 2, 83–86.
Kull, F. J., Sablin, E. P., Lau, R., Fletterick, R. J., and Vale, R. D., Crystal structure of the kinesin motor domain reveals a structural similarity to myosin. *Nature* 1996, 380, 550–555.
Lee, S. H. and Dominguez, R., Regulation of actin cytoskeleton dynamics in cells. *Mol Cells* 2010, 29, 311–325.
Lee, C. W., Han, J., Bamburg, J. R., Han, L., Lynn, R., and Zheng, J. Q., Regulation of acetylcholine receptor clustering by ADF/cofilin-directed vesicular trafficking. *Nat Neurosci* 2009, 12, 848–856.
Levitsky, D. I., Pivovarova, A. V., Mikhailova, V. V., and Nikolaeva, O. P., Thermal unfolding and aggregation of actin. *FEBS J* 2008, 275, 4280–4295.
Li, H., DeRosier, D. J., Nicholson, W. V., Nogales, E., and Downing, K. H., Microtubule structure at 8 Å resolution. *Structure* 2002, 10, 1317–1328.
Li, Y., Grenklo, S., Higgins, T., and Karlsson, R., The profilin:actin complex localizes to sites of dynamic actin polymerization at the leading edge of migrating cells and pathogen-induced actin tails. *Eur J Cell Biol* 2008, 87, 893–904.
Li, J. F. and Nebenfuhr, A., The tail that wags the dog: The globular tail domain defines the function of myosin V/XI. *Traffic* 2008, 9, 290–298.

Lin, C. T., Kao, M. T., Kurabayashi, K., and Meyhofer, E., Self-contained, biomolecular motor-driven protein sorting and concentrating in an ultrasensitive microfluidic chip. *Nano Lett* 2008, 8, 1041–1046.

Lin, S., Liu, M., Son, Y. J. et al., Inhibition of kinesin-5, a microtubule-based motor protein, as a strategy for enhancing regeneration of adult axons. *Traffic* 2011, 12, 269–286.

Lindahl, T., Instability and decay of the primary structure of DNA. *Nature* 1993, 362, 709–715.

Lindberg, U., Karlsson, R., Lassing, I., Schutt, C. E., and Hoglund, A. S., The microfilament system and malignancy. *Semin Cancer Biol* 2008, 18, 2–11.

Loring, R. H. and Salpeter, M. M., Denervation increases turnover rate of junctional acetylcholine receptors. *Proc Natl Acad Sci USA* 1980, 77, 2293–2297.

Loubery, S. and Coudrier, E., Myosins in the secretory pathway: Tethers or transporters? *Cell Mol Life Sci* 2008, 65, 2790–2800.

Lowery, L. A. and Van Vactor, D., The trip of the tip: Understanding the growth cone machinery. *Nat Rev Mol Cell Biol* 2009, 10, 332–343.

Lupa, M. T. and Caldwell, J. H., Sodium channels aggregate at former synaptic sites in innervated and denervated regenerating muscles. *J Cell Biol* 1994, 124, 139–147.

Lupa, M. T., Krzemien, D. M., Schaller, K. L., and Caldwell, J. H., Expression and distribution of sodium channels in short- and long-term denervated rodent skeletal muscles. *J Physiol* 1995, 483 (Pt 1), 109–118.

Mansson, A., Actomyosin-ADP states, inter-head cooperativity and the force–velocity relation of skeletal muscle. *Biophys J* 2010, 98, 1237–1246.

Månsson, A., Sundberg, M., Bunk, R. et al., Actin-based molecular motors for cargo transportation in nanotechnology—Potentials and challenges. *IEEE Trans Adv Pack* 2005, 28, 547–555.

McConnell, R. E. and Tyska, M. J., Leveraging the membrane–cytoskeleton interface with myosin-1. *Trends Cell Biol* 2010, 20, 418–426.

McGough, A., F-actin-binding proteins. *Curr Opin Struct Biol* 1998, 8, 166–176.

McMahan, U. J., Edgington, D. R., and Kuffler, D. P., Factors that influence regeneration of the neuromuscular junction. *J Exp Biol* 1980, 89, 31–42.

McMahan, U. J. and Slater, C. R., The influence of basal lamina on the accumulation of acetylcholine receptors at synaptic sites in regenerating muscle. *J Cell Biol* 1984, 98, 1453–1473.

Midrio, M., The denervated muscle: Facts and hypotheses. A historical review. *Eur J Appl Physiol* 2006, 98, 1–21.

Miki, H., Setou, M., Kaneshiro, K., and Hirokawa, N., All kinesin superfamily protein, KIF, genes in mouse and human. *Proc Natl Acad Sci USA* 2001, 98, 7004–7011.

Mitsui, T., Kawajiri, M., Kunishige, M. et al., Functional association between nicotinic acetylcholine receptor and sarcomeric proteins via actin and desmin filaments. *J Cell Biochem* 2000, 77, 584–595.

Murakami, K., Yasunaga, T., Noguchi, T. Q. P. et al., Structural basis for actin assembly, activation of ATP hydrolysis, and delayed phosphate release. *Cell* 2010, 143, 275–287.

Navarro, X., Vivo, M., and Valero-Cabre, A., Neural plasticity after peripheral nerve injury and regeneration. *Prog Neurobiol* 2007, 82, 163–201.

Nogales, E., Wolf, S. G., and Downing, K. H., Structure of the alpha beta tubulin dimer by electron crystallography. *Nature* 1998, 391, 199–203.

Oda, T., Iwasa, M., Aihara, T., Maeda, Y., and Narita, A., The nature of the globular- to fibrous-actin transition. *Nature* 2009, 457, 441–445.

Orlova, A. and Egelman, E. H., A conformational change in the actin subunit can change the flexibility of the actin filament. *J Mol Biol* 1993, 232, 334–341.

Otto, J. J., Actin-bundling proteins. *Curr Opin Cell Biol* 1994, 6, 105–109.

Parsons, J. T., Horwitz, A. R., and Schwartz, M. A., Cell adhesion: Integrating cytoskeletal dynamics and cellular tension. *Nat Rev Mol Cell Biol* 2010, 11, 633–643.

Patton, B. L., Miner, J. H., Chiu, A. Y., and Sanes, J. R., Distribution and function of laminins in the neuromuscular system of developing, adult, and mutant mice. *J Cell Biol* 1997, 139, 1507–1521.

Pollard, T. D. and Borisy, G. G., Cellular motility driven by assembly and disassembly of actin filaments. *Cell* 2003, 112, 453–465.

Prager-Khoutorsky, M. and Spira, M. E., Neurite retraction and regrowth regulated by membrane retrieval, membrane supply, and actin dynamics. *Brain Res* 2009, 1251, 65–79.

Prochniewicz, E., Effect of crosslinking by glutaraldehyde on interaction of F-actin with heavy meromyosin. *Biochim Biophys Acta* 1979, 579, 346–358.

Purves, D. and Sakmann, B., Membrane properties underlying spontaneous activity of denervated muscle fibres. *J Physiol* 1974, 239, 125–153.

Radtke, C. and Vogt, P. M., Peripheral nerve regeneration: A current perspective. *Eplasty* 2009, 9, e47.

Ramachandran, S., Ernst, K. H., Bachand, G. D., Vogel, V., and Hess, H., Selective loading of kinesin-powered molecular shuttles with protein cargo and its application to biosensing. *Small* 2006, 2, 330–334.

Rando, R. R., The biochemistry of the visual cycle. *Chem Rev* 2001, 101, 1881–1896.

Rayment, I., Rypniewski, W. R., Schmidt-Base, K. et al., Three-dimensional structure of myosin subfragment-1: A molecular motor. *Science* 1993, 261, 50–58.

Rayment, I., Smith, C., and Yount, R. G., The active site of myosin. *Annu Rev Physiol* 1996, 58, 671–702.

Redowicz, M. J., Regulation of nonmuscle myosins by heavy chain phosphorylation. *J Muscle Res Cell Motil* 2001, 22, 163–173.

Rieger, F., Grumet, M., and Edelman, G. M., N-CAM at the vertebrate neuromuscular junction. *J Cell Biol* 1985, 101, 285–293.

Rishal, I. and Fainzilber, M., Retrograde signaling in axonal regeneration. *Exp Neurol* 2010, 223, 5–10.

Roder, I. V., Choi, K. R., Reischl, M. et al., Myosin Va cooperates with PKA RIalpha to mediate maintenance of the endplate in vivo. *Proc Natl Acad Sci USA* 2010, 107, 2031–2036.

Roder, I. V., Petersen, Y., Choi, K. R., Witzemann, V., Hammer, J. A., 3rd, and Rudolf, R., Role of myosin Va in the plasticity of the vertebrate neuromuscular junction in vivo. *PLoS One* 2008, 3, e3871.

Rudolf, R., Bittins, C. M., and Gerdes, H. H., The role of myosin V in exocytosis and synaptic plasticity. *J Neurochem* 2011, 116, 177–191.

Saibil, H. R. and Ranson, N. A., The chaperonin folding machine. *Trends Biochem Sci* 2002, 27, 627–632.

Sancar, A., Lindsey-Boltz, L. A., Unsal-Kacmaz, K., and Linn, S., Molecular mechanisms of mammalian DNA repair and the DNA damage checkpoints. *Annu Rev Biochem* 2004, 73, 39–85.

Sandrock, A. W., Jr., Dryer, S. E., Rosen, K. M. et al., Maintenance of acetylcholine receptor number by neuregulins at the neuromuscular junction in vivo. *Science* 1997, 276, 599–603.

Sanes, J. R., The basement membrane/basal lamina of skeletal muscle. *J Biol Chem* 2003, 278, 12601–12604.

Sanes, J. R. and Lichtman, J. W., Development of the vertebrate neuromuscular junction. *Annu Rev Neurosci* 1999, 22, 389–442.

Sanes, J. R. and Lichtman, J. W., Induction, assembly, maturation and maintenance of a postsynaptic apparatus. *Nat Rev Neurosci* 2001, 2, 791–805.

Schaller, V., Weber, C., Semmrich, C., Frey, E., and Bausch, A. R., Polar patterns of driven filaments. *Nature* 2010, 467, 73–77.

Schutt, C. E., Myslik, J. C., Rozycki, M. D., Goonesekere, N. C., and Lindberg, U., The structure of crystalline profilin-beta-actin. *Nature* 1993, 365, 810–816.

Sealock, R., Butler, M. H., Kramarcy, N. R. et al., Localization of dystrophin relative to acetylcholine receptor domains in electric tissue and adult and cultured skeletal muscle. *J Cell Biol* 1991, 113, 1133–1144.

Sellers, J. R., *Myosins*, 2nd edn. Oxford University Press: Oxford, U.K., 1999.

Shah, J. V. and Cleveland, D. W., Slow axonal transport: Fast motors in the slow lane. *Curr Opin Cell Biol* 2002, 14, 58–62.

Shi, L., Butt, B., Ip, F. C. et al., Ephexin1 is required for structural maturation and neurotransmission at the neuromuscular junction. *Neuron* 2010a, 65, 204–216.

Shi, L., Fu, A. K., and Ip, N. Y., Multiple roles of the Rho GEF ephexin1 in synapse remodeling. *Commun Integr Biol* 2010b, 3, 622–624.

Smith, C. A. and Rayment, I., Active site comparisons highlight structural similarities between myosin and other P-loop proteins. *Biophys J* 1996, 70, 1590–1602.

Smith, J. W. and Thesleff, S., Spontaneous activity in denervated mouse diaphragm muscle. *J Physiol* 1976, 257, 171–186.

Son, Y. J. and Thompson, W. J., Nerve sprouting in muscle is induced and guided by processes extended by Schwann cells. *Neuron* 1995, 14, 133–141.

Soong, R. K., Bachand, G. D., Neves, H. P., Olkhovets, A. G., Craighead, H. G., and Montemagno, C. D., Powering an inorganic nanodevice with a biomolecular motor. *Science* 2000, 290, 1555–1558.

Sousa, A. D. and Cheney, R. E., Myosin-X: A molecular motor at the cell's fingertips. *Trends Cell Biol* 2005, 15, 533–539.

Spudich, J. A., Biochemistry. Molecular motors, beauty in complexity. *Science* 2011, 331, 1143–1144.

Stoll, G., Jander, S., and Myers, R. R., Degeneration and regeneration of the peripheral nervous system: From Augustus Waller's observations to neuroinflammation. *J Peripher Nerv Syst* 2002, 7, 13–27.

Sundberg, M., Bunk, R., Albet-Torres, N. et al., Actin filament guidance on a chip: Toward high-throughput assays and lab-on-a-chip applications. *Langmuir* 2006, 22, 7286–7295.

Suter, D. M., Espindola, F. S., Lin, C. H., Forscher, P., and Mooseker, M. S., Localization of unconventional myosins V and VI in neuronal growth cones. *J Neurobiol* 2000, 42, 370–382.

Svensson, A., Libelius, R., and Tagerud, S., Semaphorin 6C expression in innervated and denervated skeletal muscle. *J Mol Histol* 2008, 39, 5–13.

Tamada, A., Kawase, S., Murakami, F., and Kamiguchi, H., Autonomous right-screw rotation of growth cone filopodia drives neurite turning. *J Cell Biol* 2010, 188, 429–441.

Tani, T., Miyamoto, Y., Fujimori, K. E. et al., Trafficking of a ligand–receptor complex on the growth cones as an essential step for the uptake of nerve growth factor at the distal end of the axon: A single-molecule analysis. *J Neurosci* 2005, 25, 2181–2191.

Tilney, L. G., Bryan, J., Bush, D. J. et al., Microtubules: Evidence for 13 protofilaments. *J Cell Biol* 1973, 59, 267–275.

Vale, R. D., The molecular motor toolbox for intracellular transport. *Cell* 2003, 112, 467–480.

Vale, R. D. and Milligan, R. A., The way things move: Looking under the hood of molecular motor proteins. *Science* 2000, 288, 88–95.

van den Heuvel, M. G. L. and Dekker, C., Motor proteins at work for nanotechnology. *Science* 2007, 317, 333–336.

Vikhorev, P. G., Vikhoreva, N. N., and Mansson, A., Bending flexibility of actin filaments during motor-induced sliding. *Biophys J* 2008a, 95, 5809–5819.

Vikhorev, P. G., Vikhoreva, N. N., Sundberg, M. et al., Diffusion dynamics of motor-driven transport: Gradient production and self-organization of surfaces. *Langmuir* 2008b, 24, 13509–13517.

Vikhoreva, N. N. and Mansson, A., Regulatory light chains modulate in vitro actin motility driven by skeletal heavy meromyosin. *Biochem Biophys Res Commun* 2010, 403, 1–6.

Wells, A. L., Lin, A. W., Chen, L. Q. et al., Myosin VI is an actin-based motor that moves backwards. *Nature* 1999, 401, 505–508.

Weston, C., Gordon, C., Teressa, G., Hod, E., Ren, X. D., and Prives, J., Cooperative regulation by Rac and Rho of agrin-induced acetylcholine receptor clustering in muscle cells. *J Biol Chem* 2003, 278, 6450–6455.

Weston, C. A., Teressa, G., Weeks, B. S., and Prives, J., Agrin and laminin induce acetylcholine receptor clustering by convergent, Rho GTPase-dependent signaling pathways. *J Cell Sci* 2007, 120, 868–875.

Weston, C., Yee, B., Hod, E., and Prives, J., Agrin-induced acetylcholine receptor clustering is mediated by the small guanosine triphosphatases Rac and Cdc42. *J Cell Biol* 2000, 150, 205–212.

Wilson, M. H. and Deschenes, M. R., The neuromuscular junction: Anatomical features and adaptations to various forms of increased, or decreased neuromuscular activity. *Int J Neurosci* 2005, 115, 803–828.

Witte, H. and Bradke, F., The role of the cytoskeleton during neuronal polarization. *Curr Opin Neurobiol* 2008, 18, 479–487.

Witzemann, V., Barg, B., Nishikawa, Y., Sakmann, B., and Numa, S., Differential regulation of muscle acetylcholine receptor gamma- and epsilon-subunit mRNAs. *FEBS Lett* 1987, 223, 104–112.

Woodhoo, A. and Sommer, L., Development of the Schwann cell lineage: From the neural crest to the myelinated nerve. *Glia* 2008, 56, 1481–1490.

Woolner, S. and Bement, W. M., Unconventional myosins acting unconventionally. *Trends Cell Biol* 2009, 19, 245–252.

Wu, H. T., Xiong, W. C., and Mei, L., To build a synapse: Signaling pathways in neuromuscular junction assembly. *Development* 2010, 137, 1017–1033.

Yanagida, T., Nakase, M., Nishiyama, K., and Oosawa, F., Direct observation of motion of single F-actin filaments in the presence of myosin. *Nature* 1984, 307, 58–60.

Zhang, H., Berg, J. S., Li, Z. et al., Myosin-X provides a motor-based link between integrins and the cytoskeleton. *Nat Cell Biol* 2004, 6, 523–531.

Zhang, B., Luo, S., Wang, Q., Suzuki, T., Xiong, W. C., and Mei, L., LRP4 serves as a coreceptor of agrin. *Neuron* 2008, 60, 285–297.

Zhu, C., Zhao, J., Bibikova, M. et al., Functional analysis of human microtubule-based motor proteins, the kinesins and dyneins, in mitosis/cytokinesis using RNA interference. *Mol Biol Cell* 2005, 16, 3187–3199.

16 Fault-Tolerant Approach to the Configuration of Programmable Logic at the Nanoscale

Gianluca Tempesti, André Stauffer, and Joël Rossier

CONTENTS

The self-replication of computing systems is an idea that dates back to the very origins of electronics: in the 1950s, John von Neumann was among the first to investigate the design of processor-scale computing devices capable of self-replication [1,2] with the goal of obtaining reliability through the redundant operation of many copies of the original device.

Since von Neumann's groundbreaking work, research on self-replicating computing machines has gone through several phases, but, in general, interest in applying self-replication directly to electronic hardware waned because of technological hurdles. In recent years, the introduction of programmable logic devices such as FPGAs has revitalized the field of biologically inspired hardware by allowing (at least in theory) the runtime modification of hardware. The physical processes that underlie processes such as growth and development in nature remain unattainable in electronic devices, but they can be approximated by altering the configuration of a programmable device.

Within this revival of interest, some of the motivations that led von Neumann to study self-replication are beginning to resurface among researchers faced with design and robustness issues in nanoscale electronic devices. The vast amount of on-chip resources that will be available in the next few decades, either by further shrinking of silicon fabrication processes or by the introduction of molecular-scale devices, together with the predicted features of such devices (e.g., high fault sensitivity), will introduce layout and fault-tolerant issues that cannot be solved using current design methodologies [3–5].

The potential impact of self-replication in programmable devices is further increased by their compatibility with molecular-scale electronics: programmable logic structures, because of their regularity and their versatility, can represent a useful paradigm for the introduction of complex nanoscale devices (see, e.g., [6–8]). A mechanism that allows a complex circuit to automatically replicate within a programmable substrate can potentially be of considerable interest for the creation, within the substrate, of regular arrays of computational nodes.
In this context, the usefulness of a self-replication process seems fairly obvious:

- As biological organisms grow from an initial cell to a complete adult, large arrays of cellular computing elements could exploit self-replication to grow in the programmable substrate, rather than being completely specified at design time.
- Faced with faults in the substrate, a growth process could be able to avoid faulty areas, while the redundancy that is an automatic result of self-replication can potentially allow the circuit to self-repair in the case of online faults.

To demonstrate the feasibility of development and growth in the context of complex electronic circuits, however, it is necessary to take into account several practical issues, such as the hardware overhead and the efficiency of the mechanisms involved, and advance beyond the "toy" examples that have been traditionally used to illustrate this process. This chapter describes how a recently developed self-replication algorithm was adapted to implement self-replication within a real-world programmable device and applied to a system consisting of four dedicated processors.

While still simple compared with most real-world multiprocessor applications, this system is much more complex than any circuit to which self-replication has been applied to date and exploits mechanisms and architectures that can be easily scaled to larger systems.

16.1 SHORT BACKGROUND ON SELF-REPLICATION

The self-replication of computing machines has a long history, punctuated by relatively few milestones. The following is a brief outline of the main approaches used to study this process, ranging from von Neumann's original ideas to some of the latest results in the area.

Of course, it should be mentioned that the concept of self-replication has been applied to artificial systems in contexts other than computing. A classic example is the 1980 NASA study by Freitas and Merkle [9] (further expanded in a remarkable book [10]), where self-replication is used as a paradigm for efficiently exploring other planets. However, the self-replication of physical machines rather then computing systems is beyond the scope of this chapter and this short background will not extend to cover this kind of approaches.

16.1.1 Von Neumann's Universal Constructor

Multicellular organisms are among the most reliable complex systems known to man, and their reliability is a consequence not of any particular robustness of the individual cells, but rather of their extreme redundancy. One of the basic mechanisms that provides such reliability is cellular division, that is, self-replication at the cellular level. von Neumann, confronted with the lack of reliability of the computing systems he was designing, turned to this mechanism to find inspiration in the design of fault-tolerant computing machines.

In particular, von Neumann [2] investigated self-replication as a way to design and implement digital logic devices and attempted to develop an approach to the realization of self-replicating computing machines (which he called *artificial automata*, as opposed to natural automata, i.e., biological organisms).

Using cellular automata (CA) as a framework, von Neumann realized the first self-replicating system. Based on a 29-state CA, his approach centered on a *universal constructor* (a machine capable of building any other machine, given its description) composed of two parts: a *tape* containing the description of the cellular machine to be built and the constructor itself, a complex structure capable of reading the tape and building the corresponding machine. Given a description of itself, the machine was then able to create copies of itself, first *interpreting* the contents of the tape and then *copying* the tape to the new machine. Coupled with a (possibly universal) Turing machine, the approach conceptually allowed the self-replication of computing systems of arbitrary size and complexity (Figure 16.1a).

Never meant to be implemented in actual hardware, von Neumann's universal constructor is an extremely complex machine. An estimate by Buckley [11] places the size of the machine (without the Turing machine) at ~800 K cells, with a 12 M cells tape. Obviously, in spite of the considerable theoretical power of this approach, its complexity prevents its use from a practical standpoint.

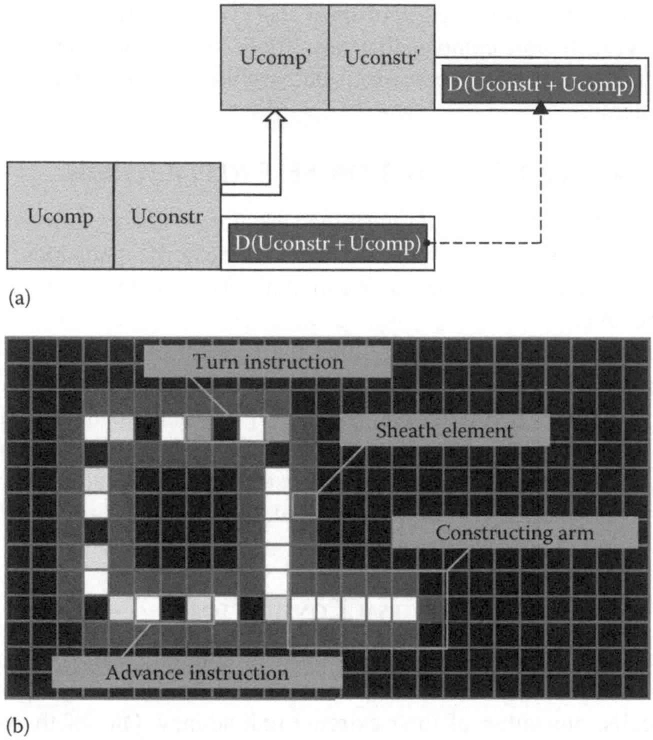

FIGURE 16.1 (a) Von Neumann's universal constructor (Uconstr) can replicate itself and an arbitrary (potentially universal) computing machine (Ucomp) given the description D of the two machines. (b) The initial configuration of Langton's loop.

16.1.2 Langton's Loop

A second stage in research on self-replication was opened by Langton in 1984 [12]. In order to reduce the complexity of the process, he dropped the universal construction and universal computation ability of the von Neumann system and proposed a simple self-replicating machine in the form of a loop (Figure 16.1b), also implemented as a CA, based on a constructing arm and on a looping replication program.

Unlike von Neumann, who was interested in self-replication from the standpoint of circuit design, Langton's research was aimed at studying the application of lifelike properties to computational structures and his goal in developing his approach was to determine the smallest automaton capable of self-replication. Further improvements to the machine led to smaller versions of the original loop [13,14] resulting in smaller self-replicating structures, but all these systems, because of the context in which they were studied, lack any computational capability.

More recently, some attempts have been made to redesign Langton's loop in order to embed calculation possibilities. Tempesti's loop [15] is thus a self-replicating automaton, with an attached executable program that is duplicated and executed in each of the copies. Perrier and Sipper [16] proposed a self-replicating loop showing universal computational capabilities. This system consists of three parts, loop, program, and data, all of which are replicated, followed by the program's execution on the given data. However, the complexity of these approaches, while considerably smaller than von Neumann's universal constructor, remains too great to be considered useful in the context of electronic hardware.

16.1.3 Other CA-Based Approaches

All of the approaches mentioned earlier share the common process of self-replicating through the interpretation of a sequence of building instructions. Some examples of self-replicating CA, however, exploit a different mechanism, that of *self-inspection*: instead of reading and interpreting a description, the self-replicating automaton inspects itself and produces a copy of what it finds. While less general than the universal constructor (obviously, the machine can only build an exact copy of itself), this approach is more versatile than Langton's loop, as structures of (almost) any size and shape can replicate. In practice, however, the best-known example of self-inspection is that of a self-replicating loop [17].

Also, while traditionally there has been a very loose connection between the kind of CA used to study self-replication and actual circuit design, some researchers have been trying to close this gap by studying automata that more closely approach some particular features of digital circuits. An example is Morita and Imai's study of self-replication in the context of reversible CA [18] (in a reversible CA, every configuration has at most one predecessor), inspired by reversible logic in digital circuits.

Similarly, Peper and coworkers [19,20] have developed self-replicating structures in self-timed cellular automata (STCA). This kind of automata do not rely on a global synchronization mechanism to update the states of the cells, but rather the state transitions only occur when triggered by transitions in neighboring cells. The basic assumption in this work is that STCA is a model that might more closely resemble molecular-scale nanoelectronic devices.

16.1.4 Self-Replication in Electronic Devices

As seen earlier, throughout its long history, CA have remained the environment of choice to study how self-replication can be applied to computing systems. However, in general, researchers in the domain (including von Neumann) have never regarded CA as the environment in which self-replication would be ultimately applied. Rather, CA have traditionally provided a useful platform to test the complexity of self-replication at an early stage, in view of eventually applying this process to real-world systems, either to electronics or, more generally, to computing systems.

Approximating a self-replication process in an electronic device, however, required the introduction of programmable circuits, where the physical construction that occurs in nature can be replaced by an information-based process. In practice, self-replication in hardware has been implemented, without exception to our knowledge, as the copy of a partial configuration of a programmable device.

One of the simplest approaches that exploit this kind of setup is *configuration cloning* [21], based on a simple replication of the configuration of part of an FPGA in order to create multiple copies of the same subsystem. In this case, of course, no *self*-replication occurs, since the configuration process is controlled by an external entity.

As the external controller still needs to sequentially program the entire circuit, most of the advantages of self-replication are lost.

A variation on the same approach, closer to self-replication, was developed in the late 1990s [22,23], with an emphasis on self-repair. In this system, applied to a very fine-grained custom FPGA, a tiny CA is used in a first step of the replication process to subdivide the programmable circuit into blocks of arbitrary size (corresponding to the circuit to be replicated). In a second step, an external entity injects a single copy of the configuration of the block, which is automatically replicated so as to completely fill the device. While more versatile than configuration cloning (the automaton does not require external control and the configuration of the device occurs in parallel rather than in series), the approach still requires a relatively complex grid of global connections and an external synchronization, which again limit the advantages of self-replication.

The system that, to date, probably best exploits self-replication in the framework of electronic devices is the cell matrix system [24,25]. Cell matrix is a fine-grained reconfigurable device composed of a collection of identical elements (referred to as *cells*) placed on the edges of a two-dimensional regular grid. Each cell in the grid is interconnected with its four cardinal neighbors and contains a lookup table, which is used as a truth table to implement logic functions. Each cell can exchange information with its four neighbors, and it uses (as a CA) its four inputs and its lookup table to define the value of its four outputs (Figure 16.2).

Unlike the systems presented earlier that have to be controlled by an external computer in order to achieve self-replication, cells of cell matrix circuit can self-replicate

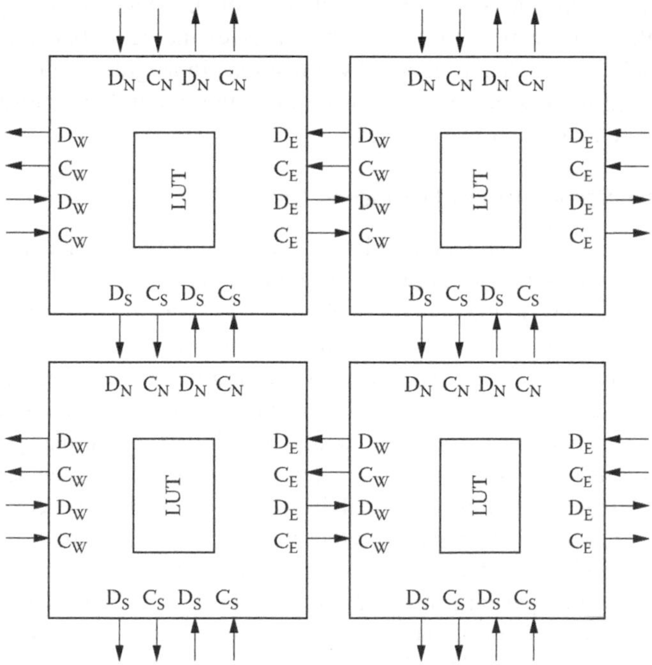

FIGURE 16.2 The 2 × 2 cell matrix grid. Each cell, which has two inputs and two outputs per edge, is connected with its four direct neighbors.

autonomously. Each cell is at the same time configurable and can configure other cells without any external input command. This feature is called *self-duality*. In order to implement self-duality, a cell has to operate in two independent modes: *D-mode* and *C-mode*. In D-mode, the cell's lookup table processes the four-input signals in order to generate output signals, whereas in C-mode the input data are used to fill up (configure) or rewrite (reconfigure) the lookup table of the cell.

Using self-duality, arbitrarily complex structures can self-replicate within the circuit. Cell matrix is an accomplished system that has been shown to work well, but is hindered both by the nonconventional structure of the cells (the approach cannot easily be generalized to arbitrary programmable logic architectures) and by the very large overhead required by self-replication, since as in von Neumann's approach the construction process relies on a description of the machine to be built that is in fact much larger than the machine itself.

The Tom Thumb algorithm [26,27] was proposed to achieve self-replication of arbitrary structures in programmable logic. The algorithm borrows from Langton and his successors the concept of loop, but was designed to be implemented in silicon. Potentially, the algorithm could be used to replicate any structure within a programmable device, following a simple systematic methodology, but so far its operation has been described only for trivial, illustrative examples. This chapter shows how the algorithm was extended and applied to a real-world programmable logic device, demonstrating how it can be used for the self-replication of complex circuits and providing indications as to its applicability to nanoscale devices.

16.2 TOM THUMB ALGORITHM

This section introduces the basic behavior of the Tom Thumb algorithm, designed to enable self-replication in programmable logic. Developed in the context of a more general bioinspired approach, the *Embryonics* project [23], which ranges from logic gates to massively parallel arrays of processors, the algorithm requires a shift in terminology compared with the traditional CA-based approaches. Notably, as illustrated in Figure 16.3, the hierarchical structure of an Embryonics system relies on four levels: the basic hardware unit is a programmable logic element; several elements are configured and connected to generate a *cell*, which represents the unit within the system that can replicate itself (in the examples used in this chapter, this will consist of a simple application-specific processor); an *organism* is a complete computing system composed of several cells working together (i.e., an application-specific array of processors); the term *population* (not used in this chapter) refers to a set of several organisms. In this context, *genome* is the term used to describe the configuration of a single cell, and hence represents a partial configuration of the FPGA used to implement the complete system.

16.2.1 Basic Structure

The basic operation of the Tom Thumb algorithm can be illustrated by means of a minimal cell composed of four elements, which grows and then divides to spawn two daughter cells. This simple example is sufficient to define the mechanisms that

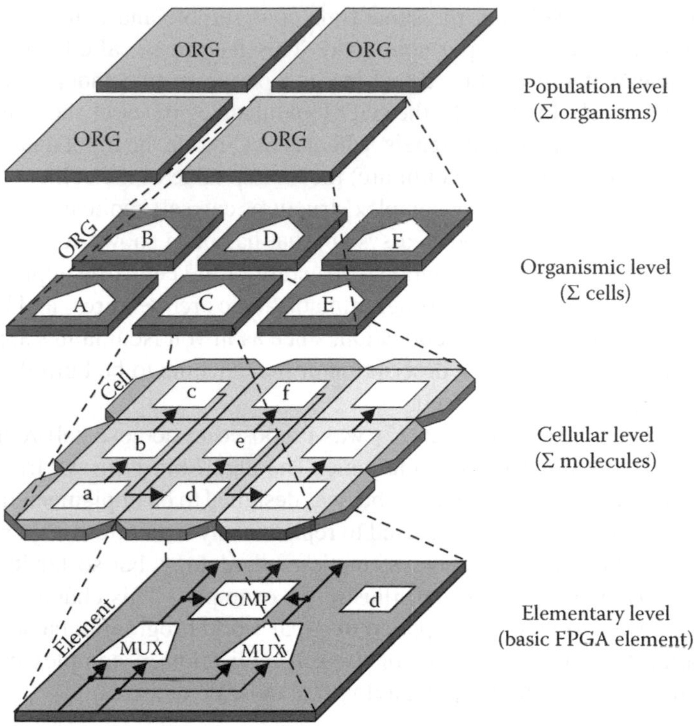

FIGURE 16.3 The four hierarchical levels of the Embryonics approach.

the basic element has to implement to enable the self-replication of the cells. Further details on the operation on the algorithms can be found in Refs. [26,27].

For this example, the basic programmable logic element has no functionality. That is, the element consists simply of a memory to store a minimal configuration, which is not used for any purpose except to store a unique number. Due to the operation of the algorithm, the configuration of the elements must be stored in a set of shift registers: while obviously more expensive in terms of surface compared with other solutions, this kind of memory storage has the fundamental advantage of allowing data to move easily within the programmable substrate, a necessary condition for a self-replicating process to take place.

The purpose of the self-replication of a mother cell in the context of a programmable logic substrate is to replicate the configuration data of its elements at a different location within the substrate, creating identical copies of the mother cell. The Tom Thumb algorithm enables such a behavior if the cell and the elements have the following characteristics: first of all, a *configuration path* has to be defined inside the cell in such a way that the configuration registers of the elements are connected in a loop that goes through each of the elements of the mother cell. A path of this type that is valid for a cell composed of four elements is shown in Figure 16.4a. This requirement is also the reason why the minimal cell that can replicate using the algorithm has to be composed of four elements, organized as a square of two rows by two columns.

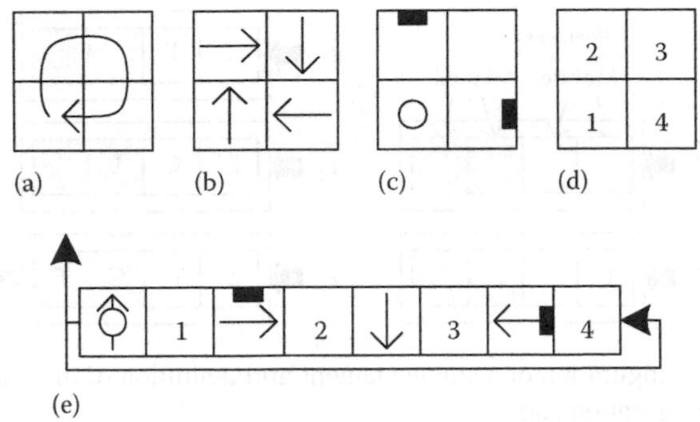

FIGURE 16.4 Basic Tom Thumb information and genome.

The second requirement of the algorithm is that each element, in addition to its configuration, must contain a *flag* indicating the direction of the configuration path (the arrows in Figure 16.4b). Additionally, the flag information must indicate which element is placed at the beginning of the loop (arbitrarily defined as the lower left corner), shown with a circle in Figure 16.4c, and which elements will be used to send out the data required for the creation of the daughter cells, shown with black rectangles in the same figure.

Taking into account that the configuration of the elements is defined by the numbers 1–4 in Figure 16.4d, it results that the minimal cell implementing the algorithm is entirely defined by the configuration bitstream shown in Figure 16.4e. This bitstream, which represents the genome of the cells, is composed of eight *packets*, four of them containing the flags and the other four used to define the configuration data.

Finally, the elements must be able to store *two* copies of their configuration data and flag. The first copy will be used as the actual configuration, while the second will circulate within the loop and be used to create the daughter cells. Roughly equivalent to the double-helix configuration of DNA in biological cells, this requirement introduces a considerable amount of overhead, but greatly simplifies the self-replication process (an alternative implementation, based on self-inspection, does away with most of the overhead at the cost of increased complexity in the control).

In the minimal example, since each element is defined by two packets of information (the configuration data and the flag), its configuration memory will have to store four packets and, in order to fully configure a cell, the bitstream shown in Figure 16.4e will have to be injected twice.

16.2.2 Constructing the Mother Cell

Note that in the genome represented in Figure 16.5, the first, third, etc., packets always contain flag information (F in figure), while the second, fourth, etc., packets contain the configuration data of the elements (C in figure). As shown in the same figure, each element contains four memory positions able to store the packets. The two positions on the right of the element will be used to store the fixed information, that is, the flag defining the role of the element in the Tom Thumb algorithm and the actual configuration of the element, whereas

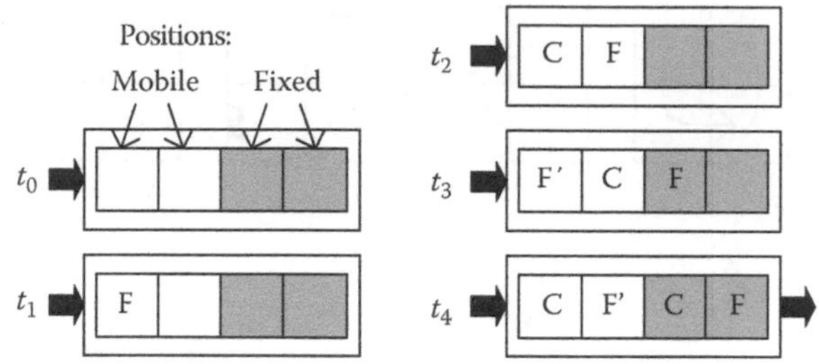

FIGURE 16.5 Configuration of a single element and definition of the direction of the next element in the configuration path.

the two left positions will be used to transmit the packets to the next element on the construction path and to contain the second copy of the genome that will be used for the replication process.

At each time step t_x, a packet of the original genome is shifted and injected in the programmable substrate. For practical reasons, the bottom-left element in the array is normally considered as the first to be accessed, but of course (the path being a loop) any element can be used to start the configuration. As the Tom Thumb algorithm relies entirely on local connections between neighbors, this initial element represents the only *injection point* where an external connection is necessary for the configuration of the entire substrate (a property that could be of interest in the context of nanoscale devices, where interfacing is complex).

When the first, empty element receives the first packets, it shifts them until its two fixed positions (on the right in figure) are filled. During this process, at time t_3 the element becomes aware of which flag (F) will be stored in its fixed position, at which point it can establish a new connection to forward the following packets of the configuration in the direction indicated by the flag in order to configure the next element on the path. This connection becomes valid one clock cycle later, that is, at time t_4. At this moment, the element has received its configuration and the flag defining in which direction the construction will proceed and has created the appropriate connection path. All further configuration packets are shifted through the two left memory positions and then out of the element in the direction indicated by its fixed flag.

The element that receives the configuration packets behaves in the same way and this process repeats itself until each element of the cell has been configured. The entire process, for the genome of Figure 16.4e, is shown in Figure 16.6. At times t_4, t_8, t_{12}, and t_{16}, new connections are established between elements. At time t_{16}, the configuration loop has closed and the four elements of the cell are configured.

Note that, during the construction process, the genome has been inserted twice. The first copy is memorized in the two right memory positions of each element (*fixed memory*) and configures the elements. In parallel, the second copy shifts indefinitely through the two left memory positions (*mobile memory*) of the elements following the configuration path. This second copy of the genome will be used to instantiate the replication of the cell, as described in Section 16.2.3.

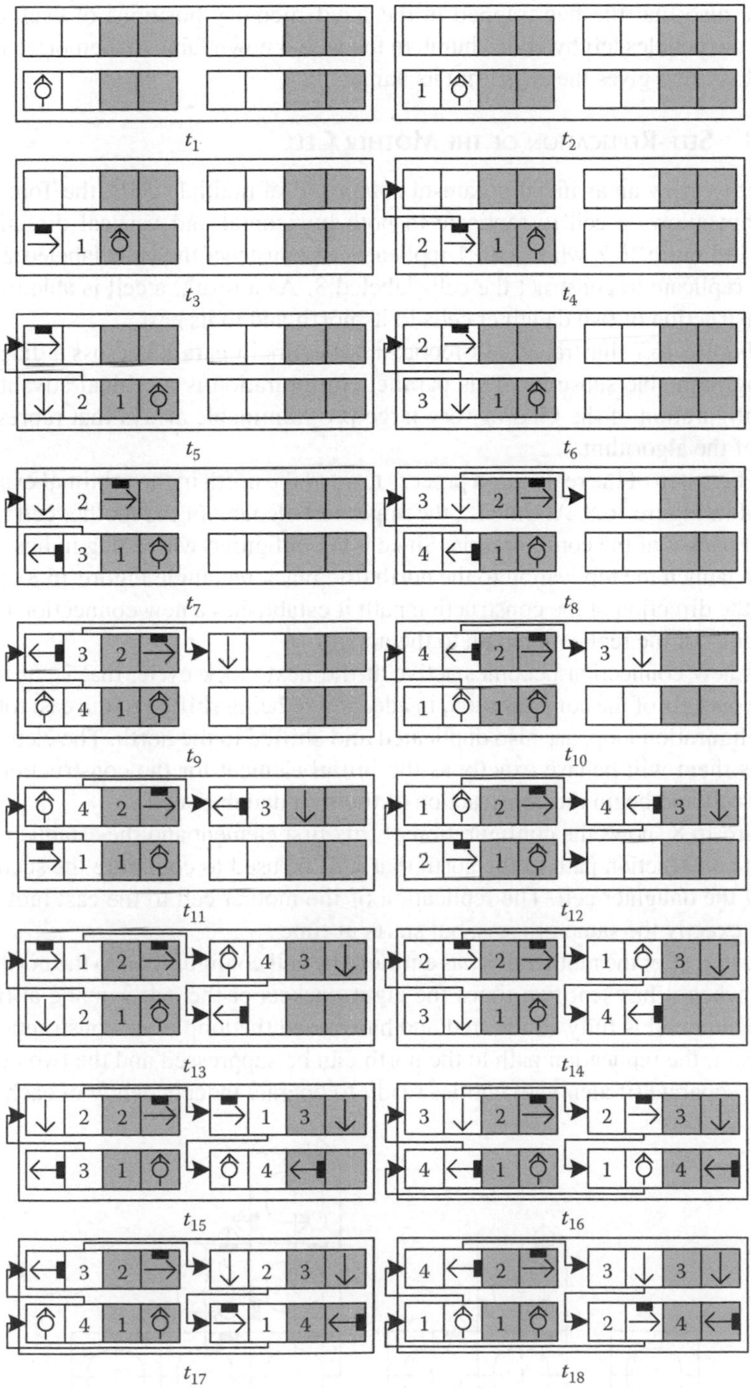

FIGURE 16.6 Construction of the first cell. Connection setup: t_4 north, t_8 east, t_{12} south, t_{16} west.

Note also that the flag trapped in the fixed memory positions of each element recalls the pebbles left by Tom Thumb in the well-known fable to memorize his way, an analogy that gives the algorithm its name.

16.2.3 Self-Replication of the Mother Cell

In order to grow an artificial organism composed of multiple cells, the Tom Thumb algorithm allows a cell to replicate in both horizontal and vertical directions, as shown in Figure 16.7, where cell 1 replicates to construct the cells labeled 2, which in turn replicate to construct the cells labeled 3. As a result, a cell is able to trigger the construction of two daughter cells to its north and to its east.

As Figure 16.7 illustrates, self-replication occurs in parallel across a diagonal of the programmable substrate. This parallel configuration is a critical advantage for the configuration of the kind of very large programmable arrays that represent the target of the algorithm.

The first step of the replication process toward the north in the minimal example is detailed in Figure 16.8. At time t_{11}, the upper left element of the mother cell receives the first packet of the configuration. Since it is configured with a flag indicating that it has to launch the replication to the north (the black rectangle Figure 16.8), in addition to the direction of the construction path it establishes a new connection that will be the start of the replication path to the north.

This new connection becomes active at the next clock cycle, that is, at time t_{12}, and the packets of the configuration, in addition to being shifted to the east following the configuration loop, are also duplicated and shifted to the north. The element that receives them will behave exactly as the initial element for the construction of the mother cell and begin the construction of the first daughter cell.

Figure 16.8 shows the configuration of this first element and the establishment of the first construction path to the north that will be used to configure the second element of the daughter cell. The replication of the mother cell to the east (not shown) follows exactly the same process, but starts at time t_{24}.

Note that after the mother cell has emitted the cell configuration to the north twice, that is, when it has sent two times the eight packets of the cell genome at time t_{28}, the daughter cell is fully configured and has closed the loop of its construction path. As a result, the replication path to the north can be suppressed and the two cells will become separate (if identical) entities ready to operate independently of each other.

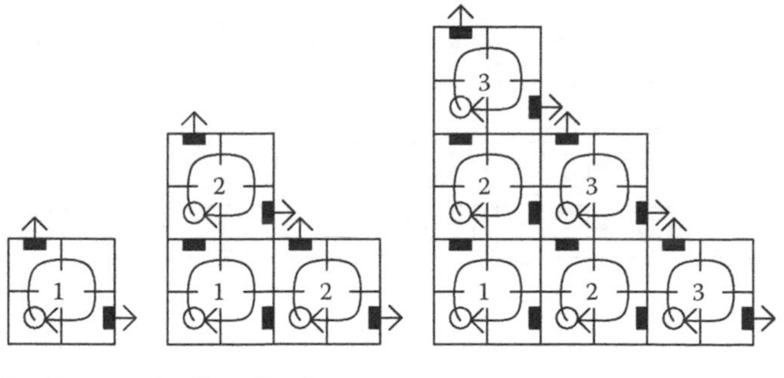

FIGURE 16.7 Pattern of cell replications.

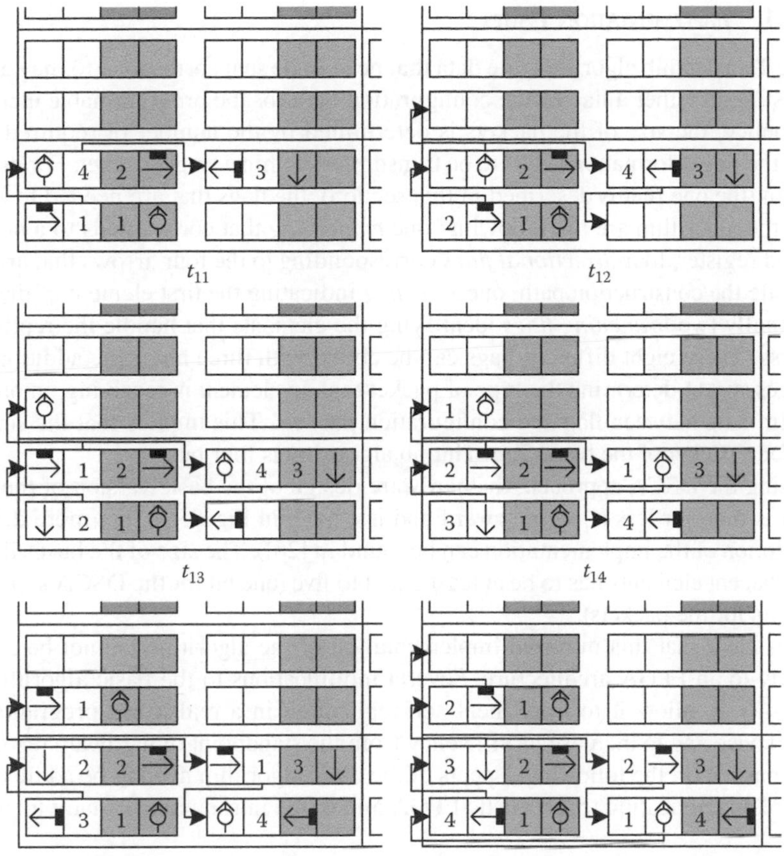

FIGURE 16.8 First steps of the cell replication to the north.

16.3 HARDWARE IMPLEMENTATION

Based on purely local interactions between neighbors, the Tom Thumb algorithm could be implemented by a canonical CA. However, while the conventional criteria used in the design of CA render them notoriously inefficient to implement in hardware, the algorithm was conceived using a design methodology that privileges ease of implementation. The Data and Signals Cellular Automata (DSCA) paradigm [28] defines the transitions of an element from one state to the next not by accessing a truth table, but rather as a consequence of a set of data and signals received from the element's neighbors. In particular, at each clock cycle information is sent from each element to its cardinal neighbors, together with a signal that indicates if the information has to be processed.

This interpretation of the behavior of the algorithm on the one hand allows a straightforward (and compact) implementation in hardware, and on the other hand simplifies the adaptation of the algorithm to the configuration of FPGAs. This adaptation involves, in practice, several minor alterations to the basic algorithm.

16.3.1 Implementation Issues

In the Tom Thumb algorithm, the data that need to be sent correspond to the packets, consisting of either a flag or the configuration data for the programmable elements. In practice, the size of the packets is determined by the number of required flags, since the flag information needs to be transmitted within a single packet. For its basic version (the one that is described in this section), the flags that are needed to implement the algorithm are the following: one *empty flag* that corresponds to a nonconfigured register, four *directional flags* corresponding to the four arrows that are used to create the construction path, one *start flag* indicating the first element of the path, and finally two *branching flags* identifying the elements that handle the replication process. These eight different flags can be coded with three bits. One additional bit is necessary to determine the type of packet that an element is receiving, in order to discriminate between flag and configuration packets. This implies that the smallest size for a packet of the basic Tom Thumb algorithm is four bits.

Using the DSCA approach, the hardware design of the basic version of the algorithm is more or less straightforward and is shown in Figure 16.9 (a more detailed description of the implementation can be found in [26]). The size of the buses linking two adjacent elements has to be at least equal to five (one bit for the DSCA signal and four bits for the packets).

It is clear that this minimal implementation of the algorithm cannot be applied directly to an FPGA architecture. Several modifications to the basic algorithm are necessary to allow it to implement self-replication in a real-world programmable logic device and to increase its efficiency from the standpoint of hardware resources.

In particular, the following aspects of the basic algorithm need to be addressed in view of an application to a realistic FPGA and of the introduction of fault tolerance:

- Only three bits of useful information are available for the configuration of an element. It is obvious that a real system must handle a greater number of configuration bits for its basic components, that is, the elements.

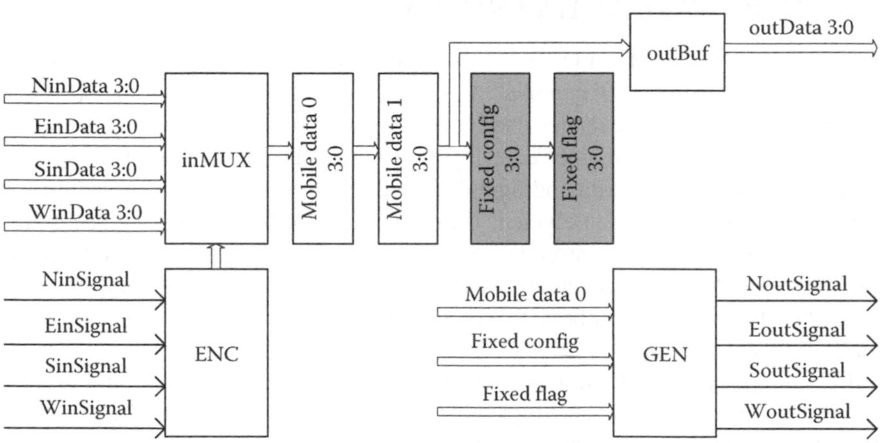

FIGURE 16.9 DSCA implementation of the Tom Thumb algorithm.

- A cell can only replicate in two directions, that is, to the north and the east. While in some cases this might be sufficient, in a real system (and particularly in one that wishes to implement fault-tolerant mechanisms), it would be an advantage to replicate in the four cardinal directions.
- The cell has no control on the replication process, which is launched at start-up and continues until the substrate is entirely configured. The algorithm could be modified to enable the cells to choose when and where to replicate, improving its capability to address runtime faults in the substrate.
- Each element in the replicated cell is activated as soon as its static configuration has been set. In a real circuit, this can lead to strange behaviors as parts of the circuit become active before the rest is configured.

Strictly speaking, only the first of these issues is crucial to achieve self-replication in an FPGA. The others are, essentially, performance issues that increase the efficiency or the versatility of the self-replication approach.

16.3.2 POEtic Tissue and MOVE Processors

To study these issues, we applied the algorithm to the configuration of a custom FPGA, the *POEtic tissue* [29,30].

Conceived for the implementation of bioinspired systems, this FPGA contains some nonstandard features, when compared with conventional off-the-shelf devices. However, the basic structure of its elements, as well as its configuration process, are essentially equivalent to their conventional counterparts: as shown in Figure 16.10,

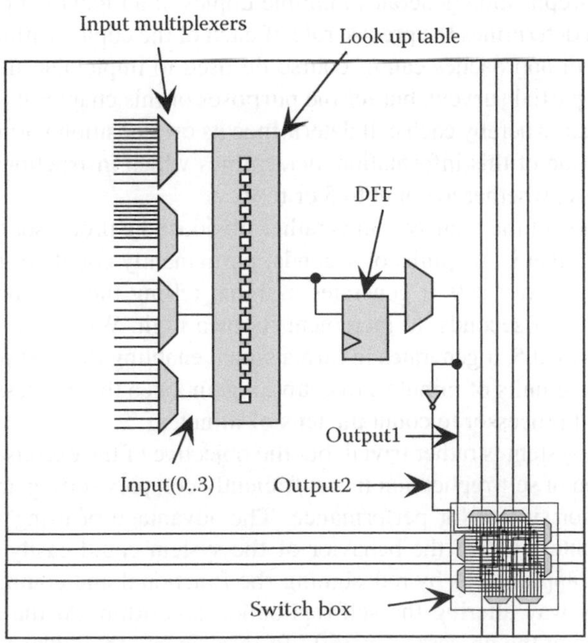

FIGURE 16.10 Basic structure of a POEtic element.

an element mainly contains a 16-bit lookup table (LUT) and a D flip-flop (DFF); its inputs are selected by a set of multiplexers and its outputs can be routed to any cardinal direction through a switch box. An element's 76 configuration bits define the content of the LUT and of the DFF, as well as the selection of the multiplexers for the inputs and the outputs of an element.

The operation and the nonstandard features of the POEtic tissue are described in some detail elsewhere [29,30]. Most of these details are not relevant to the topic of this chapter, since the POEtic tissue was selected as a test platform essentially because of the possibility to modify its hardware structure (crucial to allow the circuit to implement the Tom Thumb algorithm) and not for its detailed architecture.

To illustrate the scalability of the Tom Thumb algorithm to arbitrarily large structures, we designed a custom processor, based on the MOVE paradigm, also known as the transport-triggered architecture [31–33], originally developed for the design of application-specific dataflow processors (processors where the instructions define the flow of data, rather than the operations to be executed).

The final system (the organism) will be composed of four MOVE processors (the cells) that will form a four-digit counter that will display seconds and minutes (in practice, two connected modulus 60 counters). Each of the processors will handle one digit. Thus, two will count from 0 to 9 while the two others will count from 0 to 5. In their final configuration, they are logically organized to form a chain (Figure 16.11).

As shown in Figure 16.11, the Seed unit is used to start the counter as soon as the replication of the processors is finished and to provide the first (rightmost) processor of the final chain the information it needs to launch the *differentiation process*. This process is an important part of the setup of an organism within the bioinspired approach: self-replication generates multiple copies of an identical processor, while differentiation determines the precise role of each of the copies within the organism. Several kinds of approaches can of course be used to implement this process (see, e.g., [34] for a partial survey), but for the purposes of this chapter it can be seen as a local mechanism whereby each cell determines its own position within the organism and, as a function of this information, determines which instructions to execute (in this precise case, whether to count to 5 or to 9).

The operation of the final system is rather obvious: the processor that handles the rightmost digit, that is, the units of seconds, permanently counts from 0 to 9. When this processor arrives at 9, it generates a signal telling the next processor, which handles the tens of seconds, to increment its own digit. When the tens of seconds processor arrives at 5, it generates in turn a signal enabling the next processor on the chain, that is, the units of minutes, to count. Again, once this processor reaches 9, it signals the next processor to count the tens of minutes.

Clearly, the system is rather trivial, but the objective of the exercise was to test the implementation of self-replication in a sufficiently complex setting and not to design a multiprocessor system for performance. The advantage of using MOVE processors in this context is that the behavior of the system could easily be extended to more complex applications by redesigning the functional and communication units without in any way altering the self-replication algorithm. At the same time, the implementation of the processor on 360 FPGA elements should be a sufficient demonstration of the scalability of the Tom Thumb algorithm to realistic circuit sizes.

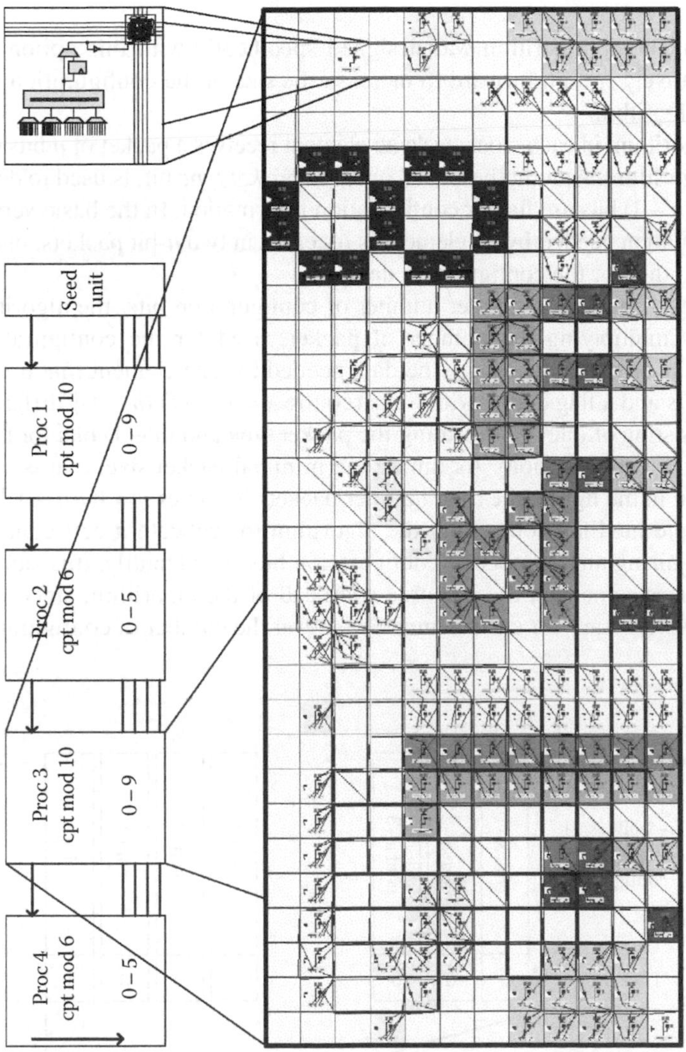

FIGURE 16.11 The three hierarchical levels of the system (cf. Figure 16.3): the organism.

16.3.3 Size of the Configuration Data

Applying the Tom Thumb algorithm to the POEtic tissue, an element would be the FPGA logic block and would require 76 bits of configuration. Generally speaking, the bits required to configure a LUT-based FPGA element (e.g., at least 16 for a four-input LUT and several more to define the behavior of the interconnection network and of the register) are many more than the three at disposal with the basic algorithm.

The Tom Thumb algorithm was designed specifically with this option in mind and it is relatively straightforward to increase the size of the configuration memory within the algorithm.

In the algorithm, in each clock cycle an element receives a packet of n bits ($n = 4$ in the basic algorithm). Among these bits, one, the *packet type* bit, is used to determine if the other $(n - 1)$ bits are flag or configuration information. In the basic version, the entire information needed by an element is received in two n-bit packets, one for the flag and the other for the configuration data.

In order to cope with a greater number of configuration bits, the algorithm was modified by multiplying the number of packets used for the configuration data (Figure 16.12). In this new design, the data needed for one element that uses c configuration bits and a flag coded with f bits require $x = \lceil (c + f)/(n - 1) \rceil$ different n-bit packets consisting of one bit indicating the packet type and $(n - 1)$ bits for the information (flag or configuration). Assuming the minimal packet size, that is, $n = f + 1$ bits as shown in the figure, the total number packets becomes $x = \lceil c/(n - 1) \rceil + 1$.

This simple modification allows the algorithm to replicate a cell composed of elements requiring any number of configuration bits. Incidentally, this same modification reduces somewhat the resource overhead of the algorithm, since only one flag packet is necessary for each element, whatever the number of configuration data packets.

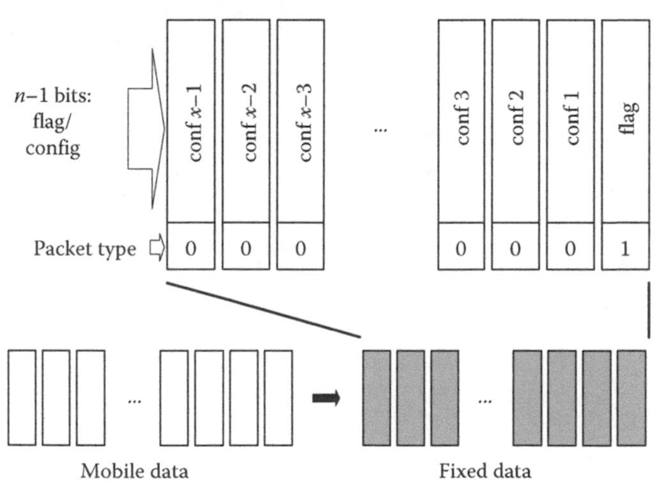

FIGURE 16.12 Configuration memory for a single element.

16.3.4 Direction of Self-Replication

The basic Tom Thumb algorithm is designed to enable the cell replication in only two directions, that is, to the north and to the east of the mother cell. Assuming that the injection point for the first configuration is placed in the south–west corner of the array, this restriction does not affect the basic operation of the system.

However, taking into account the possibility of alternate (or indeed multiple) injection points, or the application of the self-replication process to a faulty substrate, the two directions are *not* sufficient to guarantee an optimal replication pattern within the programmable circuit.

Fault tolerance, in particular, is one of the main motivations to justify the need for a self-replication process: assuming that the programmable substrate in which self-replication occurs can contain faults (a more than reasonable assumption in the kind of electronic or nanoelectronic devices that are the main targets of this approach), then it is necessary for the algorithm to be able to avoid faulty areas of the circuit and replicate only in the fault-free areas.

In Figure 16.13, each square represents the area occupied by a full cell (composed of many basic elements). The bottom-left square is the initial cell that starts the replication process, that is, the mother cell, and the gray squares are areas that contain faulty elements and should be avoided when replicating.

With the basic two-directional algorithm, the replication process will only be able to make copies of the mother cell in the squares marked with a "V," following the path shown with the black arrows. The squares labeled with question marks will not be configured, and the area will be wasted.

To avoid this potential loss of resources, the Tom Thumb algorithm has been extended to enable replication in the four cardinal directions. This extension will allow the entire fault-free surface of the circuit to be exploited, whatever the location of the faulty areas (e.g., replication will follow the configuration path indicated by the white arrows).

As detailed in Section 16.2, the algorithm works using a path spanning all the elements of the cell to be replicated, a flag marking the first element on this path and two more flags indicating the elements from which the replication has to occur and its direction. These are shown in Figure 16.14a: the direction of the path in each element is represented by the straight arrows, the start flag is the empty circle, and the

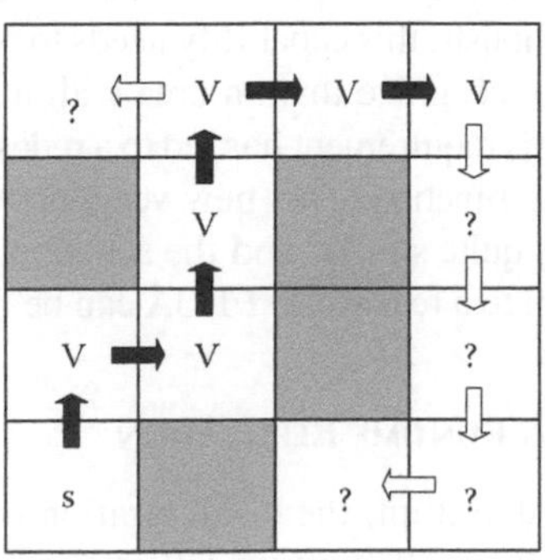

FIGURE 16.13 Replication with defaults.

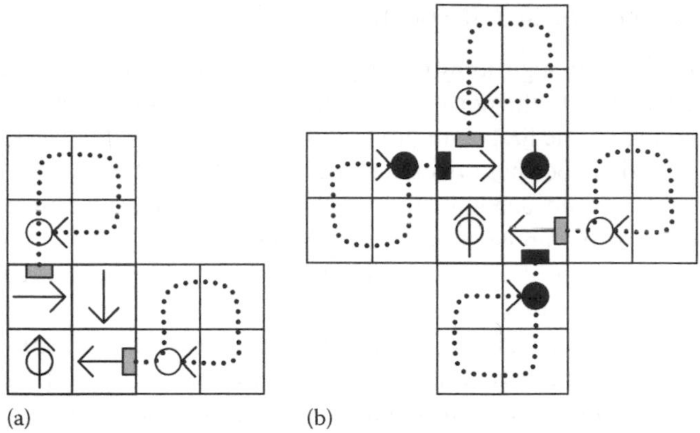

FIGURE 16.14 Basic and modified TTA directions of replication.

flags for the directions of the replication are the small gray rectangles. The dotted arrows show the new paths constructed during the replication process.

In order to make a cell replicate in four directions, the basic algorithm (Figure 16.14b) was modified by adding a new flag and modifying two existing ones. The basic concept of the path spanning all the elements of the cell, defined by the four basic directional flags, remains identical. The two branching flags were modified to indicate the two additional directions for the replication (these are shown with the small black rectangles in Figure 16.14b). Finally, while keeping the original start flag to indicate the first element in the path, a second start flag (the black circle in Figure 16.14b) had to be introduced to indicate the first element in the path for the new directions of replication: for the replication to the north and to the east directions, the first element to be configured is the one located at the bottom left of the cell, while for the west and the south directions the replication starts with the element located at the upper right corner of the cell.

Adding the empty flag, the number of flags used for the replication in the four directions is now equal to nine. This implies that the minimum packet size for this modified version of the algorithm becomes five bits (four bits for the flag plus one bit for the packet type).

It should be noted that, to allow the Tom Thumb algorithm to implement self-replication in a faulty substrate, the ability to replicate in the four directions is a necessary, but not sufficient condition. Obviously, this capability needs to be coupled with a mechanism that allows the algorithm to recognize that an area within the circuit is faulty, so as to avoid it when replicating. This requirement has led to a redesign of the Tom Thumb algorithm to implement a testing function. This new version, while more complex than the "standard" version, remains quite similar and the same procedure that was followed to adapt the standard algorithm to a real-world FPGA can be followed for the new version.

16.3.5 Start-Up versus Runtime Replication

In the basic Tom Thumb algorithm, the configuration of the cell is injected inside the circuit at start-up: the first cell (bottom left in the examples) is configured and then the replication process is automatic, filling the entire surface available on the substrate. The cells are duplicated again and again as long as there are free, not-yet-configured elements. In such a set, the cells have no influence on the replication process, that is, they cannot decide when and where to self-replicate.

While the creation of a full array of identical processors at start-up can have several practical advantages in a number of systolic applications, from a more general point of view it would be useful to give the cells the ability to choose when and where to start their replication. This capability would add the versatility, for example, to host more than one type of cell (and hence more than one organism) within the substrate, or allow a cell to create copies of itself to replace cells damaged by runtime faults.

To increase the versatility of the Tom Thumb algorithm, a possible modification would be to change the way the cells decide to replicate. In the classic algorithm, described in Section 16.2, a replication signal is emitted every time the start flag cell arrives in an element configured (with a branching flag) to initiate the replication.

A simple alteration to the algorithm would consist of disabling, by default, these replication signals. A cell would then have the ability to enable them, individually to replicate in only one direction or together to replicate in all directions at once, whenever it decides to create a copy of itself.

In such a system, as in the basic algorithm, the running genome would constantly turn within the cell following the configuration path. When the start flag arrives in an element that could initiate the replication, if the latter has not been enabled by the cell, nothing happens. On the other hand, when the cell decides to replicate, it enables one of the elements that can initiate the replication, which, when the start flag arrives, generates a replication signal and duplicates the genome in the chosen direction to create another cell, as in the original algorithm.

With this modification, the cell can now decide *when* to replicate by deciding when to enable the elements that initiate the replication. Moreover, it can also choose *where* to replicate, by enabling only the desired replication direction in the appropriate element.

16.3.6 Disabling Elements during Cell Replication

In the basic Tom Thumb algorithm, during the replication process, as soon as an element has been configured, it starts to operate according to the configuration data it has received. While not an issue for the function-less device used to illustrate the operation of the algorithm in Section 16.2, such a behavior could potentially be dangerous when the algorithm is applied to a real programmable logic device. In practice, the result of this process would be the step-by-step activation of parts of the processor (the logic gates implemented by each element) while the rest is still waiting to be configured. To achieve correct functionality, every element of the entire cell should start their normal processing at the same time, once the entire cell has been configured.

To achieve this behavior, an additional signal in each direction was added to each element. Then, when the first element of the cell is configured, it generates an active signal through this output in the direction of the replication path. This signal is forwarded combinatorially by each newly configured element and disables the normal functionality of the elements that receive it. Following the configuration path, the signal is forwarded to every element of the replicated cell and disables them as soon as they are configured.

At the end of the replication, when the last configured element of the cell closes the path, it signals that the replication process is finished. As a result, the disable signal is deactivated and this information is forwarded through the entire replication path, enabling all the elements of the cell to start their normal functionality at the same time.

16.3.7 Implementation

With the modifications outlined earlier, the Tom Thumb algorithm can be applied to the POEtic tissue to instantiate the self-replication of MOVE processors to implement a basic multiprocessor counter.

At start-up, the programmable substrate, that is, the modified POEtic tissue, contains only a small set of configured elements implementing the seed unit. This unit is located at the right side of the tissue, as shown in Figure 16.15a (the figure displays the state of the circuit through a custom interface linked to the ModelSim™ VHDL simulator). The first step of the configuration process is the injection of the configuration of the first processor (the mother cell), in the format required by the Tom Thumb algorithm. The injection point was arbitrarily set at the bottom-left element of the tissue.

A few steps of the construction and replication processes (which are essentially identical to those of the basic algorithm, illustrated in Section 16.2, expanded to take into account the additional configuration packets) are shown in Figure 16.15. In the figure, *t* represents the number of clock cycles from the beginning of the process, that is, from the injection of the first packet into the circuit.

The injected configuration automatically defines the configuration path and fixes the static configuration of each of the elements of the first processor to be implemented in the substrate. Then, since the option of using automatic replication at start-up (rather than letting the cells decide when and where to replicate) was selected, the replication process is automatically activated in every direction.

As a result, even before this first processor is fully configured, the replication of the genome according to the Tom Thumb algorithm starts to configure a copy of the processor to the north, as shown in Figure 16.15b. The third processor to be configured is also a replication of the first one to the east (see Figure 16.15c).

Finally, a fourth processor is configured on the substrate as a replication of the second one Figure 16.15d. Note that the fourth processor also tries to make a copy of itself in the east direction (Figure 16.15e), but this process halts automatically because there are not enough empty elements to contain a whole copy of the circuit.

When the four processors are configured, the whole system finds itself in an idle state (Figure 16.15e), ready for operation: the Tom Thumb algorithm has fully replicated the processors and fixed their configuration.

16.3.8 Practical Considerations

The objective of the research presented in this section was to develop a self-replication mechanism that could be integrated within a real-world programmable logic device. The hardware realization involved instantiating the modified Tom Thumb algorithm

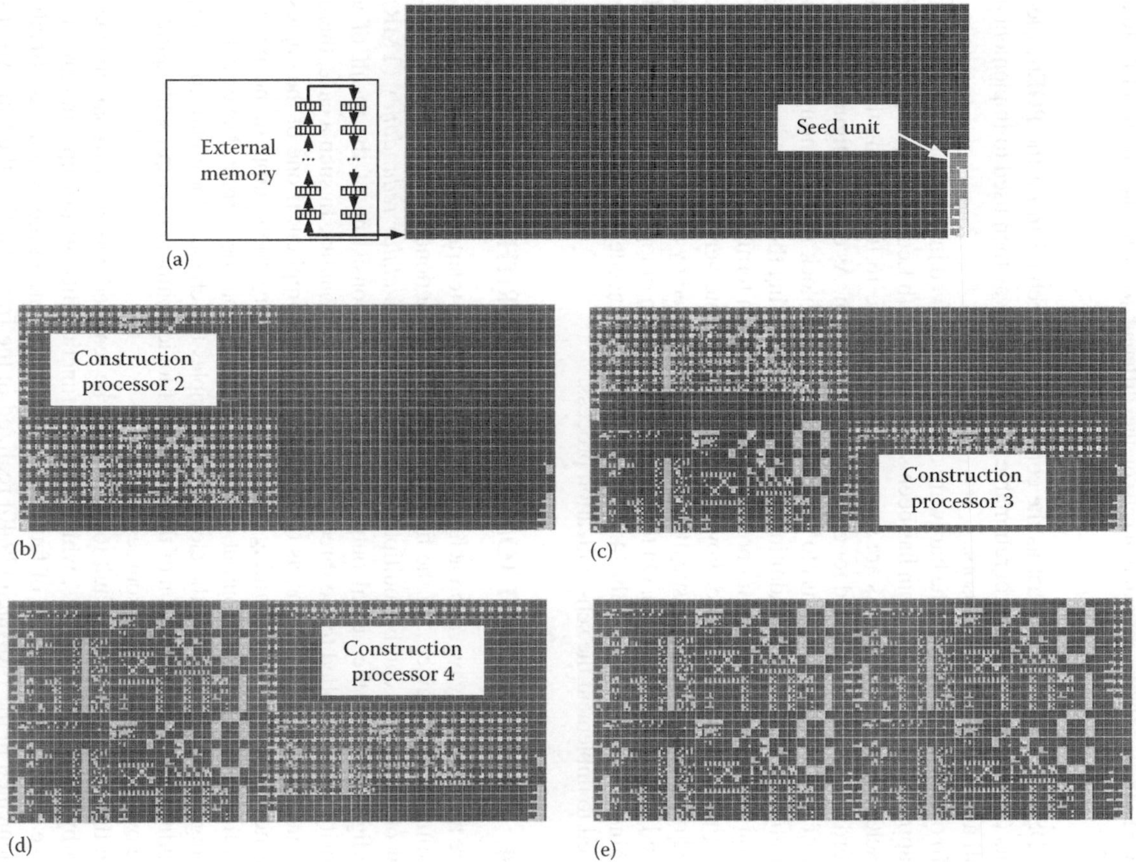

FIGURE 16.15 State of the circuit at different time steps for the test system.

	DFFs	%	Gates	%
TT	104	50.5	534	14.8
Routing	26	12.6	796	22
POEmol	76	36.9	2280	63.2

FIGURE 16.16 Hardware requirements of the different parts of the circuit (TT: Tom Thumb).

as a DSCA (Figure 16.9) to act as the configuration mechanism for the POEtic elements (Figure 16.10). This programmable logic device was then used to implement a small array of MOVE processors.

To obtain an estimate of the hardware overhead involved in the implementation of the approach, the whole system has been synthesized with Leonardo Spectrum using the scl05u library. The results are summarized in Figure 16.16, showing the amount of hardware resources needed for each part of the system. As each configuration register of the POEtic element has to be duplicated in the block implementing the Tom Thumb algorithm, with the addition of the registers storing the flags and other information related to the algorithm process, the overhead in terms of DFFs is obviously high (it should be mentioned, however, that an alternate version of the algorithm, based on self-inspection, drastically reduces this overhead by avoiding the need for a second copy of the data [35]). On the other hand, in term of logic gates, the overhead remains quite low, that is, only 14.8% of the total logic needed for the entire system is used to implement the self-replication process.

16.4 FAULT-TOLERANT TOM THUMB ALGORITHM

There are essentially two approaches to implement fault-tolerant mechanisms in programmable logic devices. In the first approach, the functionality of each basic element of the system can be multiplied as in the *triple modular redundancy* (TMR) paradigm [36–38], the actual output of the element consisting of the result of a majority voting applied to the three copies of the functionality. In such a case, there is no need for a fault detector, as the fault tolerance is implicit: if one of the copies becomes faulty, the two remaining healthy copies will give their output to the majority voting function and the output of the element will thus be correct. Nevertheless, tripling the functionality of the element also triples the chances for a fault to occur. In addition, this multiplication of the number of logic components represents quite a large overhead in term of resources.

In the second approach, fault tolerance can be realized with a self-repair mechanism by using spare elements, which have no functionality at start-up, but can take charge of the functionalities of faulty elements appearing in the system, as already implemented in the BioWatch [39,40] realized in the framework of the Embryonics project. This approach avoids the need to triplicate the functionality of each element and enables the designer of the system to choose how many spare elements must be included in the system, and thus, how many faults can be tolerated. However, in order to be able to detect the faults, to reconfigure the circuit, and to use the spare elements, the system obviously needs additional logic.

Where the Tom Thumb algorithm is concerned, however, much of the additional logic needed to implement reconfiguration using spare elements is identical to the logic used to implement the basic algorithm: as a result, the addition of reconfiguration capabilities to a self-replicating system using the TTA does not imply a large overhead.

As seen in the previous section, in its modified version, the Tom Thumb algorithm already embeds some mechanisms that target fault tolerance (Sections 16.3.4 and 16.3.6). Nevertheless, the price for such a tolerance is quite high: in the presence of a single faulty element, the area of a whole processor (or cell) is avoided, implying the waste of a large amount of resources. Consequently, we decided to implement a new low-level fault-tolerant ability with which, using spare elements, the substrate can self-repair without sacrificing a large amount of resources.

16.4.1 Global Behavior

The self-repairing version of the Tom Thumb Algorithm uses spare elements to replace faulty elements appearing in the circuit to gain a degree of fault tolerance. The global behavior of the system relies on multiple executions of the Tom Thumb Algorithm so that growth, self-replication, and self-repair can be introduced and consequently enables the cloning (cellular self-replication) and the cicatrization (cellular self-repair) of the original system.

In order to implement such mechanisms, the configuration of a cell is divided in two parts: a *structural configuration*, which defines the boundaries of the cell and the positions of its spare elements, and a *functional configuration*, which defines the configuration of the elements in order for the cell to operate following the specifications of the current application.

For a cell whose functionality requires $w \times h$ living elements, the structural configuration is defined by $(w + s) \times h$ elements, where s is the number of columns of the structure reserved to implement the spare elements. Note that s can be chosen by the designer accordingly to the level of fault tolerance required by the targeted system ($0 \leq s \leq w$). Figure 16.17 shows some examples for a cell implemented with 3×2 elements.

The construction of the cell in the substrate is then realized by the two following successive steps:

1. Using a first time the Tom Thumb algorithm, the structural configuration is injected in the substrate to create the structure of the cell, that is, to define its size $(w + s) \times h$ and the position of the columns of spare elements (Figure 16.17b).
2. When the structure has been configured, the functional configuration is in turn injected in the substrate. Once again using the Tom Thumb algorithm, the functional configuration configures the $w \times h$ living elements of the structure, that is, the elements that are not spare. Note that during this process, the spare elements are bypassed, that is, they are transparent from the functional configuration point of view.

After two executions of the TTA, the cell is fully configured and ready to start its normal operation (Figure 16.17c).

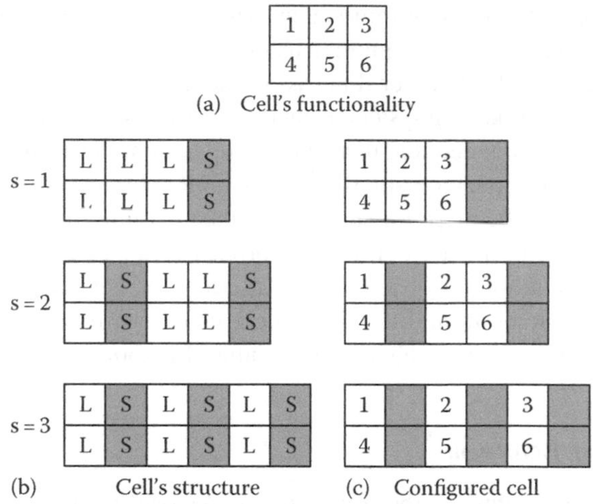

FIGURE 16.17 Functionality of a 3 × 2 elements cell (a) with different possible structures (b) and the corresponding final configured cell (c). (L = living, S = spare).

When a fault appears in one of the cell's elements, a *cicatrization* mechanism is launched (Figure 16.18). The faulty element is deactivated, isolated from the network, and replaced by the nearest healthy element to its right, which is itself replaced by the nearest right element, and so on until a spare element is reached. This process consists of the following steps:

1. The faulty element is deactivated.
2. The structural configuration of the elements at the right of the faulty one up to the spare element is changed and the elements are now considered as repair elements (Figure 16.18b).
3. The current functional configuration of the whole cell is erased.
4. The functional configuration is injected one more time in the cell using the Tom Thumb algorithm. It configures the $w \times h$ elements of the structure, which are not spare nor faulty elements, that is, the repair and normal elements. During this process, the spare elements are once again bypassed, as is the faulty element.

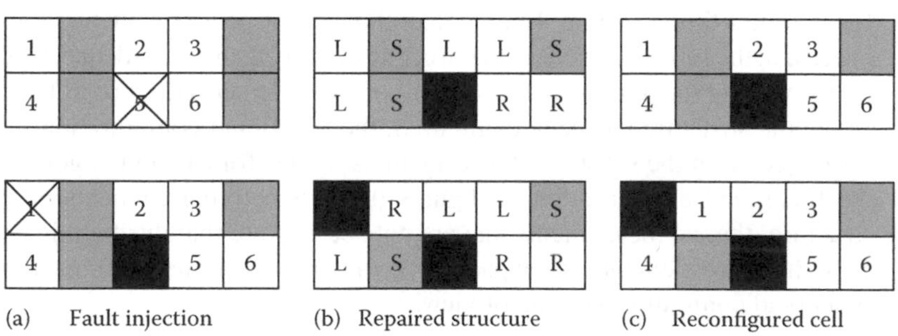

FIGURE 16.18 Cicatrization mechanism after two different fault injections. (a) Fault injection. (b) Repaired structure. (c) Reconfigured cell. L = living, S = spare, R = repair, black = faulty.

At the end of the cicatrization, the cell is fully configured again and ready to restart its normal operation (Figure 16.18c).

Note that the number of faulty elements handled by the cicatrization mechanism is necessarily limited. Indeed, the system can tolerate, per row, at most one faulty element between each pair of spare columns. If more than one element in the same row between two columns of spares is faulty, the cicatrization is impossible and the whole cell is killed, that is, deactivated.

In a multicellular point of view, as shown in Figure 16.19 where each square represents a whole cell, the global process is the following:

a. The structural configuration (S) is injected into the substrate.
b. The structure of the first cell is created. Moreover, following the TTA specification, the structural configuration is replicated and sent eastward and northward.
c. Two more cell's structures are configured, which in turn will replicate their own structural configuration. As its two neighboring cells now have their structure fixed, the first cell does not need to keep its own structural configuration running. Consequently, it stores its structure information in dedicated registers and clears its TTA memory positions: it can now accept the functional configuration (F) that is injected from outside the substrate.
d. The functional configuration of the first cell is over. The first cell will then use the TTA to replicate its functional configuration eastward and northward. As the fourth cell's structure is created, its neighboring cells do not have to keep their structural configuration running in their TTA memory positions, which are thus cleared. These two cells can then accept the functional configuration provided by the first cell.

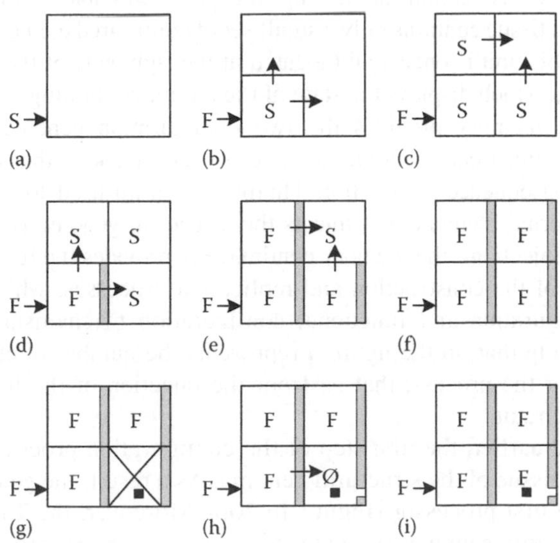

FIGURE 16.19 Multicellular behavior of the fault-tolerant Tom Thumb algorithm.

e. The functional configuration of two more cells is over. It will be replicated to functionally configure the fourth cell.
f. All four cells are configured.
g. A fault appears in the bottom right cell, which launches its cicatrization mechanism.
h. Its functional configuration is erased and its structure is repaired using one spare element. The repaired cell can now accept once again the functional configuration provided by the first cell.
i. The fourth cell has been functionally reconfigured and is ready to restart its normal operation.

Several versions of this fault-tolerant Tom Thumb algorithm have been implemented (for additional details see, e.g., [41,42]), but all share this basic approach, based on the use of spare elements within the programmable substrate.

16.4.2 Implementation

As in the previous section, the fault-tolerant Tom Thumb algorithm was integrated in the POEtic tissue. As a result, the hardware design had again to be slightly modified in order to be able to replicate the 76 bits of the functional configuration of the POEtic element. Then, as before, to prevent the step-by-step activation of parts of the system that could result in unwanted behaviors, we also implemented the signal disabling the POEtic elements during the cell's construction. Finally, to verify the functionality and efficiency of the fault-tolerant Tom Thumb algorithm applied to complex circuits, we applied it to the same multiprocessor system presented earlier, that is, a system composed of four MOVE processors (the cells) that form a four-digit counter displaying seconds and minutes, logically organized to form a chain in their final configuration (Figure 16.15).

As in the previous section, at start-up, the programmable substrate, that is, the modified POEtic tissue contains only a small set of configured elements implementing the seed unit. This unit is once again located at the right side of the tissue, as shown in Figure 16.20a, which displays the state of the circuit containing 64 × 24 elements.

The external memory provides the two configuration genomes: the structural genome and the functional genome, as a set of data packets that are continuously shifted. As a consequence of the Tom Thumb algorithm used for the configuration process, the external memory first injects the structural genome twice in the tissue. After these two injections, the memory continuously provides the functional genome.

A few steps of the construction and replication processes, which consist of the structural configuration and functional configuration mechanisms, are shown in Figure 16.20. Note that, in the figure, t represents the number of clock cycles from the beginning of the process, that is, from the injection of the first configuration packet into the circuit.

As explained earlier, the first step of the configuration process consists of the injection in the tissue of the structural genome. As a result, the genome creates the structure of the first processor (Figure 16.20b). Moreover, the Tom Thumb algorithm replicates this genome to create three supplementary processor structures (Figure 16.20c and d).

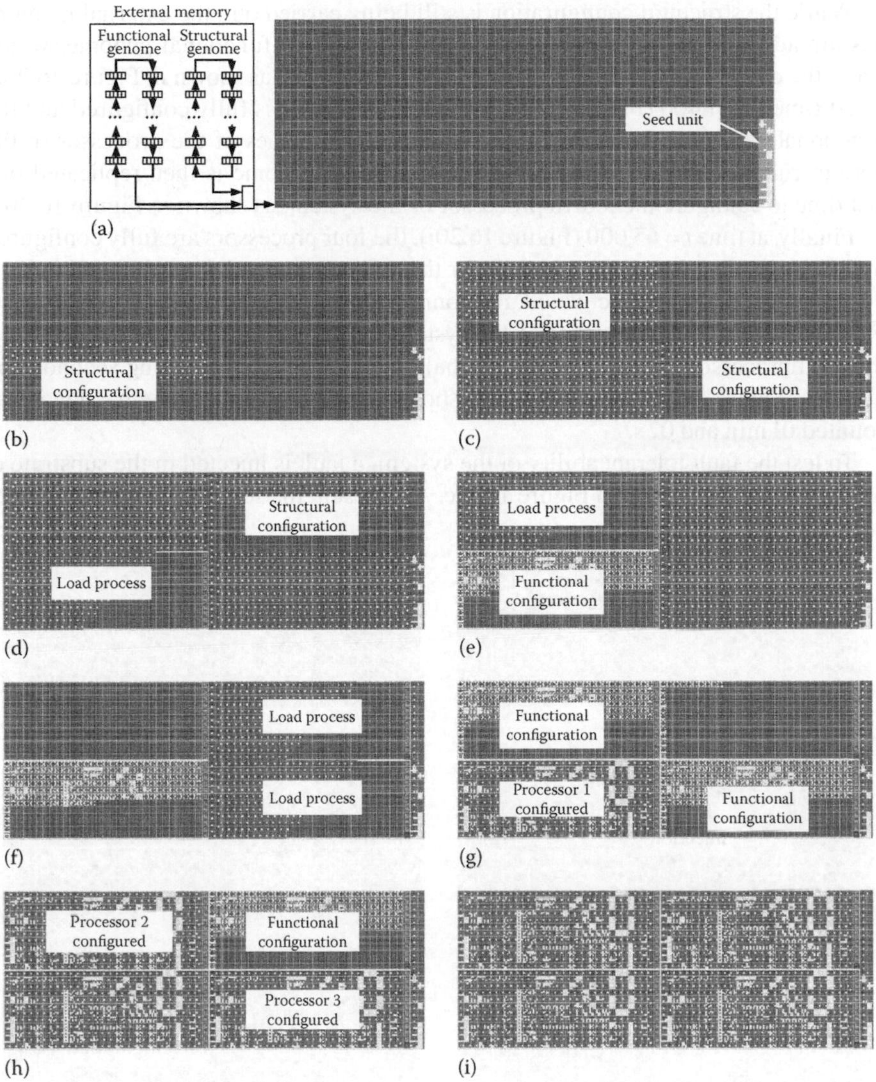

FIGURE 16.20 State of the circuit at different time steps during the self-replication process.

When the bottom right structure is configured, it sends a *close* signal to the left neighboring structure, that is, the bottom-left one, triggering the *load* process in that structure, as shown in Figure 16.20d. The top structures have the same behavior: when the top right structure is configured, it triggers the *load* process in the top left structure (Figure 16.20e).

As there is not enough place for the two right structures to replicate eastward, the *close* signals cannot be automatically generated by their right neighboring structures. As a result, these two signals are generated by the user with the button input of the elements: this triggers the *load* processes in the two right structures, as shown in Figure 16.20f.

While the structural configuration is still being carried out, the external memory has already started the injection in the substrate of the functional genome, which starts the configuration of the first processor in the tissue, as shown in Figure 16.20e.

At time $t = 51,400$ (Figure 16.20g), the first processor is fully configured and the functional genome is being replicated to create two copies of the processor in the bottom right and top left structures. The functional genome is then replicated one last time to configure the fourth processor of the system, as shown in Figure 16.20h.

Finally, at time $t = 65,000$ (Figure 16.20i), the four processors are fully configured in their respective structures and wait for the activation signal from the seed unit.

When the seed unit is activated, the connection and differentiation process starts. Figure 16.21a shows a snapshot of the system during this process. When it is over, the multiprocessor system starts its normal operation, that is, counting seconds and minutes. Figure 16.21b presents a snapshot of the state of the system after it has counted 01 min and 02 s.

To test the fault-tolerant ability of the system, a fault is injected in the substrate at time $t = 68,211$, as shown in Figure 16.21c. As a result, the selected element becomes

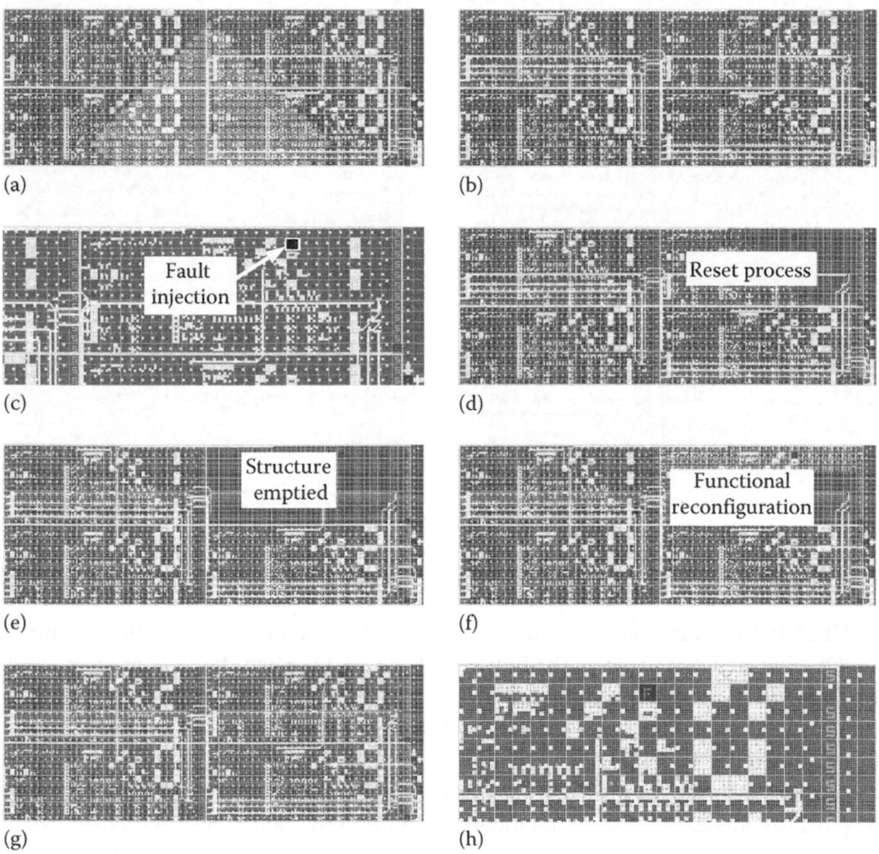

FIGURE 16.21 State of the circuit at different time steps during its operation and the cicatrization mechanism.

faulty and triggers the *reset* process of its processor (Figure 16.21d). At the end of this *reset* process, the structure of the top right processor has been repaired and its functional configuration has been erased, as shown in Figure 16.21e.

The empty structure will then receive another copy of the functional genome from one of its neighboring processors and, as shown in Figure 16.21f, it is functionally reconfigured.

At the end of this reconfiguration process, the "healed" processor can resume its role in the system. The whole chain of processors will thus continue its normal operation, that is, count seconds and minutes. Figure 16.21g shows the state of the system after it has counted 02 min and 09 s. Note that Figure 16.21h presents a zoom on the top right part of the circuit showing the *scar* that results from the structural repair: the display of zero is distorted.

16.5 PRACTICAL CONSIDERATIONS

The work presented in this section enables processor-scale digital circuit to replicate and tolerate faults in its basic constituents. Fault tolerance is achieved through a self-repair mechanism involving several successive applications of the replication and growth process of the Tom Thumb algorithm with different configurations, that is, the structural and functional genomes.

Compared with the TTA replication in the POEtic tissue presented in the previous section, which supported cellular-level fault tolerance at the cost of a potentially large amount of resources, this new fault-tolerant Tom Thumb algorithm is much more efficient. Indeed, in the presence of faulty elements, when the system is designed with just some spare elements, the algorithm can reconfigure the POEtic tissue avoiding the faulty elements. Moreover, the level of fault tolerance, that is, the maximum number of faults in the substrate that can be tolerated by the system, can be chosen by the designer of the system and simply relies on the number of columns of spare elements integrated in the structural configuration of its composing cells.

Finally, one of the key advantages of our algorithm, compared with the TMR paradigm generally used in the field of fault tolerance, is its efficiency. Indeed, with TMR, the entire circuit logic has to be multiplied by three to tolerate faults. With our algorithm, in the worst case, that is, when the designer wants the system to be able to tolerate a fault in each of its basic components, one spare column of elements per functional column of elements must be added, thus only doubling the amount of logic of the system.

16.6 CONCLUSION

The work presented in this chapter represents, to the best of our knowledge, the first example of fault-tolerant self-replication of a processor-scale digital circuit in hardware. The objective of this research was to illustrate how it might be possible to integrate self-replication in a programmable logic device using mechanisms that can scale to the thousands of elements (at least) and hence provide a substrate where developmental algorithms, with all their inherent advantages in terms of design and reliability, can be applied to hardware systems of real-world complexity.

While the algorithm was, of course, implemented and tested on conventional electronic devices, its features have been developed in an attempt to prepare the way for an eventual realization on nanoscale devices. Its completely distributed nature (no centralized control is needed for any part of the algorithm), its focus on faulty substrates, its parallel configuration, and its scalability to arbitrarily large circuits (all interactions are local and no long-distance signals are required at any stage) are all features that, we believe, could prove useful for an implementation at the molecular scale.

The algorithm presented in this chapter is obviously open to amelioration in several respects and indeed it is constantly being improved to address additional issues or environments:

- To reduce the overhead inherent in storing a second copy of the configuration in each element, a variant of the algorithm has been developed and implemented, which uses a self-inspection approach (Section 16.1.3) to generate a copy of the configuration whenever needed. Fairly similar to the basic algorithm in its replication dynamics (flags, etc.), this version drastically alters the differentiation and system dynamics.
- Fault tolerance is a crucial application area for self-replication. Several self-repairing versions of the Tom Thumb algorithm, capable of reconfiguring the programmable array to avoid faults, have been developed (e.g., [41,42] and investigations on fault-detecting logic (which was not mentioned in this chapter but is obviously a fundamental component of any self-repair mechanism) are under way.
- In the context of next-generation electronics, and particularly in that of nanoscale approaches, the global synchronization required by a CA (such as Tom Thumb) can be extremely difficult to achieve. To avoid this potential obstacle to the implementation of the algorithm, an asynchronous version has been realized, where the local interactions between the elements take place without the need for a global clock signal. Based on a custom four-phase handshaking protocol, the mechanism is an important step in the direction of a nanoscale implementation.
- One of the promises of molecular electronics (which could potentially allow a quantum leap in circuit density) is the possibility of being able to design and build circuits that exploit the three dimensions. Since self-replication is an approach designed to simplify layout and design in very complex circuits, a 3D version of the algorithm was designed and implemented [43,44]. Indeed, due to of its structure and its local interactions, the transition of the algorithm from two to three dimensions is quite simple and preserves all its properties.

Several issues connected to the practical implementation of self-replicating processors also need to be addressed: seen as part of a complex bioinspired design approach, self-replication must be integrated within a complex multicellular computing system. This context implies that several "accessory" issues have to be considered:

- The self-replication approach must be extended to real-world applications. While the test system described in this chapter is sufficient to show that the Tom Thumb algorithm can be applied to structures of arbitrary complexity, the methodology used in the design of the system (or rather, its lack, since the processors had to be designed by hand) does not scale to larger systems. Current work is under way to define a design environment for the synthesis of cellular processors [45,46].
- The design environment will have to be extended to take into account and exploit the possibilities introduced by self-replication, such as the ability of cells to replicate at will (Section 16.3.5), and to integrate fault tolerance and reconfiguration processes in the design. Specifically, the fault-tolerant possibilities offered by a developmental approach have to be verified in a real-world setting.
- The differentiation process of the cellular array will have to be closely integrated in the design as well, to allow newly replicated cells to seamlessly connect to the rest of the array. The ad hoc approach used in the test system was designed specifically for the given example, whereas a more general solution must be found to implement this process in a wider range of applications.

But even if improvements are of course possible, the self-replication approach described in this chapter represents a clear step beyond existing solutions, in terms of hardware efficiency and versatility. This latter aspect is particularly important: while the test system used to illustrate the algorithm was implemented using a custom FPGA and application-specific processors, the unconventional features of these architectures are used exclusively during their operation and are completely disjoint from the self-replication algorithm. In other words, the only requirement of the algorithm is that it must be possible to implement the configuration memory of the programmable device as a shift register (or equivalent). In turn, this implies that the algorithm is relatively technology-independent and can easily be adapted to any programmable structure, as long as the aforementioned condition is met, a crucial versatility in view of its implementation in the next generation of electronic devices at the nanoscale.

REFERENCES

1. W. Asprey, *John von Neumann and the Origins of Modern Computing*. Cambridge, MA: The MIT Press, 1992.
2. J. von Neumann, *Theory of Self-Reproducing Automata*. Urbana, IL: University of Illinois Press, 1966, edited and completed by A. W. Burks.
3. K. E. Drexler, *Nanosystems: Molecular Machinery, Manufacturing and Computation*. New York: John Wiley, 1992.
4. J. Han, J. Gao, Y. Qi, P. Jonker, and J. Fortes, Toward hardware-redundant, fault-tolerant logic for nanoelectronics, *IEEE Design and Test of Computers*, 22 (4), 328–339, 2005.

5. S. Cotofana, A. Schmid, Y. Leblebici, A. Ionescu, O. Soffke, P. Zipf, M. Glesner, and A. Rubio, Conan—A design exploration framework for reliable nano-electronics, in *Proceedings 2005 IEEE International Conference on Application-Specific Systems, Architectures and Processors (ASAP'05)*. Los Alamitos, CA: IEEE Computer Society Press, 2005, pp. 260–267.
6. C. Dong, D. Chen, S. Haruehanroengra, and W. Wang, 3-d nfpga: A reconfigurable architecture for 3-d cmos/nanomaterial hybrid digital circuits, *IEEE Transactions on Circuits and Systems I: Regular Papers*, 54 (11), 2489–2501, 2007.
7. R. Rad and M. Tehranipoor, A new hybrid fpga with nanoscale clusters and cmos routing, in *Design Automation Conference, 2006 43rd ACM/IEEE*, San Francisco, CA, 2006, pp. 727–730.
8. A. Dehon, Nanowire-based programmable architectures, *Journal of Emerging Technologies in Computing Systems*, 1, 109–162, 2005.
9. R. A. Freitas, Jr., T. J. Healy, and J. E. Long, Advanced automation for space missions, in *Proceedings 7th International Joint Conference on Artificial Intelligence (IJCAI81)*. Los Altos, CA: Morgan Kaufmann, 1981, pp. 803–808.
10. R. A. Freitas Jr. and R. C. Merkle, *Kinematic Self-Replicating Machines*. Georgetown, TX: Landes Bioscience, 2004.
11. W. R. Buckley and A. Mukherjee, Constructibility of signal-crossing solutions in von Neumann's 29-state cellular automata, in *Proceedings of 2005 International Conference on Computational Science (ICCS2005)*, ser. LNCS, Vol. 3515. Berlin, Germany: Springer Verlag, 2005, pp. 395–403.
12. C. G. Langton, Self-reproduction in cellular automata, *Physica D*, 10, 135–144, 1984.
13. J. Byl, Self-reproduction in small cellular automata, *Physica D*, 34, 295–299, 1989.
14. J. A. Reggia, S. L. Armentrout, H.-H. Chou, and Y. Peng, Simple systems that exhibit self-directed replication, *Science*, 259, 1282–1287, 1993.
15. G. Tempesti, A new self-reproducing cellular automaton capable of construction and computation, in *Advances in Artificial Life: Proceedings of 3rd European Conference on Artificial Life (ECAL95)*, ser. LNCS, Vol. 929. Berlin, Germany: Springer Verlag, 1995, pp. 555–563.
16. J.-Y. Perrier, M. Sipper, and J. Zahnd, Toward a viable, self-reproducing universal computer, *Physica D*, 97, 335–352, 1996.
17. J. Ibanez, D. Anabitarte, I. Azpeitia, O. Barrera, A. Barrutieta, H. Blanco, and F. Echarte, Self-inspection based reproduction in cellular automata, in *Proceedings of 3rd European Conference on Artificial Life (ECAL95)*, ser. LNCS, Vol. 929. Berlin, Germany: Springer Verlag, 1995, pp. 564–576.
18. K. Morita and K. Imai, Self-reproduction in a reversible cellular space, *Theoretical Computer Science*, 168, 337–366, 1996.
19. F. Peper, T. Isokawa, N. Kouda, and N. Matsui, Self-timed cellular automata and their computational ability, *Future Generation Computer Systems*, 18 (7), 893–904, 2002.
20. Y. Takada, T. Isokawa, F. Peper, and N. Matsui, Universal construction and self-reproduction on self-timed cellular automata, *International Journal of Modern Physics C*, 17 (7), 985–1007, 2006.
21. S. R. Park and W. Burleson, Configuration cloning: Exploiting regularity in dynamic DSP architectures, in *Proceedings of ACM/SIGDA International Symposium on Field Programmable Gate Arrays*. New York: ACM Press, 1999, pp. 81–89.
22. G. Tempesti, A self-repairing multiplexer-based FPGA inspired by biological processes, PhD, Ecole Polytechnique Fédérale de Lausanne (EPFL), Lausanne, Switzerland, 1998.
23. D. Mange, M. Sipper, A. Stauffer, and G. Tempesti, Towards robust integrated circuits: The Embryonics approach, *Proceedings of the IEEE*, 88 (4), 516–541, 2000.
24. L. Durbeck and N. Macias, The cell matrix: An architecture for nanocomputing, *Nanotechnology*, (12), 217–230, 2001.

25. N. Macias and P. Athanas, Application of self-configurability for autonomous, highly-localized self-regulation, in *Proceedings of 2007 NASA/ESA Conference on Adaptive Hardware and Systems (AHS2007)*. Los Alamitos, CA: IEEE Computer Society Press, 2007, pp. 397–404.
26. D. Mange, A. Stauffer, E. Petraglio, and G. Tempesti, Self-replicating loop with universal construction, *Physica D*, 191, 178–192, 2004.
27. D. Mange, A. Stauffer, L. Peparolo, and G. Tempesti, A macroscopic view of self-replication, *Proceedings of the IEEE*, 92 (12), 1929–1945, 2004.
28. A. Stauffer and M. Sipper, The data-and-signals cellular automaton and its application to growing structures, *Artificial Life*, 10 (4), 463–477, 2004.
29. A. Tyrrell, E. Sanchez, D. Floreano, G. Tempesti, D. Mange, J.-M. Moreno, J. Rosenberg, and A. Villa, POEtic tissue: An integrated architecture for bio-inspired hardware, in *Proceedings 5th International Conference on Evolvable Systems: From Biology to Hardware (ICES2003)*, ser. LNCS, Vol. 2606. Berlin, Germany: Springer Verlag, 2003, pp. 129–140.
30. Y. Thoma, E. Sanchez, J.-M.M. Arostegui, and G. Tempesti, A dynamic routing algorithm for a bio-inspired reconfigurable circuit, in *Proceedings of 13th International Conference on Field-Programmable Logic and Applications (FPL03)*, ser. LNCS, Vol. 2778. Berlin, Germany: Springer Verlag, 2003, pp. 681–690.
31. H. Corporaal and H. Mulder, MOVE: A framework for high-performance processor design, in *Proceedings of 1991 International Conference on Supercomputing*, Cologne, Germany, 1991, pp. 692–701.
32. H. Corporaal, *Microprocessor Architectures—From VLIW to TTA*. New York: John Wiley & Sons, 1998.
33. D. Tabak and G. J. Lipovski, MOVE architecture in digital controllers, *IEEE Transactions on Computers*, C-29 (2), 180–190, 1980.
34. G. Tempesti, D. Mange, E. Petraglio, A. Stauffer, and Y. Thoma, Developmental processes in silicon: An engineering perspective, in *Proceedings of 2003 NASA/DoD Conference on Evolvable Hardware (EH-2003)*. Los Alamitos, CA: IEEE Computer Society Press, 2003, pp. 255–264.
35. J. Rossier, Y. Thoma, P.-A. Mudry, and G. Tempesti, Move processors that self-replicate and differentiate, in *Proceedings of 2nd International Workshop on Biologically-Inspired Approaches to Advanced Information Technology (Bio-ADIT06)*, ser. LNCS, Vol. 3853. Berlin, Germany: Springer Verlag, 2006, pp. 328–343.
36. C. Carmichael, Triple module redundancy design techniques for Virtex series FPGA, Xilinx Corporation, Application notes 197, 2000.
37. F. Lima, L. Sterpone, L. Carro, and M. Sonza Reorda, On the optimal design of triple modular redundancy logic for SRAM-based FPGAs, in *Proceedings of the Conference on Design, Automation and Test in Europe*. Los Alamitos, CA: IEEE Computer Society Press, 2005, pp. 1290–1295.
38. B. Pratt, M. Caffrey, P. Graham, K. Morgan, and M. Wirthlin, Improving FPGA design robustness with partial TMR, in *Proceedings of the 44th IEEE International Annual Symposium on Reliability Physics*, New York, 2006, pp. 226–232.
39. A. Stauffer, D. Mange, G. Tempesti, and C. Teuscher, BioWatch: A giant electronic bio-inspired watch, in *Proceedings of 3rd NASA/DoD Workshop on Evolvable Hardware*. Los Alamitos, CA: IEEE Computer Society Press, 2001, pp. 185–192.
40. A. Stauffer, D. Mange, G. Tempesti, and C. Teuscher, A self-repairing and self-healing electronic watch: The biowatch, in *Proceedings of 4th International Conference on Evolvable Systems: From Biology to Hardware (ICES01)*, ser. LNCS, Vol. 2210. Berlin, Germany: Springer Verlag, 2001, pp. 112–127.
41. A. Stauffer and J. Rossier, Self-testable and self-repairable bio-inspired configurable circuits, in *Proceedings of 2009 NASA/ESA Conference on Adaptive Hardware and Systems (AHS09)*. Los Alamitos, CA: IEEE Computer Society Press, 2009, pp. 155–162.

42. A. Stauffer and J. Rossier, Bio-inspired self-testing configurable circuits, in *Evolvable Systems: From Biology to Hardware*, ser. LNCS, G. Tempesti, A. Tyrrell, and J. Miller, Eds., Vol. 6274. Berlin, Germany: Springer, 2010, pp. 202–213.
43. A. Stauffer, D. Mange, E. Petraglio, and F. Vannel, DSCA implementation of 3D self-replicating structures, in *Proceedings of 6th International Conference on Cellular Automata for Research and Industry (ACRI04)*, ser. LNCS, Vol. 3305. Berlin, Germany: Springer Verlag, 2004, pp. 698–708.
44. A. Stauffer, D. Mange, E. Petraglio, and G. Tempesti, Self-replication of 3D universal structures, in *Proceedings of 6th NASA/DoD Workshop on Evolvable Hardware (EH04)*. Los Alamitos, CA: IEEE Computer Society Press, 2004, pp. 283–287.
45. P.-A. Mudry, J. Ruffin, M. Ganguin, and G. Tempesti, A hardware-software design framework for distributed cellular computing, in *Proceedings of Evolvable Systems: From Biology to Hardware (ICES 2008)*. Berlin, Germany: Springer, 2008, pp. 71–82.
46. G. Tempesti, P.-A. Mudry, and G. Zufferey, Hardware/software coevolution of genome programs and cellular processors, in *Proceedings of 1st NASA/ESA Conference on Adaptive Hardware and Systems (AHS'06)*. Los Alamitos, CA: IEEE Computer Society Press, 2006, pp. 129–136.

Index

C

D

N

O

P

S

T

V

W

T - #0355 - 101024 - C16 - 234/156/26 - PB - 9781138374478 - Gloss Lamination